Electrophoretic Studies on Agricultural Pests

Proceedings of an International Symposium, held at
Rothamsted Experimental Station, Harpenden, Hertfordshire, UK,
on 6–8 April 1988

Preface

This volume represents the published proceedings of a symposium, 'Electrophoretic Studies on Agricultural Pests', held at Rothamsted Experimental Station, Harpenden, Hertfordshire, England, from 6 to 8 April 1988. The meeting was the first specifically on this topic. All lectures were given in the new lecture theatre at Rothamsted.

About 100 people attended in total and 22 papers were presented. Subjects covered ranged from studies of the genetic variation of populations of mammals and birds of agricultural importance to the use of electrophoresis to detect prey species within arthropod predators.

Organization of the meeting was not without its problems. Unfortunately, Professor M.F. Claridge, then Head of the Zoology Department, University College, Cardiff, who was to give the inaugural lecture 'Electrophoresis in agricultural pest research — a technique of evolutionary biology' was taken seriously ill some weeks before the meeting and was thus unable to attend, although fortunately he has since made a full recovery (an abbreviated form of his paper is included in this volume). Another speaker declined to attend at the last moment. Also, a ferry strike coincided with the meeting and affected many ports in Britain and Europe, preventing attendance by some would-be participants from the Continent.

Despite these difficulties, the meeting was most successful. Speakers and other participants came from all parts of the world, especially Britain, Europe, and the United States of America. Professor R.J. Berry, University College, London, kindly, and at very short notice, presented a paper in the first session on 6 April entitled, 'Electrophoretic studies: Perspectives for population biology'.

A Conference Dinner was held at the Manor House at Rothamsted on the evening of 7 April, to which all the Speakers were cordially invited and which many other participants also attended. A short speech was given after the Dinner by one of us (H.D.L.).

By good fortune, the weather was generally fine and sunny for the three days of the meeting: this was especially welcome to those participants who wished to explore the estate at Rothamsted, including the main campus area and laboratories, the Manor and its garden, and the grounds themselves, including the classic field experiments at Broadbalk and Park Grass.

Accommodation was arranged at the Hatfield Polytechnic Halls of Residence.

Four scientific manufacturing companies held exhibitions of their products at the meeting: Cambridge Electrophoresis Ltd., Burkard Scientific (Sales) Ltd., Biotech Instruments Ltd., and Bio-Rad Laboratories Ltd.

The organizers of the symposium wish to thank the following organizations and firms for generously contributing towards the running costs of the meeting: The Royal Entomological Society of London, The British Council, Shell U.K. Ltd., Imperial Chemical Industries PLC, Ciba-Geigy Ltd., British Drug Houses Ltd., Pharmacia Ltd., Biotech Instruments Ltd., Cambridge Electrophoresis Ltd., Burkard Scientific (Sales) Ltd., Bio-Rad Laboratories Ltd., Wolfe Medical Publications Ltd., Butterworths Scientific Ltd.

We should like very much to thank the respective editors of the following scientific publications for kindly advertising the meeting: *Crop Protection, Entomologia Experimentalis et Applicata, The Biologist, Antenna* (Bulletin of the Royal Entomological Society of London), *Tropical Pest Management*, and *The Bulletin of the British Ecological Society*.

We are also especially grateful to the following: Mrs Jackie Fountain, Rothamsted, for her excellent typing of the Abstracts of papers read at the meeting and a few submitted manuscripts; Mrs Moreen Budgen (The Warden, Rothamsted Manor), Mrs Pat Smith (head cook, Rothamsted canteen), as well as the catering staff at Hatfield Polytechnic for producing the array of good food; the several people at Rothamsted who kindly helped run the bar and operated the lighting and projector in the lecture theatre; and lastly, and by no means least, Cliff Brookes for his help and advice—both before and during the meeting.

Rothamsted H.D.L.
August, 1988 J.d.H.

Contents

Contributors

J. BAKKER

Department of Nematology, Wageningen Agricultural University, P.O.B. 8123, 6700 ES Wageningen, The Netherlands.

S.H. BERLOCHER

Department of Entomology, University of Illinois at Urbana-Champaign, 505 South Goodwin Avenue, Urbana, IL 61801, USA.

R.J. BERRY

Department of Biology, University College London, Gower Street, London WC1E 6BT, UK.

R.L. BLACKMAN

Department of Entomology, British Museum (Natural History), Cromwell Road, London SW7 5BD, UK.

C.P. BROOKES

Department of Entomology and Nematology, Rothamsted Experimental Station, Harpenden, Hertfordshire AL5 2JQ, UK.

P.A. BROWN

Department of Entomology, British Museum (Natural History), Cromwell Road, London SW7 5BD, UK.

L. BULLINI

Department of Genetics and Molecular Biology, University of Rome 'La Sapienza', via Lancisi 29, 00161 Rome, Italy.

M.F. CLARIDGE

School of Pure and Applied Biology, University of Wales College of Cardiff, PO Box 915, Cardiff, CF1 3TL, UK.

J.C. DALY

CSIRO Division of Entomology, Black Mountain, GPO Box 1700, Canberra, ACT 2601, Australia.

J. DEN HOLLANDER

School of Pure and Applied Biology, University of Wales College of Cardiff, PO Box 915, Cardiff CF1 3TL, UK.

A.L. DEVONSHIRE

Insecticides and Fungicides Deparment, Rothamsted Experimental Station, Harpenden, Hertfordshire AL5 2JQ, UK.

J.D. FITZGERALD

AFRC Institute of Horticultural Research, East Malling, Maidstone, Kent ME19 6BJ, UK.

C.C. FLEMING

Agricultural Zoology Research Division, Department of Agriculture for Northern Ireland, Felden, Mill Road, Newtownabbey, Co Antrim BT36 7ED, UK [and Department of Agricultural Zoology, The Queen's Univ]

C. FURK

Ministry of Agriculture, Fisheries and Food, ADAS, Harpenden Laboratory, Hatching Green, Harpenden, Hertfordshire AL5 2BD, UK.

F.J. GOMMERS

Department of Nematology, Wageningen Agricultural University, P.O.B. 8123, 6700 ES Wageningen, The Netherlands.

T.H. HSIAO

Department of Biology, College of Science, Utah State University, Logan, UT 84322–5305, USA.

M. LOUKAS

Department of Genetics, Agricultural College of Athens, 75 Iera Odos, Votanikos, Athens 118 55, Greece.

H.D. LOXDALE

Department of Entomology and Nematology, Rothamsted Experimental Station, Harpenden, Hertfordshire AL5 2JQ, UK.

S.B.J. MENKEN

Hugo de Vries-Laboratorium, University of Amsterdam, Kruislaan 318, 1098 SM Amsterdam, The Netherlands.

R.A. MURRAY

AFRC Institute of Horticultural Research, East Malling, Maidstone, Kent ME19 6BJ, UK.

G. NASCETTI

Department of Genetics and Molecular Biology, University of Rome 'La Sapienza', via Lancisi 29, 00161 Rome, Italy.

D.T. PARKIN

Department of Genetics, School of Medicine, Queen's Medical Centre, Nottingham NG7 2UH, UK.

D.P. PASHLEY

Entomology Department, Louisiana State University, Baton Rouge, LA 70803, USA.

W. POWELL

Department of Entomology and Nematology, Rothamsted Experimental Station, Harpenden, Hertfordshire AL5 2JQ, UK.

M.P. ROBINSON

Department of Entomology and Nematology, Rothamsted Experimental Station, Harpenden, Hertfordshire AL5 2JQ, UK.

A.D. SECCOMBE

Department of Entomology, British Museum (Natural History), Cromwell Road, London SW7 5BD, UK.

R.S. SINGH

Department of Biology, McMaster University, 1280 Main Street West, Hamilton, Ontario L8S 4K1, Canada.

M.G. SOLOMON

AFRC Institute of Horticultural Research, East Malling, Maidstone, Kent ME19 6BJ, UK.

J. TOMIUK

Biological Institute, University of Tübingen, Auf der Morgenstelle 28, D7400 Tübingen 1, Federal Republic of Germany.

M.P. WALTON

School of Pure and Applied Biology, University of Wales College of Cardiff, PO Box 915, Cardiff CF1 3TL, UK.

G.W. WATSON

Department of Entomology, British Museum (Natural History), Cromwell Road, London SW7 5BD, UK.

J.H. WETTON

Department of Genetics, School of Medicine, Queen's Medical Centre, Nottingham NG7 2UH, UK.

K. WÖHRMANN

Biological Institute, University of Tübingen, Auf der Morgenstelle 28, D7400 Tübingen 1, Federal Republic of Germany.

D. WOOL

Department of Zoology, George S. Wise Faculty of Life Sciences, Tel Aviv University, Tel Aviv, Israel.

1 Electrophoresis in agricultural pest research—a technique of evolutionary biology

M.F. CLARIDGE

School of Pure and Applied Biology, University of Wales College of Cardiff, Cardiff, Wales, UK

Until recently, most applied biologists would probably have doubted the relevance of evolutionary biology to problems of the control of pests of agriculture and forestry. On the other hand, the value of variability in crop plants themselves has been exploited since the earliest days of agriculture for the production of desirable cultivars. Examples of the utilization of inherited disease and pest-resistance factors by plant breeders are well known and have frequently led to successful control strategies (e.g. Russel 1978).

At a time when the control of insect pests was largely achieved by the use of broad-spectrum chemical insecticides, there was no obvious need to understand the nature of variation within and between populations of pests. The early development of resistance to chemical pesticides in populations of many species first alerted applied biologists to the need to understand the nature and evolution of such phenomena (Conway 1981). The very same problems of evolving resistance to pesticides also stimulated the present trend towards an ecologically-based system of pest control — integrated pest management in the current jargon. Such systems depend on an understanding of the interactions between pest populations, their hosts and possible biological control agents. Vital to such understanding is knowledge of the variation within and between populations of pests and their natural enemies.

The most obvious level at which an understanding of natural

Electrophoretic Studies on Agricultural Pests (ed. Hugh D. Loxdale and J. den Hollander), Systematics Association Special Volume No. 39, pp. 1-6. Clarendon Press, Oxford, 1989.

variation may be relevant is that of the species. Biological species of sexually reproducing organisms are reproductively isolated from one another and, therefore, have the possibility of maintaining distinct biological characteristics because of the lack of hybridization between them. The necessity for accurate and precise species taxonomy has often been emphasized for the understanding of pest problems. Classical taxonomy relies primarily on morphological differences between different species. The biological concept of species emphasizes reproductive isolation, so that biological, and therefore genetically distinct, species may exist that show no, or very little, morphological separation (Claridge 1988). Such sibling, or cryptic species are difficult to identify, but may be important in pest and biological control problems. In all major groups that have been studied in detail, such as the many insect vectors of disease, insect and nematode pests, and some biocontrol agents, sibling species have commonly been demonstrated (reviews in Claridge and den Hollander 1983; Diehl and Bush 1984; Claridge 1988). The solutions to these problems are simpler in principle than in practice. In many pests, populations, or even individuals within populations, have often been observed to possess distinctive properties with respect to their crop hosts. Variants of this kind have usually been, and indeed often still are, referred to as biotypes (Claridge and den Hollander 1983; Diehl and Bush 1984). Such biotypes may be distinguished by particular biological characteristics relevant to their pest status. For example, so-called host races of parasites associated with different hosts might be better termed biotypes. A wide variety of genetic entities may be included in this vague category. Sometimes even the nature of reproduction in a pest may not be well understood. For example, in aphids, which often alternate sexual and parthenogenetic reproduction, biotype development would be quite different during the different reproductive phases. When biotypes of pests have been carefully studied, they have usually been found in parthenogenetic forms to be distinct clones, and in sexual ones, to be either genetically distinct and reproductively isolated sibling species or genetic variants within a panmictic population. Any techniques that help us to understand the complexity of such variation in pest populations must be of enormous value to applied biologists.

The revolution in molecular biology in the last twenty or thirty years has not only provided a fundamental understanding of the chemical nature of the genetic material in the form of DNA, but has also provided a number of relatively simple techniques that allow new ways of looking at genetic variation. These have been excellently reviewed with respect to insect pests of agriculture by Menken and

Ulenberg (1987) and, more generally, by Ferguson (1980). Techniques now widely used include DNA sequencing, DNA hybridization, protein sequencing, and protein gel electrophoresis. Of these, gel electrophoresis is the simplest, oldest, and most widely used. The applications of this technique provided the major theme for the Rothamsted symposium and are central to the present volume. These methods make possible the demonstration of protein differences that may indicate levels of gene flow or reproductive isolation between populations that would otherwise be difficult to detect. Several authors in this volume show the existence of otherwise unrecognized sibling species in major pest groups that may necessitate fundamental changes in field control strategies (e.g. Pashley and Menken for Lepidoptera, and Robinson for nematodes). Additionally, these techniques may be used to identify markers within species populations that make possible more detailed understanding of population structure, dynamics, and migration (e.g. Berlocher and Loukas for Tephritid fruit flies, and Loxdale and Wöhrmann for aphids). The power and value of electrophoretic techniques are clearly demonstrated throughout this volume.

In the early days of electrophoresis, it seemed that there might be a quantitative correlation between degree of genetic divergence of populations and enzyme differentiation as measured by standard indices, such as Nei's genetic identity and distance (Nei 1971, 1972). It was thus assumed that taxonomic categories that represent different degrees of evolutionary divergence, such as geographic race, subspecies, sibling species, morphological species, etc., might be objectively defined by different quantitative measures of these indices. For example, the classic and beautiful work of Ayala *et al.* (1974) on the widely-sampled *Drosophila willistoni* species group in Central and South America clearly seems to conform to this hypothesis. However more recently, other instances have shown that there may often be no clear correlation between degree of morphological diversification and electrophoretic divergence. Many examples are known of morphological sibling species that are very clearly differentiated by diagnostic electrophoretic patterns (e.g. Bush and Hoy 1984). On the other hand, interesting examples are emerging in which reproductively isolated species show considerable differences in morphology and/or behaviour, but little detectable electrophoretic differentiation. Particularly well-documented cases of this are provided by some of the species of *Drosophila* in the Hawaiian archipelago. For example, the two sympatric species *D. sylvestris* and *D. heteroneura* have genetic differences of the same order of magnitude as those between local populations within one species in the *D. willistoni* group (Sene and Carson 1977; Ayala 1982). Detailed studies of polytene chromosomes of the two species also show them to be almost identical. The two species are, however, quite distinct in head shape and other characters used in courtship and pair-forming behaviour. Thus, it is clear that electrophoretically detectable differentiation is not always the first to become established during the process of speciation.

A good example of a pest for which detailed studies on variation — both within and between populations — have become necessary in order to develop appropriate control strategies, is the Brown Planthopper, *Nilaparvata lugens*. This insect is currently the major pest of irrigated rice in tropical and temperate Asia and Australasia. A major strategy for its control in tropical Asia has been the breeding of rice cultivars resistant to feeding by the pest [summaries in Wilson and Claridge (1985) and Claridge and Morgan (1987), see also den Hollander in this volume]. Planthoppers able to damage rice cultivars with different genes for resistance have often been termed biotypes. What appear to be identical insects have also been found feeding on a common tropical weed grass, *Leersia hexandra*. These insects are regarded by some as a non-virulent biotype. In fact, detailed studies on reproductive isolation show that the biotypes on different resistant rices are only variants within panmictic populations. However, those on the weed represent a very closely allied, but separate, biological species. The consequences of these findings for pest management strategies are very significant. Evolution of virulence to rice cultivars might be expected to occur rapidly in the field under intense selection. The weed-feeding species, on the other hand is most unlikely to pose any threat to rice production and may even have the advantage of maintaining a reservoir of natural enemies in the absence of rice-feeding insects. As in the example of *Drosophila sylvestris* and *D. willistoni*, so in the two sibling species of *N. lugens*, very little electrophoretically detectable genetic differentiation has been found. Behaviour and courtship-signal differentiation appear to have evolved more rapidly.

The remaining chapters in this volume demonstrate, for many different organisms, the important genetic differentiation that electrophoretic and other techniques of molecular biology now allow us to display. In the excitement, we should not be totally swept away. In all field studies, for a complete understanding, a multidisciplinary approach must be maintained. Electrophoresis will nevertheless continue to provide major contributions to such understanding.

Acknowledgments

I am pleased to thank my colleagues and co-organizers of the Rothamsted Symposium, Hugh Loxdale and Jeroen den Hollander, for taking over my responsibilities during my illness just before and during the meeting. I also thank them for inviting me to contribute this brief introductory chapter to the book.

References

Ayala, F.J. (1982). Gradualism versus punctualism in speciation: reproductive isolation, morphology, genetics. In *Mechanisms of speciation* (ed. C. Barigozzi), pp. 51–6. Alan R. Liss, New York.

Ayala, F.J., Tracey, M.L., Hedgecock, D., and Richmond, R.C. (1974). Genetic differentiation during the speciation process in *Drosophila*. *Evolution*, **28**, 576–92.

Bush, G.L. and Hoy, M.A. (1984). Evolutionary processes in insects. In *Ecological entomology* (ed. C.B. Huffaker and R.L. Rabb), pp. 247–78. Wiley, New York.

Claridge, M.F. (1988). Species concepts and speciation in parasites. In *Prospects in systematics* (ed. D.L Hawksworth), pp. 92–111. Clarendon Press, Oxford.

Claridge, M.F. and den Hollander, J. (1983). The biotype concept and its application to insect pests of agriculture. *Crop Prot.*, **2**, 85–95.

Claridge, M.F. and Morgan, J.C. (1987). The Brown Planthopper, *Nilaparvata lugens* (Stal), and some related species: a biotaxonomic approach. In *Proceedings of the 2nd International Workshop on Leafhoppers and Planthoppers of Economic Importance* (ed. M.R. Wilson and L.R. Nault), pp. 19–32. CAB International Institute of Entomology, London.

Conway, G. (1981). Man versus Pests. In *Theoretical ecology*, 2nd edition (ed. R.M. May), pp. 356–86. Blackwell, Oxford.

Diehl, S.R. and Bush, G.L. (1984). An evolutionary and applied perspective of insect biotypes. *A. Rev. Ent.*, **29**, 471–504.

Ferguson, A. (1980). *Biochemical systematics and evolution*. Blackie, Glasgow.

Menken, S.B.J. and Ulenberg, S.A. (1987). Biochemical characters in agricultural entomology. *Agric. Zool. Rev.*, **2**, 305–60.

Nei, M. (1971). Interspecific gene differences and evolutionary time estimated from electrophoretic data on protein identity. *Am. Nat.*, **105**, 385–98.

Nei, M. (1972). Genetic distances between populations. *Am. Nat.*, **106**, 283–92.

Russel, G. (1978). *Plant breeding for pest and disease resistance*. Butterworths, London.

Sene, F. M. and Carson, H. L. (1977). Genetic variation in Hawaiian *Drosophila*. IV. Allozymic similarity between *D. sylvestris* and *D. heteroneura* from the island of Hawaii. *Genetics*, **86**, 187–98.
Wilson, M. R. and Claridge, M. F. (1985). The leafhopper and planthopper faunas of rice fields. In *The leafhoppers and planthoppers* (ed. L. R. Nault and J. G. Rodriguez), pp. 381–404. Wiley, New York.

2 Electrophoretic studies: Perspectives for population biology

R.J. BERRY

Department of Biology, University College London, UK

Abstract

Variation is the core of biology, and electrophoresis is currently the best technique for quantifying inherited variation. There have been extensive gains from the wide use of electrophoresis in population biology since the method was introduced two decades ago, especially in taxonomy, phylogenetic reconstruction, and the elucidation of population structure and processes. The biggest disappointments have come from failures to relate the inherited characteristics of species to particular habitats and to determine the reasons for the adaptive behaviour of most allozymic variants. Notwithstanding, there is still much to be gained from the use of electrophoretic techniques, both in the study of pests and in learning more about populations in general.

Introduction

There is a rather painful irony in the fact that the dispute between Mendelians and biometricians that began at the same time as the rediscovery of Mendel's work in 1900, and which continued as a debate between palaeontologists and geneticists in the 1920s and 1930s, and in a less extreme form between the proponents of classical or balanced population structures in the 1950s, reignited in the 1960s when electrophoresis became a tool in population biology. The debate in the 1960s and 1970s was ostensibly about selection versus

Electrophoretic Studies on Agricultural Pests (ed. Hugh D. Loxdale and J. den Hollander), Systematics Association Special Volume No. 39, pp. 7–16. Clarendon Press, Oxford, 1989.

neutralism (or in another guise, gradualism versus punctuationism) but, throughout, the underlying problem has revolved round the nature, measurement and maintenance of variation (Lewontin 1974; Berry 1987*a*). Indeed, there is a sense in which variation can be regarded as the distinguishing characteristic of biology, separating it from 'restricted sciences' such as chemistry and physics (Pantin 1968).

The irony is that electrophoresis provided for the first time a direct way of measuring inherited variation, yet biologists have argued about variation levels at least since 1900, and continue to argue. Both theoreticians and practical geneticists (particularly animal and plant breeders) have traditionally assumed the availability of large amounts of inherited variation for adaptation (Fisher 1930), but some of the same theoreticians developed models in which variation laid a burden upon its possessors (Müller 1950; Haldane 1957). The paradox that variation could be both needful and problematical was an extension of the 1950s debate about 'classical' and 'balanced' population structures (see Mather 1974).

The advent of electrophoresis precipitated a major dispute about the value of variation to an individual, and stimulated a great deal of both experimental and theoretical work. But, above all, the possibility of determining the genotype of specific loci opened an enormous potential for solving long-running problems.

Where are we? In this paper, I examine six perspectives from a personal (and largely vertebrate) viewpoint to ascertain a sensible understanding of the current situation.

Perspectives

1. Neutralism

The predictions of neutralism about mutant accumulation and the accuracy of the molecular clock have not been borne out. Only so-called pseudogenes and the third base of codons follow the unadorned expectations of neutralist theory (Gillespie 1986; Nei 1987). For most genes, rates of gene substitution are clearly dependent on more factors than simple recurrent mutation. Both field and laboratory studies indicate that many alleles affect fitness under particular situations (Berry 1978). An interesting example of a hypothesis tested and rejected is the work of Petras, who, in an early use of electrophoresis in population work, concluded that certain allozyme variants in mice were neutral (Petras 1967) but later came to accept that strong selection was in fact acting on them (Petras and Topping 1983)

2. Taxonomy

Perhaps the most successful application of electrophoresis has been in contributing to the precision of taxonomy. For example, an important marine pollution indicator, the annelid *Capitella capitata* was shown by Grassle and Grassle (1976) to be a species complex of six different species. This was demonstrated on the grounds of allozymic differences, supported and confirmed by life-history traits. Baker (1979) distinguished between clones of the aphid *Myzus persicae* in Scotland by electrophoresis.

Electrophoresis has been particularly valuable in phylogenetic reconstruction (Avise 1983; Thorpe 1983), whilst indicating that speciation is not simply dependent upon and proportional to genic differences (Berry 1980*a*; Holmquist *et al.* 1988). Johnson *et al.* (1977) commented that their results on *Partula* speciation (which form one of the best genetic analyses of speciation so far available) 'strongly suggest that the conditions (or rules) for morphological divergence are not necessarily the same as those for divergence at the enzyme level' (see also Patterson 1987). There is no doubt that the initial euphoria about the potentialities of determining taxonomic distinctiveness by scoring individual genes has now evaporated, and has been replaced by a recognition that electrophoresis is an aid not an elixir for taxonomic problems.

3. Habitat

Perhaps the biggest disappointment from all the effort that has gone into electrophoresis has been a general failure to relate genotype to habitat. Nevo *et al.* (1984) have summarized data from 815 species showing that allozymic polymorphisms are associated with environmental heterogeneity. Notwithstanding, correlations are descriptive rather than causal and, as Nevo (1983) has himself commented, 'the unequivocal demonstration of the adaptive significance of an enzyme polymorphism depends on interrelated multidisciplinary evidence.' Certainly the connection between genotype and niche is complicated and depends on behaviour as well as physical factors (Jones *et al.* 1987). Although there have been many examinations of the relationship between habitat and variation (e.g. Felsenstein 1976; Hedrick 1976), the conclusion remains that the ability to score individual loci by electrophoresis has not produced any clear-cut answers as to why particular animals or plants exist in the habitats in which they do (Berry 1979; Hedrick 1986).

4. Stress and selection

One of the most important interfaces of ecology is with physiology. In particular, there is now a general recognition that stress, defined as an environmental stimulus that acts to disturb normal function, manifests itself as an expression of gene–environment interaction. This has been explored in particular by Southwood (1977, 1988). He envisages different habitats as templets for the moulding of life histories, and hence in principle able to provide an ecological 'periodic table'. The advantage of this concept is that stress becomes, if not synonymous with selection, at least suggestive that selection is acting (cf. Haldane 1959). This frees selection from being an idealistic perfecting agency to becoming a pragmatic, variable process. The genetics—and indeed the pesticide—literature is full of examples of the surprising and unexpectedly powerful effects of selection (Bishop and Cook 1981). Electrophoresis has a potential for dissecting complex physiological processes because so many chromosomal segments are labelled by allozymes (and even more by DNA variants) (Berry 1980*b*). The technique has not been employed seriously in this way by population biologists, although embryologists and immunologists regularly use inherited variation as a dissecting tool.

5. Allozymes as markers

If the main disappointment with electrophoresis results lies in a failure to 'explain' inherited variations in terms of habitat diversity, a second important regret stems from the poverty of reported links between enzyme function and adaptive significance, although there are a fair number of examples of these. One of the best is the association between clinal variation at the leucine aminopeptidase locus in *Mytilus edulis* and biochemical and physiological properties of the allozymes involved (Koehn and Siebenaller 1981). Another is the esterase that is electrophoretically diagnostic of organophosphate resistance in *Myzus persicae*, and is also responsible for the hydrolysis of the aphicide (Beranek and Oppenoorth 1977). But, in most cases, all that can be said of an electromorph that varies in time or space with particular environmental factors is that it is a 'marker' of an adaptive chromosome segment.

The problem is well illustrated by the *Hbb* locus in house mice. (This is the structural locus coding for the β-chain of haemoglobin.) In a number of populations, allele frequency changes with age or season have been found (Berry *et al.* 1987). The most important selective factor seems to be cold ambient temperature. Haemoglobins produced by the two common alleles differ in their oxygen-binding properties (Newton and Peters 1983) and hence in efficiency in maintaining body heat. However, laboratory selection responses, different genotypic changes to the same environmental shifts, and associations with skeletal and other characteristics seem to indicate that *Hbb* is merely labelling a key chromosome segment (Garnett and Falconer 1975; Myers 1974; Bellamy *et al.* 1973). *Hbb* is within two map units of *H-1*, which is a histocompatibility locus (although not the major one, which in the mouse is *H-2*). It is impossible to know whether the selection imputed to *Hbb* is wholly, partly, or not at all directed on the gene itself. Endler (1986) has analysed this sort of difficulty in detail.

Claims are sometimes made that levels of heterozygosity may be more important than the action of particular genes (see, e.g., Garten 1976; Garton 1984). Koehn *et al.* (1988) have examined this situation in an experimental growth study of the marine bivalve, *Mulinia lateralis*. They showed that genetic effects on growth were related to heterozygosity at particular loci, not to heterozygosity in general.

Other workers have compared amounts of variation between different groups of enzymes (e.g. Johnson 1976; Berry and Peters 1977), and have often found significant differences. However, the associations between function, flux, functional pathways, and genetical linkages are so complex that it is dangerous to generalize. For example, Bryant (1974) concluded that 83 per cent of the allozymic variation in house mice could be accounted for by temperature and rainfall data. Unfortunately, the most important climatic variable turned out to be average rainfall — but this is of little importance in the ecology of mice in most environments, quite apart from the fact that most of the data used by Bryant came from mice living in buildings and protected from the weather.

7. *Population structure*

Where electrophoresis has contributed to population biology very significantly is in revealing population structure and processes in a much more detailed and informative way than was previously possible. Estimates of gene flow and migration, social grouping and inbreeding, and differential survival can be much more precise when individuals can be genetically identified than when using traditional methods of observation, sampling or mark-recapture.

A genetical approach has added greatly to those species that have been studied from this point of view. In practice, a 'genetical approach' means electrophoresis. A few species are visibly polymorphic (see, e.g., Gershenson 1945; Barber 1954; Berry and Davis 1970), but in most cases significant genetic variation is only revealed by specialized techniques, of which electrophoresis is the simplest and commonest (e.g. for house mice *Mus musculus*, Berry *et al.* 1987; for red deer, *Cervus elephas*, Pemberton *et al.* 1988).

The number of cases in which electrophoresis work has revealed new facets of species biology are too numerous to review here. Suffice it only to note that the most valuable studies are longitudinal ones, in which the same gene pool is sampled on repeated occasions (or can be divided into different age groups: Berry *et al.* 1979). Comparison of geographically separated populations will almost certainly show that different populations are genetically distinct, but rarely provide evidence about the causes of the differentiation (Berry and Peters 1977). Particularly in work on agricultural pests, where the important problems are likely to involve changes in behaviour, population parameters, or insecticide resistance, there is no substitute for longitudinal sampling. Often, data accumulated through time can refute hypotheses, making the confirmation of the study worthwhile even without a specific hypothesis to test.

Electrophoresis and vertebrate pests

Remarkably little genetic work has been carried out on vertebrate agricultural pests, at least in Britain. The most thoroughly studied mammal is undoubtedly the house mouse (Berry 1987*b*). Extremely little work has been done on voles (*Microtus* spp.) or field mice (*Apodemus* spp.) (Berry 1985). Rats (*Rattus norvegicus*) have been studied from the point of view of warfarin resistance, but little else (Bishop *et al.* 1977). There is no published work on the genetics of either *Sciurus vulgaris* or *S. carolinensis* in Britain (Gurnell 1987). There is a small amount of electrophoretic data on rabbits *Oryctolagus cuniculus*, mostly concerned with social structure (Daly 1981). There is a fair amount of genetic work on feral cats, *Felis catus.*, but virtually all concerned with coat colour (Clark 1975). Fallow deer, *Dama dama*, are apparently allozymically monomorphic in Britain, presumably as a consequence of the small population originally introduced (Pemberton and Smith 1985). There have been a number of electrophoretic studies on red deer. (e.g. Gyllensten *et al.* 1983) and a considerable volume of work on white-tailed deer *Odocoileus virginianus* in the USA (e.g. Cothran *et al.* 1983).

This negative list could be extended considerably and unprofitably. The message that should be received from the work that has been done is that knowledge about population biology in general and some species in particular has been enormously extended by electrophoretic studies over the past two decades, and that there is much still to be learnt from pursuing electrophoresis and adding nucleic-acid investigations (which involve electrophoresis) to the present battery of techniques. But the real lesson to be digested is that electrophoresis is the best tool we have so far acquired for studying variation, and it is variation that should be the central obsession of all who profess and call themselves biologists.

References

Avise, J.C. (1983). Protein, variation and phylogenetic reconstruction. In *Protein polymorphism: Adaptive and taxonomic significance* (ed. G.S. Oxford and D. Rollinson), pp. 103–30. Academic Press, London and New York.

Baker, J.P. (1979). Electrophoretic studies on populations of *Myzus persicae* in Scotland from October to December, 1976. *Ann. appl. Biol.*, **91**, 159–64.

Barber, H.N. (1954). Genetic polymorphism in the rabbit in Tasmania. *Nature, Lond.*, **173**, 1227–9.

Bellamy, D., Berry, R.J., Jakobson, M.E., Lidicker, W.Z., Morgan, J. and Murphy, H.M. (1973). Ageing in an island population of the house mouse. *Age and Ageing*, **2**, 235–50.

Beranek, A.P. and Oppenoorth, F.J. (1977). Evidence that the elevated carboxylesterase (esterase 2) in OP resistant *M. persicae* (Sulz.) is identical with the organophosphate-hydrolyzing enzyme. *Pesticide Biochem. Physiol.*, **7**, 16–20.

Berry, R.J. (1978). Genetic variation in wild house mice: where natural selection and history meet. *Am. Scient.*, **66**, 32–60.

Berry, R.J. (1979) Genetical factors in animal population dynamics. In *Population dynamics* (ed. R.M. Anderson, B.D. Turner, and L.R. Taylor), pp. 53–80. Blackwell, Oxford.

Berry, R.J. (1980*a*). A geneticist faced with enzyme variation. In *Chemosystematics: Principles and practice* (ed. F.A. Bisby, J.G. Vaughan, and C.A. Wright), pp. 147–66. Academic Press, London and New York.

Berry, R.J. (1980*b*). Genes, pollution and monitoring. *Rapp. P.-v. Réun. Cons. int. Explor. Mer*, **179**, 253–7.

Berry, R.J. (1985). Evolutionary and ecological genetics of the bank vole and field mouse. *Symp. Zool. Soc. Lond.*, No. **55**, 1–31.

Berry, R.J. (1986). Genetics of insular populations of mammals with particular reference to differentiation and founder effects in British small mammals. *Biol. J. Linn. Soc.*, **28**, 205–30.

Berry, R.J. (1987*a*). Where biology meets; or how science advances. *Biol. J. Linn. Soc.*, **30**, 257–274.

Berry, R.J. (1987*b*). House mouse. *Biologist*, **34**, 177–86.

Berry, R.J. and Davis, P.E. (1970). Polymorphism and behaviour in the Arctic Skua (*Stercorarius parasiticus* (L.)). *Proc. R. Soc. Lond. B*, **175**, 255–67.

Berry, R.J. and Peters, J. (1977). Heterogeneous heterozygosities in *Mus musculus* populations. *Proc. R. Soc. Lond. B*, **197**, 485–503.

Berry, R.J., Bonner, W.N., and Peters, J. (1979). Natural selection in mice from South Georgia (South Atlantic Ocean). *J. Zool., Lond.*, **189**, 385–98.

Berry, R.J., Jakobson, M.E., and Peters, J. (1987). Inherited differences within an island population of the House mouse. *J. Zool., Lond.*, **211**, 605–18.

Bishop, J.A. and Cook, L.M. (eds). (1981). *Genetic consequences of man made change*. Academic Press, London.

Bishop, J.A., Hartley, D.J., and Partridge, G.G. (1977). The population dynamics of genetically determined resistance to warfarin in *Rattus norvegicus* from mid-Wales. *Heredity*, **39**, 389–98.

Bryant, E.H. (1974). On the adaptive significance of enzyme polymorphisms in relation to environmental variability. *Am. Nat.*, **108**, 1–19.

Clark, J.M. (1975). The effects of selection and human preference on coat colour gene frequencies in urban cats. *Heredity*, **35**, 195–210.

Cothran, E.G., Chesser, R.K., Smith, M.H., and Johns, P.E. (1983). Influence of genetic variability and maternal factors on fetal growth in white-tailed deer. *Evolution*, **37**, 282–91.

Daly, J.C. (1981). Effects of social organisation and environmental diversity on determining the genetic structure of a population of the wild rabbit (*Oryctolagus cuniculus*). *Evolution*, **35**, 689–706.

Endler, J.A. (1986). *Natural selection in the wild*. Princeton University Press, Princeton, N.J.

Felsenstein, J. (1976). The theoretical population genetics of variable selection and migration. *A. Rev. Genet.*, **10**, 253–80.

Fisher, R.A. (1930). *Genetical theory of natural selection*. Clarendon Press, Oxford.

Garnett, I. and Falconer, D.S. (1975). Protein variation in strains of mice differing in body size. *Genet. Res.*, **25**, 45–57.

Garten, C.T. (1976). Relationships between aggressive behavior and genetic heterozygosity in the oldfield mouse, *Peromyscus polionotus*. *Evolution*, **30**, 59–72.

Garton, D.W. (1984). Relationships between multiple locus heterozygosity and physiological energetics of growth in the estuarine gastropod, *Thais haemostoma*. *Physiol. Zool.*, **57**, 530–43.

Gershenson, S. (1945). Evolutionary studies on the distribution and dynamics of melanism in the hamster (*Cricetus cricetus* L.). *Genetics*, **30**, 207–51.

Gillespie, J. H. (1986). Rates of molecular evolution. *A. Rev. Ecol. Syst.*, **17**, 637–665.

Grassle, J.P. and Grassle, J.F. (1976). Sibling species in the marine pollution indicator *Capitella* (Polychaeta). *Science*, **192**, 567–9.

Gurnell, J. (1987). *The natural history of squirrels*. Christopher Helm, London.

Gyllensten, V., Ryman, N., Reuterwall, C. and Dratch, P. (1983). Genetic differentiation in four European subspecies of red deer (*Cervus elephas* L.). *Heredity*, **51**, 561–80.

Haldane, J. B. S. (1957). The cost of natural selection. *J. Genet.*, **55**, 511–24.

Haldane, J. B. S. (1959). Natural selection. In *Darwin's biological work* (ed. P. R. Bell), pp. 101–49. Cambridge University Press, Cambridge.

Hedrick, P. W. (1976). Genetic variation in a heterogeneous environment. *Genetics*, **84**, 145–50.

Hedrick, P.W. (1986). Genetic polymorphism in heterogeneous environments: a decade later. *A. Rev. Ecol. Syst.*, **17**, 535–66.

Holmquist, R., Miyamoto, M.M., and Goodman, M. (1988). Higher-primate phylogeny—why can't we decide? *Molec. Biol. Evol.*, **5**, 201–16.

Johnson, G.B. (1976). Genetic polymorphism and enzyme function. In *Molecular evolution* (ed. F.J. Ayala), pp. 46–59. Sinauer, Sunderland, Mass.

Johnson, M.S., Clarke, B.C., and Murray, J.J. (1977). Genetic variation and reproductive isolation in *Partula*. *Evolution*, **31**, 116–26.

Jones, J.S., Coyne, J.A., and Partridge, L. (1987). Estimation of the thermal niche of *Drosophila melanogaster* using a temperature-sensitive mutation. *Am. Nat.*, **130**, 83–90.

Koehn, R.K. and Siebenaller, J.F. (1981). Biochemical studies of aminopeptidase polymorphism in *Mytilus edulis*. II. Dependence of reaction rate on physical factors and enzyme concentration. *Biochem. Genet.*, **19**, 1143–62.

Koehn, R.K., Diehl, W.J., and Scott, T.M. (1988). The differential contribution by individual enzymes of glycolysis and protein catabolism to the relationship between heterozygosity and growth rate in the coot clam, *Mulinia lateralis*. *Genetics*, **118**, 121–30.

Lewontin, R.C. (1974). *Genetic basis of evolutionary change*. Columbia University Press, New York.

Mather, K. (1974). *Genetical structure of populations*. Chapman and Hall, London.

Müller, H.J. (1950). Our load of mutations. *Am. J. hum. Genet.*, **2**, 111–76.

Myers, J.H. (1974). Genetic and social structure of feral house mouse populations on Grizzly Island. California, *Ecology*, **55**, 747–59.

Nei, M. (1987). *Molecular evolutionary genetics*. Columbia University Press, New York.

Nevo, E. (1983). Population genetics and ecology: the interface. In *Evolution from molecules to men* (ed. D.S. Bendall), pp. 287–321. Cambridge University Press, Cambridge.

Nevo, E., Beiles, A., and Ben-Shlomo, R. (1984). The evolutionary significance of genetic diversity: ecological, demographic and life history correlates. *Lecture Notes in Biomathematics*, **53**, 13–213.

Newton, M.F. and Peters, J. (1983). Physiological variation of mouse haemoglobins. *Proc. R. Soc. Lond. B*, **218**, 443–53.

Pantin, C.F.A. (1968). *Relations between the sciences*. Cambridge University Press, Cambridge.

Patterson, C. (ed.) (1987). *Molecules and morphology in evolution: Conflict or compromise?* Cambridge University Press, Cambridge.

Pemberton, J.M. and Smith, R.H. (1985). Lack of biochemical polymorphism in British fallow deer. *Heredity*, **55**, 199–207.

Pemberton, J.M., Albon, S.D., Guinness, F.E., Clutton-Brock, T.H., and Berry, R.J. (1988). Genetic variation and juvenile survival in red deer. *Evolution*, **42**, 921–34.

Petras, M.L. (1967). Studies of natural populations of *Mus*. I. Biochemical populations and their bearing on breeding structure. *Evolution*, **21**, 259–74.

Petras, M.L. and Topping, J.C. (1983). The maintenance of polymorphism at two loci in house mouse (*Mus musculus*) populations. *Canad. J. Genet. Cytol.*, **25**, 190–210.

Southwood, T.R.E. (1977). Habitat, the templet for ecological strategies? *J. Anim. Ecol.*, **46**, 337–365.

Southwood, T.R.E. (1988). Tactics, strategies and templets. *Oikos*, **52**, 3–18.

Thorpe, J.P. (1983). Enzyme variation, genetic distance and evolutionary divergence in relation to levels of taxonomic separation. In *Protein polymorphism: Adaptive and taxonomic significance* (ed. G.S. Oxford and D. Rollinson), 131–52. Academic Press, London and New York.

3 Genetic studies of species differences and their relevance to the problem of species formation in *Drosophila*

RAMA S. SINGH,

Department of Biology, McMaster University, Hamilton, Ontario, Canada

Abstract

The genetic basis of species formation (barring sudden speciation by chromosome alteration, and by changes in breeding system, e.g. from outcrossing to selfing) remains one of the most important unsolved problems in evolutionary biology. Until the 1960s all proposed mechanisms of speciation have largely centred on geographic models of isolation and population differentiation. As far as the genetic mechanism of speciation is concerned (with the exception of Goldschmidt's discredited theory of speciation by macromutation), only general statements regarding the overall levels of genetic divergence during speciation have been made. Over the past two decades, population studies employing gene-enzymes have been productive in providing estimates of genetic divergence between closely related species. However, the only conclusion relevant to speciation drawn from these studies is that the amount of genetic change that occurs during speciation may not be as large as previously envisaged in Mayr's geographic theory of speciation by 'genetic revolution'. In our attempt to pursue this question, we have come to two conclusions. Firstly, the majority of the gene-enzyme loci normally employed in population studies are involved in intermediate metabolism and may not be relevant to speciation, even if the variation at these loci is relevant to population adaptation. Secondly, the genetic variability systems for adaptation and speciation may be separate, but in a manner quite different from

Electrophoretic Studies on Agricultural Pests (ed. Hugh D. Loxdale and J. den Hollander), Systematics Association Special Volume No. 39, pp. 17–49. Clarendon Press, Oxford, 1989.

that proposed by Goldschmidt, Wilson, and Carson. On the basis of observations that insect male genitalia show more divergence than other traits, and that species hybrids are usually phenotypically normal but sterile, we conclude that molecular studies of reproductive traits (both behaviour and reproductive organs) should provide the most relevant information about the genetic mechanism of species formation. We have made comparative studies of genic variation by employing one-dimensional gel electrophoresis for gene-enzymes and two-dimensional gel electrophoresis for male reproductive-tract proteins of *Drosophila melanogaster* and its sibling species, *D. simulans*, *D. mauritiana*, and *D. sechellia*. Two conclusions have emerged from these studies: firstly, that male reproductive-tract proteins have diverged more than some other proteins; secondly, that much of the divergence in male reproductive-tract proteins between these species may involve large changes in gene expression.

Introduction

Species hold a central position in the hierarchical organization of biological diversity, and hence mechanism of species formation has always been and still is a central problem in evolutionary biology (Mayr 1982*a*). Darwin solved the species problem by showing that species change, and in his theory the mode of change was what later came to known as *phyletic gradualism*. It took another eighty years or so before phyletic gradualism (anagenesis) and speciation (cladogenesis) became fully recognized as two separate modes of change (Huxley 1942). The next twenty years (1940s–1960s), which coincided with rapid development in population genetics, saw the debate on mechanism of speciation shift from early models framed in abstract, general terms of populations and genetically controlled phenotypic traits ('geographic' models) to formulations based on allele substitutions in gene pools ('genic' models) (Huxley 1942; Dobzhansky 1951; Mayr 1963). However, since the molecular basis of genetic variation was still unknown, very little of substance was really said about the *nature* of genetic variation for species formation, and it is better to treat all models of speciation before 1960s (barring, of course, models involving macromutations) as 'geographic' models. The last twenty years have seen the most rampant application of molecular genetic methods to the problem of species formation and a major effort has gone into confirming or denying the allopatric model of speciation proposed by Mayr (1963). We have learned a lot about the nature of genetic variation within species and genetic divergence between species (Lewontin 1974; Bush 1975; Ayala 1975; Selander

1976; Throckmorton 1977; Nevo *et al.* 1984), but the problem of speciation remains unsolved (for a most recent diversity of views, see Barigozzi 1982). A sense of disillusionment about genetic mechanisms of speciation (as much as about the evolutionary significance of gene-enzyme variation) has set in and it is not uncommon to hear that speciation is really a non-problem! Our disenchantment with the molecular approach to the problem of speciation may be due to its failure to provide a solution, but did we really know what we were looking for? Neither the genetic concept of species nor genetic mechanism(s) of speciation apply universally, and yet we speak of the mechanism of speciation as if there were indeed a universal solution to the problem. In other words, such questions as 'What is the genetic basis of speciation?' or even 'What is the central problem of speciation?' do not carry the same meanings in different disciplines of evolutionary biology. The present article deals with the nature of genetic divergence between species and what it tells us about the genetic mechanisms of speciation. But first we should define the *genetic* problems of species formation, and how genetic studies of species divergence bear on it.

The genetic problem of species formation

There are three sources of confusion that confound the genetic problem of speciation. They are *sudden versus gradual speciation*, *genic versus genomic changes* during speciation, and *genetic versus geographic* models of speciation. Sudden, catastrophic, or quantum speciation, in genetic terms, means speciation due to changes in a single major gene, chromosome, or whole genome. Evolution of inbreeders (selfers) from outbreeders, involving single gene changes in flower characteristics, fixation of choromosome translocations due to founder effect, changes in ploidy level, and species hybridization, are some well-known mechanisms of sudden speciation. We might also throw into this category all recent molecular models of speciation involving transposable elements, viruses, etc. In this class of models there is no mystery about the genetic mechanisms of speciation, as reproductive isolation is built into the genetic changes involved and they occur very rapidly if not in a single step. Also, in all these situations, reproductive isolation would obviously precede the major morphological, physiological, and ecological changes that would follow later. Thus, in those groups of organisms in which this mechanism of speciation is most likely to occur, the genetic basis of speciation is really known; what we need to know further is how common has been this mode of speciation? Evolution of selfers from outcrossers and speciation by polyploidy and by species hybridization are mechanisms whose occurrence can be determined with some certainty (Stebbins 1950; Grant 1971). With respect to the role of karyotypic changes other than polyploidy, White (1978) states that 'over 90 per cent, and perhaps 98 per cent, of all speciation events are accompanied by karyotypic change and in the majority of these cases the structural chromosomal rearrangements have played a primary role in initiating the divergence.' But all it really means is that, in 90–98 per cent of the cases, closely related species differ in their karyotype; it cannot mean (because it is unknowable) that in the majority of these cases chromosomal changes have been the initiating factor in speciation.

The second confusion is about the role of genic versus genomic changes during speciation. By genomic changes I here mean the kind of macromutational changes that involve the whole genome, first proposed by Goldschmidt (1940) and more recently resurrected by Gould and by Wilson (Gould and Eldredge 1977; Gould 1980; King and Wilson 1975), and affecting morophology, physiology, and function; I do not mean just changes in karyotype that can give rise to reproductive isolation owing to problems in chromosome pairing and segregation. It is, therefore, important not to confuse the usual kinds of genetic mechanism for sudden speciation that were mentioned above with the genetic mechanism postulated in Goldschmidt's 'saltational' model of speciation. In the former case, the karyotypic changes are genetic agents of *reproductive isolation*; in the latter they are genetic agents of change in *form and function*.

The third confusion has to do with geographic versus genetic models of speciation. Until the 1960s, all models of speciation, barring macromutational changes, were basically geographic models and have had little to say about the precise nature of the underlying genetic mechanisms. Thus, such models as the allopatric, sympatric, parapatric and stassipatric, listed in most textbooks, are all models showing how one species geographically gets split into two, and the real debate has been on the relative role of geographic *isolation versus selection* in this process. As far as the genetic basis of speciation is concerned, the main debate has been between models involving few genes (e.g. sympatric models) and those involving many genes (e.g. allopatric models). The problem of speciation has been confused unnecessarily by bringing chromosomal changes into this picture; and as I have said above, speciation by chromosomal alteration is a 'non-problem' as far as its *genetics* is concerned.

Thus, the central genetic problem of speciation consists of investigating *the genetic basis of gradual speciation involving single gene changes and the relationship between speciation and major morphological and physiological changes*. We also, of course, need to know about population structures that promote speciation, but such knowledge in the absence of the knowledge of genetic variation would not take us very far.

Constructing theoretical models of speciation: what do we need to know?

To construct models of speciation, we first need to know something about the nature of genetic variation within species, the levels of genetic divergence between species, and the relationship between the two levels of variation. Since the genetic divergence observed even between closely related species includes both speciation-related phyletic and genetic changes, some effort must be made to sort out the total genetic variation into different gene or protein 'classes', some of which might appear more relevant to speciation and worthy of more intensive study.

Secondly, we need to know about population structure and the role of various evolutionary forces that produce genetic structure within species, and about the relevance of population structure to speciation. Since genetic structure of species is also a consequence of evolution as well as a factor promoting evolution, knowledge of genetic structure without the knowledge of genetic variation is not sufficient to construct a model of speciation. It is for this reason that all models of speciation (geographic or genetic) include some feature of both population structure and genetic variation (Carson 1975, 1987; Carson and Templeton 1984; Templeton 1980, 1981; Barton and Charlesworth 1984). Thirdly, it is also important to know how much can we rely on the present genetic structure to speculate about its role in speciation and macroevolution. Because genetic structure can hardly be assumed to be stable over the evolutionary time-scale, we must have some knowledge about the role of history, i.e. the roles of population bottlenecks, colonization and range expansion, hybridization and introgression, etc., that may have affected the present genetic structure in the recent past. The role of history in evolution has now, for the first time, become amenable to genetic investigation by the use of DNA sequence and restriction analyses (see, e.g., Kreitman 1983). Finally, since both adaptation and speciation together are responsible for major morphological and physiological diversity, we should like to know whether the two processes are *causally* connected. This is a central issue in the debate on speciation and macroevolution (Gould and Eldredge 1977; Gould 1980; Wright 1982; Mayr 1982*b*; Stebbins 1982).

In the present chapter we are concerned with only the first of the above topics, i.e. genetic studies of species divergence and their relevance to the problem of speciation, to which we now return.

Genetic studies of species divergence

1. The approach

Speciation can be described in general terms as the transformation of genetic variation within populations into genetic variation between populations and species. This is of course the direction in which evolution takes place and this is also how the problem of speciation has been formally presented in population genetics. Our job is to describe the material basis of speciation and the role of evolutionary processes that give rise to new species. But, because speciation is a long-term evolutionary process and thus out of experimental reach, we cannot experimentally proceed with this problem in the direction in which it actually occurs, i.e. from within species to between species. We are forced to resort to experimental programmes that actually run in the opposite direction, and there are two such approaches.

The first approach involves genetic analysis of reproductive isolation between closely related species that can hybridize in the laborators and produce at least partially fertile hybrids. This approach was pioneered over 50 years ago by Dobzhansky (1936) and is used to determine the number and the location of genes causing sexual isolation and hybrid sterility in interspecific crosses (e.g. see Zouros 1981; Coyne 1984; Coyne and Charlesworth 1986). The advantage of this approach is that the genes discovered bear directly on the genetic basis of speciation; the disadvantage is that it can only be applied with species that are at least partially fertile and it almost certainly underestimates the number of genes, which may spuriously give credence to genetic theories of speciation involving small number of genes.

In the second approach, we start by examining species with respect to their genetic divergence and try to evaluate, by comparative analysis of species showing varying levels of reproductive isolation, the relevance (or irrelevance) to speciation of the various classes of genes and proteins. Furthermore, by simple character association–disassociation analysis, we hope to determine whether major morphological and physiological diversity is directly associated with speciation or is a consequence of phyletic divergence after the newly formed evolutionary lineages become independently established. This is essentially an approach of historical reconstruction, using existing closely related taxa, which in combination with the first approach described above can be a powerful tool for studies of species formation. This approach has been widely used with a large variety of organisms during the last twenty years and while we are still far from fully understanding the genetic mechanisms of speciation, there is a wealth of information on the genetic divergence of species.

2. *Levels of genic differentiation during speciation*

During the last twenty years the technique of gel electrophoresis has been applied to a wide variety of organisms at different stages of evolutionary divergence with a view to estimating the amount of genetic divergence that occurs during the early stages of speciation (e.g. see Ayala 1975; Powel 1975; Throckmorton 1977; MacIntyre and Collier 1986). The literature is too large to review here and therefore only a sample of average statistics for the lower stages of evolutionary divergence in *Drosophila* is provided in Table 3.1. The most important finding on species divergence is the generally low level of genic differentiation associated with the early stages of speciation. The local populations of *willistoni*, *mulleri*, and *pseudoobscura* groups appear to be very similar and show a genetic identity of over 97 per cent. The subspecies of *Drosophila* on the average show a genetic identity from 80 to 90 per cent, sibling species from 46 to 78 per cent, and non-sibling species from 16 to 35 per cent. Note that while the genetic identity values for subspecies in different *Drosophila* groups are quite comparable, the values for sibling and non-sibling species vary among groups. Thus, sibling species of *mulleri* and *bipectinata* group show higher genic similarity than those of *willistoni* and *obscura* groups; and the non-sibling species of Hawaiian *Drosophila* and *obscura* group show genic similarity that is lower than that in the *willistoni* group. The average difference of 15 per cent in genetic identity between subspecies equates to a genetic distance of $D = 0.16$, which means that 16 per cent or roughly 1 out of 6 gene loci have diverged between subspecies. Of course, the 15 per cent difference must include both relevant and irrelevant genetic changes, which leads to the unavoidable conclusion that relatively little genetic differentiation is required for speciation

Table 3.1 Summary of Nei genetic identity in populations of *Drosophila* at different stages of evolutionary divergence. (Modified from Ayala 1975.)

Species group or cluster	Local populations	Subspecies	Sibling species	Non-sibling species	Source
willistoni	0.970 ± 0.006	0.795 ± 0.013	0.563 ± 0.023	0.352 ± 0.023	Ayala *et al.* (1974)
mulleri	0.999 – 0.001	0.878 ± 0.009	0.777 ± 0.022	—	Zouros (1973)
bipectinata	—	0.907 ± 0.10	0.788	—	Yang *et al.* (1972)
pseudoobscura	0.993	0.824	—	—	Ayala and Dobzhansky (1974)
obscura	—	—	0.466 ± 0.031	0.204 ± 0.004	Lakovaara et al. (1972)
Hawaiian	—	—	0.539 ± 0.054	0.164 ± 0.019	Ayala (1975)

(Lewontin 1974; Ayala *et al.* 1974; Ayala 1975; Throckmorton, 1977).

A second feature of the species divergence data on *Drosophila* appears to be a general lack of completely differentiated loci. Even species groups showing high genetic distances (e.g. the *willistoni* group) lack fully differentiated loci. There is also generally a lack of species-specific (unique) alleles at the polymorphic loci. These feature of the data led Lewontin (1974) to conclude that 'the overwhelming preponderance of genetic differences between closely related species is latent in the polymorphisms existing within species' (Lewontin 1974, p. 179).

A third and final feature of species divergence data is the range of variation shown by the estimates of genetic distance (*D*) in different groups of organisms at similar levels of evolutionary divergence. Table 3.2. presents estimates of *D* at the species, genus, and family levels in various groups of organims. The estimates of *D* between non-sibling species vary from a minimum of 0.02 for birds to a maximum of 1.09 for sessile invertebrates. The *D* values are generally low for primates, rodents, and birds, and generally high for reptiles, fishes, *Drosophila*, and other invertebrates; the pattern is consistent at the genus and the family level where data are available. Markedly different estimates of *D* values in different groups of organisms at the same level of evolutionary divergence means that either the taxonomic categories between widely divergent groups of

Table 3.2 Summary of Nei genetic distance at the level of species and above from a random sample of studies with various groups of organisms (for primary references, see Singh 1988*a*). The numbers of pairwise comparisons made are given in parenthesis.

Group	Species (2) (non-sibling)	Genus	Family
Primates	0.12 ± 0.02 (2)	0.32 ± 0.12 (10)	0.66 ± 0.25(16)
Rodents	0.25 ± 0.17 (25)	0.83 ± 0.17 (44)	
Birds	0.02 ± 0.01 (23)	0.26 ± 0.32 (81)	0.78 ± 0.09 (8)
Reptiles	0.37 ± 0.05 (12)		
Fishes	0.37 ± 0.27 (8)	1.21 ± 0.13 (13)	
Drosophila	0.89 ± 0.48 (4)		
Other insects	0.17 ± 0.09 (97)		
Other invertebrates			
Mobile	0.34 ± 0.18 (51)	0.94 ± 0.11 (42)	
Sessile	1.09 ± 0.60 (4)		

organisms are not comparable or the population and evolutionary processes that give rise to genetic differentiation at different evolutionary stages are quite variable and hence not comparable between widely divergent groups of organisms. This would mean that different groups of organisms may show varying levels of genetic divergence at different evolutionary stages as all the divergence may not be evolutionarily meaningful.

Evolutionary constraints on species divergence

In the preceding section we summarized the main features of genetic divergence data gathered by gel-electrophoretic techniques but refrained from making any comments on the limitation or the usefulness of these data with respect to the problem they were meant to address, i.e. that of how much genetic change occurs during the early stages of speciation. A variety of criticisms related to techniques (Singh *et al.* 1976; Coyne 1982), types of genes sampled (Selander 1976; Gillespie and Kojima 1968), population history (Nei *et al.* 1975), and evolutionary relevance (Kimura 1968; King and Jukes 1969; King and Wilson, 1975) have been presented over the years. Rather than present a review and evaluation of these various criticisms, we will illustrate the various types of evolutionary factors that may constrain species divergence. We will do this with the help of summary statistics of genic divergence data on *D. melanogaster* and *D. simulans* collected in our laboratory (Singh and Rhomberg 1987; Choudhary and Singh 1987*a*; Coulthart and Singh, 1988*a*, *b*, *c*). Because of the numbers and the types of genes sampled, we think that the conclusions drawn from these data should be of general interest.

The summary statistics for genic diversity within and between *D. melanogaster* and *D. simulans* are presented in Table 3.3 and the genetic divergence data for genes on individual chromosomes are summarized in Table 3.4. There are several important aspects of these data that either extend our previous knowledge of genetic variation and species divergence or are at variance with them. (1) The abundant hemolymph proteins are more (and the non-hemolymph proteins less) polymorphic than the enzymes. This suggests that with respect to genic diversity enzymes may not provide a representative picture of the genome. (2) Seven per cent of all loci examined have become fully diverged and all are enzymes; none of the abundant protein loci show complete divergence. (3) The loci for which one species is monomorphic and the other polymorphic are also interesting. At most of these loci, one of the species apparently carries one of

Table 3.3 Summary statistics for polymorphisms in different protein classes detected by 1DE in *Drosophila melanogaster* and *D. simulans*[a].

Enzymes and proteins	Species	No. of loci studied	Proportion of loci uniformly monomorphic	Proportion of loci polymorphic (mean ± SE)	Hetero zygosity (mean ± SE)	Proportion of loci diverged ($I = 0$)
Enzymes	*D. melanogaster*	79	0.418	0.412 ± 0.049	0.127 ± 0.018	0.101
	D. simulans	79	0.443	0.329 ± 0.031	0.112 ± 0.016	
Larval (haemolymph)	*D. melanogaster*	8	0.375	0.550 ± 0.092	0.179 ± 0.069	
proteins	*D. simulans*	8	0.250	0.650 ± 0.056	0.255 ± 0.081	0
Larval (carcass)	*D. melanogaster*	30	0.667	0.193 ± 0.057	0.037 ± 0.015	
and adult proteins	*D. simulans*	27	0.926	0.059 ± 0.020	0.004 ± 0.003	0
Total (1DE)	*D. melanogaster*	117	0.470	0.420 ± 0.070	0.102 ± 0.014	0.070
	D. simulans	114	0.614	0.288 ± 0.022	0.094 ± 0.012	

[a] Data from Singh and Rhomberg (1987), and Choudhary and Singh, (1987*a*).

Table 3.4 Proportion of shared (common) and unshared (unique) alleles and mean genetic identity of allozyme loci for various chromosomes between *Drosophila melanogaster* (*D.m*) and *D. simulans* (*D.s*).[a]

Chromosome	No. loci	Total no. alleles	Proportion of alleles			Mean genetic identity	Proportion loci diverged ($I = 0$)
			Unique to *D.m*	Unique to *D.s*	Common		
X	9	31	0.29	0.13	0.58	0.709	0.11
2	17	51	0.29	0.20	0.51	0.631	0.29
3	27	97	0.23	0.22	0.56	0.774	0.04
2 + 3	57[b]	202	0.28	0.18	0.54	0.729	0.10
Total	112[c]	320	0.27	0.16	0.57	0.832	0.07

[a] Data from Choudhary and Singh (1987*a*) and Singh and Rhomberg (1987).
[b] Includes 13 additional autosomal loci that are unmapped.
[c] Includes 2 additional proteins whose chromosome location is not known.

the alleles of the polymorphic set found in the other species. Two simple logical possibilities exist for the origin of this situation: either one species (perhaps through one or more population bottlenecks) has lost all but one of the alleles of an ancestral polymorphism, or one species has acquired one or more alleles in addition to an ancestral monomorphism. A total of 27 per cent of alleles in *D. melanogaster* and 16 per cent in *D. simulans* are unique to the species. (4) More genes on chromosome II show complete divergence than on chromosome III and X (χ^2 test, $P < 0.05$). (5) The mean genetic distance between *D. melanogaster* and *D. simulans* based on 112 homologous loci is 0.179, which gives a divergence time of $t = 0.09$ Myr (Nei 1975). This is much less than the average divergence time of 2–3.5 Myr based on DNA sequence data (Stephens and Nei 1985) and suggests that functional constraints and purifying selection have played an important role in the evolution of proteins.

It is a common practice in evolutionary genetics to compare genetic divergence at a given taxonomic level across organisms and seek answers for varying rates of evolution in different organisms (e.g. King and Wilson 1975). We now turn to the various types of evolutionary constraints that can operate on species divergence and can make comparisons of *D* across widely divergent taxa rather useless for deducing rates of evolution and roles of various evolutionary forces.

1. Phylogenetic constraints

The functional significance of differences in genic polymorphisms at homologus loci between species cannot be fully understood without considering the phylogenetic aspect of species divergence at these loci. Closely related species are expected to share polymorphisms, and we need to consider joint interspecies distribution of monomorphism and polymorphism for homologous loci. With the criterion for a polymorphic locus being a frequency of 99 per cent or less for the most common allele, 47 loci are monomorphic in both species, and 37 are polymorphic in both species; 21 are polymorphic only in *D. melanogaster* and 7 polymorphic only in *D. simulans* (Table 3.5). On a more stringent criterion of polymorphism (most common allele at a frequency of 95 per cent or less), these numbers become 62, 23, 16 and 11. On either criterion, the monomorphic or polymorphic state of a locus is highly correlated between species ($\chi^2_3 = 30.29$ and 49.20, respectively, for 99 per cent and 95 per cent criteria; $P < 0.005$). This type of positive correlation is expected for two species that arose relatively recently from the same ancestral gene pool. Another explanation, and one that need not exclude the historical one, invokes

Table 3.5 Number of loci polymorphic (P) or monomorphic (M) in *Drosophila melanogaster* and *D. simulans* (*D.m*/*D.s*) and χ^2 test for independence of polymorphisms between species.[a]

Criterion for polymorphism[b]		Loci				χ^2 (Probability)
		M/M	M/P	P/M	P/P	
< 99%	Observed	47	7	21	37	30.29 ($P<0.005$)
	Expected	33	21	35	22	
< 95%	Observed	62	11	16	23	23.18 ($P<0.005$)
	Expected	51	22	27	12	

[a] Data from Choudhary and Singh (1987*a*).
[b] Frequency of the most common allele.

similar kinds of natural selection on homologous loci in the two species. However, without direct evidence to support the latter explanation, the postulation of historical effects seems at present to be the more parsimonious.

2. *Historical constraints*

By historical constraints we mean effects of population bottlenecks (affecting historical effective species size), colonization and rapid expansion into new areas, and species hybridization and introgression. These are all essentially genetic accidents that can have profound effects on species divergence.

Theoretical studies show that while the loss of alleles depends largely on the bottleneck size, the reduction in heterozygosity depends heavily on the rate of population growth following the bottleneck. If the rate of population growth is very high, a relatively high level of heterozygosity can be maintained even in the face of an extreme reduction in population size (Nei *et al.* 1975; Motro and Thompson 1982; Maruyama and Fuerst 1985*a*, *b*). Thus, allele numbers at homologous polymorphic loci can be used as indicators of historical effective size in closely related taxa. Table 3.6 presents just such a comparison between *D. melanogaster* and *D. pseudoobscura* (Choudhary and Singh 1987*b*). Multi-allelic loci would be most sensitive to loss of alleles owing to bottleneck and, as Table 3.6 shows, *D. melanogaster* has a significantly reduced number of alleles but similar heterozygosity when compared to *D. pseudoobscura*. The result is consistent with what we know about the population structures of these species. *D. melanogaster* must have gone through repeated

Table 3.6 Number of alleles (*n*) and mean heterozygosity (*H*) at some homologous allozyme loci studied by sequential gel electrophoresis in *Drosophila melanogaster* and *D. pseudoobscura*.

D. melanogaster			*D. pseudoobscura*		
Locus	*n* (>1%)	H	Locus	*n* (>1%)	H
Xdh	15	0.80	Xdh	27	0.63
Est-6	7	0.40	Est-5	33	0.85
Est-C	4	0.25	Est-4	26	0.79
Ao	7	0.42	Ao	16	0.56
Odh	2	0.14	Odh	8	0.09
Adh	2	0.23	Adh	1	0
α-Gpd	2	0.30	α-Gpd	1	0
Mean	5.57	0.36		16.0	0.42

[a] Data from Choudhary and Singh (1987*b*).

bottlenecks during its colonization and population expansion around the world. On the hand, *D. pseudoobscura*, on the basis of its ecological habitat and distribution range, can be assumed to have maintained a relatively large effective species size (Dobzhansky and Epling 1944).

The effect of population bottleneck on species divergence has not been appreciated as much as has its effect on reduction in heterozygosity (Choudhary and Singh 1987*b*). Natural populations of *D. simulans* show similar levels of heterozygosity but significantly reduced levels of geographic differentiation. From a combined analysis of allele number, heterozygosity, and geographic differentiation, we narrowed down the causes of differences in the genetic structure of these species to differences either in their selection 'strategy' (i.e. *melanogaster* being broad-niched and *simulans* being narrow-niched) or in their colonization history (i.e. recent colonization by *D. simulans* without incurring a reduction in population size) (Choudhary and Singh 1987*a*). The geographic distribution of mitochondrial DNA variants clearly shows that the recent-colonization hypothesis is clearly favoured (Hale and Singh 1987). However, since both species have expanded their range, starting from the central African region, and have become cosmopolitan, their genetic divergence to some extent must be inflated and hence not strictly in accord with the chronological time.

Hybridization between incompletely isolated taxa can lead to introgression of new alleles and elimination of old ones, creating more interlocus variance in genetic divergence than would be expected in the absence of such genetic accidents. Situations of this kind can now easily be identified with the help of mtDNA. The *Mus musculus* and *Mus domesticus* hybridization is a classic case of this kind (Selander *et al.* 1969; Ferris *et al.* 1983) and a similar situation has been proposed between subspecies *D. simulans* and *D. mauritiana* (Solignac and Monnerot, 1986).

3. *Functional constraints*

Effect of structural–functional constraints such as subunit size, subunit structure, and substrate types on number of alleles and heterozygosity have been considered in the past (see Singh 1988*b* and the references therein), but the effect of functional constraints on genetic divergence between species is as yet an unexplored area. Table 3.7 presents summary statistics on genetic identity and proportion of alleles unique or shared between *D. melanogaster* and *D. simulans* for various classes of proteins and enzymes, classified according to some commonly used structural and functional constraints. Enzymes show higher divergence (or lower genetic identity) than abundant proteins, and among enzymes the Group II enzymes show higher divergence than the Group I enzymes. The proportions of unique and shared alleles are similar between these functional groups but, in all except one case, *D. melanogaster* has higher proportion of unique alleles than *D. simulans*. The exception is in the category 'Null tolerant/non-tolerant', which can be taken as an indication of dispensable versus essential enzymes. The two species share more alleles at the 'null non-tolerant' loci than the 'null-tolerant loci', suggesting that essential enzymes are probably under purifying selection. But in this case, in contrast to the two categories mentioned above, the effect on the number of alleles does not affect the overall genetic identity between species. Subunit structure has no effect on the statistics of species divergence and subunit size appears to have some effects, but only in the large-sized protein 'class'. Thus, the data in Table 3.7 show clearly that the various functional constraints differ in their effects on *allele number* and *allele frequencies* and the two statistics of variation can be used to separate the 'neutral/purifying selection' type of functional constraints, which limit variation, from the 'heterotic selection' type of functional constraints, which promote variation within species (Singh 1988*b*). Note that both purifying and heterotic selection would put constraints on species divergence.

Because of variation in heterozygosity between different classes of proteins, the mean heterozygosity of a species would very much depend on the number and the kind of loci sampled (Singh 1988*b*).

Table 3.7 Number of loci studied, total number of alleles observed, proportion of alleles unique to *Drosophila melanogaster* (*D.m*) or *D. simulans*, (*D.s*) or common to both species, and mean genetic identity for proteins belonging to different functional classes.[a]

Chromosome	No. loci	Total no alleles	Proportion of alleles			Mean genetic identity
			Unique to *D.m*	Unique to *D.s*	Common	
Enzymes	77	277	0.279	0.188	0.532	0.771
Abundant proteins	35	77	0.246	0.065	0.688	0.964
Monomeric enzymes	31	107	0.308	0.150	0.533	0.789
Multimeric enzymes	34	115	0.278	0.191	0.530	0.750
Group I enzymes	50	147	0.258	0.177	0.567	0.826
Group II enzymes	13	63	0.333	0.174	0.492	0.591
Null tolerant	18	70	0.286	0.200	0.514	0.697
Null non-tolerant	8	15	0.133	0.133	0.734	0.708
Subunit size						
< 50 kDa	22	64	0.234	0.234	0.531	0.717
50–75 kDa	13	43	0.209	0.256	0.535	0.761
> 75 kDa	14	68	0.353	0.074	0.573	0.617

[a] From Singh (1988*b*).

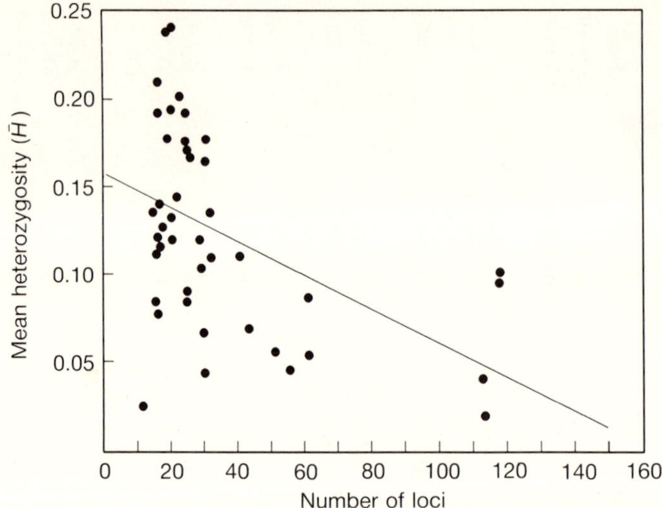

Fig. 3.1. Regression of mean heterozygosity on the number of loci studied in various species of *Drosophila*; $y = 0.154 - 0.000925x$ (regression coefficient significant: *F*-test, $P < 0.005$). The four points on the right are on four sibling species of the *D. melanogaster* complex studied in our laboratory. (Data from Singh and Rhomberg 1987.)

We have shown that in *Drosophila* the amount of heterozygosity is negatively correlated with the number of loci sampled (Fig. 3.1), which raises the question whether enzymes give us a true picture of genic variation in the genome. We have recently carried out a preliminary comparative study of proteins found in male reproductive tract tissues of *D. melanogaster* and *D. simulans* using high-resolution two-dimensional gel electrophoresis (2DE) (Coulthart and Singh, 1988*a*, *b*, *c*). In Table 3.8 we summarize some features of polymorphism within and divergence between *D. melanogaster* and *D. simulans* for the various sets of proteins that have been surveyed in the 2DE study just mentioned. Several interesting points arise from comparison of data in Tables 3.3 and 3.8.

First, the overall proportions of completely diverged loci are not very different between the soluble enzymes and other proteins surveyed by 1DE (7.0 per cent) and the abundant solubilized proteins surveyed by 2DE (13.2 per cent). The difference is reduced still further if the 1DE set is restricted to soluble enzymes (10.1 per cent). The second point concerns the differences in proportions of completely diverged loci seen between, on one hand, male reproductive-tract proteins that are found in both testis and

Table 3.8 Summary statistics for polymorphism in male reproductive tract proteins detected by 2DE in *Drosophila melanogaster* and *D. simulans*[a]

Tissue	Species	No. of loci studied	Proportion of loci uniformly monomorphic	Proportion of loci polymorphic (mean ± SE)	Heterzygosity (mean ± SE)	Proportion of loci diverged	
						Homologous (I = 0)	Non-homologous (showing + / −)
Common	*D. melanogaster*	110	0.927	0.041 ± 0.045	0.010 ± 0.008	0.098	0.183
	D. simulans	131	0.614	0.288 ± 0.022	0.094 ± 0.012		
Testis-specific	*D. melanogaster*	74	0.878	0.101 ± 0.029	0.030 ± 0.007	0.175	0.234
	D. simulans	87	0.896	0.092 ± 0.016	0.032 ± 0.010		
Gland-specific	*D. melanogaster*	26	0.769	0.173 ± 0.082	0.065 ± 0.033	0.261	0.391
	D. simulans	25	0.680	0.280 ± 0.000	0.119 ± 0.019		
Testis-elevated	*D. melanogaster*	34	0.911	0.058 ± 0.042	0.023 ± 0.019	0.035	0.291
	D. simulans	28	0.857	0.107 ± 0.051	0.028 ± 0.030		
Total (2DE)	*D. melanogaster*	244	0.893	0.080 ± 0.044	0.024 ± 0.010	0.132	0.231
	D. simulans	271	0.900	0.085 ± 0.000	0.029 ± 0.001		

[a] Summarized from Coulthart and Singh (1988*b*).

gland tissue of the tract ('common' and 'testis-elevated' proteins) and, on the other hand, those proteins found in testis but not gland tissue, or vice versa ('testis-specific' and 'gland-specific' proteins, respectively). The former classes contain 9.8 per cent and 3.5 per cent completely diverged loci, respectively, and the latter 17.5 per cent and 26.1 per cent completely diverged loci, respectively. The figure of 9.8 per cent completely diverged loci for 'common' proteins is strikingly close to the 10.1 per cent found with soluble enzymes. The question arises whether genes with relatively broad, or ubiquitous, patterns of expression between different tissues or stages of an organism ('housekeeping' genes) might have characteristically lower probabilities of complete interspecies divergence than do genes with relatively late or tissue/stage-specific patterns of expression.

Lastly, we draw attention to interspecies divergence of a different type, readily detectable with 2DE but not as readily with 1DE. Overall, 20–30 per cent of *D. melanogaster* or *D. simulans* male reproductive-tract protein loci (studied by 2DE) show 'presence/absence' differences between *D. melanogaster* and *D. simulans*. In these cases, a protein present in one of the two species lacks a detectable homologous spot in the other species; we have proposed that these seemingly species-specific proteins actually represent large interspecific differences in levels of gene expression (Coulthart and Singh 1988*b*). The proportions of loci showing such differences are given for the various tissue-distribution classes of male reproductive-tract proteins in Table 3.8. Although present/absence differences were seen frequently (18.3 per cent–39.1 per cent) in all tissue-distribution classes of proteins, the class of proteins with a 'common' expression pattern also carried the smallest proportion of loci (18.3 per cent) with differences of this type.

4. *Ontogenetic constraints*

Since the gland-specific proteins in *D. melanogaster* and *D. simulans* showed more heterozygosity within and more divergence between species than the 'common' proteins (Table 3.8), the question arises whether these observations are of general nature, and whether the differences are related to stage of development and tissue differentiation or to the organism's physiology and function.

Table 3.9 presents data on similarity measures for whole-reproductive-tract proteins and proteins of third-instar imaginal discs in randomly selected pairs of isofemale lines (one *D. melanogaster* and one *D. simulans*). As many well-defined spots are possible were compared (approximately 400 in reproductive-tracts, approximately

Table 3.9 Single-line comparisons for polypeptides of reproductive and non-reproductive tissues.[a]

Species compared	Wing discs		Reproductive tracts	
	F[b]	F_p[c]	F	F_p
D. simulans vs.	0.833 (0.811)	0.65	0.639	0.28
D. melanogaster		0.55[c]		0.27
D. simulans vs. *D. sechellia*	0.856	0.68 0.68	0.791	0.31 0.39
D. melanogaster vs. *D. sechellia*	0.700	0.51 0.56	0.664	0.37 0.45
D. simulans vs. *D. mauritiana*	(0.955)	—[d]	0.848	0.42 0.44
D. melanogaster vs. *D. mauritiana*	(0.803)	—	0.636	0.24 0.35
D. mauritiana vs. *D. sechellia*	—	—	0.781	0.43 0.33

[a] Data from Coulthart and Singh (1988*b*) and (figures in parentheses) from Ohnishi *et al.* (1983).
[b] F = shared fraction of total spots.
[c] F_p = pairable fraction of unshared spots.
[d] F_p Values refer to species listed on the same line of the table, in column 1.
[d] Data not available.

250 in imaginal discs). Reproductive-tract gels were also run for single isofemale lines of *D. mauritiana* and *D. sechellia*; wing discs were run for *D.sechellia* but not for *D. mauritiana* (Coulthart and Singh 1988*b*).

The similarity measures for imaginal discs are higher in every case than they are for reproductive-tracts. That is, more polypeptides in imaginal discs were identical (the F measure in Table 3.9) between species in each pair, and, of the non-identical polypeptides, more could be matched (the F_p in Table 3.9) with a homologous one in the other species than was possible with reproductive-tract polypeptides. The range of sampling variation in F, from random allele sampling at loci with overlapping polymorphisms, can be no greater than the total percentage of such overlapping polymorphic loci (10 per cent in *D.*

melanogaster and *simulans*) (Coulthart and Singh 1988*b*); thus, the differences in *F* values become more statistically significant. Interestingly, Ohnishi *et al.* (1983), with proteins of adult whole-body homogenates in two-dimensional gels stained with Coomassie blue, obtained *F* values that are much closer to those for imaginal discs than those for male reproductive tract data shown in Table 3.9. The proteins of testes appear to be as widely divergent between these two species as do those of whole reproductive tracts (Coulthart and Singh 1988*b*).

If these results hold, it would mean that genes and proteins that express during the early stages of development are more likely to be conserved than those that express later. The results are compatible with the observation that in most closely related species, species-distinguishing taxonomic differences first appear usually at the later stages of development. The results presented in Table 3.9 make sense in view of Haeckel's doctrine, that ontogeny recapitulates phylogeny.

5. *Selectional constraints*

Natural selection can affect species divergence in two opposite ways: it can retard divergence by eliminating alleles (purifying selection) or by maintaining balanced polymorphisms, or it can accelerate divergence by favouring different alleles in different species (adaptive genetic divergence). The purifying selection has been mentioned above and here we will concern ourself only with adaptive divergence.

Since adaptive divergence between species must use existing genetic variation within species, we should expect to find a positive correlation between genetic distance (*D*) and heterozygosity (*H*) at polymorphic loci. The same relationship is also expected under the neutral theory. The relationship between *D* and *H* for all polymorphic loci in *D. melanogaster* and *simulans* is shown in Fig. 3.2. There is indeed a strong correlation between *D* and *H*, it is negative and not positive as we would expect if genetic divergence between species were based on genetic variation within species. These results lend themselves to several interpretations. First, some of the highly heterozygous loci may indeed be locked under balanced polymorphisms and consequently their divergence is slowed down, giving low *D*. Second, genetic divergence between species may be predominantly affected by founder effect and less so by adaptive divergence, which would explain the negative correlation between *D* and *H* (Fig. 3.2). Finally, high divergence of loci with low heterozygosity may suggest a role for new mutations in adaptive divergence and speciation. I have argued elsewhere

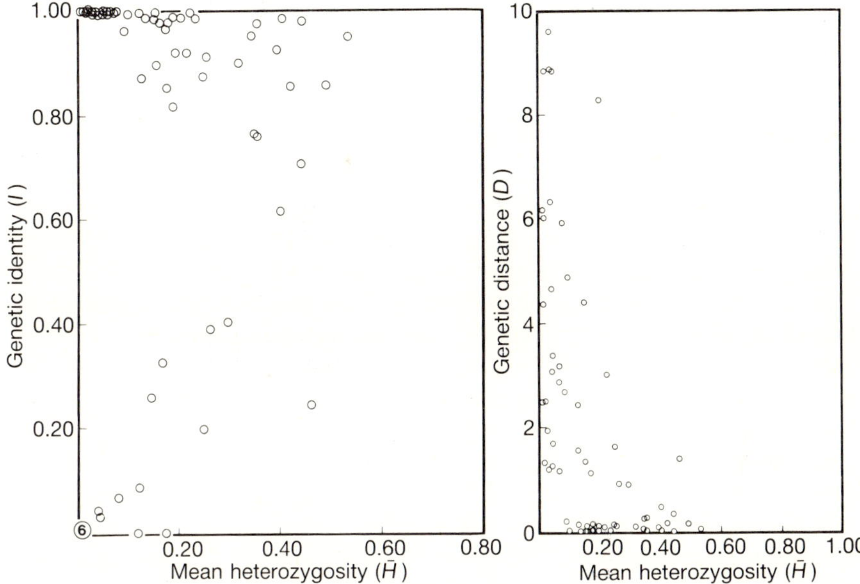

Fig. 3.2. Relationship between mean heterozygosity and genetic identity (left) and genetic distance (right) at polymorphic loci between *D. melanogaster* and *D. simulans*.

that founder effect is an important factor affecting species divergence (Singh 1988*c*) and our 2DE data suggest that role for new mutations during speciation is a real possibility (Wright 1982; Choudhary *et al.* 1988).

Genic variation and theories of speciation

Genic models of species formation make predictions of how many, or what proportion of, genes are involved in species formation. Mayr's 'peripatric' model (Mayr 1963, 1982*b*), an elaborated version of the basic allopatric model, suggested massive reorganization of the gene pool, with extensive allele substitution or fixation, in the course of achievement of reproductive isolation. Early results from electrophoretic studies, although based on fewer loci, usually fewer than 30 (Singh and Rhomberg 1987), disproved this prediction in its most extreme interpretation, and our recent results based on a larger sample of loci reinforce this conclusion (Choudhary *et al.* 1988). However, the early studies usually failed to detect complete allelic divergence at *any* loci, and thus left open the question whether such divergence occurs during species formation and, if so, at what proportion of loci.

It now appears, from our results, that in a sufficiently large sample complete divergence between closely related species can be detected at a small percentage of loci. Since *D. melanogaster* and *D. simulans* have apparently been separated for at least 2 Myr (Stephens and Nei 1985), they are probably far from their earliest stages of genetic differentiation. Thus, the figure of about 10 per cent completely diverged loci that we obtained would seem likely to include many changes attributable to side-effects of population processes (e.g. random genetic drift) accompanying species formation, or to subsequent phyletic divergence, and not causally related to the establishment of reproductive isolation *per se*. If as few as, say, 10 per cent of the completely divergent loci were actually involved in reproductive isolation, this would amount to a total of only about 50–100 genes in a genome the size of that carried by *Drosophila* [(5000–10 000 genes (Raff and Kaufman 1983)]. Although this is still a large absolute number of genes, it is perhaps not surprising that complete interspecies allelic divergence has not been widely detected by electrophoresis, especially with species more closely related than are *D. melanogaster* and *D. simulans*.

Genic models involving only small numbers of genes have been proposed periodically. The most important distinction between most of these models and the allopatric type is not, however, numbers of genes *per se*, but rather the postulation of a 'privileged' class of genes that play a key role in species formation while being constrained in their rates and/or modes of evolution within species. Two recent and rather precisely formulated examples are the 'founder-flush' model suggested by Carson (1975, 1986, 1987) and the 'genetic transilience' model formulated by Templeton (1980, 1981). Each of these entails special features of the genes involved: specifically, strong epistatic interactions in phenotype and fitness determination, involving a large number of polygenes (Carson's model) or a few major genes and their modifiers (Templeton's model). In addition, both authors have suggested that genes controlling specific phenotypic traits, especially those connected with mating behaviour, are those most likely to fit these requirements.

These models were originally inspired by patterns of species formation in the extremely speciose drosophilid fauna endemic to the Hawaiian islands. The ecological and population-genetic contexts of the differentiation of such island species are perhaps unlikely to correspond in detail to those of the *D. melanogaster*/*D. simulans* divergence, which occurred in eastern continental Africa. Nevertheless, we think that certain important formal aspects of the models may apply, and could serve to organize further investigation. Our statistical and purely structural approach does not allow us to discriminate between direct and pleiotropic effects of diverged loci on reproductive isolation, or even between those loci that have effects and those that do not. However, within these limitations, the data do contain patterns that are at least consistent with the importance of a more or less distinct class of genes in the separation between *D. melanogaster* and *D. simulans*. These genes may differ quantitatively from the rest of the genome with respect to their rates of evolution, if not qualitatively in terms of their functions or mode of evolution.

Aside from the aspect of a relatively small proportion of diverged loci, we have found evidence for the importance of selective constraint operating within species from both purifying and balancing selection (Choudhary *et al.* 1988). This suggests that the major source of genic divergence between species may be substitutions at loci that are maintained monomorphic within species. This supposition is in fact consistent with preponderance of within-species monomorphism among completely diverged loci. Thus, one distinguishing feature of loci with high potential for interspecies divergence may be tight constraint by purifying selection, perhaps combined with the presence fo appearance of rare alleles capable of fixation by directional selection. If such genetic conditions coincided with the ecological conditions favouring interpopulation divergence, allele substitutions at essentially monomorphic loci could play an important role in species formation (Wright 1982).

With respect to the functional aspects of loci involved in species formation, we are intrigued by the seeming tendency for hybrid male sterility to arise in the early stages of species differentiation in *Drosophila* (Bock 1985). This prompted us to carry out a comparison of male reproductive-tract proteins by 2DE in *D. melanogaster* and *D. simulans* (Coulthart and Singh 1988*a*, *b*, *c*). The results bear on the present discussion in two ways. First, male reproductive-tract proteins appear to be more divergent between *D. melanogaster* and *D. simulans* than are proteins of certain 'non-reproductive' tissues such as imaginal wing discs. Whether this indicates preferential genetic involvement of male reproductive-tract functions in species formation remains to be examined. Second, an unexpected consequence of our use of 2DE, with its detection of proteins by direct staining, was that a different molecular *mode* of interspecies protein divergence could be easily detected; e.g. differences in protein amount. Since such differences were very frequent among proteins of the male reproductive tract, and most frequent among proteins with organ-specific distributions within the reproductive tract, we raise the question whether the genic changes accompanying species differentiation are preferentially clustered along an axis representing the fundamental *kinds* of alterations to gene structure and function that can occur. In comparison with electrophoretic data on protein structure, very few data are available on differences in expression of homologous proteins in closely related species. Exploration of the role of such factors in genetic divergence could prove to be the most fruitful approach of all.

Speciation and macroevolution: divergence of molecules *versus* morphology

The origin of new species is an important event in the course of evolution. From the point of view of microevolution, speciation is the end result of population differentiation, a culmination of the cumulative action of the various evolutionary forces that transform genetic variation within species into genetic variation between species. From the point of view of macroevolution, speciation is clearly the first and crucial step in the origin and establishment of new evolutionary lineages that may give rise to higher taxa. Thus, speciation is clearly the link between microevolution and macroevolution and it is therefore important to understand whether the two processes, speciation and the adaptive morphological divergence, are causally connected.

In the widely accepted 'allopatric' geographical model, species formation is seen as a pleiotropic side-effect of independent general adaptive evolution during a period of reduced gene flow (Mayr 1963; Charlesworth *et al.* 1982). In recent years, however, the coupling between speciation and adaptive divergence has generally been de-emphasized, and while speciation is still seen as a gradual process, its origin is seen as a fortuitous outcome of genetic changes not necessarily accompanied by adaptation (e.g. see Gottlieb 1982). However, there are varying views on this point, with coupling between speciation and adaptation usually being emphasized in the 'genic' models and de-emphasized in the 'chromosomal' models of speciation. Gould, on the other hand, has emphasized the coupling between speciation and macroevolution to the extent that, in his view, major morphological and physiological changes mostly occur during speciation and the process occurs rapidly (geologically speaking) but at well-separated times between long periods of evolutionary stasis (Gould and Eldredge 1977; Gould 1980).

Our 2DE species-divergence results, when viewed with other genetic, chromosomal, and morphological features of *Drosophila* divergence, provide evidence to suggest that speciation and major morphological evolution (major adaptation) are two entirely different modes of evolution—one concerned with the creation of genetic *incompatibility* between evolutionary lineages and the other concerned with the maintenance of genetic *compatibility* within lineages (Singh 1988*c*). It is also interesting to note that in many organisms the ability to produce fertile interspecies hybrids has been lost much sooner than the ability to produce viable offspring. On the other hand, in those species in which reproductive isolation is of pre-mating type, laboratory crosses often produce fertile hybrids, which suggests that only the behavioural genes related to mating preference (sexual isolation) have diverged and genes related to basic physiology and function are fully compatible between species. These observations suggest that during evolution the physiological compatibility of genes involved in sexual reproduction (pre-mating or post-mating type) is lost much sooner than that of the 'housekeeping' genes involved in the development, differentiation, and maintenance of the organism. Following this reasoning, it can be hypothesized that major morphological and physiological changes do not occur with speciation but follow later, initially as a result of regulatory readjustment to the new gene pool (altered owing to speciation) and later as a result of adaptive divergence of the newly evolved lineage.

Much speculation has been offered on the importance of changes in 'gene regulation' in evolution (e.g. Britten and Davidson 1971; Zuckerkandl 1976; Wilson *et al.* 1977; Gould 1977, p. 405; Raff and Kaufman, 1983, p. 335), and measurement and characterization of interspecies divergence in patterns of gene expression have begun for pre-chosen loci (Dickinson 1980; Dickinson *et al.* 1984; Parker *et al.* 1985). Our 2DE data suggest that 2DE will not only help provide a technical base for analysis of genetic changes responsible for reproductive isolation but may also serve a unique and valuable function in the search for gene regulatory changes between species (Coulthart and Singh 1988*b*).

Conclusions

Speciation remains as an unsolved problem in evolutionary biology partly because it is not one problem but a collection of problems. The central problem of speciation consists in determining the genetic basis of gradual speciation and the relationship between speciation and major morphological and physiological changes. The genetic basis of sudden speciation involving major genes (e.g. evolution of inbreeders from outbreeders in plants), translocations, and polyploidy is known and what we need to know further is the frequency of their occurrence and the groups of organisms in which such mechanisms are most likely to occur. Population structure plays an important role in speciation but that role very much depends on the nature of genetic variation for speciation and on the breeding systems of the organisms. The genetic structure of species can not be assumed to be stable and hence the role of history versus selection in the shaping of the present genetic structure of species should be investigated.

The one- and two-dimensional gel electrophoretic studies of species differences between *D. melanogaster* and *D. simulans* have brought out three important conclusions. First, alternating fixation of different alleles at a protein-coding locus appears to be a fairly general feature of divergence between the two species, occurring at a predictable frequency of about 10 per cent for a given sample of loci. Second, there may be systematic differences between 'classes' of loci, defined on the basis of their differing roles in the economy and organization of the organism, in their probabilities of undergoing complete interspecies divergence. And thirdly, at least in certain organs and tissues, a very common mode of interspecies divergence at the protein level results in presence/absence differences, as distinct from structural differences visible as electrophoretic mobility shifts. Considering the divergence time between *D. melanogaster* and *D. simulans*, our results are consistent with the importance of a more or less distinct class of genes in speciation. These genes may differ quantitatively from the rest of the genome with respect to their rates of evolution, if not qualitatively in terms of their functions or mode of evolution.

On the basis of genic, chromosomal and morphological differences among species in the genus *Drosophila*, it is hypothesized that major morphological and physiological changes do not occur with speciation but follow later, initially as a result of regulatory readjustment to the new gene pool (altered owing to speciation) and subsequently as a result of adaptive divergence of the newly evolved lineage. Two-dimensional gel electrophoresis of proteins is a unique and valuable tool in the search for gene regulatory changes between species.

References

Ayala, F.J. (1975). Genetic differentiation during the speciation process. In *Evolutionary biology* (ed. T. Dobzhansky, M.K. Hecht, and W.C. Steere), Vol. 8, pp. 1–78. Plenum Press, New York.

Ayala, F.J. and Dobzhansky, T. (1974). A new subspecies of *Drosophila pseudoobscura*. *Pan-Pacif. Ent.*, **50**, 211–19.

Ayala, F.J., Tracey, M.L., Barr, L.G., and Ehrenfeld, J.G. (1974). Genetic and reproductive differentiation of the subspecies, *Drosophila equinoxialis carribenesis*. *Evolution*, **28**, 24–41.

Barigozzi, C. (1982). *Mechanisms of speciation*. Alan R. Liss, New York.

Barton, N.H. and Charlesworth, B. (1984). Genetic Revolutions, founder effects, and speciation. *A. Rev. Ecol. Syst.*, **15**, 133–64.

Bock, I.R. (1985). Interspecific hybridization in the genus *Drosophila*. In *Evolutionary biology* (ed. M.K. Hecht, B. Wallace, and G.T. Prance), Vol. 18, pp. 41–70. Plenum Press, New York.

Britten, R.J. and Davidson, E.H. (1971). Repetitive and non-repetitive DNA sequences and a speculation on the origins of evolutionary novelty. *Q. Rev. Biol.*, **46**, 111–38.

Bush, G.L (1975). Modes of animal speciation. *A. Rev. Ecol. Syst.* **6**, 339–64.

Carson, H.L. (1975). The genetics of speciation at the diploid level. *Am. Nat.*, **109**, 83–92.

Carson, H.L. (1986). Sexual selection and speciation. In: *Evolutionary processes and theory* (ed. S. Karlin and E. Nevo), pp. 391–409. Academic Press, New York.

Carson, H.L. (1987). The genetic system, the deme, and the origin of species. *A. Rev. Genet.*, **21**, 405–23.

Carson, H.L. and Templeton, A.R. (1984). Genetic revolution in relation to speciation phenomenon: the founding of new populations. *A. Rev. Ecol. Syst.*, **15**, 97–131.

Charlesworth, B., Lande, R., and Slatkin, M. (1982). A neo-Darwinian commentary on macroevolution. *Evolution*, **36**, 474–98.

Choudhary, M., and Singh, R.S. (1987*a*). A Comprehensive study of genic variation in natural populations of *Drosophila melanogaster* III. Variations in genetic structure and their causes between *Drosophila melanogaster* and its sibling species *Drosophila simulans*. *Genetics*, **117**, 697–710.

Choudhary, M. and Singh, R.S. (1987*b*). Historical effective size and the level of genetic diversity in *Drosophila melanogaster* and *D. pseudoobscura*. *Biochem. Genet.*, **25**, 41–51.

Choudhary, M., Coulthart, M.B., and Singh, R.S. (1988). Genetic divergence and models of species formation in *Drosophila*. *Proc. natn. Acad. Sci. USA* (submitted).

Coulthart, M.B. and Singh, R.S. (1988*a*).Low genic variation in male-reproductive tract proteins of *Drosophila melanogaster* and *D. simulans*. *Molec. Biol. Evol.*, **5**, 167–81.

Coulthart, M.B. and Singh, R.S. (1988*b*). High level of divergence of male-reproductive-tract proteins, between *Drosophila melanogaster* and its sibling species, *D. simulans*. *Molec. Biol. Evol.*, **5**, 182–91.

Coulthart, M.B. and Singh, R.S. (1988*c*). Differing amounts of genetic polymorphism in testes and male accessory glands of *Drosophila melanogaster* and *Drosophila simulans*. *Biochem. Genet.*, **26**, 153–64.

Coyne, J.A. (1982). Gel electrophoresis and cryptic protein variation. In: *Isozymes: Current topics in biological and medical research* (ed. M.C. Rattazzi, J.C. Scandalios, and G.S. Whitt), pp. 1–32. Alan R. Liss, New York.

Coyne, J.A. (1984). Genetic basis of male sterility in hybrids between two closely related species of *Drosophila*. *Proc. natn. Acad. Sci., USA*, **81**, 4444–47.

Coyne, J.A. and Charlesworth B. (1986). Location of an X-linked factor causing sterility in male hybrids of *Drosophila simulans* and *D. mauritiana*. *Heredity*, **57**, 243–46.

Dickinson, W.J. (1980). Evolution of patterns of gene expression in Hawaiian picture-winged *Drosophila*. *J. Molec. Evol.*, **16**, 73–94.

Dickinson, W.J., Rowan, R.G., and Brennan, M.D. (1984). Regulatory gene evolution: adaptive differences in expression of alcohol dehydrogenase in *Drosophila melanogaster* and *Drosophila simulans*. *Heredity*, **52**, 215–25.

Dobzhansky, T. (1936). Studies on hybrid sterility II. Localization of sterility factors in *Drosophila pseudoobscura* hybrids. *Genetics*, **21**, 113–35.

Dobzhansky, T. (1951). *Genetics and the origin of species*. Columbia University Press, New York.

Dobzhansky, T. and Epling, C. (1944). Contribution to the genetics, taxonomy and ecology of *Drosophila pseudoobscura* and its relatives. Carnegie Inst. Wash. Publ., Washington, D.C.

Ferris, S.D., Sage, R.D., Huang, C., Nielsen, J.T., Ritte, U., and Wilson, A.C. (1983). Flow of mitochondrial DNA across a species boundary. *Proc. natn. Acad. Sci. USA*, **80**, 2290–94.

Gillespie, J. and Kojima, K. (1968). The degree of polymorphism in enzymes involved in energy production compared to that in non-specific enzymes in two *D. ananassae* populations. *Proc. natn. Acad. Sci. USA*, **61**, 582–85.

Goldschmidt, R. (1940, 1982). *The material basis of evolution*. Yale University, New Haven, Conn.

Gottlieb, L.D. (1982). Does speciation facilitate the evolution of adaptation. In *Mechanism of speciation* (ed. C. Barigozzi), pp. 179–90. Alan R. Liss, New York.

Gould, S.J. (1977). *Ontogeny and phylogeny*. Harvard University Press, Cambridge, Mass.

Gould, S.J. (1980). Is a new and general theory of evolution emerging? *Paleontology*, **6**, 119–30.

Gould, S.J. and Eldredge, N. (1977). Punctuated evolution: The tempo and mode of evolution reconsidered. *Paleontology*, **3**, 115–51.

Grant, V. (1971). *Plant speciation*. Columbia University Press, New York.

Hale, L.R. and Singh, R.S. (1987). Mitochondrial DNA variation and genetic structure in populations of *Drosophila melanogaster*. *Molec. Biol. Evol.*, **4**, 622–37.

Huxley, J. (1942, 1963). *Evolution: The modern synthesis*. Allen and Unwin, New York.

Kimura, M. (1968). Genetic variability maintained in a finite population due to mutational production of neutral and nearly neutral isoalleles. *Genet. Res. Camb.*, **11**, 247–69.

King, J.L. and Jukes, T.H. (1969). Non-Darwinian evolution. *Science*, **164**, 788–98.

King, M.-C. and Wilson, A.C., (1975). Evolution at two levels in humans and chimpanzees. *Science*, **188**, 107–16.

Kreitman, M. (1983). Nucleotide polymorphism at alcohol dehydrogenase locus of *Drosophila melanogaster*. *Nature, Lond.*, **304**, 412–17.

Lakovaara, S., Saura, A., and Falk, C.T. (1972). Genetic distance and evolutionary relationship in the *Drosophila obscura* group. *Evolution*, **26**, 117–84.

Lewontin, R.C. (1974). *The genetic basis of evolutionary change*. Columbia University Press, New York.

MacIntyre, R.J. and Collier, G.E. (1986). Protein evolution in the genus *Drosophila*. In *The genetics and biology of Drosophila* (ed. M. Ashburner, H.L. Carson, and J.N. Thompson, Jr.), Vol 3e, pp. 39–146. Academic Press, New York.

Mayr, E. (1963). *Animal species and evolution*. Harvard University Press, Cambridge, Mass.

Mayr, E. (1982*a*). *The growth of biological thought: diversity, evolution and inheritance*. Harvard University Press, Cambridge, Mass.

Mayr, E. (1982*b*). Speciation and macroevolution. *Evolution*, **36**, 1119–32.

Motro, U. and Thompson, G. (1982). On heterozygosity and effective size of populations subject to size change. *Evolution*, **36**, 1059–66.

Muruyama, T. and Fuerst, P.A. (1985*a*). Population bottlenecks and nonequilibrium models in population genetics. II. Number of alleles in a small population that was formed by a recent bottleneck. *Genetics*, **111**, 675–89.

Muruyama, T. and Fuerst, P.A. (1985*b*). Population bottlenecks and nonequilibrium models in population genetics. III. Genetic homozygosity in populations which experience periodic bottlenecks. *Genetics*, **111**, 691–703.

Nei, M. (1975). *Molecular population genetics and evolution*. North-Holland, Amsterdam.

Nei, M., Muruyama, T., and Chakraborty, R. (1975). The bottleneck effect and genetic variability in populaticns. *Evolution*, **29**, 1–10.

Nevo, E., Beiles, A., and Ben-Shlamo R. (1984). Evolutionary significance of genic diversity: ecological, demographic and life-history correlates. *Lect. Notes Biomath.*, **53**, 13–213.

Ohnishi, S., Kawanishi, M., and Watanabe, T.K. (1983). Biochemical phylogenies of *Drosophila*: protein differences detected by two-dimensional electrophoresis. *Genetica*, **61**, 55–63.

Parker, H.R., Philip, D.P., and Whitt, G.S. (1985). Gene regulatory divergence among species estimated by altered developmental patterns of interspecies hybrids. *Molec. Biol. Evol.*, **2**, 217–50.

Powell, J.R. (1975). Protein variation in natural populations of animals. In *Evolutionary biology* (ed. T. Dobzhansky, M.K. Hecht, and W.C. Steere), Vol 8, pp. 79–119. Plenum Press, New York.

Raff, R.A. and Kaufman, T.C. (1983). *Embryos, genes and evolution*. Macmillan, New York.

Selander, R.K. (1976). Genic variation in natural populations. In *Molecular evolution* (ed. F.J. Ayala), pp. 21–45. Sinauer, Sunderland, Mass.

Selander, R.K., Hunt, W.G., and Yang, S.Y. (1969). Protein polymorphism and genetic heterozygosity in two European subspecies of the house mouse. *Evolution*, **23**, 379–90.

Singh, R.S. (1988*a*). Patterns of species divergence and role of genetic structure in the evolution and speciation of *Drosophila* (in preparation).

Singh, R.S. (1988*b*). A comprehensive study of genic variation in natural populations of *Drosophila melanogaster*. IV. Structural-functional constraints on genic diversity (in preparation).

Singh, R.S. (1988*c*). Patterns of species divergence and genetic theories of speciation. In *Topics in population biology and evolution* (ed. K. Wöhrmann and S.K. Jain). Springer-Verlag. (In press.)

Singh, R.S. and Rhomberg, L.R. (1987). A comprehensive study of genic variation in natural populations of *Drosophila melanogaster II*. Estimates of heterozygosity and patterns of geographic differentation. *Genetics*, **117**, 255–71.

Singh, R.S., Lewontin, R.C., and Felton, A.A. (1976). Genetic heterogeneity within electrophoretic 'alleles' of Xanthine dehygrogenase in *Drosophila pseudoobscura*. *Genetics*, **84**, 609–29.

Solignac, M. and Monnerot, M. (1986). Race formation, speciation, and introgression within *Drosophila simulans*, *D. mauritiana*, and *D. sechellia* inferred from mitochondrial DNA analysis. *Evolution*, **40**, 531–39.

Stebbins, G.L. (1950). *Variation and evolution in plants*. Columbia University Press, New York.

Stebbins, G.L. (1982). Perspectives in evolutionary theory. *Evolution*, **36**, 1109–18.

Stephens, J.C. and Nei, M. (1985). Phylogenetic analysis of polymorphic DNA sequences at the Adh locus in *Drosophila melanogaster* and its sibling species. *J. Molec. Evol.*, **22**, 289–300.

Templeton, A.R. (1980). The theory of speciation *via* the founder principle. *Genetics*, **94**, 1011–38.

Templeton, A.R. (1981). Mechanisms of speciation—A population genetics approach. *A. Rev. Ecol. Syst.*, **12**, 23–48.

Throckmorton, L.H. (1977). *Drosophila* systematics and biochemical evolution. *A. Rev. Ecol. Syst.*, **8**, 235–54.

White, M.J.D. (1978). *Modes of speciation.* Freeman, San Francisco.

Wilson, A.C., Carlson, S.S., and White, T.J. (1977). Biochemical evolution. *A. Rev. Biochem*, **46**, 573–639.

Wright, S. (1982). Character change, speciation and the higher taxa. *Evolution*, **36**, 427–43.

Yang, S.Y., Wheeler, L.L., and Bock, I.R. (1972). Isozyme variations and phylogenetic relationships in the *Drosophila bipectinata* species complex. In *Studies in Genetics* (ed. M.R. Wheeler), pp. 213–27, *Univ. Texas Publ.* 7213.

Zouros, E. (1973). Genic differentiation associated with the early stages of speciation in the *mulleri* subgroup of *Drosophila*. *Evolution*, **27**, 601–21.

Zouros, E. (1981). The chromosomal basis of sexual isolation in two sibling species of *Drosophila*: *D. arizonensis* and *D.majavensis*. *Genetics*, **97**, 703–18.

Zuckerlandl, E. (1976). Programs of gene action and progressive evolution. In *Molecular anthropology* (ed. M. Goodman and R.E. Tashian), pp. 387–447. Plenum Press, New York.

4 The complexities of host races and some suggestions for their identification by enzyme electrophoresis

STEWART H. BERLOCHER

Department of Entomology, University of Illinois, Urbana, Illinois, USA

Abstract

The phenomenon of host races remains very poorly understood, despite the fact that host races are of great theoretical and practical importance. It is at least clear that host races do exist. The best understood case is that of the apple maggot fly, *Rhagoletis pomonella*. This species, native to eastern North America, originally infested only species of hawthorn, but has in the last 130 years colonized apple, an introduced plant. Consistent differences between the apple and hawthorn host races have now been demonstrated for allozyme frequencies, and for behavioural and phenological characteristics. Confusion over host races is to be expected considering their likely mode of origin. Four possible modes of host race formation are outlined here. The modes are discussed in the light of allozyme data and its potential usefulness in distinguishing such host races.

In mode one, no genetic change occurs. Flies with genotypes that confer the ability to exist on both hawthorn and apple may have existed in some part of the geographic range of the fly, and these flies then spread throughout the range of apple. In mode two, a unique apple genotype develops once in a particular area and then spreads throughout the range of apple. Mode three is similar — except that

Electrophoretic Studies on Agricultural Pests (ed. Hugh D. Loxdale and J. den Hollander), Systematics Association Special Volume No. 39, pp. 51–68. Clarendon Press, Oxford, 1989.

the apple genotype develops many times in different geographic areas. In mode four, an apple genotype develops many times in many parts of the geographic range, but secondary adaptation to apple also occurs, perhaps originally at only one site. Secondary adaptation to apple may produce reduced gene flow between the two host races.

Most published studies of host races suffer from a variety of problems, the majority of which revolve around sampling. Insufficient sampling has impeded the acceptance of the existence of host races and at present, prevents discrimination among the four models I outline here. Four aspects of sampling are discussed. First is sampling of loci. Only a fraction of allozyme loci are involved in differentiation of host races, and sampling a small number of loci will often fail to reveal differences between host races. A second important aspect of sampling is sampling of life-history stage. If disruptive selection of flies on different host plants is occurring in the larval stage, then sampling of larvae may only provide a small amount of information about possible patterns of non-random mating. Sampling of geographic variation is critical to an improved understanding of host races. This is because geographically varying patterns of selection have the potential to confound host-plant related patterns of differentiation. Sampling over time is also critical since host races are often allochronic to varying degrees.

Suggestions for improving allozyme studies of host races are made. These include more extensive sampling of loci, individuals, and populations. Particular emphasis is placed on the desirability of sampling adult insects at sites where putative host-races co-occur on a microsympatric basis, at times when both races are active.

Introduction

The entomological literature is rife with references to biotypes, host races, strains, ecotypes, and other confusing entities. This plethora of terms reflects the biological reality that insects are capable of subtle and complex ecological specializations, which may or may not coincide with conventional taxonomic divisions. Several recent works have substantially clarified the terminological confusion surrounding host races and their ilk (Diehl and Bush 1984; Futuyma and Peterson 1985; Pashley, 1988). The work of Pashley is particularly useful because it presents a flow chart, in essence a conceptual key, for classifying an unknown case into the categories of host race, ecomorph, sibling species, etc.

If all the various terms are arranged on an imaginary scale representing the degree of reproductive isolation, it can clearly be seen that cases on the extremes of the scale are easily classified and understood.

For example, consider the following hypothetical situations. In a particular area, two different crops are fed upon by what initially appears to be a single species. Yet, despite the morphological uniformity, suspicions arise that differences exist between the insects from different host plants. If research reveals that insects associated with the two plants never mate in nature, then reproductive isolation is probably complete, and one can conclude that two sibling species are present (see Diehl and Bush 1984 for examples). On the other hand, research might reveal that the population is composed of several genotypes that specialize (as immatures) on one or the other host plant, but that adult insects nevertheless mate totally at random. In this case, there is no reproductive isolation. Futuyma and Peterson (1985) provide examples of such genetic polymorphism for food choice in insects, although one of the most striking examples of a trophic polymorphism is a cichlid fish population from northern Mexico that contains genotypes specializing on either snails or algae (Kornfield *et al.* 1982).

In the middle of this scale of reproductive isolation, however, things are much less clear-cut. If our hypothetical population turns out to be composed of genotypes that specialize on one or the other of the two host plants, and mating between the genotypes is neither random nor absent, then we are dealing with host races. Note that I am using the term in agreement with Diehl and Bush (1984) and Pashley (1988), but not in complete agreement with Janeike (1981), who imposes restrictions that I consider unrealistic upon the definition of host race.

Host races, as populations that display only partial reproductive isolation, are by their very nature bound to be complex. In what follows, I shall address two aspects of host races. First, I will discuss some ideas about how host races originate in sexually reproducing species, and how their origin may effect allozyme differentiation between host races. This discussion will focus on the case of the apple and hawthorn races of *Rhagoletis pomonella* (Walsh), the apple maggot fly. Secondly, I will discuss how electrophoretic studies should be carried out to maximize their ability to reveal host races.

Host races of *R. pomonella*: what we do and do not know

The host-associated populations (Pashley 1988; Smith 1988) of *Rhagoletis pomonella* have been widely discussed (Diehl and Bush 1984). In brief, this North American species originally infested the fruits of hawthorns (*Crataegus* spp.), but in historical time has established itself on apples, cherries (Bush, 1966), and rose (Prokopy and Berlocher 1980). Although the case has been discussed for over 100 years (Walsh 1867), it is due to the efforts of Bush (1966, 1969, 1974, 1975; Bush and Diehl 1982) that *R. pomonella* has been brought to the attention of entomologists, population geneticists, and evolutionary biologists. Bush has argued that the various host-associated populations of *R. pomonella* represent genetically differentiated host races, and that host races can gradually evolve more and more complete reproductive isolation until they become species. Bush's persuasive case for sympatric host race formation and speciation has, together with the economic importance of the fly, generated a substantial body of research on *R. pomonella*. Recently, three independent lines of evidence have bolstered the case for host races in *R. pomonella*.

1. *Genetic differences at allozyme loci*

Two studies, one in Illinois (McPheron *et al.* 1988*b*) and one in Michigan (Feder *et al.* 1988*b*) have demonstrated allozyme frequency differences between the hawthorn and apple populations. In the Illinois study there are consistent, statistically significant differences at two loci between five pairs of microsympatric apple and hawthorn populations. In the Michigan study there are consistent, significant differences at six loci between three pairs of microsympatric apple and hawthorn populations. The results are largely consistent across the two studies, in that the two loci of interest in the Illinois study are also involved in the Michigan study, which examined several loci not studied in the Illinois study.

2. *Differences in oviposition behaviour*

Prokopy *et al.* (1987) have demonstrated significant differences between apple and hawthorn flies with respect to choice of fruit for oviposition. The results are in part counter to expectations, in that both apple and hawthorn flies prefer hawthorn fruits in cage experiments. However, apple flies chose apples significantly more often than did hawthorn flies.

3. *Differences in post-diapause emergence time*

Very recent work (Smith 1988; McPheron *et al.* 1988b) has demonstrated that apple flies are genetically programmed to emerge sooner after diapause than hawthorn flies. When apple and hawthorn flies taken from the field are reared for one generation under uniform laboratory conditions, the mean post-diapause emergence time for the hawthorn flies is approximately 3 weeks later than that for the apple flies. This difference corresponds to that in the field in Illinois, where apple flies are active earlier than hawthorn flies.

4. Conclusion

These three lines of evidence indicate that host races of *R. pomonella* do exist. That is, the apple and hawthorn populations are genetically differentiated, and are unlikely, on the electrophoretic evidence, to be completely randomly mating with one another. Of course, much more work needs to be done (see caveats about allozyme evidence in the final section), but the evidence is strong enough that the terms apple race and hawthorn race can be used with some confidence.

The work described above provides a foundation for an understanding of host races, but raises as many questions as it answers. Are the other host-associated populations, such as the rose and cherry populations, also genetically differentiated? Is it possible for host races to be differentiated at loci controlling behavioural or phenological characteristics, but not differentiated for allozyme loci? Do host races arise independently several times, or only once?

Answering such questions will require very carefully thought-out experiments, experiments best planned in the context of understanding the origin of host races. The basic model for the origin of host races is that of Bush (1969, 1974, 1975; Bush and Diehl 1982), who proposed that recombination between alleles of a host selection gene that determines where a phytophagous insect mates and oviposits, and alleles of a host survival gene that determines the ability of the insect to feed upon different host plants, will produce new genotypes that can colonize new plants. While this process is almost certainly at the core of host-race formation, much detail remains to be filled in. In the following section, I consider how geography and history can produce complex origins of host races, and how this complexity may affect electrophoretic analysis of host races.

Four possible ways in which a host race could arise

I can at present identify four sequences of events that could give rise to host races of *R. pomonella*. All of these sequences involve geographic variation, for which there is abundant evidence in *R. pomonella*. Northern and southern populations in eastern North America differ in size (Bush 1966), and at least some of these differences have a genetic basis (Berlocher, unpublished). More to the point, a large amount of allozyme variation exists, with an F_{ST} of 0.22 over eastern North America (McPheron, 1987). Southern populations are very distinct from northern ones, but significant variation among populations at the same latitude also exists. Significant allozyme variation exists even on a microgeographic scale (McPheron *et al.* 1988a).

1. Sequence one: The 'lack of genetic change' model

The first possible sequence of events is one in which, oddly enough, little or no genetic change accompanies the production of an apple race from the ancestral hawthorn race. To demonstrate how this could occur, I use a diagram (Fig. 4.1) in which the horizontal axis represents a transect through hypothetical geographical space, while the vertical axis is an ecological transect with two niche spaces, apple and hawthorn.

Along the geographical transect are arrayed four genotypes, each of which is adapted to local conditions and is the most common genotype in the area. Areas of genetic intergradation, where mixtures of common genotypes occur in populations, are shown as shaded areas. Of course, in reality, populations are never organized as neatly and consistently as this, but the simplification facilitates discussion. The genotypes are named for the geographic area to which they are adapted, and the host plant to which they are adapted, so that the genotype predominating in area 1, adapted to area 1 and to hawthorn, is genotype 1^H.

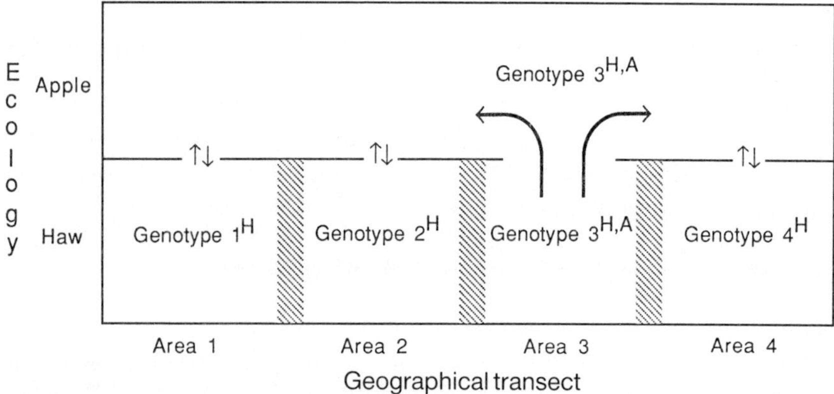

Fig. 4.1. Origin of apple race of *R. pomonella* by sequence one, the 'lack of genetic change' model (see text).

The essence of sequence one is that one of the original hawthorn genotypes, genotype $3^{H,A}$, is preadapted to existing on apples. This could be because the hawthorn fruits in area 3 happened to be large, approaching apples in size, or that they were chemically similar to apples. When apples were introduced the flies of genotype $3^{H,A}$ started using apples as well as hawthorn and then spread throughout the transect. Partial reproductive isolation between races could be attained (except in area 3) if flies of genotype $3^{H,A}$ did not prefer the hawthorn fruits in the other areas. (Partial reproductive isolation between hawthorn flies in different areas need not occur if the transition from one hawthorn fruit type to another is gradual).

Allozyme frequency differences would exist between apple and hawthorn races in most areas if allozyme loci were involved in the preadaptation *or*, more likely, if they were linked to loci involved in the preadaptation. Gene flow between the host races (small arrows in Fig. 4.1) would act to reduce differentiation, but differentiation could be maintained by selection, aided by host specificity of the genotype. Overall, the allozyme pattern expected from sequence one is substantially greater genetic homogeneity of the 'apple' race compared to the hawthorn race, and at least one area of very similar allele frequencies in the two races. Since the numbers of flies colonizing apple would be large, no genetic bottleneck effects should be expected in the apple race.

2. *Sequence two: The 'single origin' model*

In sequence two, a genetic change (denoted Δp in Fig. 4.2) is required somewhere to produce a novel genotype that prefers and can feed upon apple instead of hawthorn. This new genotype, 3^{A} in Fig. 4.2, then spreads as in sequence one to occupy the rest of the geographic area.

Sequence two is the Bush (1969, 1974, 1975) model of host-race formation. The alleles that make up genotype 3^{A} were probably already in existence at low frequency in area 3, and genotype 3^{A} was recombining into existence sporadically even before apple was introduced. The arrival of apple provided an ecological opportunity for a new population of apple flies to increase and expand.

In both sequences one and two, the genetic variation to produce an apple race was pre-existent. The difference is that in sequence one a genotype that could use either host was predominant in area 3 and little or no genetic change in the population was required to colonize apple, while in sequence two the genotype that comes to dominate

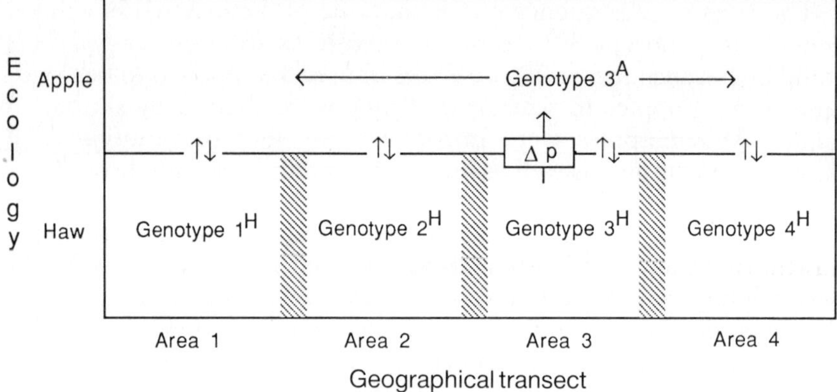

Fig. 4.2. Origin of apple race of *R. pomonella* by sequence two, the 'single origin' model (see text).

the apple race is created out of low-frequency alleles in the ancestral hawthorn population.

With respect to allozymes, sequence two should result, like sequence one, in uniformity of the apple race relative to the hawthorn race. However, allozyme frequency differences between races should exist in all areas and, in addition, at least some loss of rare alleles might be expected owing to the smaller founding size of the apple race in sequence two.

3. *Sequence three: The 'multiple origin' model*

Any adaptation can arise independently several times if ecological opportunity and genetic variation allow it. Hybridization of *R. pomonella* and a closely related species suggests that the genetic control of certain aspects of oviposition behaviour and post-diapause emergence time is polygenic, producing a continuous distribution of the traits (Smith 1986), in which case it might be possible for an apple population to be selected out of almost any hawthorn population. I know of no direct evidence for multiple colonization of apples (and Bush's 1969 historical map suggests a single colonization), but given that we know very little of the nature and extent of genetic variation in relation to host-plant usage in natural insect populations (Futuyma and Peterson 1985), there is no *a priori* reason to expect host races to arise only once.

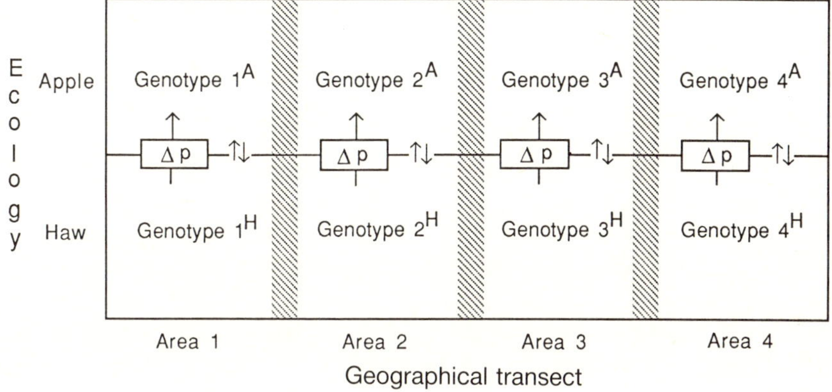

Fig. 4.3. Origin of apple race of *R. pomonella* by sequence three, the 'multiple origin' model (see text).

Figure 4.3 illustrates multiple origin of the apple race in my hypothetical case. In each of the four areas, recombination and selection have formed a new population on apple that differs genetically from the local hawthorn population at loci directly involved with host adaptation. At the same time, however, the new apple populations (which will intergrade with one another more or less as do the ancestral hawthorn populations) will be like their ancestral populations at loci related to local climate, etc. Consequently, the new apple populations comprising the apple race will be genetically similar at some loci and dissimilar at others.

Sequence three can result in a complex pattern of allozyme differentiation. An allozyme locus directly involved in adaptation to apple (or linked to such a locus) might show a pattern of consistent frequency differences between apple and hawthorn races, while a locus involved with adaptation to local climate might retain in the apple race the original geographic pattern of the hawthorn race. Multiple colonizations, either with or without polygenic origination of each apple population, would assure that little loss of rare alleles occured during host-race formation. Overall, conflicting and problematic allozyme results are to be expected from sequence three.

4. *Sequence four: The 'multiple origin–secondary adaptation' model*

I have intentionally kept the first three sequences as simple as possible to focus attention on their differences. In reality, host-race information is likely to involved combinations of these sequences of events. For example, sequences one and two might begin independently in different geographic areas. In lieu of attempting to consider all conceivable combinations of sequences, I propose for my final case a sequence that I believe is most likely, based upon my experience with *Rhagoletis* flies.

The starting point for sequence four is sequence three under the assumption that the local colonizations of apple involve polygenic, continuous variation. In the absence of direct data for this sequence, I would offer the observation that multiple adaptation in the same species is common; for example the several types of resistance to the same insecticide in the housefly (Plapp 1976), and the fact that the apple maggot fly has colonized several different host plants in a very short time.

The next step of sequence four is the selection of a new genotype, produced either by mutation or recombination, that is much better adapted to apple than the predominant genotype of the initial colonists. In Fig. 4.4, I show the production of a 'super-apple' genotype, 1^{AA}, that has arisen from genotype 1^{A}, the genotype characteristic of the original colonists of apple in area 1. Genotype 1^{AA} will then eventually spread by gene flow to other geographic areas, replacing or altering the population of the original colonists. In Fig. 4.4, genotype 1^{AA} has already replaced the original apple genotype of area 2.

Sequence four is inspired in part by observations on the different types and levels of insecticide resistance achieved by laboratory colonies and field populations of insects. Resistance in laboratory

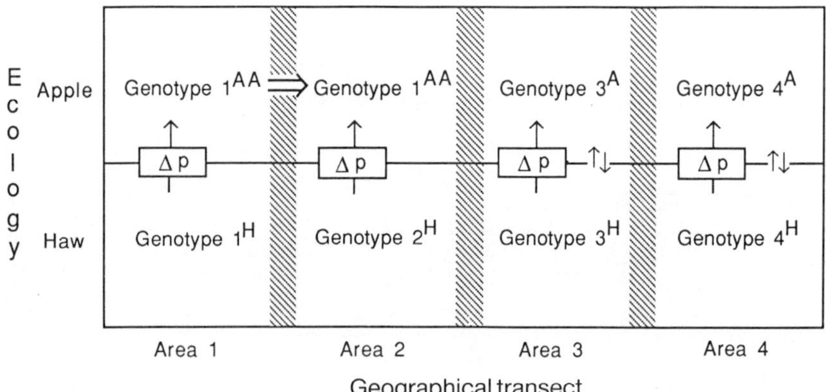

Fig. 4.4. Origin of apple race of *R. pomonella* by sequence four, the 'multiple origin–secondary adaptation' model (see text).

colonies is frequently polygenic and limited; resistance in nature is often monogenic and quite effective. As discussed by Rousch and McKenzie (1987), the reason for this is that laboratory colonies have a limited gene pool, so that resistance must be attained by selecting upon the cumulative effects of many alleles with limited individual effects, while, in field populations of millions or billions, selection has the opportunity to find those extremely rare alleles that have major effects.

An analogous process may be acting to produce the apple race of *R. pomonella*. The establishment of multiple populations, perhaps poorly adapted to apple at first, would provide large populations likely to contain rare but especially effective 'apple genotypes', in an environment in which those genotypes would be rapidly selected. Bush (1969) briefly discusses secondary adaptation to a new host in a somewhat different context, and implies that such adaptation could be involved in curtailing gene flow between host races. In Fig. 4.4 this process is indicated by the elimination of gene flow between the host races in areas 1 and 2.

Whether host races in actuality arise through an initial local, polygenic adaptive process followed by secondary adaptation remains to be seen. In the context of this chapter, the primary value of sequence four is to demonstrate a feasible process in which allozyme differentiation could be exceedingly complex. Strong selection for the 'super apple' genotype might produce homozygosity for linked allozyme loci within the selected populations, and genetic uniformity for such allozyme loci among those populations, superimposed upon the patchwork allozyme pattern characteristic of sequence three.

Some suggestions for improving allozyme studies of host races

The point of the preceding discussion is not to promote a particular scheme for the origin of a host race, but to focus attention upon the need for carefully designed studies. None of the small but growing number of allozyme studies relating to host races (Tabachnik *et al.* 1980; Jaenicke and Selander 1979, 1980; Pashley and Bush 1979; Menken 1981; Malavasi and Morgante 1983; Pashley 1986; Sturgeon and Mitton 1986), including those on *R. pomonella* (Feder *et al.* 1988*b*; McPheron *et al.* 1988*b*), have dealt with all of the possible complexities of host races.

To a large extent, the task of maximizing the information content of an allozyme study of a host race in nature becomes a sampling problem. At least four aspects of sampling need to be considered.

1. Sampling of loci and individuals

Newcomers to electrophoretic analysis often ask 'How many loci are enough?' The answer to this excellent question is, of course, 'As many as you can get' (an answer seldom viewed as satisfactory). If one wishes to find allozyme loci that are directly involved in host adaptation, or that are linked to important loci, then the ideal number is several hundred. I am all too familiar with the difficulty of attaining this goal; the number of polymorphic loci in *R. pomonella* has increased from 9 in 1976 (Berlocher and Bush 1982) to a current total of only 17 (Feder *et al.* 1988*a*). The ultimate answer is to augment one's set of allozymes with restriction-fragment-length polymorphism sites in the nuclear genome (e.g. Bernatzky and Tanskley 1986), for which a set of hundreds of 'loci' may be feasible.

However, there is no reason to abandon allozyme work; it remains an inexpensive and simple way of studying genetic variation. What I am suggesting, though, is that studies of host races based upon few loci are very unlikely to be informative, and that researchers should resist the temptation to begin population analysis with the first, say, six loci that resolve well. (I note that none of the nine allozyme loci resolved in *R. pomonella* in 1976 are significantly involved in differentiation of the host races.)

Also to be considered as an aspect of genetic sampling is mapping of loci. Even a crude genetic map may be of use in deciding, for example, whether loci that appear to be differentiating along host-plant lines are doing so independently or as part of a block of genes.

No set of guidelines has been published for determining adequate sample sizes for studying population structure. Some empirical guidance may be taken from work on host races and microgeographic variation in *R. pomonella* (McPheron *et al.* 1988*a*, *b*), where samples of approximately 100 flies were necessary to reveal any pattern. It is clear that very large sample sizes are desirable to permit the most sensitive scrutiny of pest populations; in general, entomologists have not availed themselves of the tremendous population sizes of most pests when carrying out allozyme work.

2. Sampling of life-history stage

The decision to electrophorese insects as larvae, pupae, or adults is often made on the basis of such considerations as ease of collecting and rearing insects, and electrophoretic resolution. Such considerations are important, but the most information about the existence of host races will be obtained by sampling a life stage that will reveal whether mating is random. Usually this means sampling adults, ideally when they are actually mating and ovipositing on the host plants.

Recent work on microgeographic variation in *R. pomonella* (McPheron *et al.* 1988*a*) illustrates how conclusions about population structuring are affected by the stage analysed. Sampling from nine large hawthorn trees in a park was performed by collecting infested fruits and rearing the larvae to adulthood in the laboratory. Allozyme analysis revealed significant inter-tree allele frequency heterogeneity, but this could have come about in at least two ways. The adult population in the park could be panmictic, with the frequency differences being due to disruptive selection in the larval stage, related to some aspect of the tree environment (see also Jaenicke 1981). In other words, the population structuring would develop during the larval stage and then be eradicated in each generation by random mating of adults. Alternatively, the adult population might be sustructured, with subpopulations (centring on host trees) mating non-randomly with one another (The frequency differences might still be due to selection acting in the larval stage, but the differences would be maintained rather than eradicated during the adult stage.) As discussed by McPheron *et al.* (1988*a*), direct sampling of mating and ovipositing adults needs to be done to distinguish between the two possibilities.

The same two explanations will often apply to studies of host races. The most information about patterns of mating is obtained by sampling individuals capable of mating (Feder *et al.* 1988b; McPheron *et al.* 1988*b*). Sampling of as many life stages as is possible is of course desirable in gaining an understanding of how and when selection is acting to produce differentiation.

3. *Sampling of geographic variation*

Inadequate geographic sampling can lead to two kinds of problems. The first arises if all samples of one putative host race are reared from a particular plant in one geographic area while the other putative host race is sampled from a different host plant in another area. The difficulty here is that geographic variation is hopelessly confounded with potential host-related variation. This is a very significant problem, since geographic variation in insects is ubiquitous, ranging from very slight in migratory insects such as the Monarch butterfly (Eanes and Koehn 1978) to substantial in insects with more limited dispersal ability, such *Euphydryas* butterflies (McKechnie *et al.* 1975).

Furthermore, geographic variation can occur on a very small geographic scale (McCauley and Eanes 1987; McPheron *et al.* 1988*a*; Barker *et al.* 1986).

The second problem concerns the intrinsic complexity of host races I have discussed, complexity in which geographic variation plays a key part. Quite simply, none of the sequences described (nor any other conceivable sequences) will be distinguishable without careful geographic sampling. Unfortunately, it is impossible at present to put any quantitative estimates on the number of samples needed. For example, the statistical problem of rejecting sequence one in a particular case hinges upon the size of the area in which the original bi-host genotypes occurred.

Generally speaking, there are three possible patterns of geographic sampling of two putative host races. Allopatric sampling, sampling one host from one area and the other host from another area, is highly undesirable. However, allopatric sampling may be unavoidable if the host plants do not occur in sympatry. In such cases, electrophoretic analysis of the populations will reveal nothing about whether host races are present (although a very large genetic distance would indicate that the populations are species — see Thorpe 1982). Allopatric populations can only be evaluated on the basis of thorough behavioural and ecological work, which might allow a reasonable projection of how mating would proceed in sympatry.

Sympatric geographic sampling is much more informative than allopatric sampling, but still presents problems. For example, if one host plant occurs in dry uplands while the other occurs in wet bottomlands, and if certain allozymes are selected for in dry and others in wet conditions, then host races might be incorrectly diagnosed. In fact, this is simply geographic heterogeneity due to selection, and indicates nothing about patterns of mating. If host plants never occur in the same habitat, sympatric sampling might be unavoidable, and in these cases allozyme differences associated with hosts are suggestive but not conclusive. The decision that the populations represent host races still depends as much upon behavioural and ecological work as allozyme data.

Microsympatric geographic sampling maximizes the value of allozyme data with respect to host races. If individuals of the different host plants are intermixed and touching, or at least within the typical cruising range of the insect in question, then allozyme frequency differences between samples of adult insects from the two plants strongly suggest the presence of host races. With such samples, from many parts of the range of the host plants, maximally informative analysis of the pattern of differentiation can be carried out.

4. Sampling over time

Time is related to the detection and understanding of host races in two ways. First, many putative host races, especially in univoltine insects, are at least partially allochronic (Diehl and Bush, 1984). This introduces problems analagous to those associated with geographic variation. For example, if one host-associated population is active in spring, and the other in midsummer, selection for allozyme frequency differences related to climate may occur. A classic example of seasonal selection is the cyclical change in chromosome inversion frequencies in *Drosophila pseudoobscura* (Dobzhansky 1943), and examples involving molecular polymorphism are known (Templeton and Johnston 1982). It is ironic that allochronic separation, which may aid the formation of host races (Bush 1969, 1974, 1975), may also diminish the ability of allozyme data to prove the existence of host races. The simplest solution to the allochrony problem, if feasible, is to sample at a time when at least some members of both host-associated populations are actively mating and ovipositing. Otherwise, inferences about host races must be based primarily upon ecological and behavioural work.

Time must also be considered on a longer scale. The sequences described earlier may occur over periods of several years, and sequence four might extend for decades. Obviously, studies of host races should ideally begin as soon as host races are suspected, and continue until the pattern of change is clear. Unfortunately, the political realities of research funding usually prohibit such an investment in time, although the fact that a few long-term studies of genetic change have been carried out (e.g. Anderson *et al.* 1975) proves that such studies are possible.

Conclusion

The ideal study of a host race, as I have described it, would sample thousands of individuals for hundreds of loci, across dozens of geographic sites, and over many years. Such a study will never occur. However, only by pursuing this goal will we substantially better our understanding; to do otherwise is to invite continued confusion about host races.

Acknowledgements

I thank my colleagues in 'rhagoletology', G.L. Bush, B.A. McPheron, D.C. Smith, J. Feder, G. Steck, D. Howard, M. Huettel, and S. William, for many hours of discussion of these problems over the years, but I take sole credit for any errors of thinking or fact. J.M. Berlocher prepared the figures.

References

Anderson, W., Dobzhansky, Th., Pavlovsky, O., and Yardley, D. (1975). Genetics of natural populations XLII. Three decades of genetic changes in *Drosophila pseudoobscura*. *Evolution*, **29**, 24–36.

Barker, J.S.F., East, P.D., and Wier, B.S. (1965). Temporal and microgeographic variation in allozyme frequencies in a natural population of *Drosophila buzzatii*. *Genetics*, **112**, 577–611.

Barton, J.S., Jones, J.S., and Mallet, J. (1988). Evolution: no barrier to speciation. *Nature Lond.*, **336**, 13–14.

Berlocher, S.H. and Bush, G.L. (1982). An electrophoretic analysis of *Rhagoletis* (Diptera: Tephritidae) phylogeny. *Syst. Zool.*, **31**, 136–155.

Bernatzky, R. and Tanksley, S.D. (1986). Toward a saturated linkage map in tomato based upon isozymes and random cDNA sequences. *Genetics*, **112**, 887–98.

Bush, G.L. (1966). The taxonomy, cytology, and evolution of the genus *Rhagoletis* in North America (Diptera: Tephritidae). *Bull. Mus. Comp. Zool.*, **134**, 431–562.

Bush, G.L. (1969). Sympatric host race information and speciation in frugivorous flies of the genus *Rhagoletis* (Dipera: Tephritidae). *Evolution*, **23**, 237–51.

Bush, G.L. (1974). The mechanisms of sympatric host race formation in the true fruit flies (Tephritidae). In *Genetic mechanisms of speciation in insects* (ed. M.J.D. White), pp. 3–23. Australia and New Zealand Book Co., Sidney.

Bush, G.L. (1975). Sympatric speciation in phytophagous parasitic insects. In *Evolutionary strategies of parasitic insects and mites* (ed. P.W. Price), pp. 187–206. Plenum Press, New York.

Diehl, S.R., and Bush, G.L. (1982). Host shifts, genetic models of sympatric speciation and the origin of parasitic insect species. In *Proceedings 5th International Symposium on Insect–Plant Relationships*, pp. 297–306. Pudoc, Wageningen.

Diehl, S.R. and Bush, G.L. (1984). An evolutionary and applied perspective of insect biotypes. *A. Rev. Ent.*, **29**, 471–504.

Dobzhansky, Th. (1943). Genetics of natural populations: IX. Temporal changes in the composition of populations of *Drosophila pseudoobscura*. *Genetics*, **28**, 162–86.

Eanes, W.P. and Koehn, R.K. (1978). An analysis of genetic structure in the Monarch butterfly, *Danaus plexippus* L. *Evolution*, **32**, 784–97.

Feder, J.L., Chilcote, C.A., and Bush, G.L. (1988*a*). Genetic differentiation between sympatric host races of the apple maggot fly, *Rhagoletis pomonella*. *Nature, Lond.*, **336**, 61–64.

Feder, J.L., Chilcote, C.A., and Bush, G.L. (1988*b*). Inheritance and linkage relationships of allozymes in the apple maggot fly. *J. Hered.* (In press.)

Futuyma, D.J., and Peterson, S.C. (1985). Genetic variation in the use of resources by insects. *A. Rev. Entomol.*, **30**, 217–38.

Jaenike, J. (1981). Criteria for ascertaining the existence of host races. *Am. Nat.*, **117**, 830–4.

Jaenike, J. and Selander, R.K. (1979). Ecological generalism in *Drosophila falleni*: genetic evidence. *Evolution*, **33**, 741–8.

Jaenike, J. and Selander, R.K. (1980). On the question of host races in the fall webworm, *Hyphantria cunea*. *Entomologia exp. appl.*, **27**, 31–7.

Kornfield, I., Smith, D.C., Gagnon, P.S., and Taylor, J.N. (1982). The cichlid fish of Cuatro Cienegas, Mexico: Direct evidence of conspecificity among distinct trophic morphs. *Evolution*, **36**, 658–64.

Malavasi, A. and Morgante, J.S. (1983). Population genetics of *Anastrepha fraterculus* (Diptera: Tephritidae) in different hosts: genetic differentiation and heterozygosity. *Genetica*, **60**, 207–11.

McCauley, D.E. and Eanes, W.F. (1987). Hierarchical population structure analysis of the milkweed beetle, *Tetraopes tetraophthalmus* (Forster). *Heredity*, **58**, 193–201.

McKechnie, S.W., Ehrlich, P.R., and White, R.B. (1975). Population genetics of *Euphydryas* butterflies. I. Genetic variation and the neutrality hypothesis. *Genetics*, **81**, 571–94.

McPheron, B.A. (1987). The population genetics of the colonization of the western United States by the apple maggot, *Rhagoletis pomonella* (Walsh) (Diptera: Tephritidae). Unpublished D. Phil. thesis, University of Illinois at Urbana-Champaign.

McPheron, B.A., Smith, D.C. and Berlocher, S.H. (1988*a*). Microgeographic genetic variation in the apple maggot *Rhagoletis pomonella*. *Genetics*, **119**, 445–51.

McPheron, B.A., Smith, D.C., and Berlocher, S.H. (1988*b*). Genetic differences between host races of *Rhagoletis pomonella*. *Nature, Lond.*, **336**, 64–66.

Menken, S.B.J. (1981). Host races and sympatric specialization in small ermine moths, Yponomeutidae. *Entomologia exp. appl.*, **30**, 280–92.

Pashley, D.P. (1986). Host associated genetic differentiation in fall armyworm: a sibling species complex? *Ann. ent. Soc. Am.*, **79**, 898–904.

Pashley, D.P. (1989). Causes of host-associated variation in insect herbivors: an example from fall armyworm. In *Evolution in insect pests: The pattern of variation* (ed. K.C. Kim). John Wiley, New York. (In press.)

Pashley, D.P. and Bush, G.L. (1979). The use of allozymes in studying insects movement with special references to the codling moth, *Laspeyresia pomonella* (L.) (Olethreutidae). In *Movement of highly mobile insects: Concepts and methodology in research* (ed. R.L. Rabb and C.G. Kennedy), pp. 33–41. University Graphics, Raleigh, North Carolina.

Plapp, F.W., Jr. (1976). Biochemical genetics of insecticide resistance. *A. Rev. Ent.*, **21**, 179–197.

Prokopy, R.J. and Berlocher, S.H. (1980). Establishment of *Rhagoletis pomonella* (Diptera: Tephritidae) on rose hips in southern New England. *Can. Ent.*, **112**, 1319–20.

Prokopy, R.J., Diehl, S.R., and Cooley, S.S. (1987). Behavioral evidence for host races in *Rhagoletis pomonella* flies. *Oecologia*, **76**, 138–47.

Rousch, R.T. and McKenzie, J.A. (1987). Ecological genetics of insecticide and acaride resistance. *A. Rev. Ent.* **32**, 361–80.

Smith, D.C. (1986). Genetics and reproductive isolation of *Rhagoletis* flies. Unpublished D. Phil. thesis, University of Illinois at Urbana-Champaign.

Smith, D.C. (1988). Heritable divergence of *Rhagoletis pomonella* host races by seasonal asynchrony. *Nature, Lond.*, **336**, 66–68.

Sturgeon, K.B. and Mitton, J.B. (1986). Allozyme and morphological differentiation of mountain pine beetles *Dendroitonus ponderosae* Hopkins (Coleoptera: Scolytidae) associated with host tree. *Evolution*, **40**, 290–302.

Tabachnick, W.J., Munsterman, L.E., and Powell, J.R. (1980). Genetic distinctness of sympatric forms of *Aedes aegypti* in East Africa. *Evolution*, **33**, 287–95.

Templeton, A.R. and Johnston, J.S. (1982). Life history evolution under pleitropy and K-selection in a natural population of *Drosophila mercatorum*. In *Ecological genetics and evolution: The cactus-yeast-drosophila model system* (ed. J.S.F. Barker and W.T. Starmer), pp. 25–239. Academic Press, Australia.

Thorpe, J.P. (1982). The molecular clock hypothesis: Biochemical evolution, genetic differentiation and systematics. *A. Rev. Ecol. Sys*, **12**, 139–168.

Walsh, B.D. (1867). The apple-worm and the apple-maggot. *Am. J. Hort.*, **2**, 338–43.

5 Population genetics studies of fruit flies of economic importance, especially medfly and olive fruit fly, using electrophoretic methods

MICHAEL LOUKAS

Department of Genetics, Agricultural University of Athens, Votanikos, Athens, Greece

Abstract

The olive fruit fly, *Dacus oleae*, is an extremely monophagous species totally dependent on olive fruit. Electrophoretic analysis of natural populations from almost all areas of its distribution revealed a substantial amount of genetic variation ($\overline{H}$ = 0.188). When field-collected flies were maintained on artificial substrate, profound changes were observed in the allozyme frequencies of two enzyme loci, alcohol dehydrogenase (ADH) and 6-phosphogluconate dehydrogenase (6-PGD), out of ten loci studied. These changes, which occur within a few generations of artificial rearing, coincide with other changes in physiology and behaviour. When fruit flies collected from the wild were reared artificially under three different daily temperature regimes, the frequency of ADH alleles also changed strikingly in all cases within four generations. Wild flies kept on olives for four generations showed no change in frequency of the ADH alleles. When flies reared in the laboratory for about 2 years were provided olives for oviposition, the ADH allozyme frequencies changed substantially within only one generation towards the frequencies of the wild populations, remaining stable thereafter. Lastly, when a new laboratory colony was established from other laboratory colonies with known initial ADH allozyme frequencies, the changes in the allozyme

Electrophoretic Studies on Agricultural Pests (ed. Hugh D. Loxdale and J. den Hollander), Systematics Association Special Volume No. 39, pp.69–102. Clarendon Press, Oxford, 1989.

frequencies in the new colony were similar to those observed in colonies established from field-collected flies.

Numerous wild populations of medfly originating from different parts of the world and from different host trees were analysed for 25 enzyme systems. All the populations proved to be highly monomorphic ($\overline{H} \simeq 0.05$) except of those coming from Kenya, S. Africa and Reunion Island. This worldwide low genetic heterogeneity seems to be 'real', since the application of the urea denaturation method did not detect any further genetic variability. Experiments in the field and in the laboratory failed to detect any pattern of preference for ovipositing sites associated either with the taxonomic status of the host fruit or with the size of the fruit. By estimating the genetic distances between the populations and/or by using the allozymes as genetic markers, we could roughly trace the route of dispersion of the fly from its geographic centre of origin, which is placed in Africa. No systematic changes in allele frequencies were observed in populations reared on artificial substrate.

Introduction

Following the demonstration by Hubby and Lewontin (1966) and Harris (1966) that many enzyme loci segregate for several electrophoretically-detectable variants in natural populations, population biologists have used the electrophoretic technique to answer questions of paramount importance in theoretical and applied genetics (see Nevo *et al.* 1984 for a recent review).

The present study focuses on the application of the electrophoretic technique in agricultural entomology and, especially, in two of the most harmful insect pests, the olive fruit fly, *Dacus oleae* (Diptera: Tephritidae) and the Mediterranean fruit fly (medfly), *Ceratitis capitata* (Diptera: Tephritidae). *D. oleae* is a monophagous species totally dependent on the olive fruit, whereas *C. capitata* is an extremely polyphagous species infesting more than 253 fruit trees, nut trees and vegetables, 40 of which are considered 'heavily or general infested' (Hagen *et al.* 1981).

Allozyme analysis has been used in these two species

(1) to study the levels of genetic variation and the distribution of allele frequencies in natural populations;
(2) to study the processes accounting for the maintenance of genetic variability in natural populations and the role of this variability to species' adaptation;
(3) to trace the introduction and subsequent dispersal of each species;

(4) to estimate the genetic distances between different populations;
(5) to study the creation of host races that may occur by the adaptation of species to new plants (Bush 1969);
(6) to monitor the genetic structure of natural and artificially-reared populations;
(7) to develop a genetic sexing system (Riva and Robinson 1986; Robinson *et al.* 1986);
(8) to establish linkage groups (Malacrida *et al.* 1987), for genetic manipulations (Saul 1986);
(9) to study development differences and tissue specificity (Malacrida *et al.* 1983);
(10) to study sperm precedence and multiple matings (Zouros and Krimbas 1970);
(11) to study the effects of insecticides on polymorphisms (Tsakas and Krimbas 1970).

In the following sections, I shall discuss topics 1 to 7, not only because of their importance from the population genetics point of view, but also because many data concerning these applications are presented here for the first time.

Allozyme polymorphisms in natural populations

Zouros *et al.* (1968) described the first two enzyme polymorphisms in the olive fruit fly *Dacus oleae*, namely, Est A and Est B. Zouros and Krimbas (1969) detected 12 alleles for Est A and 11 for Est B and demonstrated that each locus segregates for a 'null' (or silent) allele. A survey of 12 populations, extending over most of the known geographical distribution of the species, increased the number of active alleles of Est A to 20 (Tsakas and Zouros 1980). In a preliminary study, Bush and Kitto (1979) scored about 50 individuals of *D. oleae* from two neighbouring populations from Attica, Greece for 23 enzyme loci. Five loci exhibited substantial amounts of variation, six were moderately polymorphic and 12 were monomorphic. In a recent report, Loukas *et al.* (1985) analysed 10 loci (two of which, hexokinase and leucine-aminopeptidase, were not included in the study of Bush and Kitto (1979)) in a natural population of *D. oleae* (Isle of Aghia Triada, Evia, Greece). The general profile for the loci involved was similar in both studies, although Loukas *et al.* (1985) detected more rare alleles because they used a larger sample size. Table 5.1 gives the present information on enzyme polymorphisms in natural populations of *D. oleae*.

The mean heterozygosity was estimated to be 0.188 — which is

Table 5.1 Electrophoretic enzyme polymorphisms in natural populations of *Ducus oleae* (Zouros and Loukas (1988).

Enzyme locus	K^a	n^b	$n_e{}^c$	$\bar{H}^d$	Source[e]
Esterase A (Est A)	14	20	3.7	0.728	1,2,3,4
Esterase B (Est B)	14	11	2.3	0.561	1,2,3,4
6-Phospho gluconate dehydrogenase (6-PGD)	3	5	2.1	0.530	5,6
Leucine aminopeptidase (LAP)	1	4	2.1	0.514	6
Alkalinc phosphatase (APH)	3	2	2.0	0.496	3
Hexokinase (HK)	1	6	1.9	0.467	6
Alcohol dehydrogenase (ADH)	8	3	1.8	0.444	5,3,7.6
Octanol dehydrogenase (ODH)	3	4	1.5	0.345	3
Mannose phosphate isomerase (MPI)	2	5	1.5	0.325	5
Malate dehydrogenase-2 (MDH-2)	2	3	1.3	0.222	5
Peptidase-3 (PEP-3)	1	3	1.2	0.196	6
Phosphoglucomutase (PGM)	3	4	1.2	0.185	5,6
Glutamate-oxaloacetate transaminase-1 (GOT-1)	2	3	1.1	0.064	5
Glucose-6-phosphate dehydrogenase (G6PD)	1	3	1.1	0.060	6
Malate dehydrogenase-1 (MDH-1)	2	2	1.0	0.039	5
Acetaldehyde oxidase (AO)	3	2	1.0	0.037	3
Tetrazolium oxidase (TO)	3	2	1.0	0.032	3
Isocitric dehydrogenase (IDH)	3	3	1.0	0.028	5,6
Peptidase-1 (PEP-1)	3	3	1.0	0.006	5,6
Aconitase-1 (ACON-1)	2	1	1.0	0	5
Aconitase-2 (ACON-2)	2	1	1.0	0	5
Aldolase (ALD)	2	1	1.0	0	5
Malic enzyme (ME)	2	1	1.0	0	5
Fumarase (FUM)	2	1	1.0	0	5
Peptidase-2 (PEP-2)	2	1	1.0	0	5
Glycerol-3-phosphate dehydrogenase (G-3-PD)	2	1	1.0	0	5
Malate dehydrogenase-3 (MDH-3)	2	1	1.0	0	5
Glutamate-oxaloacetate transaminase-2 (GOT-2)	2	1	1.0	0	5
Mean	—	3.467	1.2	0.188	—

[a] K = number of populations surveyed.
[b] n = number of alleles observed.
[c] n_e = effective number of alleles.
[d] $\bar{H}$ = average heterozygosity.
[e] Source: 1. Zouros and Krimbas (1969); 2. Krimbas and Tsakas (1971); 3. Tsakas and Krimbas (1975); 4. Tsakas and Zouros (1980); 5. Bush and Kitto (1979); 6. Loukas *et al.* (1985); 7. Zouros *et al.* (1982).

higher than the average heterozygosity for *Drosophila*, estimated at 0.140, and twice as high as the average heterozygosity for insects, estimated at 0.074 (Nevo 1978). This is especially notable because *D. oleae* is a monophagous species totally dependent on the olive fruit.

Krimbas and Tsakas (1971) tested the hypothesis that the polymorphisms in Est A and B loci are not under the influence of natural selection. By using the theoretical expectation of the gene-frequency variance among subpopulations, which is a function of the time elapsed since the separation of the populations and the effective population size, they showed that temporal variation in gene frequencies was compatible with the neural theory. On the contrary, application of the same test to six loci showed that Est B and ADH may be under the influence of natural selection (Tsakas and Krimbas 1975).

The uniformity of the gene frequency distribution observed by Tsakas and Zouros (1980) (alleles A_2 and B_4 of the loci Est A and B, respectively, are the most common ones in all populations) can be explained either by a neutral hypothesis or by selection. Under the first hypothesis, even a small rate of gene flow among populations will be enough to cancel out the effects of random drift, resulting in gene frequency uniformity, while under the second the action of selection would be the same throughout the species range.

Huettel *et al.* (1980) studied the genetic variability of four natural populations of medfly from South Africa, Israel, Hawaii, and Costa Rica, by examining 23 allozyme loci and the relationship between genetic structure and the geographic distribution of natural populations. Morgante *et al.* (1981) studied the allozymic variation encoded by 13 loci in four natural Brazilian populations of medfly from different host-fruits, while Gasperi *et al.* (1986) studied the genetic variability in samples of medfly from such geographically and ecologically distinct localities as Kenya, Libya, and the Italian Islands of Procida and Sardinia. Finally, Kourti and Loukas (in prep.) studied the allozymic variation of 14 natural populations of medfly from different host fruits and distinct geographic areas (one each from S. Africa, Reunion Island, Hawaii, Israel, Island of Procida (Italy), Spain, Guatemala, Cyprus, and two from continental Greece, the Green Islands Chios, and Crete, and Egypt) by examining 25 enzyme loci.

Tables 5.2 and 5.3 summarize the information on enzyme variability in natural populations of medfly. In Table 5.3 the degree of genetic variation of each population is expressed by the expected mean heterozygosity ($\bar{H}$), by the proportion of polymorphic loci (P) and by the mean actual number of alleles per locus ($\bar{n}$). The

Table 5.2 Electrophoretic enzyme polymorphisms in natural populations of medfly[a]

Enzyme locus	S. African population				Reunion population				Introduced populations[b]			
	n^c	$n_e{}^d$	N^e	H^f	n^c	$n_e{}^d$	N^e	H^f	$\bar{n}^g$	$\bar{n}_e{}^h$	N^e	$\overline{H}^i$
Mannose phosphate isomerase (MPI)	14	5.7	120	0.815	6	2.0	126	0.497	2	1.3	97–178	0.259
Glutamate-oxaloacetate transaminase-2 (GOT-2)	2	1.2	107	0.162	2	1.1	112	0.102	1	1.0	80	0
Phosphoglucomutase (PGM)	5	1.6	120	0.358	3	1.2	110	0.136	2	1.0	77–140	0.014
Peptidase-1 (PEP-1)	3	1.8	142	0.456	2	1.8	112	0.453	2	1.7	101–404	0.427
Peptidase-2 (PEP-2)	6	2.4	142	0.583	3	1.9	112	0.482	2	1.0	80–282	0.035
Peptidase-3 (PEP-3)	4	1.5	142	0.330	2	1.2	112	0.177	2	1.2	101–387	0.155
6-Phosphogluconate dehydrogenase (6-PGD)	3	1.1	118	0.051	1	1.0	110	0	2	1.0	80	0.001
Pupal esterase	—	—	—	—	4	2.1	91	0.516	5	1.7	80	0.412
Phosphoexose isomerase (PHI)	3	1.3	120	0.232	1	1.0	110	0	1	1.0	80	0
Hexokinase-1 (HK-1)	3	2.3	120	0.561	3	1.7	115	0.427	3	1.0	80	0.002
Hexokinase-3 (HK-3)	2	1.6	120	0.391	2	1.9	115	0.461	1	1.0	80	0
Glocose-6-phosphate dehydrogenase (G-6-PD)	3	1.9	118	0.464	1	1.0	110	0	3	1.1	80–154	0.077
Isocitric dehydrogenase (IDH)	3	1.4	80	0.275	3	1.4	115	0.288	2	1.1	80–140	0.113
Octanol dehydrogenase (ODH)	2	1.2	80	0.180	1	1.0	80	0	1	1.0	80	0
Leucine aminopeptidase (LAP)	3	2.4	72	0.591	2	1.2	76	0.199	3	1.0	60–140	0.044

Table 5.2 *Continued*

Enzyme locus	S. African population				Reunion population				Introduced populations[b]			
	n^c	$n_e{}^d$	N^e	H^f	n^c	$n_e{}^d$	N^e	H^f	$\bar{n}^g$	$\bar{n}_e{}^h$	N^e	$\bar{H}^i$
Tetrazolium oxidase (To)	1	1.0	80	0	1	1.0	80	0	1	1.0	80	0
Glutamate-oxaloacetate transaminas-1 (GOT-1)	1	1.0	80	0	1	1.0	80	0	1	1,.0	80	0
Diaphorase-1 (DIAPH-1)	1	1.0	80	0	1	1.0	80	0	1	1.0	80	0
Diaphorase-2 (DIAPH-2)	1	1.0	80	0	1	1.0	80	0	1	1.0	80	0
Alcohol dehydrogenase (ADH)	1	1.0	80	0	1	1.0	80	0	1	1.0	80	0
Adenylate Kinase (AK)	1	1.0	80	0	1	1.0	80	0	1	1.0	80	0
Fumarase (FUM)	1	1.0	80	0	1	1.0	80	0	1	1.0	80	0
Malate dehydrogenase (MDH)	1	1.0	80	0	1	1.0	80	0	1	1.0	80	0
Malic enzyme (ME)	1	1.0	80	0	1	1.0	80	0	1	1.0	80	0
α-Glycerophosphate dehydrogenase (α-GPD)	1	1.0	80	0	1	1.0	80	0	1	1.0	80	0

[a] From Kourti and Loukas (in prep.)
[b] Except Reunion and South Africa populations
n^c = number of alleles observed
$n_e{}^d$ = effective number of alleles
N^e = number of individuals analysed per population
H^f = heterozygosity
$\bar{n}^g$ = mean actual number of alleles
$\bar{n}_e{}^h$ = mean effective number of alleles
$\bar{H}^i$ = average heterozygosity

Table 5.3 Genetic variation in natural populations of medfly. (Further explanations in the text.)

Population	Host fruit-tree	No. of loci	$\overline{H}$	P	$\bar{n}$	Source[a]
Costa Rica	—	23	0.036	—	—	1
Guatemala	Coffee	23	0.063	0.13	1.3	2
Brazil	See note b	13	0.030[b]	—	—	3
Libya	Oranges	27	0.021	0.15	1.2	4
S. Africa	Guavas	24	0.234	0.54	2.7	2
S. Africa	—	23	0.167	—	2.3	1
Kenya	Coffee	25	0.164	0.56	1.8	4
Reunion	Coffee	25	0.153	0.44	1.8	2
Egypt	Apricots	23	0.068[c]	0.17[c]	1.3[c]	2
Hawaii	—	23	0.034	—	—	1
Hawaii	Guavas	24	0.055	0.20	1.2	2
Israel	—	23	0.071	—	—	1
Israel	Apricots	25	0.066	0.16	1.2	2
Cyprus	Tangerines	25	0.039	0.08	1.2	2
Crete (island, Greece)	Grapefruit	23	0.046	0.17	1.3	2
Attiki (Greece)	Oranges	25	0.057	0.16	1.3	2
Kalamata (Greece)	Figs	25	0.046	0.12	1.3	2
Chios (island, Greece)	Figs	25	0.042	0.16	1.2	2

Table 5.3 *Continued*

Population	Host fruit-tree	No. of loci	$\overline{H}$	P	$\bar{n}$	Source[a]
Procida (island, Italy)	Figs	23	0.045	0.17	1.2	2
Procida (island, Italy)	See note c	23–25	0.046[d]	0.19[d]	1.2[d]	4
Sardinia (island, Italy)	Figs	15	0.115	0.47	1.6	4
Sardinia (island, Italy)	Plums	23	0.114	0.30	1.4	4
Spain	Peaches	23	0.065	0.17	1.2	2

[a] Source: 1. Huettel *et al.* (1980); 2. Kourti and Loukas (in prep.); 3. Morgante *et al.* (1981); 4. Gasperi *et al.* (1986).
[b] Data based on four populations
[c] Data based on two populations
[d] Data based on 16 samples collected in the years 1983, 1984, 1985 from bitter oranges (3), apricots (3), figs (3), peaches (6) and prickly-pears (1).

main feature of Table 5.3 is the extremely low genetic variability, ranging from 0.021 to 0.071, of all the populations (irrespectively of the host fruit-tree), with the exception of those coming from S. Africa, Kenya, and Reunion Island. This observation is especially notable because the medfly is an extremely polyphagous species. (The comparably high genetic variability detected in the two samples from Sardinia may be due to a founder effect.)

According to theory (Gillespise and Langley 1974), and some experimental studies (Powell 1971; Levinton 1973), one would expect a positive correlation between a species genetic variation and the degree of its environmental diversity. However, comparison of genetic variation levels between fruit-fly species that have a single host, and the medfly, which has multiple hosts, shows that the genetic variation of medfly populations does not follow these expectations. For example, *Dacus oleae*, *Rhagoletis pomonella*, *R. completa*, and *Delia antiqua* are monophagous species infesting olive trees, species of the plant families *Rosaceae* and *Juglandaceae* (walnut trees) and cultivated *Allium* species (onions), respectively. Their respective degrees of mean heterozygosity were found to be 0.188 (see Table 5.1), 0.181, and 0.075 (Berlocher and Bush, 1982) and 0.09 (M. Harris, pers, comm.), which are considerably higher than the heterozygosity values reported for the Mediterranean populations of medfly.

It could be argued, however, that the low genetic variation observed is a technical artifact: electrophoretic 'alleles' or 'electromorphs', group together many real alleles that cannot be differentiated by such a crude technique as simple gel electrophoresis. Often, by employing a large number of techniques coupled with electrophoresis (thermostability, different gel concentrations and buffer systems, and sensitivity to urea), a considerable amount of additional 'hidden' variation can be revealed (Singh *et al.* 1975, 1976; Coyne 1976; Coyne and Felton 1977; Coyne *et al.* 1978; Singh 1979; Loukas *et al.* 1981).

Kourti *et al.* (in prep.) utilized the urea denaturation method to detect hidden genetic variability. They constructed 20 'isogenic' strains of medfly using 'brother–sister' matings for 14 successive generations. Urea denaturation was performed before electrophoresis for three enzyme systems, two of which were monomorphic (ODH and ME) and one highly polymorphic (PEP-1) in all populations studied. None of the 20 'isogenic' strains displayed either quantitative or qualitative differences from standard electrophoresis. The results indicate that the low genetic variability detected in medfly populations by the usual electrophoretic method is real.

Intra- and interpopulational variation

Host acceptance patterns

Allozyme variability patterns have also been used to indicate the formation and occurrence of host races (Diehl and Bush, 1984), defined as 'populations of a species partially reproductively isolated from other conspecific populations as a direct consequence of adaptation to a specific host'. The phenomenon of host-plant shift in phytophagous insects may lead to the evolution of a new host race and sympatric speciation (Bush 1975). It is obvious that the ability to recognize distinct populations can have important consequences for pest-control management (Rosen 1978; Bush and Hoy, 1983). Knowledge and understanding of genetic variability within and between target populations provide evidence on population structure, dispersal, and gene flow and are a primary goal of integrated pest management (Pashley and Bush 1979; Florence *et al.* 1982; Pashley 1986).

Prokopy *et al.* (1982) reported that some, if not most, of the observed interpopulational variation in host-acceptance patterns of *Rhagoletis pomonella* is attributable to genetic differences between populations of that species. Jaenike and Grimaldi (1983) found substantial genetic variation for host preference among strains of the polyphagous *Drosophila tripunctata* species. Prokopy *et al.* (1984) reported significant interpopulational differences among medfly females for oviposition sites. Their data suggest that the fruit size and not its taxonomic status has a strong influence on the acceptance pattern of each population and that at least a portion of the interpopulational variation has a genetic basis.

The striking similarities in the genetic structure of the populations of medfly infesting different host fruits (Table 5.2, for the Mediterranean populations) is in apparent contradiction to a host-acceptance pattern associated with genetic differences. On the other hand, it could be argued that the genetic changes for host preference, if they do exist, may be too small to be detected by the usual population samples, and/or they could be detected only if samples from different kinds of fruit-tree species cultivated in the same area and collected at the same time, were studied.

To study the last possibility, Kourti *et al.* (in prep.) analysed two pairs of populations coming from two different localities from the Island of Crete (Greece) for 23 enzyme loci. The first pair of populations (Crete-I and Crete-II) derived from infested grapefruit

(Crete-I) and oranges (Crete-II) collected in May 1986 from two adjacent orchards. The second pair derived from infested grapefruit (Crete-III) and peaches (Crete-IV) collected in April 1986 (Crete-III) and July 1986 (Crete-IV), again from two adjacent orchards. The mean diameters of the infested grapefruit (production of the previous year maintained on the trees), oranges, and peaches, from which the populations were collected, were 11.7 ± 0.9 cm, 6.3 ± 0.5 cm, and 6.8 ± 0.6 cm, respectively. The mean heterozygosity of each population was 0.046 for Crete-I, 0.044 for Crete-II, 0.049 for Crete-III, and 0.050 for Crete-IV. The genetic identity values (Nei 1972) in all pair-wise comparisons of the four populations, were estimated to be 1.000 for the pairs 'Crete-I–Crete-II', 'Crete-I–Crete-III', 'Crete-I–Crete-IV' and 'Crete-III–Crete-IV', and 0.999 for the pairs 'Crete-II–Crete-III' and 'Crete-II–Crete-IV'. Therefore, all four populations displayed extremely high coincidence in their genetic structures, irrespective of the taxonomic status and the size of the host fruit. Moreover, numerous single pair crosses on artificial food between individuals coming from grapefruit and oranges (♀ 'grapefruit' × ♂ 'grapefruit', ♀ 'orange' × ♂ 'orange', ♀ 'grapefruit' × ♂ 'orange', ♀ 'orange' × ♂ 'grapefruit') did not yield significantly different mean numbers of (adult) offspring (Kourti *et al.*, in prep.).

Of course, the possibility remains that the host-preference patterns that Prokopy *et al.* (1984) reported for the medfly are due to relatively few genetic changes or that they may not involved the genetic loci studied by Kourti and Loukas (see Table 5.2). Kourti and Loukas (in prep.) studied the preference for oviposition sites of the medfly females, in relation to size or the kind of the fruit in a laboratory experiment. They established two medfly strains, one homozygous for allele PEP-$1^{1.26}$ and the other homozygous for allele PEP-$1^{1.00}$. The first was from flies that had emerged from grapefruit and the second from those from peaches. The two strains were maintained in the laboratory for two consecutive generations on grapefruit and peaches respectively. In the third generation, 300 pupae from each strain were transferred to separate cages containing artificial medium. Ten days after emergence, 100 females of each strain (probably already inseminated) were transferred to two new cages (50 females in each cage), so that each new cage contained 100 females. The females in each cage oviposited on four grapefruit and four peaches placed in each cage. These fruits were changed daily for ten concecutive days. Kourti and Loukas collected a total of 449 pupae from the peaches and 285 from the grapefruit. All the adults that emerged (411 from peaches and 245 from grapefruit) were electrophorezed for PEP-1. Among the 411 adults coming from peaches, 187 were found to be homozygous for allele PEP-$1^{1.26}$ and 224 homozygous for allele PEP-$1^{1.00}$. Among the 245 adults coming from grapefruit, 131 were found to be homozygous for allele PEP-$1^{1.26}$ and 114 homozygous for allele PEP-$1^{1.00}$. Although the females showed some tendency to oviposit on the fruit in which were fed as larvae, the χ^2 homogeneity tests (to test the deviation from the 1 : 1 ratio) were not statistically significant ($0.05 < P < 0.10$ in the first case and $0.20 < P < 0.30$ in the second).

Again, these results, which are independent of the larval viability in the different host fruits, failed to detect any pattern of preference for oviposition sites in the medfly.

From the previous experiments, it seems that medfly may experience the different host plants as a fine-grained environment. It is possible for the different host fruits not to represent different environments, because these fruits do not differ in the chemical compounds which are the substrates for the enzymes studied. The fact that the same kind of fruit is not available throughout the year forces the medfly to shift from host to host. Such a shift does not seem to be connected with any apparent action of selection (unpub. data). Since the fine-grained species utilize many alternate food resources they require functional flexibility (Valentine 1976). Thus, selection would increase the frequency of the allele that, averaged over the different environments, provides the highest fitness.

In conclusion, we agree with Menken and Ulenberg (1987), who stated that 'host races are extremely difficult to identify positively and careful ecological, genetic and behavioural studies are indispensable at this low level of divergence. Allozyme data must be evaluated, together with results from host-plant choice and larval-survival experiments as well as pheromone attraction experiments in the laboratory and the field before conclusions can be drawn concerning host races.'

2. *Differences in genetic structure between native and introduced populations*

The low levels of genetic variation observed in the 'introduced' (see Table 5.2) populations of medfly may reflect historical reasons, namely, the time elapsed since colonization and the number of flies that formed the founder populations.

Tropical west Africa is regarded as the centre of origin of the medfly. The biologically effective time of separation for species of Dacinae and Trypetinae is usually assigned to the beginning of the Oligocene epoch, about 36 million years ago (Myr) (see Tarling

(1980) for discussion of the geological setting; Sarich and Cronin (1980) and Beverley and Wilson (1984) for other examples of Old and New World groups that diverged at this time through separation of Africa and South America).

The high degrees of heterozygosity of the medfly in the populations of S. Africa and Kenya (Table 5.3) in combination with the high number of alleles detected, many of which are rare (Table 5.2), support the hypothesis that the centre of origin of the species is in Africa. Therefore, the low genetic variability observed, e.g. in the Mediterranean populations, may have arisen from the introduction of a small number of individuals (the founder effect—Mayr 1954) that later spread to different geographic areas.

The reduction in average heterozygosity depends on both the size of the bottleneck and the time the bottleneck persists, whereas the reduction in average number of alleles per locus depends primarily on bottleneck size (Nei *et al.* 1975; Chakraborty and Nei 1976). Such a reduction both in average heterozygosity and in numbers of alleles compared with the source populations, has been found in the gypsy moth, *Lymantria dispar* (Harrison *et al.* 1983).

For the Mediterranean populations of medfly, it has been proposed (Hagen *et al.* 1981) that the insect first invaded Spain in 1842, and subsequently spread into France, Italy, Greece, and the Middle East. Of course, there is no way to determine the amount of genetic variation contained in the founder population, simply because there is no unified theory for predicting colonization success at the species level (Levins and Wilson, 1980). The allele frequency distribution in the native populations is U-shaped, which appears to be typical of distributions of populations assumed to have maintained large population sizes in the recent past. The distributions of the derived populations are effectively bimodal, with peaks in the 0.40–0.60 and 0.91–1.00 frequency classes and an almost complete absence of rare and low-frequency alleles. Such distributions have been predicted for populations that have gone through a genetic bottleneck (Nei *et al.* 1975); Berlocher 1976; Chakraborty *et al.*1980). This is exactly the pattern found in medfly from Kenya and S. Africa (the probable source areas) and from the Mediterranean countries, Guatemala, Costa Rica, and Hawaii (Huettel *et al.* 1980; Kourti and Loukas, in prep.).

Table 5.4 gives the genetic identity values, I, as well as the genetic distances, D (Nei 1972) between all the pairwise comparisons of 13 populations of medfly (Kourti and Loukas, in prep.) based on 23 loci. The values of I can be classified into three groups. One group includes combinations of any two populations except the populations Reunion and S. Africa; there are 55 such values. The second group includes all the pairs 'any population (except S. Africa)–Reunion' and 'any population (except Reunion)–S. Africa'; there are 22 such values. The third group consists of a simple value, the index of identity of the pair 'S. Africa–Reunion'. The non-parametric two-sample rank test (Brownlee 1967, p. 251) for this pattern gives a standard normal deviate $U_p = -6.8$. The two-tailed probability that a value as large as this or larger is due to chance is approximately equal to zero. Therefore, there are statistically significant differences in the genetic structure between the introduced and the native populations of medfly.

From the distance matrix of Table 5.4, the dendrogram of Fig. 5.1 was constructed, according to the unweighted pair group mean analysis (UPGMA) method of Sneath and Sokal (1973). Although a small number of loci contribute to the differentiation of the *D* values between populations (only the polymorphic loci of the introduced populations), the most significant features of the dendrogram depicted in Fig. 5.1 are the following. (1) There is a clear-cut separation of the introduced and the source populations. (2) The population of Reunion belongs to the same lineage as the population of S. Africa, a fact indicating that this population derived from a different route of dispersion than, e.g. the Mediterranean populations. (3) The lineage leading to Mediterranean populations splits into two lineages. One leads to the population of Spain and the other to the remaining Mediterranean populations. The latter splits into two sublineages: the sublineage of the populations Israel and Egypt and the sublineage of the populations Attiki, Cyprus, Kalamata, Procida, Chios, and Crete.

The dendrogram of Fig. 5.1, based on quantitative data, seems to be in agreement with the proposal of Hagen *et al.* (1981) concerning the dispersion of the fly from its geographic centre of origin. On the other hand, the qualitative data (by using the alleles as genetic markers to trace the routes of dispersion of the fly) support such a hypothesis in some cases but not in others. Table 5.5 gives the allele frequencies at three loci in the populations from S. Africa and Guatemala (supposedly 'introduced' from Spain), as well as in the Mediterranean populations. While the kinds of alleles and their frequencies at the G-6-PD locus support the hypothesis of Hagen *et al.* (1981), the allele distribution of the IDH and PGM loci supports another route of dispersion of the fly, from Africa to the Mediterranean countries, not through the Iberian peninsula but through the Nile valley to Egypt. Of course, many more populations from different countries need to be analysed to answer this question.

Table 5.4 Genetic identify values I (above the diagonal) and genetic distances, D, (below the diagonal) between 13 natural populations of medfly (given as multiples of 10^3). (Further explanations in the text.)

	1	2	3	4	5	6	7	8	9	10	11	12	13
1	—	999	996	1,000	997	997	995	989	986	985	974	942	932
2	1 ± 7	—	999	999	998	995	995	987	993	980	981	943	932
3	4 ± 13	1 ± 7	—	997	998	990	993	985	982	977	983	936	925
4	0	1 ± 7	3 ± 11	—	998	997	995	988	988	987	980	949	938
5	3 ± 11	2 ± 9	2 ± 9	2 ± 9	—	994	995	989	994	988	987	949	942
6	3 ± 11	5 ± 15	10 ± 21	3 ± 11	6 ± 16	—	993	988	989	992	970	951	941
7	5 ± 15	5 ± 15	7 ± 18	5 ± 15	5 ± 15	7 ± 18	—	988	983	986	972	939	933
8	11 ± 22	13 ± 24	15 ± 26	12 ± 23	11 ± 22	12 ± 23	2 ± 9	—	977	989	965	939	938
9	14 ± 25	7 ± 17	18 ± 28	12 ± 24	6 ± 16	11 ± 22	17 ± 28	23 ± 32	—	981	974	953	942
10	15 ± 26	20 ± 30	23 ± 31	13 ± 23	12 ± 23	8 ± 19	14 ± 24	11 ± 22	19 ± 29	—	957	941	939
11	26 ± 34	19 ± 29	17 ± 27	20 ± 30	13 ± 24	30 ± 37	28 ± 35	36 ± 40	26 ± 34	43 ± 44	—	961	948
12	60 ± 52	59 ± 51	66 ± 55	52 ± 48	52 ± 48	50 ± 47	63 ± 53	63 ± 53	48 ± 46	60 ± 52	40 ± 42	—	979
13	70 ± 52	70 ± 56	78 ± 59	64 ± 54	60 ± 52	61 ± 52	69 ± 56	64 ± 54	59 ± 52	62 ± 53	53 ± 49	21 ± 31	—

1, Cyprus; 2, Kalamata (Greece); 3, Chios (island, Greece); 4, Attiki (Greece); 5, Procida (island, Italy); 6, Crete (island, Greece); 7, Israel; 8, Egypt-I; 9, Spain; 10, Guatemala; 11, Hawaii; 12, Reunion (island, France); 13, South Africa.

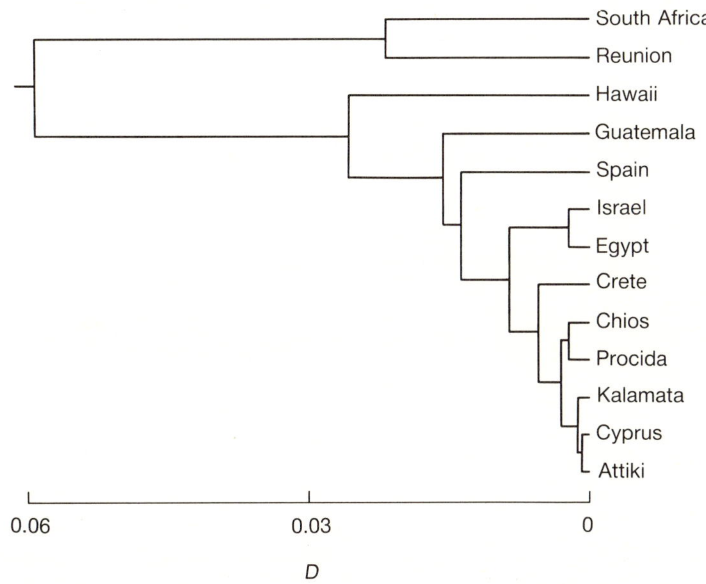

Fig. 5.1. Dendrogram of 13 natural populations of medfly constructed from the distance matrix shown in Table 5.4.

Genetic changes in artificially-reared colonies

Most control programmes for harmful insects involve the rearing of the insect under artificial conditions. This is particularly true for biological programmes of control or programmes that combine biological and traditional techniques. The conditions that prevail in such colonies are often quite different from those experienced by natural populations of the species. Food, oviposition site, temperature and photoperiod regimes, population density, and presence or absence of preditors and parasites are some of the most important factors that can be radically different between the artificial and natural environments. Therefore, a new regime of selective forces is expected to act on the colony. Morever, maintenance under a uniform environmental laboratory regime without any gene flow from outside the population will undoubtedly lead to changes in gene frequencies and loss of alleles (especially rare alleles) owing to genetic drift and

Table 5.5 Allele frequencies at three enzyme loci in the Mediterranean populations of medfly as well as in the populations from S. Africa and Guatemala (Kourti and Loukas, in prep.)

Locus	Guatemala	Spain	S. Africa	Egypt	Mediterranean populations[a]
IDH alleles					
1.10	—	—	0.012	—	—
1.00	1.000	1.000	0.838	0.670	0.955
0.83	—	—	0.150	0.330	0.045
G-6-PD alleles					
1.02	0.103	0.031	0.051	—	0.001
1.00	0.661	0.802	0.678	1.000	0.993
0.98	0.236	0.167	0.271	—	0.006
PGM alleles					
1.26	—	—	0.004	—	—
1.13	—	—	0.087	—	—
1.00	1.000	1.000	0.788	0.956	1.000
0.89	—	—	0.117	0.044	—
0.79	—	—	0.004	—	—

[a] Excepting the populations from Spain and Egypt.

inbreeding depression. This loss of genetic variability is of paramount concern to workers in biological control (Mackauer 1976). The only way to counteract the negative effects of drift and inbreeding is to keep the population sizes large enough (in this case we need information on the effective population size) and/or to introduce genes from outside, from relatively unrelated populations, to reduce the frequency of unfit homozygous recessives.

The adaptation to artificial conditions can be studied either by examining changes in the life-history traits (i.e. alterations in viability, fertility, sexual maturity, ethology, body size, and other such characters), or by looking at the level of the gene and its primary product. Usually, one is faced with the problem of establishing an insect colony with low maintenance and production costs, whose individuals exhibit the 'typical' physiology and behavior of the species. When these requirements are not met, it is essential to know whether this reflects a purely phenotypic response, which can be restored with appropriate improvements of the rearing conditions, or a more permanent genetic change. The simplest and the most direct way to decide between these two alternatives is to employ a technique that looks at the genetic material itself or its immediate products. Enzyme gel electrophoresis is by far the easiest, fastest and least costly of these techniques. Thus, allozyme analysis can be used to assess and monitor the quality of such artificially produced populations.

The method was successfully applied to factory colonies of the screwworm fly, *Cochliomyia hominivorax*, in which significant changes at the α-glycerophosphate dehydrogenase locus owing to a high constant rearing temperature (Singh 1984) produced laboratory strains with reduced flight activities. Hence, irradiated sterile laboratory males were unable to compete with wild males and the control management failed (Bush and Neck 1976; Bush *et al.* 1976).

In the olive fruit fly *Dacus oleae*, Zouros *et al.* (1982) studied the allele frequency distribution of the alcohol dehydrogenase (ADH) locus both in natural populations and laboratory populations maintained on an artificial substrate. Each laboratory colony was established from large numbers of wild adults. Samples of about 200 flies from the founder populations were used to obtain estimates of the ADH allele frequencies at the start of the colonies. The gene frequencies in the colonies were monitored by taking samples every generation of artificial rearing.

Three alleles were detected in natural populations: a 'slow' (*S*) with frequency of about 0.65, a 'fast' (*F*) with frequency of about 0.35, and an 'intermediate' (*I*) with frequency less than 0.01. After four generations of artificial rearing, the frequency of *S* dropped to 0.35 or 0.40, that of *F* remained practically unchanged and that of *I* increased to about 0.30. This increase of the *I* allele occured in all cases in which the *I* allele was present in the founding population and proved to be independent of the geographical origin of the founders. After four generations of artificial rearing, the ADH frequencies stabilized and the colony remained at a polymorphic equilibrium that is quite different from the equilibrium characteristics of natural populations.

Following DuMouchel and Anderson (1968), Zouros *et al.* (1982) estimated relative fitness for the six ADH genotypes. The two heterozygotes for the *I* allele displayed the highest fitness, whereas the three homozygous genotypes displayed very low fitness values of about 5 per cent of the heterozygotes. This accounts for the abrupt increase of the *I* allele in the colonies and for the maintenance of all three alleles in a stable equilibrium. These are probably the largest selection differentials reported for allozymes and are comparable only to differentials for tolerance to heavy metals in plants (Jain and Bradshaw 1966).

In another study (Loukas *et al.* 1985) in which the founder sample of the olive fruit flies contained no *I* alleles (a fact occurring completely by chance), the absence of allele *I* has not rendered the

ADH polymorphism neutral to forces of natural selection. The *S* allele again suffered a dramatic decrease, but the new two-allele equilibrium point was different and could not be predicted from the dynamics of the three-allele system. The final outcome is dependent on the genetic make-up of the founder populations, even if environmental conditions are identical.

Economopoulos and Loukas (1986) attempted to identify the specific factors responsible for the selection pressure on ADH variants under artificial rearing. Three newly established colonies maintained on artificial medium were kept under three temperature regimes: constant 17°C; alternating daily between 17°C and 25°C; and constant 25°C. In all cases, the general pattern of ADH allozyme frequency changes was the same and similar to that described by Zouros *et al.* (1982). In another experiment they established four new colonies, two of which were maintained on artificial medium and two on fresh ripe olives, and scored the ADH allele frequencies for four consecutive generations. In colonies maintained on olive fruits no allele frequency change was observed, whereas in colonies maintained on artificial diet the changes were again similar to those described by Zouros *et al.* (1982). It follows that the cause of selection on ADH polymorphism lies with the artificial medium, and not the general 'indoor' laboratory conditions.

In the same report, Economopoulos and Loukas (1986) tried to answer the question 'whether the natural ADH allele frequencies could be restored when artificially-reared flies, kept in the laboratory for several generations, are provided with olives instead of artificial larval medium'. This 'reverse' experiment proved to be very difficult, because larvae from artificially reared adults carry no symbionts and therefore are apparently unable to survive on olive fruits that in the laboratory are heavily infested with bacteria and fungi. In spite of this, three generations on olive fruits were completed. In the first generation, the frequency of the *I* allele decreased from 0.28 to 0.08. This decrease was accompanied by an increase of the frequency of the *S* allele from 0.22 to 0.48. The frequency of the *F* allele remained practically unchanged. In the second and third generations, the frequencies of all three alleles remained more or less stable. The most likely explanation for the stability observed in the allele frequencies in the second and third generations was that the olives were picked at an advanced stage of maturity and stored at low temperature for several weeks. During this time they may undergo physiological changes that result in changes in the selection pressures on ADH in such a way that the *I* alleles are no longer selected against. For example, during the treatment and subsequent washes with water of olive fruits, a marked decrease was observed in certain constituents, such as sugars, mannitol, and alcohol-precipitable matter (Diez 1971).

These findings concerning the selective response of ADH allozymes to artificial rearing bear upon another question. Is this profound response in ADH unique to this enzyme, or is it shared by other polymorphic enzymes? To answer this question, Loukas *et al.* (1985) established a colony of olive fruit flies and scored the allele frequencies for 10 loci in the founding sample as well as in the first five generations of artificial rearing. They also scored for the same loci two old colonies that had been kept on artificial diet for 30 and for 240 generations, respectively. Systematic changes were observed for three loci, HK, 6-PGD, and of course ADH. For HK, the most common allele increased steadily from 0.69 in the founder sample to 0.82 after five generations of artificial rearing. The increase was steeper in the early generations and levelled off as the frequency of the favoured allele came closer to the upper limit of unity. For 6-PGD, the natural population of *D. oleae* displayed four alleles, 6-PGD^1 with frequency 0.19, 6-PGD^2 with frequency 0.64, 6-PGD^3 with frequency 0.15, and finally a null (silent) allele with frequency 0.02. After four generations of artificial rearing, allele 6-PGD^2 decreased to 0.22, whereas allele 6-PGD^1 increased to 0.78. Allele 6-PGD^3 declined from 0.15 to almost zero. The same gene frequency distribution for the 6-PGD locus (displayed by the new colony after only four generations of artificially rearing), was also observed in the two older colonies, a fact suggesting that the response of the 6-PGD locus is characteristic of all artificially-reared colonies. Since the shift from one equilibrium state to the other is completed within four generations (as was the case for ADH locus), it is clear that the selection pressure on 6-PGD variants is as strong as for the variants of ADH.

In conclusion, the response of enzyme polymorphisms surveyed in the study of Loukas *et al.* (1985) can be summarized as follows. Out of 10 loci studied 6 can be considered polymorphic with the 95 per cent criterion for monomorphism. Two of the 6 polymorphic loci, 6-PGD and ADH, appear to be influenced by powerful selective forces acting against the allele that is most common under natural conditions. Selection may also occur at a third locus, HK, but here it favours the allele that is most common under natural conditions. The evidence for selection at the other loci, especially the nearly monomorphic ones, is less clear. The variation at these loci may be truly neutral, the selection differentials of the various genotypes may be small, or strong selection pressures may be operating but gene frequencies may have reached a point beyond which selective changes are difficult to detect. When all loci are considered collectively, one detects a reduction in the mean actual number of alleles with colony age (from $\bar{n}_0 = 3.2$ in the founder sample to $\bar{n}_5 = 2.3$ after five generations of artificial rearing), owing to loss of alleles that are rare in the founder sample. A similar trend is seen for heterozygosity (from $\bar{H}_0 = 0.25$ to $\bar{H} = 0.18$), which can be explained partly by the loss of rare alleles (owing to random drift) and partly by the selective changes at 6-PGD, ADH, and HK loci.

According to the theory of population genetics, the intensity of selection on such a 'costly' polymorphism (Crow and Kimura, 1970) as the one observed at the ADH locus will declined with time, possibly through the selective accumulation of modifiers. Let us assume that the increase of the *I* allele is caused by selection acting not on *I* itself, but on a gene with which *I* is in linkage association. Because of the strong selection and the small population size, in the first few generations selection will overcome the effect of recombination and appear to be acting on the *I* allele itself. After a few generations, the population will reach equilibrium with all three ADH alleles in approximately equal frequencies. At this stage, recombination will prevail and, given enough time, a state of random association between ADH alleles and the undetected selected genes in the background genotype will be established. After this stage, the ADH polymorphism will be neutral to the forces of natural selection and its frequencies will be subject only to random drift, a process whose results will be apparent only after long periods of time.

To test this possibility, Zouros *et al.* (1986) established a new colony by mixing individuals from two laboratory colonies with different histories and different ADH allozyme frequency distributions in such a way that the expected gene frequency input was 0.75, 0.05, and 0.20 for alleles *F*, *I*, and *S*, respectively. Since one of the founder laboratory colonies was 30 generations old, the three allele frequencies were at approximately equilibrium and, therefore, the supposed non-random associations between ADH alleles and undetected variants in the background genotype were expected to have ceased to exist or to be greatly reduced. The other founder laboratory colony was six generations old but the ADH locus in this colony was segregated only for alleles *F* and *S*. Zouros *et al.* (1986) scored the allele frequencies of the new colony for 10 consecutive generations of artificial rearing as well as in the 15th and 20th generations, and found that the frequency of allele *I* rebounded and quickly returned to the levels present in the founder colony. The increase was as dramatic as in colonies established afresh from animals collected in the wild. These results suggest that no genetic changes have occurred in the founder colony that would have 'neutralized' the selection forces acting on the ADH locus. Thus, the possibility that selection acts on a hidden genetic polymorphism in linkage disequilibrium with the ADH alleles appears unlikely. Therefore, either selection acts directly on the ADH alleles, or the linkage with the selected genes is permanent (not subject to decoupling through crossing-over).

The same conclusions were also drawn by Vergini *et al.* (in prep.), who performed a similar experiment. These authors constructed laboratory strains, homozygous for the *F*, *I*, and *S* alleles of the ADH locus, coming from old laboratory colonies in which the three allele frequencies were at equilibrium. Then, by using individuals of these strains, they established two replicate colonies in which the initial frequencies of the alleles *F*, *I*, and *S* were 0.45, 0.10, and 0.45, respectively. Again, in both colonies, the allele frequencies changed towards equilibrium levels described by Zouros *et al.* (1982).

As far as life-history parameters are concerned, artificially reared flies

(1) reach sexual maturity faster and exhibit mating activity at the age of 2–3 days, whereas flies reared on olive fruits show such activity at 4–6 days (Economopoulos *et al.* 1971);
(2) produce 3–4 times more eggs during the first month after mating, and 1–2 times more thereafter (Economopoulos *et al.* 1976);
(3) search for mates 4 hours before scotophase, compared to 2 hours for wild insects (Zervas and Economopoulos 1982);
(4) produce higher amounts of sex-pheromone (Economopoulos *et al.* 1971);
(5) are smaller in size — during the first two generations of artificial rearing the average pupae weight is 4.1 mg, which then increases in succeeding generations and stabilizes at about 5.3 mg, compared to 6.6 mg for pupae reared on olive fruit (Manoukas 1983);
(6) mate less often than olive-reared flies (Economopoulos 1972);
(7) have a third to half the life-span of olive-reared flies (Economopoulos *et al.* 1976);
(8) have much lower flight ability than olive-reared flies, which in field experiments are recovered at longer distances from release points (Remund *et al.* 1977; Economopoulos *et al.* 1978);
(9) show reduced sensitivity to different colours (Remund *et al.* 1981). It is possible that when artificially-reared flies are released in the field they do not disperse satisfactorily, either because their ability for flight is low and/or because they are behaviourally

predisposed against long flights. This limited dispersal activity together with the restricted visual sensitivity reduces the fly's ability to search for food or to escape predators. Another probable factor accounting for the lower performance of the artificially reared flies in the field is their lack of the symbiotic bacteria in olive-reared examples.

Although there is no direct evidence, without doubt these differences between artificially and naturally reared flies are to a large extent genetically determined. The only relevant information is that the changes occur within four to five generations of artificial rearing. During these generations, the colony's size is dramatically reduced because of limited oviposition on paraffin cones (the artificial oviposition substrate) and high larval mortality. During this period, the artificially reared insects also lose their symbiotic bacteria.

The cumulative work on the ADH locus in *Drosophila* has made it possible to use this locus as a model to trace fitness responces to a single locus level. This claim can be made only for a few other electrophoretically detected enzyme loci, such as amylase (De Jong *et al.* 1972) and α-glycerophosphate dehydrogenase (Kitto and Briggs 1962; O'Brien and MacIntyre 1972). King *et al.* (1976) and Richmond and Gerking (1979) have studied oviposition preference between sites containing different amount of alcohol in various species of *Drosophila*. Some correlation appears to exist between ADH activity in the fly and the level of alcohol in the preferred (or avoided) oviposition site, but neither the nature nor the extent of this correlation is understood at present. Also, several studies (McKenzie and Parson 1972; Starmer *et al.* 1977) have shown that environmental alcohols may have different effects on the dispersal and longevity of related species and may, ultimately, contribute to the spatial distribution of competing species. Gelfand and McDonald (1980) studied alcohol preference behaviour in larvae of *Drosophila melanogaster* and observed that alcohol concentrations above 10 per cent caused a differential response among ADH genotypes in such a way that genotypes with lower alcohol dehydrogenase levels exhibited a higher degree of avoidance of the alcohol-rich medium. They also showed that the three homozygous genotypes (for fast, slow, and null alleles) had different alcohol dehydrogenase levels (fast > slow > null).

With regard to 6-PGD, the observation by Cavener (1984) that the 6-PGD polymorphism of *Drosophila melanogaster* responds to presence of ethanol in the food suggests that ADH and 6-PGD polymorphisms may indeed be affected by the same factor.

These data on the physiological role of the enzymes in question, together with the fact that both the dramatic genetic changes and the major changes in life-history traits appear within the first four generations of artificial rearing, suggest a strongly correlated response.

In the medfly, Kourti and Loukas (in prep.) scored the changes in allele frequencies of the polymorphic loci PEP-1, PEP-3, and EST in three populations during the first five generations of artificial rearing. Out of the nine χ^2 homogeneity tests performed to test the changes in allele frequencies of each locus between generations zero (founder sample) and five, four were statistically significant. Since the changes at each locus were not of the same magnitude and in the same direction in all cases, the previous authors regarded them as resulting through the process of random drift. Of course, a neutrality hypothesis cannot easily be tested without knowledge of the effective sizes of the colonies. The fact that the very rare alleles of the EST locus continued to be present even after five generations of artificial rearing seems to support some form of balancing selection that prevents the extinction of rare alleles. However, the changes observed, if they are not selectively neutral, do not seem to affect severely (or to be connected to) the field performance of the fly. The eradication of medfly from southern Mexico, by the release of mass-reared sterilized flies (Hendrichs *et al.* 1983) supports such an assertion.

Electrophoretic data and the control of the olive fruit fly

The objective of current experimental work on the insect is to develop a methodology for the ultimate replacement of the chemical control of the fly by an ecologically less damaging integrated system that will rely heavily on non-insecticidal methods (Economopoulos 1979). For this we need an efficient method for culturing the insect in the laboratory and genetic markers that will facilitate comparisons between laboratory reared and wild-reared animals for such characters as dispersal, longevity, sexual drive, sensitivity to heat shock, etc.

The striking alterations in the genetic make-up of laboratory-produced flies in a few generations following colonization coincide with significant changes in the physiology and behaviour of the fly. All these changes are the results of powerful selective processes that operate on the very first generations of colonization. Observations suggest that the key problems wild-bred flies encounter when reared in the laboratory are ovipositing in a ceresin or paraffin shell (or surface) and surviving in the artificial larval diet. The later problem is probably related to the monophagous habit of the larva which, under natural conditions, feeds only on olive fruits. Economopoulos and Loukas (1986) have shown that the profound changes in the allozyme frequencies of ADH in laboratory colonies are exclusively due to the artificial larval diet. The morphological, physiological, or behavioural differences betwen wild and artificially reared flies have serious consequences when laboratory insects are produced for control applications, e.g. in the sterile insect technique. There is now good reason to suspect that these differences have a genetic basis.

Because of this correlation between changes at certain loci and the colony's performance, these allozyme changes might be used as indicators of the colony's quality, even if the causal relationship between the two sets of characters remains unknown. If the investigator wishes to improve the quality of the colony by selection from within the colony, then allozymes could be useful in choosing the individuals from which new lines will be established and for monitoring the response of these lines. Also, by knowing which enzymes are most affected by artificial rearing, it may be possible to identify the factors that must be modified in order to improve the rearing. However, the final decision on the stability of laboratory or factory-produced insects used in behavior or control experiments should still be primarily based on studies that directly measure any deviation of artificially reared insects from the typical behaviour and physiology of the species.

The reinforcement of the hypothesis that selection acts on the ADH locus (Zouros *et al.* 1986) provides valuable clues for search for the factor or factors that are responsible for this response. Identification of these factors would suggest modifications in the rearing protocol that may improve the yield and quality of laboratory colonies. These factors are most probably chemical compounds in the larval food that act as substrates, inhibitors, or activators of alcohol dehydrogenase. These chemicals may be ingredients of the medium or may be produced secondarily during larval growth. Specifically, it is known that the main role of ADH is alcohol detoxification (Van Delden 1982). Experiments with strains homozygous for different ADH alleles should show which allozyme is more effective in this role, and many lead to research into the alcohol content of the larval diet as a way of improving the rearing protocol. Similar insights can be gained by considering the physiological role of the 6-PGD locus (Cavener and Clegg 1981).

On the other hand, the fact that the genetic changes occurred within four generations of artificial rearing and could not easily be reversed (Economopoulos and Loukas 1986) may have serious implications on the applicability of insect quality restoration procedures.

The availability of a *Dacus oleae* stock containing a genetic variant (allele *I*) not found in appreciable frequencies in natural populations opens the way for a comprehensive study of the competitiveness of artificially reared insects with wild-caught flies in the laboratory as well as in the field. Such a colony may have a number of interesting properties. Because the *I* allele is favoured under laboratory conditions, one may expect that colonies containing only this allele will have higher yields under the method of artificial rearing currently in use. Also, because the *I* allele is apparently selected against in the natural habitat, a colony homozygous for this allele may be most appropriate to provide insects for sterilization and release. Low survival of larvae carrying the *I* allele on olive fruit may mean that, in the event a small fraction of the released insects are sterile, the offspring of these flies will die before eclosion. This in turn may allow for lower radiation levels for sterilization and thus increase the longevity and mating competitiveness of released flies. Clearly, a long series of laboratory and field experiments is needed before the profound response of ADH allozymes to the shift from natural to artificial larval substrate can be of use in the insects' control.

Furthermore, the ADH locus holds the promise for the development of a 'sexing' system — a technique for obtaining unisexual populations. Not only does the production and release of females add to the cost of the operation, sterile females retain the habit to open oviposition holes on the olive fruit, thus rendering it liable to bacterial and fungal infection. Such a technique has already been developed in *Drosophila melanogaster* by Robinson and van Heemert (1981) and consists of a Y-chromosome translocation carrying an 'active' ADH allele, whereas autosomes carry a 'null' allele. In a food supplemented with 4 per cent alcohol, only larvae carrying the active allele, i.e. males, survive.

Thus, another possibility suggested by the availability of stocks homozygous for the ADH alleles is to devise a sexing technique similar to that developed in *Drosophila melanogaster* by Robinson and van Heemert (1981). This will involve the creation of a stock in which females will be homozygous for a ADH allele other than *I* and males will be identical with the females for the autosomal ADH locus but carry, in addition, a translocation containing the *I* allele on their Y chromosome. It will then be easy to obtain an artificial medium in which only allele-*I*-carrying larvae will survive.

Acknowledgements

Part of this work was supported by the FAO/IAEA Joint Division, Vienna, and by the Greek Ministry of Agriculture.

References

Berlocher, S.H. (1976). *The genetics of speciation in Rhagoletis (Diptera: Tephritidae)*. Ph. D. thesis. University of Texas. Austin.

Berlocher, S.H. and Bush, G.L. (1982). An electrophoretic analysis of *Rhagoletis* (Diptera: Tephritidae) phylogeny. *Sys. Zool.*, **31**, 136–55.

Beverley, S.M. and Wilson, C.A. (1984). Molecular evolution in *Drosophila* and the higher Diptera II. A time scale for fly evolution. *J. Molec. Evol.*, **21**, 1–13.

Brownlee, K.A. (1967). In *Statistical theory and methodology (in science and engineering)*, 2nd edn. Wiley, New York and London.

Bush, G.L. (1969). Sympatric host race formation and speciation in frugivorous flies of the genus *Rhagoletis* (Diptera, Tephritidae). *Evolution*, **23**, 237–51.

Bush, G.L. (1975). Models of speciation. *A. Rev. Ecol. Syst.*, **6**, 339–64.

Bush, G.L. and Hoy, MA. (1983). Evolutionary processes in insects. In *Ecological entomology* (ed. C.B. Huffaker and R.L. Rabb), pp. 247–78. Wiley, New York.

Bush, G.L. and Kitto, G.B. (1979). Research on the genetic structure of wild and laboratory strains of the olive fly. F.A.O. Report: *Development of pest management systems for olive culture program.* Food and Agriculture Organization of the United Nations, Rome.

Bush, G.L. and Neck, R.W. (1976). Ecological genetics of the screwworm fly *Cochliomyia hominivorax* (Diptera: Calliphoridae) and its bearing on the quality control of mass-reared insects. *Environ. Ent.*, **5**, 821–26.

Bush, G.L., Neck, R.W., and Kitto, G.B. (1976). Screwworm eradication: inadvertent selection for noncompetitive ecotypes during mass rearing. *Science*, **193**, 491–3.

Cavener, D.R. (1984). Response of the G6pd and 6Pgd polymorphisms of *Drosophila melanogaster* to diety selection. *Genetica*, **63**, 81–3.

Cavener, D.R. and Clegg, M.T. (1981). Evidence for biochemical and physiological differences between enzyme genotypes in *Drosophila melanogaster*. *Proc. natn. Acad. Sci. USA*, **78**, 4444–7.

Chakraborty, R. and Nei, M. (1976). Hidden genetic variability within electromorphs in finite populations. *Genetics*, **84**, 385–93.

Chakraborty, R., Fuerst, P.A., and Nei, M. (1980). Statistical studies on protein polymorphism in natural populations. III. Distribution of allele frequencies and the number of alleles perlocus. *Genetics*, **94**, 1039–63.

Coyne, J. A. (1976). Lack of genetic similarity between two sibling species as revealed by varied techniques. *Genetics*, **84**, 593–607.

Coyne, J. A. and Felton, A. A. (1977). Genic heterogeneity at two alcohol dehydrogenase loci in *Drosophila pseudoobscura* and *Drosophila persimilis*. *Genetics*, **87**, 285–304.

Coyne, J. A., Felton, A. A., and Lewontin, R. C. (1978). Extent of genetic variation at a highly polymorphic esterase locus in *Drosophila pseudoobscura*. *Proc. natn. Acad. Sci. USA*, **75**, 5090–3.

Crow, J. F. and Kimura M. (1970). *An introduction to population genetics theory*. Harper and Row, New York.

De Jong, G., Hoorn, A. J. W., Thörig, G. E. W., and Scharloo, W. (1972). Frequencies of amylase variants in *Drosophila melanogaster*. *Nature, Lond.*, **238**, 452–3.

Diehl, S. R. and Bush, G. L. (1984). An evolutionary and applied perspective of insect biotypes. *A. Rev. Ent.*, **29**, 471–504.

Diez, M. J. F. (1971). The olive. In *The biochemistry of fruits and their products*, Vol. 2 (ed. A. C. Hulme), pp. 255–79. Academic Press, New York.

DuMouchel, W. H. and Anderson, W. W. (1968). The analysis of selection in experimental populations. *Genetics*, **58**, 435–49.

Economopoulos, A. P. (1972). Sexual competitiveness of γ-ray sterilized males of *Dacus oleae*. Mating frequency of artificially reared and wild females. *Environ. Ent.*, **1**, 490–7.

Economopoulos, A. P. (1979). Prospects for the control of *Dacus oleae* (Gmelin) (Diptera, Tephritidae) by methods that do not involve insecticides. The sterile insect release technique and oflactory and visual traps, integrated approach. *International Organization for Biological Control West Palearctic Regional Section (IOBC/WPRS)*, Bulletin (1979) 2/1, pp. 42–9.

Economopoulos, A. P. and Loukas, M. (1986). ADH allele frequency changes in olive fruit flies shift from olives to artificial larval food and vice versa, effect of temperature. *Entomologia exp. appl.* **40**, 215–21.

Economopoulos, A. R., Giannakakis, A., Tzanakakis, M. E., and Voyatzogloy, A. V. (1971). Reproductive behaviour and physiology of the olive fruit fly. I. Anatomy of the adult vectum and odor emitted by adults. *Ann. ent. Soc. Am.*, **64**, 1114–6.

Economopoulos, A. P., Voyatzogloy, A. V., and Giannakakis, A. (1976). Reproductive behaviour and physiology of *Dacus oleae*. Fecundity as affected by mating, adult diet and artificial rearing. *Ann. ent. Soc. Am.* **69**, 725–9.

Economopoulos, A. P., Haniotakis, G. E., Mathioudakis, J., Missis, N., and Kinigakis, P. (1978). Long-distance flight of wild and artificially reared *Dacus oleae* (Gmelin) (Diptera, Tephritidae). *Z. angew. Ent.*, **87**, 101–108.

Florence, L. Z., Johnson, P. C., and Coster, J. E. (1982). Behavioral and genetic diversity during dispersal: analysis of a polymorphic esterase locus in southern pine beatle *Dendroctonus frontalis*. *Environ. Ent.*, **11**, 1014–8.

Gasperi, G., Malacrida, A. R., and Milani, R. (1986). Protein variability and population genetics of *Ceratitis capitata*. In *Fruit flies: Proceedings of the Second International Symposium* (ed. A. P. Economopoulos), pp. 148–57.

Gelfand, L. J. and McDonald, J. F. (1980). Relationship between ADH activity and behavioral response to environmental alcohol in *Drosophila*. *Behav. Genet.*, **10**, 237–49.

Gillespie, J. H. and Langley, C. H. (1974). A general model to account for enzyme variation in natural populations. *Genetics*, **76**, 837–48.

Hagen, K. S., William, W. W., and Tassan, R. L. (1981). Mediterranean fruit fly: The worst may be yet to come. *California Agriculture* (University of California, Division of Agricultural Sciences, Reports of progress in research), March–April 1981, **35**, 5–7.

Harris, H. (1966). Enzyme polymorphism in man. *Proc. R. Soc. B*, **164**, 298–310.

Harrison, R. G., Wintermeyer, S. F., and Odell, T. M. (1983). Patterns of genetic variation within and among gypsy moth *Lymantria dispar* (Lepidoptera: Lymantriidae) populations. *Ann. Ent. Soc. Am.*, **76**, 652–6.

Hendricks, J., Ortiz, G., Liedo, P., and Schwarz, A. (1983). Six years of successful medfly program in Mexico and Guatemala. In *Fruit flies of economic importance: Proceedings of the CEC/IOBC International Symposium* (ed. R. Cavalloro), pp. 353–65.

Hubby, J. l. and Lewontin, R. C. (1966). A molecular approach to the study of genic heterozygosity in natural populations. I. The number of alleles at different loci in *Drosophila pseudoobscura*. *Genetics*, **54**, 577–94.

Huettel, M. D., Fuerst, P. A., Maruyama, T., and Chakraborty, R. (1980). Genetic effects of multiple population bottlenecks in the Mediterranean fruit fly (*Geratitis capitata*). *Genetics*, **94**, s47.

Jaenike, J. and Grimaldi, D. (1983). Genetic variability for host proference within and among populations of *Drosophila tripunctata*. *Evolution*, **37**, 1023–3.

Jain, S. K. and Bradshaw, A. D. (1966). Evolutionary devergence among adjacent plant populations. I. The evidence and its theoretical analysis. *Heredity*, **21**, 407–41.

King, S. B., Rockwell, R. F., and Grossfield, J. (1976). Oviposition response to ethanol in *Drosophila melanogaster* and *Drosophila simulans*. *Genetics*, **83**, 5–39.

Kitto, G. B. and Briggs, M. H. (1962). Relationship between locomotory habits and enzyme concentration in insects. *Science*, **135**, 918.

Krimbas, C. B. and Tsakas, S. (1971). The genetics of *Dacus oleae*. Changes of esterase polymorphism in a natural population following insecticide control. Selection or drift? *Evolution*, **25**, 454–60.

Levins, R. and Wilson, M. (1980). Ecological theory and pest management. *A. Rev. Ent.*, **25**, 287–308.

Levinton, J. (1973). Genetic variations in a gradient of environmental variability: marine Bivalvia (Mollusca). *Science*, **180**, 75–6.

Loukas, M., Vergini, Y., and Krimbas, C. B. (1981). The genetics of

Drosophila subobscura populations. XVII. Further genic heterogeneity within electromorphs by urea denaturation and the effect of the increased genic variability on linkage disequilibrium studies. *Genetics*, **97**, 429–41.

Loukas, M., Economopoulos, A.P., Zouros, E., and Vergini, Y. (1985). Genetic changes in artificially reared colonies of the olive fruit fly (Diptera: Tephritidae). *Ann. ent. Soc. Am.*, **78**, 159–65.

Mackauer, M. (1976). Genetic problems in the production of biological control agents. *A. Rev. Ent.*, **21**, 369–85.

Malacrida, A., Gasperi, G., and Milani, R. (1983). Recent developments in the study of enzyme polymorphisms in *Ceratitis capitata*. *Atti XIII Congresso nationale italiano di Enomologia, Sestriere*, 1983, pp. 511–7.

Malacrida, A.R., Gasperi, G., Biscaldi, G.F., and Milani, R. (1987). Updating of the genetics of *Ceratitis capitata*. *FAO/IAEA (Symposium on Modern Insect Control. Nuclease Techniques and Biotechnology, Vienna 16–20 Nov., 1987)* (in press).

Manoukas, A.G. (1983). The adaptation process and the biological efficiency of *Dacus oleae* larvae. In *Fruit flies of economic importance: Proceedings of the CEC/IOBC International Symposium* (ed. R. Cavalloro), pp. 91–5.

Mayr, E. (1954). Change of genetic environment and evolution. In: *Evolution as a process* (ed. J. Huxley, A.C. Hardy, and E.B. Ford), pp. 167–80. George, Allen and Unwin Ltd., London.

McKenzie, J.A. and Parson, P.A. (1972). Alcohol tolerance: An ecological parameter in the relative success of *Drosophila melanogaster* and *Drosophila simulans*. *Oecologia*, **10**, 373–88.

Menken, J.B.S. and Ulenberg, A.S. (1987). Biochemical characters in agricultural entomology. *Agric. Zool. Rev.*, **2**, 305–60.

Morgante, J.S., de Souza, H.M.L., de Conti, E., and Cytrynowicz, M. (1981). Allozymic variability in an introduced fruit fly pest *Ceratitis capitata* (Wiedeman), 1824, (Diptera-Tephritidae), *Rev. Brasilian Genet.*, **2**, 183–91.

Nei, M. (1972). Genetic distance between populations *Am. Nat.*, **106**, 283–92.

Nei, M., Maruyama, T., and Chakroborty, R. (1975). The bottleneck effect and genetic variability in populations. *Evolution*, **29**, 1–10.

Nevo, E. (1978). Genetic variation in natural populations: Patterns and theory. *Theor. Pop. Biol.*, **13**, 127–77.

Nevo, E., Beiles, A., and Ben-Shlomo, R. (1984). The evolutionary significance of genetic diversity: ecological, demographic and life history correlates. *Lecture Notes in Biomathematics*, **53**, 13–213.

O'Brien, S.J. and McIntyre, R.J. (1972). The α-glycerophosphate cycle in *Drosophila melanogaster*. I. Biochemical and developmental aspects. *Biochem. Genet.* **7**, 141–61.

Pashley, D.P. (1986). The use of population genetics in migration studies: a comparison of three noctuid species. In *The movement and dispersal of agriculturally important biotic agents* (ed. D.R. MacKenzie *et al.*), pp. 305–24. Claitor's Publishing Division. Baton Rouge, Fla.

Pashley, D.P. and Bush, G.L. (1979). The use of allozymes in studying insect movement with special reference to the codling moth, *Laspeyresia pomonella* (L.) (Olethreutidae). In *Management of highly mobile insects: concepts and methodology in research. Proceedings of a Conference: Movement of Selected Species of Lepidoptera in the South-eastern Unites States, Raleigh, North Carolina, 9–11 April 1979* (ed. R.L. Rabb and G.G. Kennedy), pp. 333–41. University Graphics, Raleigh, N.C.

Powell, J.P. (1971). Genetic polymorphisms in varied environments. *Science*, **174**, 1035–1036.

Prokopy, R.J., Averill, A.L., Cooley, S.S., Roitberg, C.A., and Kallet, C. (1982). Variation in host acceptance pattern in apple maggot flies. *Proceedings of the 5th International Symposium on Insect–Plant Relations. Wageningen (PUDOC)*, pp. 123–9.

Prokopy, R.J., McDonald, P.T., and Wong, T.T.Y. (1984). Interpopulation variation among *Ceratitis capitata* flies in host acceptance pattern. *Entomologia exp. appl.*, **35**, 65–9.

Remund, U., Boller, EF., Economopoulos, A.P., and Tsitsipis, J.A. (1977). Flight performance of *Dacus oleae* reared on olives and artificial diet. *Z. angew. Ent.*, **82**, 330–9.

Remund, U., Economopoulos, A.P., Boller, E.F., Agee, H.R., and Davis, J.C. (1981). Fruit fly quality monitoring: The spectral sensitivity of field collected and laboratory reared olive flies, *Dacus oleae* (Gmel.) (Diptera, Tephritidae), *Bul. Soc. Ent. Suisse*, **54**, 221–7.

Richmond, R.C. and Gerking, J.L. (1979). Oviposition site preference in *Drosophila. Behav. Genet.*, **9**, 233–41.

Riva, M.E. and Robinson, A.S. (1986). Induction of alcohol dehydrogenase null mutants in the Mediterranean fruit fly *Ceratitis capitata. Biochem. Genet.*, **24**, 765–74.

Robinson, A.S. and Van Heemert, C. (1981). Genetic sexing in *Drosophila melanogaster* using the alcohol dehydrogenase locus and a Y-linked translocation. *Theor. appl. Genet.*, **59**, 23–5.

Robinson, A.S., Riva, M.E., and Zapater, M. (1986). Genetic sexing in the Mediterranean fruit fly *Ceratitis capitata. Theor. appl. Genet.*, **72**, 455–7.

Rosen, D. (1978). The importance of cryptic species and species identification as realted to biological control. In *Beltsville Symposia in Agricultural Research 2. Biosystematics in Agriculture* (ed. J.A. Romberger), pp. 23–35. Wiley, New York.

Sarich, Y.M. and Cronin, E.J. (1980). South American mammal molecular systematics, evolutionary clocks, and continental drift. In *Evolutionary biology of the New World monkeys and continental drift* (ed. L.R. Ciochon and B.A. Charelli), pp. 399–421. Plenum Press, New York.

Saul, S.H. (1986). Genetics of the Mediterranean fruit fly (*Ceratitis capitata*) (Wiedemann). In *Agricultural zoology reviews* (ed. G.E. Russell), Vol. 1, pp. 73–108. Intercept, Ponteland, Newcastle upon Tyne.

Singh, R.S. (1979). Genetic heterogeneity within electrophoretic 'alleles' and the pattern of variation among loci in *Drosophila subobscura*. *Genetics*, **93**, 997–1018.

Singh, P. (1984). Insect diets: Historical developments, recent advances and future prospects. In *Advances and Challenges in insect rearing* (ed. E.C. King and N.C. Leppla), pp. 32–44. Agricultural Research Series. USDA, Washington, DC.

Singh, R.S., Hubby, J.L., and Throckmorton, L.H. (1975). The study of genic variation by electrophoresis and heat denaturation techniques at the octanol dehydrogenase locus in members of the *Drosophila virilis* group. *Genetics*, **80**, 637–50.

Singh, R.S., Lewontin, R.C., and Felton, A.A. (1976). Genic heterogeneity within electrophoretic 'alleles' of xanthine dehydrogenase in *Drosophila pseudoobscura*. *Genetics*, **84**, 609–29.

Sneath, P.H.A. and Sokal, R.R. (1973). In *Numerical taxonomy*. W.H. Freeman, San Francisco.

Starmer, W.T., Heed, W.B., and Rockwood-Sluss, E.S. (1977). Extension of longevity in *Drosophila melanogaster* by environmental ethanol: Differences between substrates. *Proc. natn. Acad. Sci. USA*, **74**, 387–91.

Tarling, D.H. (1980). The geological evolution of South America with special reference to the last 200 million years. In *Evolutionary biology of the New World monkeys and continental drift* (ed. L.R. Ciochon and B.A. Charelli), pp. 1–41. Plenum Press, New York.

Tsakas, S. and Krimbas, C.B. (1970). The genetics of *Dacus oleae*. IV. Relation between adult esterase genotypes and survival to organophosphate insecticides. *Evolution*, **24**, 807–15.

Tsakas, S. and Krimbas, C.B. (1975). How many genes are selected in populations of *Dacus oleae*? *Genetics*, **79**, 675–9.

Tsakas, S. and Zouros, E. (1980). Genetic differences among natural and laboratory-reared populations of the olive fruit fly *Dacus oleae* (Diptera: Tephritidae). *Ent. exp. appl.*, **28**, 268–76.

Valentine, J.W. (1976). Genetic strategies of adaptation. In *Molecular evolution* (ed. F.J. Ayala), pp. 78–94. Sinauer, Sunderland, Mass.

Van Delden, W. (1982). The alcohol dehydrogenase in *Drosophila melanogaster*: Selection at an enzyme locus. *Evol. Biol.*, **15**, 187–222.

Zervas, G.A. and Economopoulos, A.P. (1982). Mating frequency in caged populations of wild and artificially reared (normal or sterilized) olive fruit flies. *Environ. Ent.*, **11**, 17–20.

Zouros, E. and Krimbas, C.B. (1969). The genetics of *Dacus oleae*. III. Amounts of variation at two esterase loci in a Greek population. *Genet. Res.*, **14**, 249–58.

Zouros, E. and Krimbas, C.B. (1970). Frequency of female digamy in a natural population of the olive fruit fly *Dacus oleae* as found by using enzyme polymorphism. *Ent. exp. appl.*, **13**, 1–9.

Zouros, E. and Loukas, M. (1988). Biochemical and colonization genetics of *Dacus oleae*. In *Fruit flies, their biology, natural enemies and control* (ed. G.H.S. Hooper). Elsevier, Amsterdam. (In press.)

Zouros, E., Tsakas, S., and Krimbas, C. B. (1968). The genetics of *Dacus oleae*. II. The genetics of two adult esterases. *Genet. Res.*, **12**, 1–9.

Zouros, E., Loukas, M., Economopoulos, A., and Mazomenos, B. (1982). Selection at the alcohol dehydrogenase locus of the olive fruit fly *Dacus oleae* under artificial rearing *Heredity*, **48**, 169–85.

Zouros, E., Loukas, M., Economopoulos, A. P., and Vergini, Y. (1986). The alcohol dehydrogenase locus (ADH) of *Dacus oleae*: Further evidence for selection under artificial rearing. In *Fruit flies: Proceedings of the Second International Symposium* (ed. A. P. Economopoulos), pp. 341–7.

6 Host-associated differentiation in armyworms (Lepidoptera: Noctuidae): An allozymic and mitochondrial DNA perspective

DOROTHY P. PASHLEY

Entomology Department, Louisiana State University, Baton Rouge, Louisiana 70803, USA

Abstract

The fall armyworm, *Spodoptera frugiperda* (J. E. Smith), is a major pest of corn, rice, and forage grasses throughout the Western Hemisphere. Long considered to be a single, polyphagous species, it has recently been determined to include two host-associated strains that differ at five allozyme loci. Genetic analysis of individuals reared on both sets of hosts indicated that natural selection in the host environment was not the cause of genetic differentiation. Instead, strains were either partially or completely reproductively isolated. Subsequent studies that aimed at determining the taxonomic status of the strains and their biological differences have involved the use of both protein electrophoretic and mitochondrial DNA (mtDNA) markers.

Intra- and interstrain variation in mtDNA from individual fall armyworms was low. Of 35 restriction enzymes examined, only five differed within strains and four between strains. Within the corn strain, identical mtDNA genotypes were found in individuals from Central America, the Caribbean, and the south-eastern USA with only a few variants. Although mtDNA variation appears to be of limited use for population studies within strains, it will be useful for studying interstrain hybridization when used in conjunction with allozymes.

Electrophoretic Studies on Agricultural Pests (ed. Hugh D. Loxdale and J. den Hollander), Systematics Association Special Volume No. 39, pp.103–14. Clarendon Press, Oxford, 1989.

Introduction

One of the most commonly sought after parameters of population genetic studies is an estimate of gene flow among populations. The point at which gene flow ceases is a key question in systematic biology. Determination of levels and cessation of gene flow have remained difficult tasks even with the widespread use of protein electrophoresis. Generally, if species do not interbreed and have not been exchanging genes for some time, genetic differentiation will be detected. Below this interspecific level, there are two difficulties associated with studies of gene flow.

First, it is nearly impossible to separate levels of gene flow; large amounts, very little, and recent cessations in gene flow cannot easily be differentiated from one another. If no differences between taxa are detected, high levels of gene flow cannot be assumed, because markers may not be refined enough to distinguish current gene flow from gene flow that ceased perhaps hundreds or thousands of years ago.

A second difficulty exists when differentiation is detected among populations. There are numerous potential causes of differentiation, the most obvious being environmental induction and natural selection, and it is not possible to conclude that differences are due to reduced gene flow until other possibilities are ruled out (see Pashley 1988*b* for a review). Molecular studies have an advantage over quantitative traits such as morphology and behaviour in that a genetic basis can either be assumed or easily be verified and ecotypic variation can be ruled out. Unfortunately, natural selection cannot be discounted so readily.

Because of refinement problems mentioned above and the possibility that natural selection acts on allozyme variation, evolutionary biologists eagerly embraced mitochondrial DNA (mtDNA) technology about a decade ago when it became streamlined enough to be used in population studies. Its rapid rate of evolution (Brown *et al.* 1979) suggested that mtDNA might finally allow accurate estimates of gene flow (see Avise *et al.* 1987 and Moritz *et al.* 1987 for reviews). Whether it has lived up to this expectation is debatable. However, when used in combination with other characters or markers, it has clearly resolved certain aspects of the biology of insects (Simon 1988).

The work presented in this chapter is centred on two taxa of the fall armyworm (*Spodoptera frugiperda* (J.E. Smith)) that exhibit genetic differentiation associated with host usage in sympatry. Their taxonomic status is currently undefined. They are so similar morphologically that they were not detected until recently (Pashley 1986), even though they are strains of a major agricultural pest that has been the subject of hundreds of studies. My primary objective has been to quantify the level of genetic distinctiveness between these strains and, in so doing, to estimate levels of gene flow and determine the causes of differentiation.

Natural history and background

Although the fall armyworm feeds on numerous plants (Luginbill 1928; Pashley 1988*c*), its primary hosts are grasses, including many crops such as corn, rice, and bermudagrass. It is distributed throughout the Western Hemisphere but, because it does not have a diapausing stage, adults must reinvade temperate regions of North and South America each spring from overwintering grounds in the neotropics (Sparks 1979).

Major genetic differences were detected between Puerto Rican and mainland moths (Pashley *et al.* 1985) in an allozyme study of populations from winter and summer ranges. Because the Puerto Rican population had been collected on rice and all others were from corn, a host effect was possible. Genetic studies of populations from different hosts in Louisiana and Puerto Rico and cross-rearing of larvae on different hosts revealed the presence of two genetic host-associated strains (referred to subsequently as the rice and corn strains) whose differences were probably due to reduced gene flow among hosts (Pashley 1986).

The degree of reproductive isolation between the strains has been examined in two ways. First, laboratory interstrain matings indicated unidirectional incompatibility in the parental crosses (rice females mated with corn males but corn females did not mate with rice males) (Pashley and Martin 1987). F1 males mated with parental females, whereas F1 females did not mate. Second, studies in nature of the attraction of males to females (which emit a sex-attractant pheromone) indicated a strong preference among males for females of their own strain (Pashley 1988*b*). In both strains, about 75 per cent of the males preferred females of their own strain, whereas 25 per cent chose females of the other strain.

It was desirable to determine whether interstrain matings actually occurred in these frequently occurring 'mistakes'. Two direct approaches are currently underway to determine whether interstrain matings occur in nature: (1) strain identification of wild individuals captured in copula in corn and bermudagrass fields, and (2) identification of wild males that copulate with laboratory females tethered to a structure in the field. A third, indirect, method that can more easily be applied across a broad geographic area involves the combined analysis of mtDNA and allozymes. The principles behind this approach, procedures, and preliminary results are the focus of the remainder of this chapter.

Use of molecular markers to study hybridization

1. Mitochondrial DNA principles

Comparisons of nuclear and mtDNA genotypes have facilitated the identification of hybrids and their parental types in several organisms (Brown and Wright 1979; Powell 1983; Avise *et al.* 1984; Avise and Saunders 1984; Spolsky and Uzzell 1984). The utility of this approach depends upon differing modes of inheritance for the two molecules. Male and female parental nuclear genes are equally represented in progeny; however, the fact that mitochondria are maternally inherited results in progeny mtDNA genotypes identical to the mother. Hybrids in nature possess intermediate nuclear genotypes and mtDNA genotypes of the female parent. Several possible scenarios for interstrain hybridization in fall armyworm are presented in Fig. 6.1.

To use this method, it was necessary first to determine whether the mtDNA of the two strains differed. If so, it was necessary then to survey individuals at both mtDNA and allozyme genotypes to resolve a potential problem of overlap in allozyme genotypes, the latter of which are not diagnostic for each strain (Fig. 6.2). Certain multilocus allozyme genotypes are found in individuals collected from both hosts (e.g. rightmost column in Fig. 6.2). There are two possible explanations for this result. Strains could overlap genetically and a small number of the rice-strain individuals collected feeding on rice could have genotypes that are more common in corn strain (i.e. host fidelity could be complete). Alternatively, strains may overlap in host usage and individuals collected on rice that have corn-strain genotypes could actually be members of the corn strain feeding on the 'wrong' host. In the latter case, allozymes are nearly diagnostic between strains, whereas in the former they are not and therefore cannot be used to identify hybrids. One way to resolve this problem is to examine individuals at other markers (e.g. mtDNA).

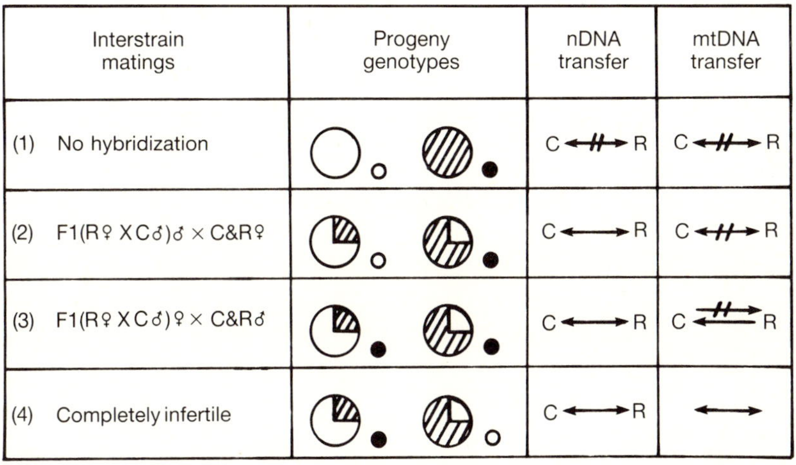

Interstrain matings	Progeny genotypes	nDNA transfer	mtDNA transfer
(1) No hybridization		C ←//→ R	C ←//→ R
(2) F1(R♀ X C♂)♂ × C&R♀		C ←→ R	C ←//→ R
(3) F1(R♀ X C♂)♀ × C&R♂		C ←→ R	C —//→ ←— R
(4) Completely infertile		C ←→ R	←→

Fig. 6.1. Genetic expectations for nuclear genes (nDNA, large circles) and mtDNA (small circles) in certain progeny of interstrain fall armyworm crosses. Parental strain genotypes are diagnostic as indicated by different patterns in the nucleus and mitochondrion. Only four types of backcrosses are presented. The genetic combination in number 2 is expected if the laboratory data are correct (Pashley and Martin 1987); number 3 is expected if lab data are only partially correct in that F1 females also mate.

2. *Methods*

Total DNA was isolated from individual moths (weighing about 100 mg each) according to the following procedure. Individuals were homogenized as in Coen *et al.* (1982) and the homogenate was subjected to extraction with phenol, followed by chloroform, and then ethanol-precipitated prior to resuspension in buffer. Approximately 0.5 μg total DNA was obtained per individual and was sufficient for digestion with at least 20 restriction enzymes. Digested samples were electrophoresed on agarose gels and transferred to nitrocellulose paper (Southern 1975). A ^{32}P nick-translated probe of *Drosophila silvestris* mtDNA was hybridized to fragments on the filter and bands visualized by autoradiography (Maniatis *et al.* 1982).

Band mobilities were determined using molecular-weight standards, and individual fall armyworms were scored for fragment patterns at each restriction enzyme. Each pattern was given a letter designation and multienzyme letter genotypes were then assigned to each individual.

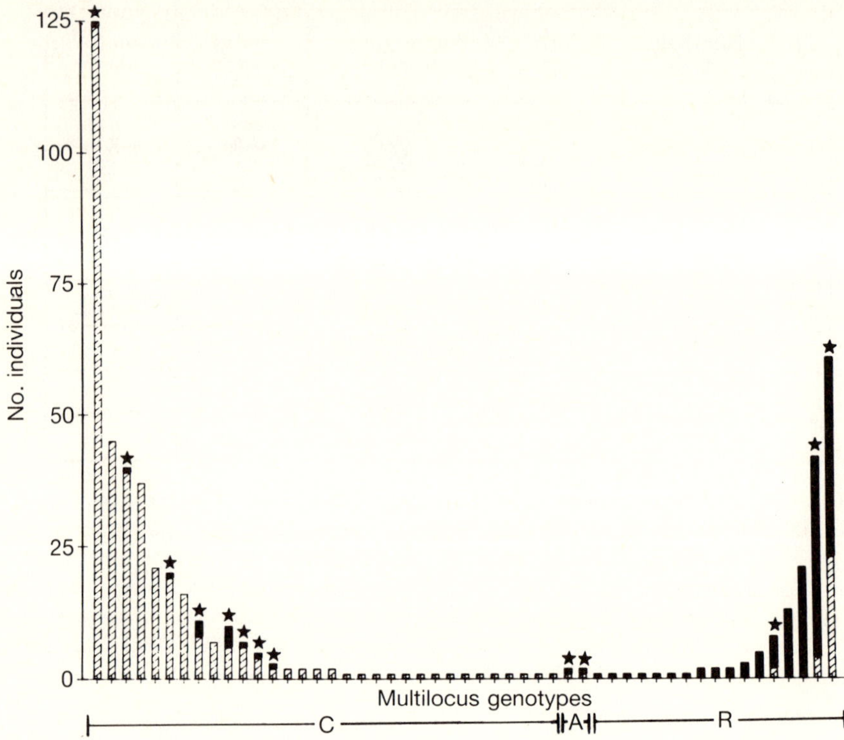

Fig. 6.2. Genotypic composition of individual fall armyworms collected from rice or bermudagrass (solid bars; N = 152) and corn (hatched bars; N = 388) throughout the Western Hemisphere. The 52 multilocus genotypes are based on combined allelic states at esterase and hydroxybutyrate dehydrogenase and are arranged according to the most common genotypes in corn (on left) to the most common genotypes in rice/bermudagrass (on right). Stars denote genotypes observed on both hosts. Strain affinities of genotypic classes indicated on the X axis are based on the host plants that represent the majority for a given genotype (C = corn strain; A = ambiguous; R = rice strain).

3. Results

Of the 35 restriction enzymes surveyed, most were uninformative (Table 6.1). Four were potentially diagnostic between strains and five exhibited minor intrastrain variation of limited use in population studies (i.e. single individuals had variant types but no population differences were apparent).

Approximately 20 individuals collected from both corn and rice in

Table 6.1 Summary of restriction enzymes surveyed for variation in mtDNA of fall armyworm populations. Asterisks indicate differences between strains; pluses indicate intrastrain variation.

Minimum number of restriction fragments[a]						
0/1	2	3	4	6	7	>8
AatI	NdeI	HindIII	BanII +	DraI	AluI +	BstNI*
ApaI	NruI	NcoI	EcoRI +	MboI*	RsaI +	HinfI*
BamHI	PstI	XbaI			SspI	MboII*
BanIII	SnaBI +					
BclI	TthIII(I)					
BglI						
BglII						
CfoI						
ClaI						
EcoRV						
HpaI						
PvuII						
SalI						
SpeI						
SphI						
StuI						
XhoI						

[a] A minimum estimate is given because small fragments might not be detected with a heterologous probe.

Table 6.2 MtDNA genotypes at two restriction enzymes (*HinfI* and *BstNI*, individually, and composite two-enzyme genotypes) in individual fall armyworms collected on rice (N = 15) and corn (N = 18) in Puerto Rico in January 1985. Low overlap (composite genotype BB on corn and AA on rice) is probably due to individuals of one strain feeding on the host of the other strain.

	HinfI		*BstNI*				*HinfI–BstNI*			
	A	B	A	B	C	D	AA	AC	AD	BB
Corn	16	2	14	2	1	1	14	1	1	2
Rice	2	13	2	13	0	0	2	0	0	13

Puerto Rico were initially examined at two of the four distinctive restriction enzymes (Table 6.2). The apparent small amount of overlap in genotypic categories can best be explained by overlapping host usage (the second hypothesis noted above for all allozyme data). For example, in Table 6.2, the two corn-collected individuals with the B genotype at *HinfI* were the same two individuals with the B genotype at *BstNI*. These were both predominantly rice-strain genotypes, strongly suggesting that these two individuals were really members of the rice strain feeding on corn. Likewise, two rice-feeding individuals with the A genotype at *HinfI* were the same two individuals with the A genotype at *BstNI*. Using similar rationale, they were probably members of the corn strain feeding on rice.

A second study was performed on individuals from various geographic localities and the two remaining restriction enzymes were also examined (Table 6.3). Again, there were primarily two genotypic classes, ABAA associated with corn and BABB associated with bermudagrass. Only one other genotype was apparent in individual 2 from Ecuador [based on allozyme data, the corn strain feeds on cotton (Pashley 1988*b*)]. Three individuals, two from Santiago Tuxtla and one from Trinidad, show a discrepancy between the host of collection and the strain assignment. Allozyme data indicated that these populations were composed of a mixture of strains (Table 6.4). The probable diagnostic nature of mtDNA and the likelihood that the allozymes are also strain-specific (or nearly so) indicates that the combined use of the two molecules will be useful in resolving interstrain hybridization in nature.

Conclusions

The two fall armyworm taxa are clearly not biotypes of a single species, because they exhibit too many genetic differences. In addition to allozyme and mtDNA, there are differences in development on rice and corn (Pashley 1988*a*) and on bermudagrass varieties (Pashley *et al.* 1987*a*), as well as differences in insecticide resistance (Pashley *et al.* 1987*b*). The life-history characteristics of fall armyworm suggest that the host strains are more likely to be separate species than host races (see Diehl and Bush 1984 for a distinction). If the two strains are interbreeding, introgression may be occurring throughout the range of the species because of their high degree of sympatry.

Because this study is probably an interspecific rather than intraspecific one, the difficulty of quantifying gene flow levels is

Table 6.3 MtDNA restriction-enzyme genotypes that exhibit host-associated variation in individual fall armyworms from various geographical populations.

Collection site Date (host)	Individual no.	BstNI	HinfI	MboI	MboII	Strain
Isabela, Puerto Rico, USA	1	—	B	A	A	C
VII-1984 (corn)	2	A	B	A	A	C
Baton Rouge, La., USA	1	B	A	B	B	R
VIII-1984 (bermudagrass)	2	B	A	B	B	R
Mocya, Trinidad	1	B	A	B	B	R
VII-1984 (corn)	2	—	—	—	A	C
Palm Co., Fla., USA	1	A	B	A	A	C
III-1983 (corn)	2	A	B	A	A	C
Santiago Tuxtla, Ver., Mexico	1	B	A	B	B	R
VIII-1984 (corn)	2	B	A	B	B	R
Guayaquil, Ecuador	1	A	B	A	A	C
V-1984 (cotton)	2	A	B	B	A	C/R
Hammond, La. USA VIII-1985 (corn)	1	A	B	A	A	C

somewhat reduced. There are significant genetic differences at both allozymes and mtDNA that are almost certainly due to a near absence of gene flow. After an extensive survey using both markers, it should be possible to quantify precisely whether or how much interbreeding is occurring.

An ancillary issue addressed by this study was the utility of mtDNA for population studies. MtDNA was slightly better at distinguishing the two fall armyworm strains and in this regard aided in establishing the realiability of the allozyme markers. Using the two markers together will allow studies of hybridization that would not be

Table 6.4 Strain affinities of individual fall armyworms based on multilocus allozyme genotypes (*Est, Hbdh, Pep*) and mtDNA restriction patterns (*BstNI, HinfI, MboI, and MboII*).

Collection site, date (host)	Allozymes Corn/Rice Strain	mtDNA Corn/Rice Strain
Isabela, Puerto Rico, VII–1984 (corn)	34/ 0	2/0
Baton Rouge, La., VIII–1984 (bermudagrass)	2/38	0/2
Mocya, Trinidad, VII–1984 (corn)	23/ 5	1/1
Palm Co., Fla., IV–1984 & III–1983 (corn)	29/ 0	2/0
Santiago Tuxtla, Mexico, VIII–1984 (corn)	11/12	0/2
Guayaquil, Ecuador, V–1984 (cotton)	26/ 3	1/0
Hammond, La., VII–1984 & VIII–1985 (corn)	33/ 4	1/0

possible with either method alone. However, use of mtDNA within strains may be compromised by limited variation. In the 60 or so individuals surveyed at 35 restriction enzymes, only a handful of individuals departed in their nucleotide sequences from the two strain-specific types. Within the corn strain, the same genotypes were common to populations from Central America, the Caribbean, and the south-eastern USA. More restriction enzymes should be examined before it is concluded that mtDNA will not be useful for intraspecific studies of fall armyworm, but preliminary data are not promising. As technology advances, a preferred method may be analysis of rapidly evolving stretches of single-copy nuclear DNA as opposed to or in combination with organellar (e.g. mt) DNA. There is still no universal, best method for studying gene flow.

Acknowledgements

S.J. Johnson, J. Harper, S. Pair, and A.N. Sparks aided in collection of fall armyworm. Technical help with mtDNA was provided by R. DeSalle and E.A. Zimmer. J.A. Martin assisted in the laboratory work. The manuscript was improved by comments from D.N. Pashley and R.M. Zink. Research was supported by USDA Grant No. 86-CRCR-1-2027.

References

Avise, J. C. and Saunders, N. C. (1984). Hybridization and introgression among species of sunfish (*Lepomis*): analysis by mitochondrial DNA and allozyme markers. *Genetics*, **108**, 237–55.

Avise, J. C., Bermingham, E., Kessler, L. G., and Saunders, N. C. (1984). Characterization of mitochondrial DNA variability in a hybrid swarm between subspecies of bluegill sunfish (*Lepomis macrochirus*). *Evolution*, **38**, 931–41.

Avise, J. C., Ball, R. M., Bermingham, E., Lamb, T., Neigel, J. E., Reeb, C. A., and Saunders, N. C. (1987). Intraspecific phylogeography: the mitochondrial DNA bridge between population genetics and systematics. *A. Rev. Ecol. Sys.*, **18**, 489–522.

Brown, W. M. and Wright, J. W. (1979). Mitochondrial DNA analyses and the origin and relative age of parthenogenetic lizards (Genus *Cnemidophorus*). *Science*, **203**, 1247–9.

Brown, W. M., George, Jr., M., and Wilson, A. C. (1979). Rapid evolution of animal mitochondrial DNA. *Proc. natn. Acad. Sci. USA*, **76**, 1967–71.

Coen, E. S., Thoday, J. N., and Dover, G. (1982). Rate of turnover of structural variants in the rDNA gene family of *Drosophila melanogaster. Nature, Lond.* **295**, 564–8.

Diehl, S. R. and Bush, G. L. (1984). An evolutionary and applied perspective of insect biotypes. *A. Rev. Ent.*, **29**, 471–504.

Luginbill, P. (1928). The fall armyworm. *USDA Technical Bulletin* No. 34.

Maniatis, T., Fritsch, E., and Sambrook, J. (1982). *Molecular cloning*. Cold Spring Harbor Lab. Publ., New York.

Moritz, C., Dowling, T. E., and Brown, W. M. (1987). Evolution of animal mitochondrial DNA: relevance for population biology and systematics. *A. Rev. Ecol. Sys.*, **18**, 269–92.

Pashley, D. P. (1986). Host associated genetic differentiation in fall armyworm: a sibling species complex? *Ann. ent. Soc. Am.*, **79**, 898–904.

Pashley, D. P. (1988*a*). Quantitative genetics, development and physiological adaptation in sympatric host strains of fall armyworm. *Evolution*, **42**, 93–102.

Pashley, D. P. (1988*b*). Causes of host-associated variation in insect herbivores: an example from fall armyworm. In *Evolution of insect pests: The pattern of variation* (ed. K. C. Kim). Wiley, New York (in press).

Pashley, D. P. (1988*c*). Current status of fall armyworm host strains. *Florida Entomologist*, **71**, 227–34.

Pashley, D. P. and Martin, J. A. (1987). Reproductive incompatibility between host strains of fall armyworm (Lepidoptera: Noctuidae). *Ann. ent. Soc. Am.*, **80**, 731–3.

Pashley, D. P., Johnson, S. J., and Sparks, A. N. (1985). Genetic population structure of migratory moths: the fall armyworm (Lepidoptera: Noctuidae). *Ann. ent. Soc. Am.*, **78**, 756–62.

Pashley, D. P., Quisenberry, S. S., and Jamjanya, T. (1987*a*). Impact of fall armyworm (Lepidoptera: Noctuidae) host strains on the evaluation of Bermudagrass resistance. *J. econ. Ent.*, **80**, 1127–30.

Pashley, D. P., Sparks, T. C., Quisenberry, S. S., Jamjanya, T., and Dowd, P. (1987*b*). Two fall armyworm strains feed on corn, rice and Bermudagrass. *Louisiana Agriculture*, **30**, 8–9.

Powell, J. R. (1983). Interspecific cytoplasmic gene flow in the absence of nuclear gene flow: evidence from *Drosophila*. *Proc. natn. Acad. Sci. USA*, **80**, 492–5.

Simon, C. (1988). Evolution of 13- and 17-year periodical cicadas (Homoptera: Cicadidae: *Magicicada*). *Bull. ent. Soc. Am.*, **34**, 163–76.

Southern, E. M. (1975). Detection of specific sequences among DNA fragments separated by gel electrophoresis. *J. Molec. Biol.*, **98**, 503–17.

Sparks, A. N. (1979). A review of the biology of the fall armyworm. *Florida Entomologist*, **62**, 82–7.

Spolsky, C. and Uzzell, T. (1984). Natural interspecies transfer of mitochondrial DNA in amphibians. *Proc. natn. Acad. Sci. USA*, **81**, 5802–5.

7 The use of electrophoretic data in a study of gene flow in the pest species *Heliothis armigera* (Hübner) and *H. punctigera* Wallengren (Lepidoptera: Noctuidae).

JOANNE C. DALY

CSIRO Division of Entomology, Canberra, Australia.

Abstract

Gene flow results when an immigrant organism breeds successfully and its genes persist in the new area. In pest species, gene flow can be important in a number of ways—such as the evolution of pesticide resistance, the development of host races, and adaptation to localized environments. Although gene flow results from dispersal, it is not an inevitable outcome. Gene flow can, however, influence the genetic structure of a population even when it occurs infrequently or at low frequency, which is also when it is most difficult to study within and/or between natural populations.

Electrophoresis is a useful tool with which to study the genetic structure of a population. It is a relatively cheap and quick method for screening genetic variation at different loci in a number of populations. Inferences can be drawn about gene flow from this structure. However, as a given genetic structure can be generated by a variety of contrasting population structures, electrophoretic data must be supplemented by independent observations of the species' population biology (population size, sex ratio, breeding structure, and so on).

The patterns of electrophoretic variation in the Noctuid moth species *H. armigera* and *H. punctigera* throughout eastern Australia are similar. In both species, the differences in gene frequencies between

Electrophoretic Studies on Agricultural Pests (ed. Hugh D. Loxdale and J. den Hollander), Systematics Association Special Volume No. 39, pp.115–41. Clarendon Press, Oxford, 1989.

regions are small and the distribution of rare alleles is consistent with high levels of gene flow between populations. Thus, electrophoretic data have not been able to detect any qualitative differences in gene flow in the two pest species. Yet gene flow may have led to the rapid spread of insecticide resistance in *H. armigera*, and retarded the evolution of resistance in *H. punctigera.* Differences in the outcome of gene flow in these two moths is thought to reflect the effective sizes of populations within, and outside, sprayed cropping areas.

Introduction

This chapter discusses how the technique of allozyme electrophoresis has been applied to a study of gene flow in insect pest species. Gene flow can be difficult to measure directly, so inferences about its relative magnitude have been drawn, with a knowledge of population genetics theory, from the genetic structure of a population. The first part of the chapter will consider the limitations of this approach. Part two discusses the problems in obtaining a random sample of a population. Part three describes a number of techniques that can be used to measure genetic differentiation between populations and evaluates their usefulness in predicting the amount of gene flow that occurs between populations. Examples will largely be taken from the study of the genetic structure of the moths *Heliothis armigera* and *H. punctigera* (Lepidoptera: Noctuidae) (Daly and Gregg 1985; Daly, unpublished).

The definitions of the terms *movement*, *dispersal*, *gene flow*, and *migration* vary among disciplines. In this chapter, the movement of individuals between populations will be defined as dispersal. Gene flow occurs when the dispersers reproduce successfully and contribute to the gene pool (i.e. when they leave descendants). The term migration will be avoided because, although it can be used in population genetics to mean gene flow, it can also be used to describe long-distance movement or seasonal movement.

The measurement of gene flow in pest species

An understanding of gene flow is important because the pest status of many insects is, in part, a consequence of their ability rapidly to invade and reproduce in new habitats. The pattern of agriculture in which a variety of crops is cultivated and each reaches maturity at a different time, and the expansion of cropping into new regions, have benefited those insects that are good colonizers (Southwood 1971;

Farrow and Daly 1987). Also, the development and spread of insecticide resistance is dependent on the rate of gene flow between populations (Comins 1977).

1. *Direct observations of gene flow*

Despite its importance, very little is known about gene flow in insect pest species because, for a number of reasons, it is difficult to observe and measure directly. First, the amount of gene flow occurring between populations is not necessarily proportional to the amount of dispersal. Dispersal may be common in an insect species but dispersers may contribute little to the new gene pool if, for example, they disperse at a time of year unfavourable to reproduction or to an unfavourable habitat (e.g. Farrow 1979).

Secondly, populations may be genetically homogeneous as a result of the successful reproduction of only a few dispersers per generation (Lewontin 1974). Thus, the study may lack the power to detect gene flow in species if dispersal, particularly over long distance, is an occasional event or if dispersers are only a small part of the population.

Thirdly, dispersal may be difficult to measure even in species in which it is common. For example, individuals may be too small to observe easily, such as aphids; species may be hard to identify if dispersal occurs at night, such as in *Heliothis* spp. (Farrow and Daly 1987); individuals may be hard to identify if dispersal occurs in the geostrophic winds, 0.1–1 km above the ground, because the density of individuals can be too low for individuals to be collected by conventional trapping techniques (Farrow and Daly 1987). Drake and Farrow (1985) collected only four *H. punctigera* moths in aerial traps in a 9-day period, yet this represented 100 000 moths crossing a 1 km front during the 9-day period (Farrow and Daly 1987).

2. *Electrophoretic studies of gene flow in insects.*

A second approach for estimating gene flow is an indirect one: inferences about the level of gene flow can be drawn from the genetic structure of a species, which is defined as the spatial and temporal pattern of gene and genotype frequencies (Pashley 1985; Slatkin 1985*a*). Allozyme electrophoresis is a useful technique with which to detect genetic variants (Harris and Hopkinson 1976; Richardson *et al.* 1987) because heterozygotes can be distinguished from homozygotes and each individual can be classified for a number of loci cheaply and quickly. Allozyme studies have been used widely to study gene flow in plants (e.g. Hamrick *et al.* 1979; Soltis and Soltis 1987; Goldenberg 1987) and vertebrates (e.g. Patton and Feder 1981; Larson *et al.* 1984; Waples 1987). This chapter will focus on studies with insects.

A variety of hypotheses about gene flow in insects have been examined using electrophoretic data:

1. Genetic differentiation between populations will be less in a migratory (or vagile) species compared with a closely related but more sedentary species. Daly and Gregg (1985) tested this hypothesis in the two sympatric species of Australian pests, *H. punctigera* and *H. armigera.* Hypotheses about the evolution of insecticide resistance in *H. armigera*, but not in *H. punctigera*, could not be evaluated because of lack of information about the population structure of either species (Daly and Murray 1988), although *H. armigera* was considered to be more sedentary than *H. punctigera* (Wardhaugh *et al.* 1980; Morton *et al.* 1981; Farrow and Daly 1987).

   Liebherr (1986) also examined this hypothesis in two species of carabid beetles of differring vagility. He compared the genetic structure between populations of the winged *Platynus tenuicollis* with that between populations of the wingless *P. angustatus.*

2. Large genetic differences will not occur between populations of the same species if there is gene flow between populations, e.g. in the horseflies *Tabanus nigrovittatus* and *T. conterminus* (Sofield *et al.* 1984), in European populations of the gypsy moth, *Lymantria dispar* (Harrison *et al.* 1983; George 1984), in the boll weevil, *Anthonomus grandis* (Bartlett *et al.* 1983), in the spruce budworm *Choristoneura occidentalis* (Willhite and Stock 1983), in the mosquito *Culex quinquefasciatus* (Urbanelli *et al.* 1985), in the horn fly, *Haematobia irritans* (McDonald *et al.* 1987), in eleven species of cave arthropods (Caccone 1985), and in *Drosophila pseudoobscura* (Jones *et al.* 1981).

   In particular, genetic differentiation between populations will be small in highly migratory species, e.g. *Spodoptera exempta* (den Boer 1978); within a species, genetic differentiation will be greater between populations taken from different geographic races than between populations within the same geographic race, e.g. in the southern pine beetle, *Dendroctonus frontalis* (Anderson *et al.* 1979; Namkoong *et al.* 1979), and the corn rootworm, *Diabrotica barberi* (McDonald *et al.* 1985).

3. In migratory species, the genetic structure of immigrant populations will more closely resemble source rather than non-source

populations, e.g. in the velvetbean caterpillar *Anticarsia gemmatalis* (Pashley and Johnson 1986).

4. Colonization of new habitat will be accompanied by genetic differentiation as the population goes through bottlenecks, e.g. in the walnut husk fly *Rhagoletis completa* (Berlocher 1984) and the face fly *Musca autumnalis* (Bryant *et al.* 1981). Pashley and Bush (1979) discuss the use of rare allozymes to draw inferences about the origin of derived populations and illustrate this approach with codling moth, *Cydia pomonella*.
5. The degree of genetic differentiation will be related to the size and distribution of habitat patches, e.g. in species of carabid beetles, *Agonum* spp. and *Platynus* spp. (Liebherr 1988).

Each of these studies makes an important assumption in the interpretation of electrophoretic data: that the genetic structure of a species is a consequence largely of the pattern of gene flow between populations. This is true only if the following four conditions are met. First, gene flow is random with respect to the genotypes studied. Second, the rate of gene flow (m) exceeds the effects of selection, which certainly will be true for neutral polymorphisms (polymorphisms for which there is no difference in relative fitness of the different genotypes). Third, the rate of gene flow exceeds the effects of genetic drift described mathematically as, $m >> 1/4N_e$ (Wright 1931). N_e, the effective size of a population, summarizes the deviations of real populations from the ideal model (see Crow and Kimura 1970 for details). The magnitude of genetic drift is inversely proportional to the effective size of a population (N_e). Fourth, the populations are at equilibrium.

Most allozymes are considered to be either neutral or under weak selection (Kimura 1983), so that assumptions 1 and 2 are usually met. Assumptions 3 and 4, however, present more problems. In most insect pest species, the magnitude of genetic drift is largely undetermined, so the pattern of genetic variation by itself provides insufficient evidence about gene flow. For example, populations may be genetically homogeneous in species with either high gene flow or low gene flow, as long as the effect of gene flow exceeds that of genetic drift. Conversely, geographic differentiation may occur in species that are highly mobile if each population is founded from a small number of individuals or if population size fluctuates (i.e. if N_e is small; Slatkin 1977).

Even in pest species, numbers may be very low in some generations as a result of pest management, seasonal factors, or natural catastrophes. Such bottlenecks can be important in generating changes in gene frequencies so that the genetic structure of a population may reflect historical events not measurable within the period of study even in species where gene flow is occurring (e.g. Larson *et al.* 1984). It may take many generations for the populations to reach the equilibrium gene frequency after a population crash. In such species the effects of gene flow may not exceed that of genetic drift, even if population sizes appear large and gene flow is common.

As a consequence, a given genetic structure is not the result of a unique level of gene flow. This presents problems in interpreting the results of the electrophoretic studies outlined above. Consider the studies in hypothesis (1). Daly and Gregg (1985) concluded that gene flow was common in both species of *Heliothis* because no regional variation was detected in either species. In contrast, marked regional variation in both *Playtnus* spp. was consistent with low gene flow in both species (Liebherr 1986). These results do not mean that there are no qualitative differences in gene flow within either genera. For example, little genetic differentiation would be expected if dispersal occurred either every generation in one species compared with once every five generations in the other, or if 10 per cent or 90 per cent of the population dispersed. Yet in pest species, such as *Heliothis* spp., such differences could be important to population dynamics, to pest control, and to the evolution of insecticide resistance.

Another problem arises with studies given in hypothesis (2). Subdivisions of a population can differ in gene frequency yet still be part of the same gene pool, e.g. in the brown snail, *Helix aspersa* (Selander and Kaufman 1975), in the pocket gopher *Thomomys bottae* (Patton and Feder 1981) and in the mouse *Mus musculus* (Selander 1970; Baker 1981). Differences in gene frequency do not necessarily mean that gene flow is restricted. Studies in which there are no *a priori* reasons for concluding that gene flow is limited can only suggest this as one of a number of alternative hypotheses if genetic divergence between populations is observed.

Despite these problems, electrophoresis is a useful technique because it can be used to test hypotheses about population structure derived from independent data (e.g. *ad hoc* observations of migration, subdivision of the species into geographic races, apparent barriers to dispersal). Alternatively, if little or nothing is known about dispersal in a species, an electrophoretic study is a relatively quick and cheap method of developing hypotheses about population structure to be tested in other ways. In either case, it is an example of how theory developed in one discipline, population genetics, can be applied to problems that are intractable or difficult to study in another discipline, population ecology.

Intra-population variation

Most allozyme studies of insects focus on the genetic differences between populations. The analyses assume that the individuals collected are a random sample of the local population. This assumption is not necessarily correct if the sampling procedure cannot detect any microspatial differences in gene frequencies (e.g. due to intrapopulation subdivision) or if the effective size of the sample, N_e, is considerably less than the number of individuals collected, N (e.g. if individuals are the progeny of only a few females; see below). Non-random samples from within populations can exaggerate the magnitude of genetic differences observed between populations. Thus, homogeneity of samples from within a population needs to be established before variation among populations can be considered biologically meaningful (Richardson *et al.* 1987).

In mobile lepidopterous pest species, microspatial variation in gene frequencies is probably minimal in adults. For example, adult *H. armigera* and *H. punctigera* have been observed with night-vision devices to move between fields separated by hundreds of metres (R.A. Farrow and G.P. Fitt, pers. comm.). Collections of eggs or larvae, however, can be non-random samples of the population if more than two progeny of the same female are sampled ($N_e < N$). In some species, individual females lay egg masses, e.g in the gypsy moth, *Lymantria dispar* (George 1984). If the sperm of only one male is used to fertilize the mass, the effective sample size of the egg mass is 4, irrespective of the number of eggs sampled from each mass (see Crow and Kimura 1970). Similarly, $N_e < N$ if the sample consists of many eggs collected from adult females, e.g. in the Colorado potato beetle, *Leptinotarsa decemlineata* (Jacobsen and Hsiao 1983).

Field collections of larvae may also be non-random samples, particularly if the population is sampled from only one field (or part of a field) in species in which females lay many eggs in a small area. This is a potential source of bias in studies of insect pests of trees if many individuals are sampled from only a few trees and the relatedness of individuals within a tree is not known (e.g. Willhite and Stock 1983; Anderson *et al.* 1979).

Statistical tests are available that can detect local population structuring: deviations from Hardy–Weinberg proportions, significance tests for Wright's F_{IS} (Wright 1978) and Smith's H (Smith 1970) statistics. Applications of these tests are discussed elsewhere (Wright 1978; Pashley 1985; Richardson *et al.* 1987).

A study of the microspatial variation of gene frequencies was undertaken in *Heliothis* spp. to determine whether larvae taken from different parts of a field were random sample of the local population (Daly, unpublished).

1. Microspatial variation in Heliothis *spp.*

In broadacre agriculture, pest control is the responsibility of each farmer, and within a farm, each field or crop may be managed independently. These divisions of pest populations into farms, fields, and crops are convenient to farmers but may not reflect the population structure of the pest on a fine scale. One situation in which microspatial variation will be important is in monitoring for insecticide resistance in eggs or larvae. It is necessary to know whether the frequency of resistance varies within a field, or between fields, so that sampling can be minimized and yet still give unbiased estimates.

In *Heliothis* spp., the eggs represent a suitable life-stage for examination because, for the first half of embryo development, the diploid genotype detected for many isoenzymes is that of the female parent (Fisk and Daly 1989). Thus, all eggs laid by a female initially will have an identical genotype to her, so that the allele frequencies in the eggs are a direct measure of the non-randomness of oviposition.

In a preliminary study, Daly (unpublished data) sampled a cotton field and a maize field on a farm in the Namoi Valley, New South Wales (Fig. 7.1, site 2 from Daly and Gregg 1985). Samples from within each field were regularly spaced (Fig. 7.2). Within each sample, eggs were collected from three adjacent rows over a length of 50 m. Only one egg was collected per plant. Electrophoresis techniques are given in Daly and Gregg (1985). Eggs were scored for the isoenzyme isocitrate dehydrogenase-1 (ICD-1) to identify the species of *Heliothis*, and for phosphoglucomutase (PGM), which is a variable locus. All eggs in the maize sample were *H. armigera*, while 94 per cent of the eggs in the cotton sample were *H. punctigera*.

The results are illustrated in Fig. 7.2. Gene frequency differences were compared by contingency χ^2 testing. Limitations of this test are discussed in the next section. There were significant differences in allele frequencies between samples taken from different parts of the same field in both *H. armigera* ($\chi^2 = 37$, 15 df, $P < 0.001$) and *H. punctigera* ($\chi^2 = 24$, 12 df, $P < 0.05$). These results suggest that eggs are not randomly distributed with respect to genotype within a field.

Similar microspatial variation has been observed for the distribution of phenotypes resistant or susceptible to synthetic pyrethroids

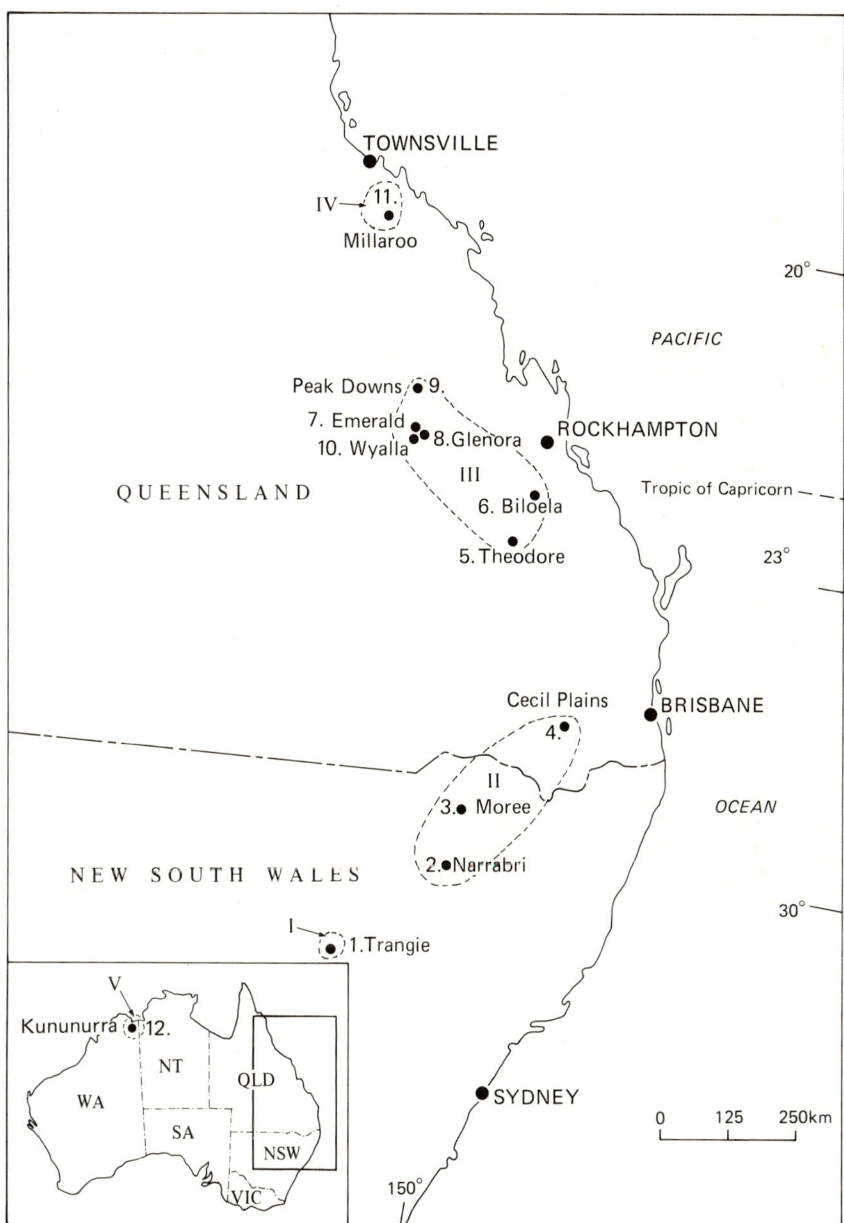

Fig. 7.1. Areas where samples were collected. Samples 1–12 were grouped into five regions I–V indicated by enclosure in broken lines. (From Daly and Gregg 1985.)

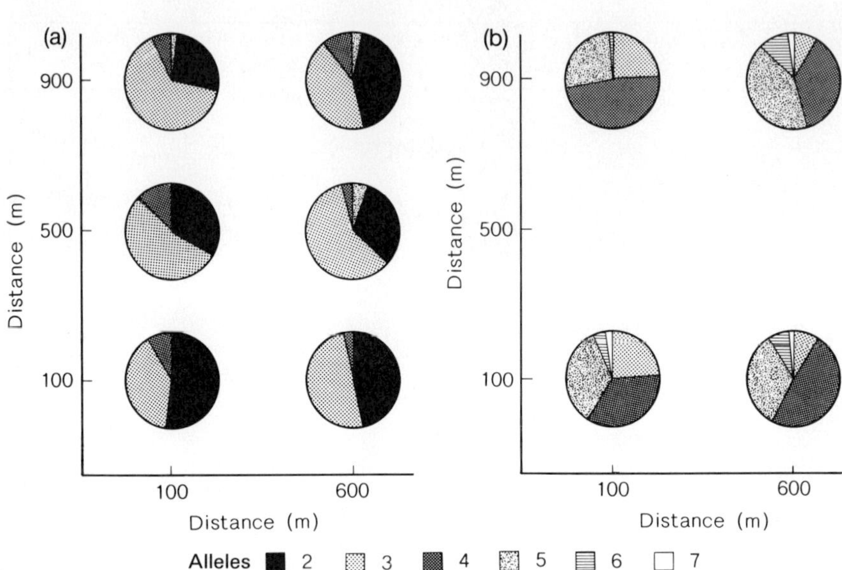

Fig. 7.2. The frequency of *Pgm* alleles in two fields: (a) six samples of *H. armigera* from a maize crop; (b) four samples of *H. punctigera* from a cotton crop.

in *H. armigera* (Daly, unpublished). Eggs were collected from either three or four locations, approximately 200 m apart, within four sites in the Emerald Irrigation Area, Queensland (site 7 in Daly and Gregg 1985). There was significant heterogeneity in the frequency of resistance in the samples taken from site A but not from the other crops (Table 7.1). The magnitude of the differences between samples from site D could be important in pest management even though they were not statistically significant.

These results are only preliminary and need to be repeated over different nights, fields crops, and stages of crop growth before we have a full description of microspatial variation in either species. Nevertheless, it may be inferred from the data that samples taken from different parts of a field are not necessarily a random genetic sample of the population.

Genetic variation between populations

A number of techniques are available for analysing the genetic structure among populations. Two of these, contingency tables and genetic distance measurements, provide only a measure of the degree

Table 7.1 The percentage of individuals resistant to synthetic pyrethroids in subsamples taken from localities A–D in the Emerald Irrigation Area.

Sample location	Percent resistant in subsamples				*P*
	1	2	3	4	
A	3(65)[a]	9(55)	21(38)	4(27)	<0.05
B	13(56)	15(73)	10(48)		>0.05
C	3(34)	0(46)	6(47)		>0.05
D	14(52)	25(48)	14(21)		>0.05

[a] Figures in parentheses are sample sizes.

of differentiation. Three further techniques, *F*-statistics (Wright 1978), conditional allele frequencies (Slatkin 1981), and private allele frequencies (Slatkin 1985*b*), also provide a measure of the amount of gene flow between populations. Each of these will be evaluated with specific reference to the study of the genetic structure of *Heliothis* spp. (Daly and Gregg 1985).

Daly and Gregg (1985) sampled 11 *H. armigera* and 6 *H. punctigera* populations in eastern Australia and one population of each species from northern Western Australia from a variety of crops and weeds (Fig. 7.1). A number of populations were sampled in two regions so that intra- and interregional variation could be compared. Details of collections and electrophoretic techniques are given by Daly and Gregg (1985). Individuals were identified to species using the three diagnostic isoenzymes: ICD-1, phoshogluconate dehydrogenase (PGD), and glucose-6-phosphate dehydrogenase (GD). Considerable misidentification of larvae had occurred when morphological characters had been used (Daly and Gregg 1985). The polymorphic isoenzymes examined were β-hydroxybutyric dehydrogenase (HBDH), phosphoglucose isomerase (PGI), mannose phosphate isomerase (MPI), guanine deaminase (GDA), and PGM.

1. Contingency χ^2 tests and log-linear models

Gene frequencies in different populations were compared using contingency χ^2 tests. Heterogeneity was partitioned into a nested analysis using log-linear models and computed with the statistical package GLIM (Payne 1985). These analyses are also called G-tests (Sokal and Rohlf 1969).

The tests examine the null hypothesis that there is no significant difference between populations. If sample sizes are inadequate, the χ^2 tests lack power to detect significant differences because of the chance of type 1 errors (rejecting the null hypothesis when it should be accepted, usually set at 5 per cent) and type 2 errors (accepting the null hypothesis when it should be rejected). If the probability of a type 2 error is set at 10 per cent, then at least 50 individuals are required to detect a gene frequency difference of 0.20 between two populations and 100 individuals if the difference is 0.1 (Richardson *et al.* 1987). It is unusual for sample sizes in electrophoretic studies to exceed 50 individuals per population. Some studies have used considerably fewer (e.g. 10–12 individuals in Liebherr 1986).

Daly and Gregg (1985) observed significant heterogeneity between populations for four of the seven loci in *H. armigera* and two of the four loci in *H. punctigera* (Table 7.2). However, they concluded that there was no distinct regional differentiation in eastern Australia, because the interregional variation was not significantly greater than the

Table 7.2 Tests of heterogeneity of gene frequencies in *H. armigera* and *H. punctigera* populations. (From Daly and Gregg 1985.)

	Significance levels (*P*)		
Locus	Between all population	Within regions	Between regions
H. armigera			
Pgi, Icd-1	NS	NS	NS
Pgm	<0.05	NS	NS
Pgd	NS	NS	<0.05
Icd-2	<0.05	<0.01	NS
Mpi	<0.001	NS	<0.001
Gda	<0.01	<0.05	NS
H. punctigera			
Pgm, Mpi	NS	NS	NS
Pgi	<0.05	NS	<0.01
Hbdh	<0.01	NS	<0.01

NS = not significant.

intraregional variation (Table 7.2; see Daly and Gregg 1985 for further discussion of this point).

No evidence for regional variation in gene frequencies was obtained in either *H. virescens* (Sluss and Graham 1979) or in *H. zea* (Sell *et al.*, 1974, 1975), although there was indication of intraregional variation in both species.

2. *Genetic distance measurements*

The differences between populations can also be measured using a variety of genetic distance measurements. The common ones are Nei's identity (*I*) and distance (*D*) (Nei 1978) and Rogers' similarity (*I*) and distance (*R*) measurements (Rogers 1972). These coefficients can be summarized in a dendogram that reflects genetic affinities of different populations. There are a variety of methods of constructing the dendograms (Farris 1972; Felsenstein 1982; Rogers 1984).

Genetic distance measurements are not always appropriate for looking at population subdivision within a species (Richardson *et al.* 1987). Both Rogers' and Nei's distance measurements can have large standard errors, particularly for small distance values. The standard errors depend mainly on the number of loci sampled but also on the population sample size (Richardson *et al.* 1987). Thus, small differences in genetic distance values may be meaningless. Few authors quote standard errors in their genetic distance trees, yet dendograms are used to interpret genetic subdivision when genetic distances are of the order of 0.1 or less, which is normally the upper limit for differences between populations of the same species (Ayala 1975). In the absence of statistical tests, it is difficult to determine whether the preferred dendrogram is a much better fit to the data than other dendograms. For example, in the spruce budworm, *Choristoneura occidentalis*, Willhite and Stock (1983) allocated populations into regions using genetic distances and the distribution of rare alleles, yet the genetic distances between populations within regions (0.001–0.072) were of the same order of magnitude as between populations in different regions, and the rare alleles were not unique to a given region.

Only loci that are polymorphic within or between populations yield information on the extent of population divergence. Genetic distance measurements fail to detect population differences in species when most of the loci are monomorphic yet populations differ markedly for some of the poylmorphic loci. The tables of genetic distances given in Daly and Gregg (1985) do not contain useful information about population subdivision in *Heliothis* spp. because most loci in both *H.*

armigera and *H. punctigera* were either monomorphic or nearly so. Thus it is not surprising that all genetic distance measurements were small (<0.01 for all population comparisons).

More polymorphic loci were examined in *H. virescens* (14 loci) than in the two Australian *Heliothis* species (9 loci). All Nei's distances for pairwise comparisons between populations were less than 0.073, and the average genetic distance between populations was 0.034 (C.I. $\pm$ 0.002) (Sluss and Graham 1979).

Genetic distance measurements may also fail to detect population subdivision even in species where barriers to dispersal would suggest that gene flow should be very limited, e.g. in the white pine weevil, *Pissodes strobi* (Phillips and Lanier 1985).

3. *F-statistics*

F-statistics are three related statistics that can be used to interpret population differentiation. These were originally defined by Wright as 'inbreeding coefficients' but they have been generalized to describe the genetic structure of populations that have been subdivided (Nei 1977; Wright 1978). *F*-statistics can also be used to examine the hierarchical nature of populations, although few studies of insects attempt this (e.g. McCauley and Eanes 1987). In the present context, the species range of *H. armigera* will be considered to be a population and the samples from different localities will be considered to be subdivisions of that population.

F_{IS} and F_{ST} measure the deviations from Hardy–Weinberg proportions within each subpopulation, and within the total population, respectively. In the absence of selection, these departures are the result of non-random mating within a subpopulation or within the whole population because it is subdivided. Deviations of *F*-statistics from expected values can be compared with the χ^2 distribution, although the test for the significance of F_{ST} must be modified if sample sizes from the different subpopulations are not the same (see Eanes and Koehn 1978).

F-statistics have been used widely in the literature to interpret electrophoretic data. Weir and Cockerham (1984) warn that the estimates of *F*-statistics made by different authors may be sensitive to the numbers of populations sampled, sample sizes, and heterozygote frequencies. Estimates from different species must therefore be interpreted with caution.

Table 7.3 summarizes the *F*-statistics for eight loci from the seven populations of Narrabri, Moree, Cecil Plains, Biloela, Emerald (2), Glenora, and Kununurra in *H. armigera* (Daly and Gregg,

Table 7.3 Wright's *F*-statistics for populations of *H. armigera* (Daly and Gregg, unpublished).

Locus	F_{IS}	F_{ST}
Pgi	– 0.029	0.010
Pgm	0.066	0.017
Pgd	0.060	0.011
Icd-1	0.065	0.009
Icd-2	0.057	0.012
Gd	– 0.023	0.017
Hbdh	– 0.027	0.014
Mpi	0.070	0.039[a]
Mean	0.053	0.023[a]

[a] $P < 0.05$, otherwise all $P > 0.05$.

unpublished). The mean F_{IS} (0.053) was not significantly different from zero, which indicates populations were in Hardy–Weinberg equilibrium. The mean F_{ST} (0.023) was just significant ($P < 0.05$), which indicates that, overall, there were significant genetic differences between populations. When individual loci were analysed separately, the F_{ST} (0.039) was significant only for MPI ($P < 0.05$). In general, the results are consistent with those obtained in the tests of heterogeneity (Table 7.2). However, the *F*-statistics lacked the power to detect significant heterogeneity between populations for *Pgm* and *Icd-2* (Table 7.2, cf. Table 7.3).

F_{ST} values for a variety of pest species are given in Table 7.4. Many of these values have been calculated by Pashley *et al.* (1985) or McCauley and Eanes (1987) from published data. The F_{ST} value for *S. frugiperda* (0.032) excludes the Puerto Rican population, as it probably is a sibling species (Pashley 1986). It is clear that sedentary species such as *Euphydryas* have high values of F_{ST} (high genetic differentiation between populations), while highly migratory species such as *S. exempta* and *A. gemmatalis* have low values of F_{ST} (little genetic differentiation between populations).

McCauley and Eanes (1987) note that there is general agreement between the relative value of F_{ST} and the proposed vagility of the species among all animal groups. An exception is in the ground beetles. Extreme population subdivision was observed in both the wingless species *P. tenuicollis*, and in the winged species *P. angustatus* (Liebherr 1986), while little population subdivision was observed in the winged species *Agonum elongatulum* (Liebherr 1986, 1988).

Table 7.4 F_{ST} values and calculated values of gene flow (Nm) for pest and non-pest species of insects.

Insect	Family[a]	F_{ST}	Nm	Reference
Pest species				
Spodoptera exempta	L	0.006	41.4	Pashley *et al.* (1985)[b]
Pieris rapae	L	0.014	17.6	Pashley *et al.* (1985)[b]
Anticarsia gemmatalis	L	0.021	11.7	Pashley (1985)
Heliothis armigera	L	0.023	10.6	Daly and Gregg (1985)
Dendroctonus ponderosae	C	0.030	8.1	McCauley and Eanes (1987)[b]
Spodoptera frugiperda	L	0.032	7.6	Pashley *et al.* (1985)
Heliothis virescens	L	0.048	5.0	McCauley and Eanes (1987)[b]
Cydia pomonella	L	0.066	3.5	Pashley *et al.* (1985)
Dendroctonus frontalis	C	0.068	3.4	McCauley and Eanes (1987)[b]
Leptinotarsa decemlineata	C	0.068	3.4	Jacobson and Hsiao (1983)
Diabrotica barberi	C	0.098	2.3	McCauley and Eanes (1987)[b]
Pissodes strobi	C	0.098	2.3	Phillips and Lanier (1985)
Non-pest Species				
Agonum elongatulum	C	0.003	83.1	Liebherr (1988)
Drosophila willistoni	D	0.022	11.1	McCauley and Eanes (1987)[b]
Agonum decorum	C	0.08	2.9	Liebherr (1988)
Euphydryas editha	L	0.118	1.9	Pashley *et al.* (1985)
Drosophila pavani	D	0.126	1.7	McCauley and Eanes (1987)[b]
Tetraopes tetraopthalmus	C	0.154	1.4	McCauley and Eanes (1987)
Agonum extensicolle	C	0.16	1.3	Liebherr (1988)
Platynus tenuicollis	C	0.26	0.7	Liebherr (1986)
Platynus angustatus	C	0.27	0.7	Liebherr (1986)

[a] L = Lepidoptera; C = Coleoptera; D = Diptera.
[b] Calculated by the authors from published data.

Liebherr (1988) concluded that genetic heterogeneity as measured by F_{ST} was correlated not with the degree of flight-wing development but with the degree of habitat patchiness.

Conversely, F_{ST} alone is not sufficient to indicate the level of gene flow that is occurring in a species. Gene-frequency differences may exist even if gene flow is common in species in which population extinctions and recolonizations are occurring or sample sizes are small (see Slatkin 1985*a*).

4. *Conditional allele frequencies*

Most analyses of gene frequency data lack power with respect to alleles that are not common ($P < 0.05$) in any population. Yet these alleles provide a measure of gene flow. Slatkin (1981) showed that the average frequency of an allele found only in some populations, $\bar{p}(\mathrm{i})$, depended strongly on the overall level of gene flow. He simulated the distribution of these allele frequencies for different levels of gene flow among populations for two models of population structure: the 'island model', in which immigration from all populations is equally likely; and the 'one-dimensional stepping-stone model', in which immigration is more likely from adjacent populations. Gene flow was given as the number of immigrants per generation, *Nm*. These distributions are given in Fig. 7.3 for a one-dimensional stepping-stone

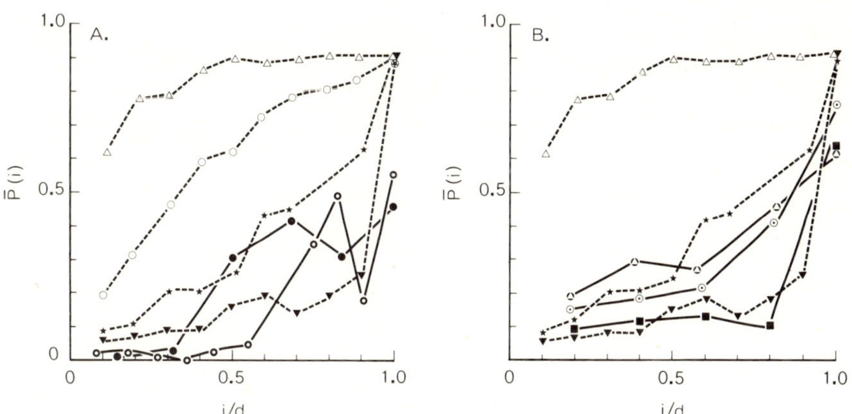

Fig. 7.3. A comparison of the conditional allele frequencies $\bar{p}$ (i) from Slatkin's simulation models (1981) (dashed lines) for high (▲), medium (★), low-medium (○) and low (△) gene flow compared with (A) *H. armigera* (**O**) and *H. punctigera* (●) (Daly and Gregg (1985), (B) *D. willistoni* (■, from Slatkin 1981), *P. angustatus* (**O**) and *P. tenuicollis* (⊙, Liebherr 1988).

model for high (Nm = 2.5), medium (Nm = 1.25), medium-low (Nm = 0.25) and low (Nm = 0.025) gene flow.

Actual distributions for *Heliothis* spp. (Daly and Gregg 1985), the two carabid beetles, *Playtnus tenuicollis* and *P. augustatus* (Liebherr 1988) and *Drosophila pseudoobscura* (quoted in Slatkin 1981) are also given in this figure. The distributions for *H. armigera*, *H. punctigera*, and *D. pseudoobscura* are consistent with medium-to-high gene flow, and with medium gene flow for the two *Platynus* species.

There are three reasons for suspecting that comparisons between actual distributions for a species and the simulated distributions are crude, particularly for intermediate distributions. First, in the absence of information about the pattern of movement in a species, a choice must be made if the population structure of the species is that of the 'island' or 'the stepping-stone' model. The simulations for these two models are similar for the extreme cases of high and low gene flow but differ for intermediate values (Slatkin 1981).

Secondly, there is no way of comparing the simulated and actual distributions other than by eye. Thus, the interpretation of the results is subjective, particularly when an actual distribution (e.g. that for *H. punctigera*) resembles one simulated line over part of the distribution and another simulated line over the rest (see Fig. 7.3A). Liebherr (1988) suggests that the distribution for *P. tenuicollis* is typical of a species with medium gene flow while that for *P. angustatus* is typical of a species with low gene flow. This is not apparent from the distributions that he gives (see Fig. 7.3B) and his interpretation is at variance with his calculations of Nm in both species, 0.68–0.70 (Liebherr 1988).

Thirdly, the distributions do not always agree with independent data on movement. For example, the distribution for *H. armigera* would suggest that gene flow was higher in this species than in *H. punctigera*, although *H. punctigera* appears from ecological data to be more vagile than *H. armigera* (Farrow and Daly 1987).

Caccone (1985) applied Slatkin's (1981) method to 11 species of cave arthropods. Although the patterns of gene flow were consistent with the dispersal abilities and ecological requirements of each species, Caccone concluded that Slatkin's models would have difficulty distinguishing between historical patterns and the effects of gene flow among extant populations.

5. *Private alleles*

Slatkin (1985*b*) further refined his analysis of low-frequency alleles by focusing on 'private alleles', which are alleles found only in one population. The logarithm of average frequency of private alleles $\bar{p}(1)$ is approximately related to the logarithm of the number of immigrations per generation, *Nm*:

$$ln(\bar{p}(1)) = a\ ln(Nm) + b,$$

where *a* and *b* are constants. The estimate of *Nm* is adjusted for size of the samples in the electrophoretic study by the ratio of sample size/25 (Slatkin 1985*b*).

Estimates of *Nm* for a number of insect species were calculated from published work (Table 7.5). In many studies there was considerable variation in the number of individuals screened per locus within each population. The average sample sizes were computed using the largest number of alleles sampled per population, so that the values of *Nm* in Table 7.5 would be underestimates. Low values of *Nm* will be obtained if a private allele is relatively common (> 0.1) in the population in which it occurs.

In general, the estimates of *Nm* are consistent with the life history of the insect. Highly mobile insects (e.g. *Spodoptera* spp., *A. gemmatalis*) have values of $Nm >> 1$ while in the more sedentary species, $Nm \leqslant 1$ (see references in Table 7.5). The values for *H. armigera* and *H. punctigera* suggest that gene flow is common in both species.

Estimates of *Nm* calculated this way are sensitive to the number of populations sampled and the number and type of loci screened. In the case of the European populations of the gyspy moth, there is considerable discrepancy between the estimate of $Nm = 0.3$ from George (1984) compared with that of 47.0 from Harrison *et al.* (1983). The two studies have only three loci in common out of the combined 25 loci screened. Harrison *et al.* (1983) observed seven private alleles in their six European populations all at frequency less or equal to 0.01. George (1984), on the other hand, monitored six populations within France. Only one private allele was observed, at frequency of 0.12. There is some doubt as to whether this allele is a true polymorphism, as it was observed only in a homozygous condition; no heterozygotes were observed although 8.9 were expected.

Two estimates are also available for another species, *Dendroctonus frontalis*. Although the relative estimates differ as 75.5 versus 29.5 (Namkoong *et al.* 1979; Anderson *et al.* 1979, respectively), both values are consistent with high gene flow.

6. *A comparison of estimates of Nm*

Nm can also be estimated from F_{ST} values (Table 7.4):

$$F_{ST} = \frac{1}{4Nm + 1}$$

Table 7.5 Estimates of rates of gene flow (*Nm*) from the average frequency of private alleles ($\bar{p}(1)$, Slatkin 1985*b*).

Species	$\bar{p}(1)$	*Nm*	Average sample size	Reference
Pest species				
Dendroctonus frontalis	0.005	75.5	95	Namkoong *et al.* (1979)
Lymantria dispar	0.008	47.0	59	Harrison *et al.* (1983)
Heliothis armigera	0.010	43.3	42	Daly and Gregg (1985)
Dendroctonus frontalis	0.004	29.5	378	Anderson *et al.* (1979)
Tabanus nigrovittatus	0.015	28.8	28	Sofield *et al.* (1984)
Anticarsia gemmatalis	0.012	28.2	NA[c]	Pashley and Johnson (1986)
Heliothis punctigera	0.015	27.0	30	Daly and Gregg (1985)
Choristoneura occidentalis	0.010	26.4	69	Willhite and Stock (1983)
Tabanus conterminus	0.017	18.9	35	Sofield *et al.* (1984)
Pissodes strobi	0.016	17.0	44	Phillips and Lanier (1985)
Spodoptera frugiperda	0.014	16.8	54	Pashley *et al.* (1985)[a]
Dendroctonus ponderosae	0.018	14.6	41	Stock *et al.* (1984)
Anthonomus grandis	0.010	14.4	126	Bartlett *et al.* (1983)
Heliothis virescens	0.025	7.5	41	Sluss and Graham (1979)
Lymantria dispar	0.12	0.3	39	George (1984)[b]
Non-pest species				
Agonum elongatulum	NA[c]	11.7	NA[c]	Liebherr (1988)
Drosophila willistoni	0.014	9.9	94	Slatkin (1985b)
Agonum decorum	NA[c]	2.4	NA[c]	Liebherr (1988)
Agonum extensicolle	NA[c]	2.0	NA[c]	Liebherr (1988)
Drosophila pseudoobscura	0.075	1.0	32	Slatkin (1985b)
Platynus tenuicollis	0.145	0.9	11	Liebherr (1988)
Platynus angustatus	0.144	0.5	11	Liebherr (1988)
Speonomus hydrophilus	0.134	0.1	102	Crouau-Roy (1986)

[a] These data exclude the *Laps-c* allele from population 5 as only 3 individuals were sampled.

[b] French populations only.

[c] NA = not available.

(Wright 1951). Estimates of *Nm* calculated from F_{ST} and from private alleles are given in Table 7.6. Although these two methods appear to differ, Barton and Slatkin (1986) showed that both methods are simply different ways of estimating the same essential properties of gene-frequency distributions.

The absolute values of the estimates do differ but, with one exception, both methods give a consistent view of gene flow, both in mobile and sedentary species. Estimates of *Nm* for *Dendroctonus frontalis* from the private allele analysis suggest that gene flow is very common ($Nm > 30$) in this species relative to other pest insects. However, estimates of *Nm* from F_{ST} suggest that gene flow, while common, is less than in other pest species. Discrepancies such as this indicate that both techniques (private allele analysis and F_{ST}) provide only a crude estimate of gene flow. Although it appears to be possible to allocate species into general categories (low, medium, and high gene flow), relative rankings of *Nm* within each category may not be very meaningful.

7. *Comparison of allozyme and ecological data for* Heliothis *spp.*

All the genetic data presented for *Heliothis* species (*armigera*, *punctigera*, *virescens*, and *zea*) indicate that gene flow occurs between populations throughout the appropriate species' ranges in eastern Australia or North America. Thus, Daly and Gregg (1985) concluded that, for the

Table 7.6 Comparisons of *Nm* values from F_{ST} and private allele analysis. Data are from Tables 7.4 and 7.5.

Species	*Nm*	
	F_{ST}	Private alleles
A. elongatulum	83.1	11.7
A. gemmatalis	11.7	28.2
H. armigera	10.6	43.3
P. strobi	2.3	17.0
S. frugiperda	7.6	16.8
D. ponderosae	8.1	14.6
H. virescens	5.0	7.5
D. frontalis	3.4	75.5, 29.5
A. decorum	2.9	2.4
A. extensicolle	1.3	2.0
P. tenuicollis	0.7	0.9
P. angustatus	0.7	0.5

Australia species, populations in different regions formed a common gene pool and the effective size of *Heliothis* populations was large. This was contrary to the notion that *H. armigera* was a relatively more sedentary species than *H. puntigera* (see Farrow and Daly (1987) for review).

These conclusions agree with independent data on gene flow in all species. For example, in Australia, *H. punctigera* is known to disperse long distances (Morton *et al.* 1981; Farrow and Daly 1987) and pest populations in the cotton-growing area of the Namoi Valley, New South Wales, do not overwinter there (G.P. Fitt, pers. comm.). Evidence for long-distance movement in *H. armigera* in Australia is slowly accumulating.

1. Long-distance movement is periodically observed and is associated with synoptic winds (Farrow 1985; Farrow and Daly 1987).
2. Insecticide resistance is widespread even in areas in which insecticides are not used or only used infrequently (e.g. Gunning *et al.* 1984). More recently, Gunning and Easton (1989) have demonstrated that the frequency of pyrethroid resistance is high throughout the unsprayed maize growing areas in northern New South Wales.
3. *H. armigera* has been detected regularly in a light-trap in a forested area of northern New South Wales on nights with northerly winds (Gregg and Daly 1988; Gregg, pers. comm.).

Thus, although there may be qualitative differences in dispersal of both Australian *Heliothis* species, gene flow in both is sufficient to minimize genetic differences between populations.

Summary

Allozyme electrophoresis is a useful tool for studying gene flow between populations. A variety of analytical methods applied to allozyme data give a consistent view of this phenomenon. Species known to be migratory or vagile do, in general, have less regional differentiation than those species that are more sedentary. Gene flow appears to be common in many insect pest species.

However, genetic structure is still only a crude measure of gene flow. Historical events, such as population bottlenecks, or an inadequate sampling strategy may lead to estimates of gene flow that are at variance with the true level. Thus, in the absence of independent observations of the species biology, electrophoretic data only provide hypotheses about gene flow in a species.

Acknowledgements

This work was supported, in part, by funds from the Cotton Research Council of Australia. The manuscript benefited from comments by John Dearn, Rod Mahon, and Barry Richardson. Jenny Fisk gave technical assistance. I wish to thank C.A.B. International, Wallingford, for permission to reproduce parts of Fig. 7.1 and 7.3 and Table 7.2 from *Bulletin of Entomological Research;* and *Genetics* and M. Slatkin for permission to reproduce data from Dr Slatkin's 1981 paper.

References

Anderson, W.W., Berisford, C.W. and Kimmich, R.H. (1979). Genetic differences among five populations of the southern pine beetle. *Ann. ent. Soc. Am*, **72**, 323–7.

Ayala, F.J. (1975). Genetic differentiation during the speciation process. In *Evolutionary biology* (ed. T. Dobzhanzky, M.K. Hecht, and W.C. Steere), Vol. 8, pp 1–78. Plenum Press, New York.

Baker, A.E.M. (1981). Gene flow in house mice: introduction of a new allele into free-living populations. *Evolution*, **35**, 243–58.

Bartlett, A.C., Randall, W.C. and May, J.E. (1983). Allozyme variation among populations of boll weevils in Arizona and Mexico. *Southwest. Ent.*, **8**, 118–29.

Barton, N.H. and Slatkin M. (1986). A quasi-equilibrium theory of the distribution of rare alleles in a subdivided population. *Heredity*, **56**, 409–15.

Berlocher, S.H. (1984). Genetic changes coinciding with the colonization of California by the walnut husk fly, *Rhagoletis completa. Evolution*, **38**, 906–18.

Bryant, E.H., Van Dijk, H., and Van Delden, W. (1981). Genetic variability in the face fly, *Musca autumnalis* De Geer, in relation to a population bottleneck. *Evolution*, **35**, 872–81.

Caccone, A. (1985). Gene flow in cave arthropods: a qualitative and quantitative approach. *Evolution*, **39**, 1223–35.

Comins, H.N. (1977). The development of insecticide resistance in the presence of migration. *J. theor. Biol.*, **64**, 177–97.

Crouau-Roy, B. (1986). Population studies on Pyrenean troglobitic beetles: local genetic differentiation and microgeographic variations in natural populations. *Biochem. Syst. Ecol.*, **14**, 521–6.

Crow, J. F. and Kimura, M. (1970). *An introduction to population genetics theory.* Harper and Row, New York.

Daly, J. C. and Gregg, P. (1985). Genetic variation in *Heliothis* in Australia: species identification and gene flow in the two pest species *H. armigera* (Hübner) and *H. punctigera* Wallengren (Lepidoptera: Noctuidae). *Bull. ent. Res.*, **75**, 169–84.

Daly, J. C. and Murray D. A. H. (1988). Evolution of resistance to synthetic pyrethroids in *Heliothis armigera* (Hübner) (Lepidoptera: Noctuidae) in Australia. *J. econ. Entomol*, **81**, 984–8.

Drake, V. A. and Farrow R. A. (1985). A radar and aerial-trapping study of an early spring migration of moths (Lepidoptera) in inland New South Wales. *Aust. J. Ecol.*, **10**, 223–35.

Eanes. W. F. and Koehn R. K., (1978). An analysis of genetic structure in the monarch butterfly, *Danaus plexippus* L. *Evolution*, **32**, 784–97.

Farris, J. S. (1972). Estimating phylogenetic trees from distance matrices. *Am. Nat.*, **106**, 645–68.

Farrow, R. A. (1979). Population dynamics of the Australian plague locust, *Chortoicetes terminifera* (Walker), in central western New South Wales. I. Reproduction and migration in relation to weather. *Aust. J. Zool.*, **27**, 717–45.

Farrow, R. A. (1985). Detection of transoceanic migration of insects to a remote island in the Coral Sea, Willis Island. *Aust. J. Ecol.*, **9**, 253–72.

Farrow, R. A., and Daly, J. C. (1987). Long-range movements as an adaptive strategy in the genus *Heliothis* (Lepidoptera: Noctuidae): a review of its occurrence and detection in four pest species. *Aust. J. Zool.*, **35**, 1–24.

Felsenstein, J. (1982). Numerical methods for inferring evolutionary trees. *Q. Rev. Biol.*, **57**, 379–404.

Fisk, J. H. and Daly, J. C. (1989). Electrophoresis of *Helicoverpa armigera* (Hübner) and *H. punctigera* Wallengren (Lepidoptera: Noctuidae): genotype expression in eggs and allozyme variations between life stages. *J. Aust. ent. Soc.* (in press).

George, C. (1984). Allozyme variation in natural populations of *Lymantria dispar* (Lepidoptera). *Génét. Sél. Evol.*, **16**, 1–14.

Goldenberg, E. M. (1987). Estimation of gene flow and genetic neighborhood size by indirect methods in a selfing annual, *Triticum dicoccoides*. *Evolution*, **41**, 1326–34.

Gregg, P. and Daly, J. C. (1988). The Australian species of *Heliothis*: identification, genetic variation and migration. *Acta phytopath. ent.*, (in press)

Gunning, R. V. and Easton, C. S. (1989). Pyrethroid resistance in *Heliothis armigera* (Hübner) collected from unsprayed maize crops in New South Wales 1983–1987. *J. Aust. ent. Soc.*, **28**, 57–61.

Gunning, R. V., Easton, C. S., Greenup, L. R., and Edge, V. E. (1984). Pyrethroid resistance in *Heliothis armigera* (Hübner) (Lepidoptera: Noctuidae) in Australia. *J. econ. Entomol.*, **77**, 1283–7.

Hamrick, J.L., Linhart Y.B., and Mitton, J.B. (1979). Relationships between life history characteristics and electrophoretically detectable genetic variation in plants. *A. Rev. Ecol. Syst.*, **10**, 173–200.

Harris, H. and D.A. Hopkinson (1976). *Handbook of enzyme electrophoresis in human genetics.* North-Holland, Oxford.

Harrison, R.G., Wintermeyer S.F., and Odell, T.M. (1983). Patterns of genetic variation within and among gypsy moth, *Lymantria dispar* (Lepidoptera: Lymantriidae), populations. *Ann. ent. Soc. Am.*, **76**, 652–6.

Jacobson, J.W. and Hsiao, T.H. (1983). Isozyme variation between geographic populations of the Colorado potato beetle, *Leptinontarsa decemlineata* (Coleoptera: Chrysomelidae). *Ann. ent. Soc. Am.*, **76**, 162–6.

Jones, J.S., Bryant, S.H., Lewontin, R.C., Moore, J.A., and Prout, T. (1981). Gene flow and the geographical distribution of a molecular polymorphism in *Drosophila pseudoobscura. Genetics*, **98**, 157–78.

Kimura, M. (1983). *The neutral theory of evolution.* Cambridge University Press, Cambridge.

Larson, A., Wake D.B., and Yanev, K.P. (1984). Measuring gene flow among populations having high levels of genetic fragmentation. *Genetics*, **106**, 293–308.

Lewontin, R.C. (1974). *The genetic basis of evolutionary change.* Columbia University Press, New York.

Liebherr, J.K. (1986). Comparison of genetic variation in two carabid beetles (Coleoptera) of differing vagility. *Ann. ent. Soc. Am.*, **79**, 424–33.

Liebherr, J.K. (1988). Gene flow in ground beetles (Coleoptera: Carabidae) of differing habitat preference and flight-wing development. *Evolution* **42**, 129–37.

McCauley, D.E. and Eanes, W.F. (1987). Hierarchical population structure analysis of the milkweed beetle, *Tetraopes tetraophthalmus* (Forster). *Heredity*, **58**, 193–201.

McDonald, I.C., Krysan, J.L., and Johnson, O.A. (1985). Genetic variation within and among geographic populations of *Diabrotica barberi* (Coleoptera: Chrysomelidae). *Ann. ent. Soc. Am.*, **78**, 271–8.

McDonald, P.T., Hilburn, L.R., and Kunz, S.E. (1987). Genetic similarities among natural populations of the horn fly (Diptera: Muscidae). *Ann. ent. Soc. Am.*,**80**, 288–92.

Morton, R., Tuart L.D., and Wardhaugh, K.G. (1981). The analysis and standardisation of light-trap catches of *Heliothis armigera* (Hübner) and *H. punctigera* Wallengren (Lepidoptera, Noctuidae). *Bull. ent. Res.*, **71**, 207–25.

Namkoong, G., Roberds, J.H. Nunnally, L.B., and Thomas, H.A. (1979). Isozyme variations in populations of southern pine beetles. *Forest Sci.*, **25**, 197–203.

Nei, M. (1977). F-statistics and analysis of gene diversity in subdivided populations. *Ann. hum. Genet., Lond.*, **41**, 225–33.

Nei, M. (1978). Estimation of average heterozygosity and genetic distance from a small number of individuals. *Genetics*, **89**, 583–90.

Pashley, D. P. (1985). The use of population genetics in migration studies: a comparison of three noctuid species. In *The movement and dispersal of agriculturally important biotic agents* (ed. D. R. MacKenzie, C. S. Barfield, G. C. Kennedy, and R. D. Berger), pp. 305–24. Louisiana State Univeristy, Baton Rouge.

Pashley, D. P. (1986). Host-associated genetic differentiation in fall armyworm (Lepidoptera: Noctuidae): A sibling species complex? *Ann. ent. Soc. Am.*, **79**, 898–904.

Pashley, D. P. and Bush, G. L. (1979). The use of allozymes in studying insect movement with special reference to the codling moth, *Laspeyresia pomonella* (L.) (Olethreutidae). In *Movement of highly mobile insects: concepts and methodology in research* (ed. R. L. Rabb and G. G. Kennedy), pp. 333–41, North Carolina State University, Raleigh.

Pashley, D. P. and Johnson, S. J. (1986). Genetic population structure of migratory moths: the velvetbean caterpillar (Lepidoptera: Noctuidae). *Ann. ent. Soc. Am.*, **79**, 26–30.

Pashley, D. P., Johnson, S. J., and Sparks, A. N. (1985). Genetic population structure of migratory moths: the fall armyworm (Lepidoptera: Noctuidae). *Ann. ent. Soc. Am.*, **78**, 756–62.

Patton, J. L. and Feder, J. H. (1981). Microspatial genetic heterogeneity in pocket gophers: non-random breeding and drift. *Evolution*, **35**, 912–20.

Payne, C. D. (1985). *The GLIM system release 3.77 manual.* Numerical Algorithms Group, Oxford.

Phillips, T. W. and Lanier, G. N. (1985). Genetic divergence among populations of white pine weevil, *Pissodes strobi* (Coleoptera: Curculionidae). *Ann. ent. Soc. Am.*, **78**, 744–50.

Richardson, B. J., Baverstock P. R., and Adams, M. (1987). *Allozyme Electrophoresis.* Academic Press, Sydney.

Rogers, J. S. (1972). Measures of genetic similarity and genetic distance. *Studies in genetics VII: Univ. Texas Publ. 7213,* pp. 145–53.

Rogers, J. S. (1984). Deriving phylogenetic trees from allele frequencies. *Syst. Zool.*, **33**, 52–63.

Selander, R. K. (1970). Behaviour and genetic variation in natural populations. *Zoologist*, **10**, 53–66.

Selander, R. K. and Kaufman, D. W. (1975). Genetic structure of populations of the brown snail (*Helix aspersa*). I. Microgeographic variation. *Evolution*, **29**, 385–401.

Sell, D. K., Whitt, G. S., and Luckman, W. H. (1975). Esterase polymorphism in the corn earworm, *Heliothis zea* (Boddie): a survey of temporal and spatial allelic variation in natural populations. *Biochem. Genet.*, **13**, 885–98.

Sell, D. K., Whitt, G. S., Metcalf, R. L., and Lee, L. K. (1974). Enzyme polymorphism in the corn earworm, *Heliothis zea* (Lepidoptera:

Noctuidae). Hemolymph esterase polymorphism. *Can. Ent.*, **106**, 701-9.

Slatkin, M. (1977). Gene flow and genetic drift in a species subject to frequent local extinctions. *Theor. Pop. Biol.*, **12**, 253-62.

Slatkin, M. (1981). Estimating levels of gene flow in natural populations. *Genetics*, **99**, 323-35.

Slatkin, M. (1985*a*). Gene flow in natural populations. *A. Rev. Ecol. Syst.*, **16**, 393-430.

Slatkin, M. (1985*b*). Rare alleles as indicators of gene flow. *Evolution*, **39**, 53-65.

Sluss, T.P. and Graham, H.M. (1979). Allozyme variation in natural populations of *Heliothis virescens*. *Ann ent. Soc. Am.*, **72**, 317-22.

Smith, C.A.B. (1970). A note on testing the Hardy-Weinberg law. *Ann. hum. Genet.*, **33**, 377-83.

Sofield, R.K., Buroker, N.E., Hansens E.J., and Vrijenhoek, R.C. (1984). Genetic diversity within and between sibling-species of salt-marsh horseflies (Diptera: Tabanidae). *Ann. ent. Soc. Am.*, **77**, 663-8.

Sokal, R.R. and Rohlf, F.J. (1969). *Biometry*. Freeman, San Francisco.

Soltis, P.S. and Soltis D.E. (1987). Population structure and estimates of gene flow in the homosporous fern *Polystichum munitum*. *Evolution*, **41**, 620-9.

Southwood, T.R.E. (1971). The role and measurement of migration in the population system of an insect pest. *Trop. Sci.*, **13**, 275-8.

Stock, M.W., Amman G.D. and Higby P.K. (1984). Genetic variation among mountain pine beetle (*Dendroctonus ponderosae*) (Coleoptera: Scolytidae) populations from seven western states. *Ann. ent. Soc. Am.*, **77**, 760-4.

Urbanelli, S., Bullini L., and Villani F. (1985). Electrophoretic studies on *Culex quinquefasciatus* Say from Africa: genetic variability and divergence from *Culex pipiens* L. (Diptera: Culicidae). *Bull. ent. Res.*, **75**, 291-304.

Waples, R.S. (1987). A multispecies approach to the analysis of gene flow in marine shore fishes. *Evolution*, **41**, 385-400.

Wardhaugh, K.G., Room P.M. and Greenup L.R. (1980). The incidence of *Heliothis armigera* (Hübner) and *H. punctigera* Wallengren (Lepidoptera: Noctuidae) on cotton and other host plants in the Namoi Valley of New South Wales. *Bull. ent. Res.*, **70**, 113-31.

Weir, B.S. and Cockerham C.C. (1984). Estimating F-statistics for the analysis of population structure. *Evolution*, **38**, 1358-70.

Willhite, E.A. and Stock M.W. (1983). Genetic variation among western spruce budworm (*Choristoneura occidentalis*) (Lepidoptera: Tortricidae) outbreaks in Idaho and Montana. *Can. Ent.*, **115**, 41-54.

Wright, S. (1931). Evolution in Mendelian populations. *Genetics*, **16**, 97-159.

Wright, S. (1951). The genetical structure of populations. *Ann. Eugen.*, **15**, 323-54.

Wright, S. (1978). *Evolution and the genetics of populations*, Vol. 4, *Variability within and among natural populations*. University of Chicago, Chicago.

8 Estimation of genetic variability amongst Coleoptera

T.H. HSIAO

Department of Biology, Utah State University, Logan, Utah, USA

Abstract

The Coleoptera, which is the largest insect order, comprises an estimated 300 000 species. Many phytophagous species of this order are major pests of agricultural crops. Their economic importance has stimulated a variety of physiological and biochemical studies, performed in order to attempt to improve pest-management strategies. Gel electrophoresis of proteins and enzymes has been used to study Coleoptera since the early 1970s. Of some 50 articles concerning the subject, over 70 per cent have appeared since 1980. A total of 43 enzymes has been examined by gel electrophoresis in 64 beetle species belonging to nine families. Major pests species include the cotton boll weevil, corn rootworms, bark beetles, flour beetles, the Colorado potato beetle, and the alfalfa weevil, while beneficial species include ground beetles and ladybird beetles. Genetic variability has now been estimated for 30 species. A mean heterozygosity ($\bar{H}$) of 0.168 is among the highest recorded for insects. Genetic distances (Nei's and Rogers') between beetle species also appear larger than for other insects. Allozyme data have clarified the often confusing taxonomic status of several agricultural and forest pests. Gel electrophoresis has been used successfully in the study of genetic inheritance, species identification, physiological function of isoenzymes, and reproductive behaviour. The technique is presently under-utilized in Coleoptera research. Advantages and prospects of allozyme electrophoresis in biosystematic studies are discussed.

Electrophoretic Studies on Agricultural Pests (ed. Hugh D. Loxdale and J. den Hollander), Systematics Association Special Volume No. 39, pp.143–80. Clarendon Press, Oxford, 1989.

Introduction

The Coleoptera (beetles) comprise nearly one-third of all species in the animal kingdom. This insect order contains an estimated 300 000 species and members are characterized by a rigid exoskeleton with a pair of hard elytra, and a complete metamorphosis. Beetles feed on diverse organic matter as specialists or generalists and are found in all terrestrial and many aquatic habitats. The abundant beetle species are a rich resource for several types of scientific inquiry. Taxonomy has traditionally been an integral part of many investigations and the rich species diversity of the Coleoptera, especially in the New World tropics, will continue to occupy taxonomists and systematists. Most biological studies of coleopteran species are motivated by their importance as pests of food crops or fibres, or as beneficial organisms. The ecological roles of phytophagous and predaceous beetles have received increasing attention. Many peculiar adaptations of coleopteran species are facinating topics of biological research (Crowson 1981; Jolivet *et al.* 1988).

Because of their long life cycle, and difficulties in rearing and handling, beetles are generally not amenable to genetic studies. Chromosome numbers are known for about 2500 species (Smith and Vikki 1978; Petitpierre 1988). Most formal genetic analyses are restricted to a few stored-product species, such as *Tribolium* (Sokoloff 1977). The development of gel electrophoresis since the early 1970s provided the means for assessing genetic diversity, helped to define intra- and interspecific genetic variability, and resolved taxonomic problems of selected insect taxa, including Coleoptera. Several recent reviews have discussed the concept and uses of biochemical methodologies in insect systematics (Bush and Kitto 1978; Ayala 1978; Berlocher 1984; Buth 1984) and agricultural entomology (Diehl and Bush 1984; Menken and Ulenberg 1987). This discussion surveys the use of allozyme electrophoresis in Coleoptera.

Survey of gel electrophoretic application in Coleopteran studies

Gel electrophoresis of enzymes has been available since the mid-1960s. However, the technique was first applied to Coleoptera by Avise and Selander (1972), who examined three enzymes in two populations of a troglobitic carabid beetle, *Rhadine subterranea*, during a study of cave-dwelling fishes. Prior to 1980, 15 papers dealt with the subject (Table 8.1). Since 1980, an additional 45 papers have

Table 8.1 Survey of beetle species for which gel electrophoresis of enzymes has been conducted.

Taxa	Gel used[a]	No. enzymes	No. loci[b]	Areas of study[c]	enzymes used[d]	Reference
Anobiidae						
Lasioderma serricorne (F.)	A	4	4	Ia	EST,GP	Frankel and Duffield (1984)
Carabidae						
Abax ater Villa	A	1	5	Ia, Ib	EST	Schneider *et al.* (1984)
A. parallelus Dft.	A	1	5	Ia, Ib	EST	Schneider *et al.* (1984)
Agonum decorum (Say)	S	15	17	Ia	ACON,ACP, ADA,ALDH,EST, ENOL,G6PD, GAPD,MDH,IDH, MPI,PGI,PGM, PEP,SOD	Liebherr (1983)
	S	15	17	Ia, Ib	Same as above	Liebherr (1986*a*)
A. elongatulum (Dej.)	S	15	17	Ia, Ib	Same as above	Liebher (1986*a*)
A. extensicolle (Say)	S	15	17	Ia, Ib	Same as above	Liebherr (1986*a*)
A. extimum Liebherr	S	15	17	Ia, Ib	Same as above	Liebherr (1986*a*)
A. taxanum Liebherr	S	15	17	Ia, Ib	Same as above	Liebherr (1986*a*)
Carabus convexus F.	A	1	5	Ia, Ib	EST	Schneider *et al.* (1984)
C. nemoralis Muell.	A	1	5	Ia, Ib	EST	Schneider *et al.* (1984)

Table 8.1 *Continued*

Taxa	Gel used[a]	No. enzymes	No. loci[b]	Areas of study[c]	enzymes used[d]	Reference
C. violaceus Muell.	A	1	5	Ia, Ib	EST	Schneider *et al.* (1984)
Harpalus rufipes Dej.	A	1	5	Ia, Ib	EST	Schneider *et al.* (1984)
Neaphaenops tellkampfii	A	4	6	Ia	AP,EST,MDH, GP	Giuseffi *et al.* (1978)
	A	6	13	Ia	AP,EST,MDH, PGI,GP,XDH	Turanchik and Kane (1979)
	A	6	9	Ia	AP,EST,MDH, PGM,PGI,GP	Kane and Brunner (1986)
Platynus angustatus (Dej.)	S	20	22	Ia, Ib	ACP,GAM, HBDH,GAPD, IDH,GDA,MDH, ME,MPI,PGI, PGDH,SOD, PEP,PGK,GP, TPI	Liebherr (1986*b*)
P. tenuicollis (LeConte)	S	20	22	Ia, Ib	Same as above	Liebherr (1986*b*)
Pterostichus niger Schall.	A	1	5	Ia, Ib	EST	Schneider *et al.* (1984)
P. oblongopunctatus F.	A	1	5	Ia, Ib	EST	Schneider *et al.* (1984)
Rhadine subterranea	S	3	3	Ia	MDH,PGI,PGM	Avise and Selander (1972)

Catopidae						
Leptodirus hohenwarti Schmidt	S	9	14	Ia	ACP,AP,EST, HK,LAP,NDH, PGI,GP,SOD	Sbordoni *et al.* (1980)
Orostygia doderoi bucciarellii Tamanini	S	9	12	Ia, Ib	Same as above	Sbordoni *et al.* (1980)
O. meggiolaroi Agazzi	S	9	12	Ia, Ib	Same as above	Sbordoni *et al.* (1980)
Ptomaphagus hirtus (Tellkampf)	A	12	13	Ia	ACP,AP,EST, CA,HBDH, SOD,HK,MDH, ME,ODH,PGI, PGM	Laing *et al.* (1976)
Speonomus hydrophilus	S,A	12	17	Ia	AP,EST,HK, LAP,ME,ACP, FUM,G6PD, HBDH,MDH, PGI	Crouau-Roy (1986)
Cerambycidae						
Tetraopes tetraophthalmus (Foster)	S	10	10	Ia	HK,G6PD,GOT, IDH,LAP,LDH, ME,PEP,PGI, PGM	McCauley and Eanes (1987)

Table 8.1 *Continued*

Taxa	Gel used[a]	No. enzymes	No. loci[b]	Areas of study[c]	enzymes used[d]	Reference
Chrysomelidae						
Bromius obscurus (L.)	S	11	16	Ia	ACP,AK,AMY, EST,HK,IDH, LAP,MDH,ME, SOD,TPI	Lokki *et al.* (1976)
Chrysolina aurichalcea	S	7	11	Ia	AMY,CA,G6PD, GOT,LAP,ME, GP	Sakanoue and Fujiyama (1987)
Diabrotica barberi Smith and Lawrence	S	20	>30	Ic	ACON,ACP, ADH,AK,EST, FUM,HK,GOT, G6PD,α-GPD, IDH,LDH, MDH,ME,MPI, PGI,PGM,PEP, GP,SOD	Piedrahita *et al.* (1985)
	A	4	10	II	EST,HK,MDH, XDH	McDonald *et al.* (1982)
	A	10	10	Ia	AMY,EST, HBDH,HK, GOT,IDH, MDH,PGM, SOD,XDH	McDonald *et al.* (1985)
D. longicornis (Say)	A	4	10	II	EST,HK,MDH, XDH	McDonald *et al.* (1982)

D. virgifera virgifera (LeConte)	S	20	>30	Ic	ACON,ACP, ADH,AK,EST, FUM,HK,GOT, G6PD,α-GPD, IDH,LDH, MDH,ME,MPI, PGI,PGM,PEP, GP,SOD	Piedrahita *et al.* (1985)
Leptinotarsa decemlineata (Say)	S	10	11	Ia, Ib	ACP,EST,FUM, GOT,G6PD, α-GPD,IDH, MDH,ME,SOD	Jacobson and Hsiao (1983)
Leptinotarsa haldemani (Rogers)	S	10	11	Ib	Same as above	Jacobson and Hsiao (1983)
Plagioderma versicolora Baly	S	1	1	IV	PGM	McCauley and O'Donnell (1984)
Coccinellidae						
Anatis ocellata L.	A	1	5	Ia, Ib	EST	Schneider *et al.* (1984)
Coccinella septempunctata L.	A	1	5	Ia, Ib	EST	Schneider *et al.* (1984)
Exochomus quadripustulatus L.	A	1	5	Ia, Ib	EST	Schneider *et al.* (1984)
Harmonia quadripunctata Pont.	A	1	5	Ia, Ib	EST	Schneider *et al.* (1984)
Henosepilachna vigintioctomaculata complex	A	2	5	Ia	ACP,EST	Kuboki (1978)
Myrrha octodecimguttata L.	A	1	5	Ia, Ib	EST	Schneider *et al.* (1984)

Table 8.1 *Continued*

Taxa	Gel used[a]	No. enzymes	No. loci[b]	Areas of study[c]	enzymes used[d]	Reference
Propylaea quatuordecimpunctata L.	A	1	5	Ia, Ib	EST	Schneider *et al.* (1984)
Thea vigintiduopunctata L.	A	1	5	Ia, Ib	EST	Schneider *et al.* (1984)
Curculionidae						
Anthonomus grandis Boh.	S	9	9	Ia	ACP,ADH,EST, HK,LAP,PGI, PGM,SOD, SORDH	Bartlett (1981)
	S	9	9	Ia	Same as above	Bartlett *et al.* (1983)
	A	1	5	II, III	EST	Bancroft and Jones (1977); Biggers and Jones, (1976); Jones and Bancroft (1986)
	A	24	35	Ia	ACP,ADH, AP,AK,AMY,AO, EST,FUM,HK, HEXD,GOT, G6PD,α-GPD, IDH,LDH,LAP, MDH,ME, ODH,PGI,PGM, SOD,TYR,XDH	Terranova (1981)

	A	3	4	II	AO,GOT,PGM	Terranova (1980)
	A	1	5	II	EST	Terranova (1982)
	A	2	3	II	ADH,ODH	Terranova and North (1984*a*)
	A	2	2	II	G6PD,PGI	Terranova and North (1984*b*)
	A	1	1	II	IDH	Terranova and North (1984*c*)
	A	6	6	II	AAT,AMY,FUM, LAP,ME,SOD	Terranova and North (1985)
Cleonus piger Scop.	A	1	5	Ia, Ib	EST	Schneider *et al.* (1984)
Conotrachelus nenuphar (Herbst)	S	4	4	IV	ADH,AK,IDH, PGM	Huettel *et al.* (1976)
Hylobius abictis L.	A	1	5	Ia, Ib	EST	Schneider *et al.* (1984)
Hypera postica (Gyll.)	S	4	5	Ia	AAT,EST, α-GPD,MDH	Sell *et al.* (1978)
	A	12	22	Ia	ACP,ADH,AMY, AO,EST,GOT, G6PD,MDH, ME,SOD,TYR, XDH	Hsiao and Stutz (1985)
	A	3	3	III	AO,G6PD,LDH	Romero-Severson *et al* (1986)
Otiorrhynchus scaber	S	16	23	Ia	ACP,AK,AMY, AO,EST, α-GPD,HK, IDH,LAP,MDH, ME,PGDH, PGM,SOD,TPI, XDH	Suomalainen and Saura (1973)

Table 8.1 *Continued*

Taxa	Gel used[a]	No. enzymes	No. loci[b]	Areas of study[c]	enzymes used[d]	Reference
Strophosomus capitatum Deg.	A	1	5	Ia, Ib	EST	Schneider *et al.* (1984)
Strophosomus capitatus	S	12	17	Ia, Ib	ACP,AK,AMY, EST,α-GPD, HK,IDH,LAP, MDH,PGM, SOD,TPI	Suomalainen and Saura (1973)
Sitophilus granarius (L.)	S	6	9	Ia, Ib	ADH,AMY,EST, α-GPD,LAP, MDM	Beiras and Petitpierre (1981)
	A	1	2	Ia	AMY	Baker (1987)
S. oryzae (L.)	S	6	9	Ia, Ib	ADH,AMY,EST, α-GPD,LAP, MDH	Beiras and Petitpierre (1981)
	A	1	2	Ia	AMY	Baker (1987)
S. zeamais Motsch.	S	6	9	Ia, Ib	ADH,AMY,EST, α-GPD,LAP, MDH	Beiras and Petitpierre (1981)
	A	1	2	Ia	AMY	Baker (1987)
Pissodes strobi (Peck)	S	10	11	Ia	AAT,ADH,AK, EST,GAPD, IDH,MDH, PGI,SOD	Phillips and Lanier (1985)

Scolytidae						
Dendroctonus brevicomis Leconte	S	5	5	Ia, Ib	ACP,EST,MDH, PGI,PGM	Namkoong *et al.* (1979)
D. frontalis Zimmerman	S	5	5	Ia, Ib	Same as above	Namkoong *et al.* (1979)
	S	6	6	Ia	HK,GOT,MDH, PGI,PGM,SOD	Anderson *et al.* (1979)
D. jeffreyi Hopkins	S	11	17	Ia, Ib	AAT,ACP,CK, EST,α-GPD, IDH,LAP, MDH,PGI,PEP, SOD	Higby and Stock (1982)
D. ponderosae Hopkins	S	13	18	Ia	AAT,ACP,CK, EST,α-GPD, IDH,LDH, MDH,ME,PGI, MPI,PEP,SOD	Stock and Guenther (1979)
	S	8	15	Ia	AAT,EST, α-GDP,LDH, MDH,PGM,PEP, SOD	Stock and Amman (1980)
	S	11	17	Ia, Ib	AAT,ACP,CK, EST,α-GPD, IDH,LAP, MDH,PGI, PEP,SOD	Higby and Stock (1982)
	S	4	5	Ia	EST,GOT,MPI, PEP	Sturgeon and Mitton (1986)
	S	11	18	Ia	AAT,ACP,CK, EST,α-GPD, IDH,LAP, MDH,PGI,PEP, SOD	Stock *et al.* (1984)

Table 8.1 *Continued*

Taxa	Gel used[a]	No. enzymes	No. loci[b]	Areas of study[c]	enzymes used[d]	Reference
D. pseudotsugae Hopkins	S	12	13	Ia	AAT,ACP,EST, α-GPD,IDH, LAP,MDH,ME, PEP,PER,PGM, SOD	Stock *et al.* (1979)
D. terebrans (Olivier)	S	8	12	Ia	ADH,EST,GOT, IDH,LDH,MDH, ME,PGM	Anderson *et al.* (1983)
Ips calligraphus Germar	S	5	8	Ia	ADH,GOT,IDH, MDH,PGM	Anderson *et al.* (1983)
Pityogenes chalcographus L.	A	6	NA	Ia	ACP,AMY,EST, GOT,LAP,PER	Ritzengruber and Führer (1986)
Tenebrionidae						
Tribolium brevicornis	A	5	NA	Ia, Ib	ACP,AMY,AO, EST,MDH	Wool (1982)
T. castaneum (Herbst)	A	1	4	III	EST	Cohen *et al.* (1977)
	A	5	NA	Ia, Ib	ACP,AMY,AO, EST,MDH	Wool (1982)
	A	1	1	III	AMY	Bremner *et al.* (1983)
	A	5	NA	I	ADH,G6PD, IDH,LDH, PGDH	Yeh and Scheinberg (1974)
	S	2	4	II	HK,ME	Samollow *et al.* (1983)

	S	1	1	II	PGM	Riddle *et al.* (1983)
	S	1	1	II	G6PD	Dawson and Hollingsworth (1982)
T. confusum J. & V.	A	1	4	III	AMY	Bremner *et al.* (1983)
	A	5	NA	Ia, Ib	ACP,AMY,AO, EST,MDH	Wool (1982)
	S	2	4	II	HK,ME	Samollow *et al.* (1983)
	S	2	2	II	HK,ME	Dawson and Jost (1983)
T. destructor	A	1	2	III	AMY	Bremner *et al.* (1983)
	A	5	NA	Ia, Ib	ACP,AMY,AO, EST,MDH	Wool (1982)

[a] A = Acrylamide; S = Starch.
[b] NA = not available.
[c] Areas of study: I, Biochemical genetics; Ia, Intraspecific variation; Ib, Interspecific variation; Ic, species identification; II, Allozyme inheritance; III, Developmental and physiological function; IV, Mating and reproduction.
[d] See Table 8.2 for detail of abbreviation.

appeared. The starch gel system was used initially. The polyacrylamide gel system was first adopted about 15 years ago (Yeh and Scheinberg 1974). The literature indicates that starch and acrylamide gel systems receive equal use. The advantages and disadvantages of the two systems were recently compared by Menken and Ulenberg (1987). The resolving power of each system varies with the enzyme, the buffer system, and the species under investigation. Menken and Ulenberg (1987) advocate the use of the starch gel system. However, the polyacrylamide gel system has been used as frequently in coleopteran studies (Table 8.1). With one exception, all investigators appear to chose a single gel system for their electrophoretic analyses. This is probably a matter of convenience, since other insect studies have shown that, at many loci, there is generally good agreement between the two systems.

A total of 43 enzymes have been examined by gel electrophoresis in 64 beetle species belonging to 9 families (Tables 8.1 and 8.2). Several non-enzymatic proteins were included in these analyses. The number of enzymes used varied greatly with the type of the study. Of the 98 entries in Table 8.1, only 26 studies involved more than 10 enzymes. A single enzyme was used in 31 entries and the remaining 41 entries examined 2–9 enzymes. The number of putative gene loci surveyed also varied. Only 22 entries examined more than 15 loci. The majority of the entries (85 per cent) examined intra- and interspecific genetic variation; the remaining 15 per cent concerned allozyme inheritance, developmental and physiological function, and mating behaviour.

The enzymes selected for coleopteran studies are typical of those used in insect studies in general. Of the 43 enzyme systems selected, 15 were used in more than 10 studies, while the remaining 28 enzymes have been used in 1–10 studies (Table 8.2). The 15 enzymes selected most often, arranged in a descending order of use, include EST, MDH, ACP, SOD, PGI, IDH, PGM, ME, HK, LAP, G6PD, GOT, AMY, α-GPD, and ADH. The number of scorable loci per enzyme in coleopteran studies varies (Table 8.2). Twenty-one enzymes were scored only for one locus, whilst 22 enzymes were scored for more than one locus. Esterase was the enzyme most often selected and had the greatest number of scorable loci (~3.0). It is a variable substrate enzyme displaying more than five loci in several studies. In general, the number of scorable loci may be influenced by the electrophoretic procedures. Nevertheless, data on scorable loci of each enzyme system under specific conditions may be helpful in selecting enzymes suitable for beetle studies, as can data from other insect studies. Several investigators have advocated the use of at least

Table 8.2 Specific enzymes examined by gel electrophoresis in beetle studies.

Abbreviation[a]	Enzyme (EC no.)	No. studies used	Mean no. scorable loci
AAT	Aspartate aminotransferase (2.6.1.1)	8	1.22
ACON	Aconitase (4.2.1.3)	3	1.75
ACP	Acid phosphatase (3.1.3.2)	22	1.29
ADA	Adenosine deaminase (3.5.4.4)	2	1.00
ADH	Alcohol dehydrogenase (1.1.1.1)	11	1.30
AK	Adenylate kinase (2.7.4.3)	6	1.25
ALDH	Aldehyde dehydrogenase (1.2.1.5)	2	1.00
AP	Alkaline phosphotase (3.1.3.1)	7	1.11
AMY	Amylase (3.2.1.1)	12	1.27
AO	Aldehyde oxidase (1.2.3.1)	6	1.10
CA	Carbonic anhydrase (4.2.1.1)	2	1.00
CK	Creatine kinase (2.7.3.2)	3	1.00
ENOL	Enolase (4.2.1.11)	2	1.00
EST	Esterase (3.1.1.1)	38	3.04
FUM	Fumerase (4.2.1.2)	5	1.00
GAM	Galactosaminidase	1	1.00
GAPD	Glyceraldehyde-3-phosphate dehydrogenase (1.2.1.12)	4	1.00
GDA	Guanine deaminase (3.5.4.3)	1	1.00
GOT	Glutamate-oxaloacetate transaminase (2.6.1.1)	12	1.43
α-GPD	α-Glycerophosphate dehydrogenase (1.1.1.8)	11	1.53
G6PD	Glucose-6-phosphate dehydrogenase (1.1.1.49)	13	1.07
GP	General proteins (nonenzymatic)	8	1.00
HBDH	β-hydroxybutyrate dehydrogenase (1.1.1.30)	4	1.40
HEXD	Hexanol dehydrogenase	1	1.00
HK	Hexokinase (2.7.1.1)	15	1.47
IDH	Isocitrate dehydrogenas (1.1.1.42)	21	1.15
LAP	Leucine aminopeptidase (3.4.11.1)	15	1.58
LDH	Lactate dehydrogenase (1.1.1.27)	6	1.14
MDH	Malate dehydrogenase (1.1.1.37)	29	1.43
ME	Malic enzyme (1.1.1.40)	16	1.14
MPI	Mannose phosphate isomerase (5.3.1.8)	6	1.00
NDH	Nothing dehydrogenase	1	1.00
ODH	Octanol dehydrogenase (1.1.1.73)	3	1.00
PEP	Peptidases (3.4.11–13)	10	1.17

Table 8.2 *Continued*

Abbreviation[a]	Enzyme (EC no.)	No. studies used	Mean no. scorable loci
PER	Peroxidase (1.11.1.7)	2	1.00
PGDH	Phosphogluconate dehydrogenase (1.1.1.43)	3	1.00
PGI	Phosphoglucose isomerase (5.3.1.9)	21	1.03
PGK	Phosphoglycerate Kinase (2.7.2.3)	1	1.00
PGM	Phosphoglucomutase (2.7.5.1)	20	1.03
SOD	Superoxide dimutase/Tetrazolium oxidase (1.15.1.1)	21	1.00
SORDH	Sorbic acid dehydrogenase (1.1.1.14)	2	1.00
TPI	Triosephosphate isomerase (5.3.1.1)	3	1.00
TYR	Tyrosinase (1.14.18.1)	2	2.00
XDH	Xanthine dehydrogenase (1.2.3.2)	6	1.00

[a] Abbreviations follow the usage of Richardson *et al.* (1986).

20 putative loci for a meaningful allozyme analysis (Ayala 1978; Menken and Ulenberg 1987). Such a criterion can easily be met by selecting the 15 most frequently used enzymes listed above (Table 8.2).

Studies on intraspecific variation

Gel electrophoresis is valuable in the study of genetic variability at the population level as well as the species level. Amongst Coleoptera, genetic diversity has been analysed in 30 species belonging to five families (Table 8.3). Whilst the majority of studies involved agricultural pests or beneficial species, several studies have examined the influence of geographic isolation on genetic diversity of cave-dwelling species. The number of loci examined in these studies ranged from 6 to 35. Of the 37 entries (representing 30 species) included in Table 8.3, only 5 examined more than 20 loci. However, 17 entries (14 species) examined at least 15 loci. There was a great deal of variation between species and, in several instances, even within species surveyed by different investigators. For instance, there were rather wide ranges (over all loci examined) in the mean heterozygosity ($\overline{H}$) and polymorphic loci (P) reported in three separate studies for the cave carabid, *Neaphaenops tellkampfii*, and the cotton boll weevil,

Table 8.3 Mean heterozygosity ($\overline{H}$) and polymorphic loci (P) of coleopteran species.

Species	No. loci	$\overline{H}$	P	Reference
Carabidae				
Agonum decorum (Say)	17	0.204	0.710	Liebherr (1983)
	17	0.220	0.563	Liebherr (1986*a*)
A. elongatulum (Dej.)	17	0.233	0.602	Liebherr (1986*a*)
A. extensicolle (Say)	17	0.232	0.462	Liebherr (1986*a*)
A. extimum Liebherr	17	0.145	0.470	Liebherr (1986*a*)
Neaphaenops tellkampfii	6	0.121	0.360	Giuseffi *et al.* (1978)
	13	0.154	0.470	Turanchik and Kane, (1979)
	9	0.094	0.302	Kane and Brunner (1986)
Platynus angustatus (Dej.)	22	0.162	0.710	Liebherr (1986*b*)
P. tenuicollis (LeConte)	22	0.195	0.620	Liebherr (1986*b*)
Catopidae				
Leptodirus hohenwarti Schmidt	14	0.195	0.571	Sbordoni *et al.* (1980)
Orostygia doderoi bucciarellii Tamanini	12	0.185	0.667	Sbordoni *et al.* (1980)
O. meggiolaroi Agazzi	12	0.133	0.583	Sbordoni *et al.* (1980)
Ptomaphagus hirtus (Tellkampf)	13	0.048	0.154	Laing *et al.* (1976)
Speonomus hydrophilus	17	0.049	0.410	Crouau-Roy (1986)
Chrysomelidae				
Bromius obscurus (L.)	16	0.180	0.563[a]	Lokki *et al.* (1976)
Chrysolina aurichalcea	11	0.139	0.439	Sakanoue and Fujiyama (1987)
Diabrotica barberi Smith	10	0.081	0.710[a]	McDonald *et al.* (1985)
Leptinotarsa decemlineata (Say)	11	0.206	0.515	Jacobson and Hsiao (1983)
Leptinotarsa haldemani (Rogers)	11	0.164	0.364	Jacobson and Hsiao (1983)
Curculionidae				
Anthonomus grandis Boh.	9	0.383	0.910[a]	Bartlett (1981)
	9	0.357	0.870[a]	Bartlett *et al.* (1983)
	35	0.236	0.600	Terranova (1981)

Table 8.3 *Continued*

Species	No. loci	$\overline{H}$	P	Reference
Hypera postica (Gyll.)	22	0.231	0.536	Hsiao and Stutz (1985)
Otiorrhynchus scaber	23	0.309	0.830	Suomalainen and Saura (1973)
Strophosomus capitatus	17	0.157	0.410	Suomalainen and Saura (1973)
Sitophilus granarius (L.)	9	0.029	0.100	Beiras and Petitpierre (1981)
S. oryzae (L.)	9	0.061	0.140	Beiras and Petitpierre (1981)
S. zeamais Motsch.	9	0.067	0.106	Beiras and Petitpierre (1981)
Pissodes strobi (Peck)	11	0.172	0.420	Phillips and Lanier (1985)
Scolytidae				
Dendroctonus jeffreyi Hopkins	17	0.131	0.350[a]	Higby and Stock (1982)
D. ponderosae Hopkins	15	0.144	0.300[a]	Stock and Amman (1980)
	17	0.143	0.350[a]	Higby and Stock (1982)
	18	0.120	0.330[a]	Stock *et al.* (1984)
D. pseudotsugae Hopkins	13	0.288[a]	0.540[a]	Stock *et al.* (1979)
D. terebrans (Olivier)	12	0.123	NA[b]	Anderson *et al.* (1983)
Ips calligraphus Germar	8	0.114	NA[b]	Anderson *et al.* (1983)
Mean ± SE		(37) 0.168 ± 0.013	(35) 0.487 ± 0.034	

[a] Calculated from published result.
[b] NA = not available.

Anthonomus grandis. On the other hand, there were only minor differences in the $\overline{H}$ and P values reported in separate studies of *Agonum decorus* and *Dendroctonus ponderosae* (Table 8.3). The latter studies were conducted by the same researchers using the same loci. Two cave-dwelling beetles, *Ptomaphagus hirtus* and *Speonomus hydrophilus*, and three granary weevils, *Sitophilus granarius*, *S. oryzae*, and *S. zeamais* had the lowest $\overline{H}$ values. These data seem to support the hypothesis that habitat specialists, such as the species cited above, have lower $\overline{H}$

values than habitat generalists because of inbreeding and/or founder effects (Beiras and Petitpierre 1981). However, three other cave-dwelling catopids, *Leptodirus hohenwarti*, *Orostygia doderoi bucciaqrellii*, and *O. meggiolaroi*, and the cave carabid, *Neaphaenops tellkampfii*, did not have low $\bar{H}$ values (Table 8.3). The cotton boll weevil, *Anthonomus grandis*, had the highest $\bar{H}$ value, which ranged from 0.236 to 0.383. The lowest value (0.236) was based on 35 loci (the largest number of loci examined in a study of any coleopteran species) and from a laboratory strain (Terranova 1981). The $\bar{H}$ value for Coleoptera is the highest of the insect orders studied. Using all 37 entries of Table 8.3, $\bar{H} \pm$ SE is 0.168 ± 0.013. For the 14 species for which at least 15 loci have been examined, $\bar{H}$ is 0.187 ± 0.017. This value is the highest among six insect orders compiled by Graur (1985), who reported a $\bar{H}$ value of 0.107 ± 0.005 for 188 insect species. The three largest insect orders, Coleoptera, Diptera, and Lepidoptera, appeared to exhibit the most genetic variation (Graur 1985).

The proportion of polymorphic loci is found to vary considerably among coleopteran species (Table 8.3). *P* values range from 0.100 to 0.910 with a mean of 0.487 ± 0.034 (SE) for 35 entries. This mean value for Coleoptera is similar to that reported for *Drosophila* (0.480 for 39 species), but is larger than the mean for other insects (0.370 for 179 species) (Wood and Guttmann, 1987).

Nei's genetic identity (*I*) and distance (*D*) and Wright's fixation index (F_{ST}) are used to assess genetic differentiation between and within populations. Mean values (i.e. $\bar{I}$ and $\bar{D}$) for these indices are now available for 28 beetle species belonging to six families (Table 8.4). $\bar{I}$ values for beetles range from 0.810 to 0.998 (mean of 0.935 ± 0.013), whilst $\bar{D}$ values, which reflect genetic dissimilarity, range from 0.002 to 0.245 (mean of 0.069 ± 0.015). The difference represented by these $\bar{I}$ and $\bar{D}$ values is larger than those for geographic populations of *Drosophila* (Ayala 1978). Perhaps, the intraspecific relationships at the subspecies or semispecies levels are not as well-defined among Coleoptera as in *Drosophila* and other insects. Among beetle species whose taxonomy is well-defined, $\bar{I}$ and $\bar{D}$ values are comparable to those for other insects. The fixation index (F_{ST} measures genetic differentiation between populations. A value above 0.05 indicates substantial differentiation among populations (Wright 1978). Data compiled from 15 beetle species shows that F_{ST} ranges between 0.011 and 0.270 with a mean of 0.103 ± 0.020 (Table 8.4). These values are not greatly different from those reported from other insects (McCauley 1987; Sweeney *et al.* 1987).

The following three examples are presented to illustrate how allozyme data can help clarify intraspecific relationships of beetles.

Table 8.4 Mean values of Nei's genetic identity (I), genetic distance (D), and fixation index (F_{ST}) between populations of beetle species.

Species	No. populations	Mean I	Mean D	Mean F_{ST}	Reference
Carabidae					
Agonum decorum (Say)	10	0.917[a]	0.087	0.080	Liebherr (1983)
	17	0.970	0.030[a]	0.080	Liebherr (1986*a*, 1988)
A. elongatulum (Dej.)	5	0.998	0.002[a]	0.030	Liebherr (1986*a*, 1988)
A. extensicolle (Say)	8	0.960	0.041[a]	0.160	Liebherr (1986*a*, 1988)
Neaphaenops tellkampfii	5	0.972	0.029	NA[b]	Guiseffi (*et al.*, 1978)
	8	0.980	0.020	NA[b]	Turanchik and Kane (1979)
Platymus angustatus (Dej.)	5	0.840	0.180[a]	0.270	Liebherr (1986*b*, 1988)
P. tenuicollis (LeConte)	5	0.890	0.121	0.260	Liebherr (1986*b*, 1988)
Catopidae					
Ptomaphagus hirtus (Tellkampf)	6	0.810	0.183	NA[b]	Laing *et al.* (1976)
Speonomus hydrophilus	5	0.949	0.052	0.112	Crouau-Roy (1986)
Cerambycidae					
Tetraopes tetraophthalmus (Foster)	5	NA[b]	NA[b]	0.154	McCauley and Eanes (1987)
Chrysomelidae					
Chrysolina aurichalcea	9	0.966[a]	0.035	NA[b]	Sakanouce and Fujiyama (1987)
Diabrotica barberi Smith	15	0.987[a]	0.013	0.098	McDonald *et al.* (1985)
Leptinotarsa decemlineata (Say)	11	0.978	0.022	0.068	Jacobson and Hsiao (1983)

Curculionidae					
Anthonomus grandis grandis Boh.	2	0.934[a]	0.068	NA[b]	Bartlett (1981)
A. grandis thurberiae	3	0.842[a]	0.172	NA[b]	Bartlett (1981)
A. grandis grandis	15	0.803[a]	0.220	NA[b]	Bartlett *et al.* (1983)
Hypera postica (Gyll.)	6	0.968	0.033	0.024	Hsiao and Stutz (1985)
Pissodes strobi (Peck)	17	0.963	0.038[a]	0.098	Phillips and Lanier (1985)
Scolytidae					
Dendroctonus frontalis (Zimm.)	5	NA[b]	NA[b]	0.068	Anderson *et al.* (1979)
D. jeffreyi Hopkins	4	0.992	0.005	NA[b]	Higby and Stock (1982)
D. ponderosae Hopkins	6	0.947	0.054[a]	NA[b]	Stock and Guenther (1979)
	4	0.996	0.004	NA[b]	Stock and Amman (1980)
	3	0.992	0.003	NA[b]	Higby and Stock (1982)
	15	0.981	0.006	0.030	Stock *et al.* (1984)
D. terebrans (Olivier)	4	0.986[a]	0.014	NA[b]	Anderson *et al.* (1983)
Ips calligraphus Germar	4	0.998[a]	0.002	0.011	Anderson *et al.* (1983)
Tenebrionidae					
Tribolium castaneum (Herbst)	19	0.836	0.181	NA[b]	Wool (1982)
T. confusum	7	0.788	0.245	NA[b]	Wool (1982)
Mean ± SE		(27) 0.935 ± 0.013	(27) 0.069 ± 0.015	(15) 0.103 ± 0.020	

[a] Calculated from published data, $D = -\ln I$.
[b] NA = not available.

1. Bark beetle studies

The bark beetles are destructive forest pests in North America. Eight species have been examined by gel electrophoresis (Table 8.1). Of these, five species (*Dendroctonus jeffreyi*, *D. ponderosae*, *D. pseudotsugae*, *D. terebrans*, and *Ips calligraphus*) have been investigated extensively to assess genetic diversity between species and host-adapted populations (Table 8.3 and 8.4). As many as 18 enzyme loci have been examined in several populations of each species. The $\overline{H}$ values of geographically widely separated populations are similar and range between 0.12 and 0.14, although the Douglas-fir beetle, *D. pseudotsugae*, has a high $\overline{H}$ value of 0.288. The $\overline{H}$ value of 0.153 for the five species is somewhat lower than the $\overline{H}$ value (0.168) for all coleopteran species (Table 7.3). The proportion of polymorphic loci (0.374) is also lower than other beetle species. Mean I and D values between the bark beetles populations are 0.923 and 0.014, respectively — somewhat lower than for other beetle species (Table 8.4). The values for *D. ponderosae* are similar in four studies. Marked genetic divergence was detected in two populations of *D. pseudotsugae* (Stock *et al.* 1979). A similarity coefficient (Rogers') of $\overline{S}$ = 0.63 was found between a coastal Oregon population and a population from inland Idaho. Since the similarity coefficients of conspecific populations (including races and subspecies) exceed 0.75 (on a scale of 0–1), the low value for the two populations suggests they belong to different races or subspecies.

Molly Stock and her colleagues have conducted some interesting studies of the mountain pine beetle complex of the western United States. This scolytid beetle has been assigned at various times to three species (*Dendroctonus ponderosae*, *D. monticolae*, and *D. jeffreyi*). In the taxonomic revision of Wood (1963), these 'species' were synonmized as one species, although *D. jeffreyi* was later reinstated as a separate species. The 'mountain pine beetle' attacks 16 host tree species of *Pinus*, whilst lodgepole pine, western white pine, limber pine, sugar pine, ponderosa, and whitebark pine serve as its preferred hosts. The 'Jeffrey pine beetle' is specialized on Jeffrey pine. Both 'species' are widely distributed throughout 11 western states and in western Canada.

UPGMA (unweighted pair-group arithmetic average) dendrograms of Nei's genetic distance have, in two studies, clarified the genetic relationships between the mountain pine and Jeffrey beetles (Higby and Stock 1982; Stock *et al.* 1984). The 1982 study examined 12 populations from Northern California. The $\overline{D}$ value obtained between the mountain pine beetle and Jeffrey pine beetle was 0.168. In the 1984 study, 15 populations from six states were analysed. The 14 populations of the mountain pine beetle that fed on lodgepole pine were clustered. The population that fed on Jeffrey pine was separated from the other populations by a $\overline{D}$ value of 0.163, which is almost identical to the value found from the 1982 study. Stock and co-workers consider the two bark beetles to be sibling species that specialize on different hosts but occur sympatrically in some areas. The two species are reproductively isolated and do not produce viable progeny in the laboratory (Lanier and Wood, 1968). Thus, allozyme studies have identified two distinct species that could not be distinguished by morphological criteria. Two diagnostic loci (IDH and MDH) can also separate the two species.

2. Colorado potato beetle studies

The Colorado potato beetle (*Leptinotarsa*) is now a pest on two continents—North America and Europe. This indigenous North American insect became a pest when it adapted to the cultivated potato some 130 years ago. In North America, the species is found from southern Mexico and nearly all parts of the United States except California and in several provinces of Canada (Hsiao 1985). In different regions, the beetles utilize different native as well as introduced solanaceous hosts. The geographic diversity of the beetle provides an opportunity to investigate the genetic variability between geographic populations (Jacobson and Hsiao 1983). Twelve populations from North America and Europe were examined at 11 putative enzyme loci. $\overline{H}$ values of 11 populations from cultivated potatoes and egg plants were similar with a mean over all populations of 0.206. The $\overline{H}$ value (0.092) of a Mexican population was, however, much lower. A highly inbred laboratory strain maintained for over 150 generations has retained the same average heterozygosity as the field populations. The proportions of polymorphic loci is identical among populations. The F_{ST} for these populations (0.068) indicates that 6.8 per cent of the total variation is attributable to genetic differences among populations, while 93.2 per cent of the total variation occurs within populations (Table 8.4).

The mean D value was 0.022 for the 11 populations. The value between the Mexican population and the other 11 populations was 0.212, which is comparable to the $\overline{D}$ value for subspecies of *Drosophila*. The $\overline{D}$ value between *Leptinotarsa decemlineata* and its congener *L. haldemani* was 0.439. The Mexican beetle population is presently considered to be a subspecies rather than a distinct species because it interbreeds readily with other beetle populations under laboratory conditions (Hsiao 1982). It is not known whether the two subspecies hybridize in nature. One diagnostic locus (SOD) can be used to distinguish between the two subspecies (Jacobson and Hsiao 1983). Again, allozyme data have clarified the taxonomic relationships among geographic populations of pest beetles.

3. *Alfalfa weevil studies*

The alfalfa weevil is a destructive pest of the alfalfa crop in North America and in many parts of Europe and Asia. It is native to Europe and Asia and was introduced to the United States on three separate occasions. The first introduction, found in Salt Lake City, Utah, in 1904 was named *Hypera postica.* The second, introduction found in Yuma, Arizona in 1939, was named the Egyptian alfalfa weevil *H. brunneipennis*. A third introduction, discovered in Maryland in 1951, was called the 'eastern strain' of the alfalfa weevil to distinguish it from the western strain from Utah. The three weevil introductions have now spread to all parts of the United States as well as to Mexico and Canada (Hsiao 1989). These weevils are morphologically indistinguishable but exhibit considerable differences in biology and ecology. Formerly, they were identified by geographic origin, but this is no longer feasible because populations now overlap in several parts of the USA. A unidirectional reproductive incompatibility first reported by Blickenstaff (1965) also makes it difficult to distinguish populations. Infertile eggs are produced when females of the eastern weevil are crossed with males of the western weevil. The reciprocal cross is compatible but produces a high ratio of females. Such a unidirectional incompatibility also occurs in the cross between females of the Egyptian weevil and males of the western weevil (Hsiao and Hsiao 1985*a*). However, there is no reproductive incompatibility between the eastern and Egyptian weevils (Hsiao and Hsiao 1985*a*). Hence, the taxonomic status of alfalfa weevils in North America is rather confusing (Hsiao 1989).

As with the two aforementioned beetle groups, allozyme electrophoresis is very useful in order to clarify the taxonomic relationships and genetic divergence between these weevils (Hsiao and Stutz 1985). Two populations from each of the three weevil types were analysed using 12 enzymes representing 22 putative loci. The H and P values were very similar in all six populations with a mean of 0.231 and 0.536, respectively (Table 8.3). These means were slightly higher than those reported for other beetle species. The mean I and D values calculated for the six weevil populations were 0.968 and 0.033, respectively, suggesting that there was only intraspecific variation between these weevils (Table 8.4). Also, allozyme electrophoresis did not identify any diagnostic loci that could reliably differentiate these weevil types. Cytogenetic studies of the three weevil types also failed to detect chromosomal differences (Hsiao and Hsiao 1984). Consequently, the three alfalfa weevil introductions in North America are now regarded as strains of the same species rather than two separate species.

However, allozyme and cytogenetic data do not explain the reproductive incompatibility. The cause of the unidirectional reproductive incompatibility was eventually traced to a rickettsia, *Wolbachia postica*, found only in the western weevil but not in the eastern and Egyptian weevils (Hsiao and Hsiao 1985*b*). By experimentally eliminating the rickettsia with antibiotics, reproductive incompatibility was fully restored between the three weevil strains (Hsiao 1989). The rickettsia thus provides a reliable diagnostic tool for distinguishing these strains.

As the above examples demonstrate, the study of genetic differentiation of insect populations often involves several approaches. Basic biological data such as host relationships, life-history traits, reproductive behaviour, and genetic information, as well as biosystematic and taxonomic data, are needed to interpret intraspecific relationships. Allozyme analysis is a novel approach that allows the rapid gathering of genetic data not available by classical genetic approaches. The technique provides a new dimension enabling genetic variation to be assessed, information that is indispensable in the study of evolutionary processes.

Studies on interspecific relationships

Allozyme electrophoresis has been used to assess interspecific relationships in six beetle families (Table 8.5). Nei's genetic identity and distance coefficients were calculated in most instances. There was a wide range of values in the 23 entries. The D values ranged from 0.098 to 2.207 with a mean of 0.703 $\pm$ 0.123 (SE); the I values from 0.110 to 0.907 with a mean of 0.563 $\pm$ 0.049. These means for Coleoptera are comparable to those reported for *Drosophila* (Ayala 1978) and other insects (Menken and Ulenberg 1987).

Three species groups, including two stored-product genera, *Sitophilus* and *Tribolium*, and the carabid genus *Agonum*, have been studied in some detail (Table 8.5). Both *Sitophilus* and *Tribolium* display rather large mean D values (1.76 $\pm$ 0.447 and

Table 8.5 Nei's genetic identity (I) and distance (D) between coleopteran species.

Species pair	Mean I	Mean D	Reference
Carabidae			
Agonum decorum vs. *A. elongatulum*	0.907[a]	0.098	Liebherr (1986*a*)
A. decorum vs. *A. extensicolle*	0.801[a]	0.222	Liebherr (1986*a*)
A. decorum vs. *A. extimum*	0.700[a]	0.356	Liebherr (1986*a*)
A. decorum vs. *A. texanum*	0.759[a]	0.276	Liebherr (1986*a*)
A. elongatulum vs. *A. extensicolle*	0.825[a]	0.192	Liebherr (1986*a*)
A. elongatulum vs. *A. extimum*	0.662[a]	0.413	Liebherr (1986*a*)
A. elongatulum vs. *A. teranum*	0.739[a]	0.304	Liebherr (1986*a*)
A. extensicolle vs. *A. extimum*	0.736[a]	0.307	Liebherr (1986*a*)
A. extensicolle vs. *A. texanum*	0.724[a]	0.323	Liebherr (1986*a*)
A. extimum vs. *A. texanum*	0.728[a]	0.318	Liebherr (1986*a*)
Platynus angustatus vs. *P. tenuicollis*	0.574	0.555[a]	Liebherr (1986*b*)
Catopidae			
Orostygia doderoi vs. *O. meggiolaroi*	0.388	0.947	Sbordoni *et al.* (1980)
Chrysomelidae			
Leptinotarsa decemlineata vs. *L. haldemani*	0.645	0.439	Jacobson and Hsiao (1983)
Curculionidae			
Sitophilus oryzae vs. *S. zeamais*	0.420	0.867[a]	Beiras and Petitpierre (1981)
S. oryzae vs. *S. granarius*	0.110	2.207[a]	Beiras and Petitpierre (1981)
S. granarius vs. *S. zeamais*	0.110	2.207[a]	Beiras and Petitpierre (1981)
Scolytidae			
Dendroctonus jeffreyi vs. *D. ponderosae*	0.830	0.186	Higby and Stock (1982)

Table 8.5 *Continued*

Species pair	Mean *I*	Mean *D*	References
Tenebrionidae			
Tribolium castaneum vs. *T. confusum*	0.310	1.199	Wool (1982)
T. castaneum vs. *T. brevicornis*	0.418	0.878	Wool (1982)
T. castaneum vs. *T. destructor*	0.273	1.339	Wool (1982)
T. confusum vs. *T. brevicornis*	0.474	0.750	Wool (1982)
T. confusum vs. *T. destructor*	0.355	1.042	Wool (1982)
T. brevicornis vs. *T. destructor*	0.470	0.754	Wool (1982)
Mean ± SE	0.563 ± 0.049	0.703 ± 0.123	

[a] Calculated from published data, $D = -\ln I$.

0.994 ± 0.099, respectively) compared to values for other insects. In fact, *D* values of 2.207 between *S. granarius* and *S. oryzae*, and *S. granarius* and *S. zeamais* exceed those separating insect genera (Menken and Ulenberg 1987). The *D* value of 0.8867 between the two sibling species *S. oryzae* and *S. zeamais* is considerably higher than that for sibling species of the *Drosophila willistoni* group (0.581) (Ayala 1978). Among the four *Tribolium* species, the mean *D* value of 0.994 ± 0.099 (Wool 1982) is not substantially different from the value of 1.056 ± 0.068 for non-sibling species of *D. willistoni* (Ayala 1978). Further discussion concerning the assessment of genetic variation among *Tribolium* species is given by Wool (this volume).

Allozyme variation of five *Agonum* species has been investigated in considerable detail by Liebherr (1986*a*). The study was based on both electrophoretic and morphological data and thus provides a more realistic estimation of phylogenetic relationships. The mean values of *I* and *D* for the five species were 0.758 ± 0.022 and 0.281 ± 0.028, respectively. These values are within the range reported for congeneric species of *Drosophila* and other insects (Ayala 1978; Menken and Ulenberg 1987). The electrophoretic and morphological data produced compatible estimates of phylogeny in these carabid species (Liebherr 1986*a*). As discussed in a previous section, allozyme data separated the sibling species of *Dendroctonus* *jeffreyi* and *D. ponderosae*. The marked genetic dissimilarity (large *D* values) between congeneric beetle species, especially among stored-product species, is intriguing. More data on other beetle groups must be compiled before it can be determined whether the pattern is unique to Coleoptera.

Allozyme variation and inheritance

Ideally, the allelic status of electromorphs should only be assigned after genetic crosses of individuals bearing different mobility variants. However, the difficulties in rearing and crossing mean that the majority of allozyme studies with Coleoptera involve the interpretation of allozyme patterns without associated genetic testing. These interpretations are based upon the overall variability patterns and the agreement of allozyme genotype frequencies with Hardy–Weinberg expectations at the population level (Menken and Ulenberg 1987). Several types of variation have been described. Non-genetic enzyme variations are especially troublesome, such as those introduced by modifications occurring during storage and sample preparation. Another type of variation is due to the fact that some enzymes may be active or inactive at different developmental stages. For example, the expression of three enzymes (G6PD, AO, and LDH) of the alfalfa weevil, *Hypera postica*, was found to vary with the age of the adult beetle (Romero-Severson *et al.* 1986). Similarly, esterases are known to show major ontogenetic changes in *Tribolium castaneum* (Cohen *et al.* 1977) and *Anthonomus grandis* (Jones and Bancroft 1986). Analysing enzymes in insects of a certain physiological age would minimize such variations. To detect rare electromorphs, a large population sample is often necessary. Several hundred individuals were routinely used to assess genotype frequencies in bark beetle studies (Stock and Amman 1980; Anderson *et al.* 1983). Other non-genetic variations that may interfere with interpretation of allozyme patterns include diseases, parasites, or gut contents.

As stated above, analysis of allozyme inheritance can confirm the allelic status of electromorphs, although such an approach, whilst desirable, is seldom done because it is difficult to rear insects for genetic crosses. Amongst Coleoptera, allozyme inheritance has been investigated with stored-product and agricultural pests, especially species that can be maintained on artificial diets. In *Tribolium castaneum* and *T. confusum*, genetic inheritance has been examined on PGM (Riddle *et al.* 1983), ME, HK (Samollow *et al.* 1983; Dawson and Jost 1983) and G6PD (Dawson and Hollingsworth 1982). A comprehensive study of isoenzyme inheritance has been conducted with the cotton boll weevil, *Anthonomus grandis*, including ADH and ODH (Terranova and North 1984*a*); EST (Biggers and Bancroft 1977; Terranova 1982); AO, GOT, and PGM (Terranova, 1980; AMY, FUM, LAP, MDH, and SOD (Terranova and North 1985); GPI and G6PD (Terranova and North 1984*b*). Inheritance of EST, HK, MDH, and XDH has also been examined in two subspecies of *Diabrotica longicornis* (McDonald *et al.* 1982).

Other applications of gel electrophoresis

Gel electrophoresis of enzymatic and non-enzymatic proteins has been used in several other ways. In population studies, allozyme techniques are widely used to assess levels of inter- and intraspecific hybridization and to study the dynamics of hybrid zones. Among sibling species distinguished by diagnostic enzyme loci, the detection of heterozygotes proves the existence of natural hybridization. Allozyme analysis thus provides valuable information on reproductive behaviour and isolating mechanisms. In coleopteran studies, allozyme data more frequently verify the lack of genetic exchange between sibling species (Table 8.1). A unique application was the use of allozyme markers to determine sperm precedence and multiple matings; Huettel *et al.* (1976) crossed two strains of the plum curculio, *Conotrachelus nenuphar* with different frequencies of an IDH allele to assess the incidence of sperm precedence and the refractory period between matings. Gel electrophoresis is also useful in taxonomic identification. Immature stages of congeneric beetle species are sometimes difficult to identify morphologically and diagnostic allozyme loci facilitate identification. Electrophoretic differences in IDH, EST, ACP, and HK have been used to accurately identify second and third instar larvae of *Diabrotica barberi* and *D. virgifera virgifera* (Piedrahita *et al.* 1985). Gel electrophoresis has also been used to detect EST allozymes in the frass of the cotton boll weevil (Bancroft and Jones 1977). This procedure makes genotyping possible without sacrifice of experimental organisms and should be useful in genetic selection studies. Gel electrophoresis is also routinely used to monitor genetic changes and for quality control when mass-rearing insects (Terranova 1981). These examples illustrate that the potential application of gel electrophoresis is limited only by the ingenuity of the researcher.

Table 8.6 Comparison of genetic variability among major insect orders.

Insect order	*Estimated species*[a] No.	Percentage	*Species used*[b] No.	Percentage	Mean heterozygosity ($\overline{H}$) ± SE[b]
Orthoptera	20 000	2.8	25	12.7	0.076 ± 0.009
Hemiptera	56 000	7.9	24	12.2	0.104 ± 0.017
Coleoptera	300 000	41.8	14	7.1	0.187 ± 0.017
Lepidoptera	113 000	15.7	34	17.3	0.145 ± 0.014
Diptera	120 000	16.7	70	35.5	0.124 ± 0.007
Hymenoptera	108 000	15.1	30	15.2	0.036 ± 0.004
Total	717 500		197		0.112 ± 0.010

[a] Data from Borror *et al.* (1981).
[b] Data from Graur (1985), except for Coleoptera, which were taken from this survey.

Conclusion

Using gel electrophoresis, genetic variability has been examined in nearly 200 insect species from six major orders (Table 8.6). Only studies based on at least 15 gene loci from each species have been included in this table. Even though Coleoptera comprises over 40 per cent of the total estimated number of species, it comprises only 7.1 per cent of the species examined for genetic variability. The majority of the 37 entries (30 species) cited in the present survey involved the examination of relatively few loci. Only in five species were more than 20 loci surveyed, whilst in another 14 species, more than 15 loci were examined (Table 8.6). Thus, allozyme data from Coleoptera may not be as accurate as those from Diptera (especially *Drosophila*) and other insect orders. Nevertheless, the data do reveal several interesting aspects of the order. The high average heterozygosity of Coleoptera is striking. Values calculated from 30 species (0.168) or from 14 species (0.187) are much higher than those for many other insect orders (Table 8.3 and 8.6). Thus Coleoptera, Lepidoptera, and Diptera are the most genetically variable insect orders. There is also considerable genetic divergence between congeneric species. Nei's coefficient of genetic distance, which measures genetic dissimilarity, is usually quite large between species. The D values for stored-product species exceed the values usually used to separate genera (Table 8.5). In contrast, the D and F_{ST} values between beetle populations are not significantly different when compared to those of other insect groups (Table 8.4). From the available allozyme data it may be inferred that there is much more genetic differentiation in Coleoptera above the species level than at the intraspecific level. The large allozyme variation observed between beetle species suggests that the technique may not be suitable for assessing genetic diversity above the genus level.

Allozyme eleoctrophoresis is a versatile tool for many types of population and biosystematic studies. Agricultural pests are often introduced into new habitats by human activity and agricultural practices and are therefore well-suited for analysis by this technique. The genetic variability of these pest species is largely unknown. Many insect pests are mobile and adaptable and can rapidly form biotypes or races that are morphologically indistinguishable. Failure to recognize biotypes and genetic variants often leads to inaccurate taxonomic identification and can impede pest management. Several examples cited in this chapter have shown that the allozyme technique is especially suitable for studying pest complexes, particularly in defining intraspecific relationships and sibling species.

Lastly, as mentioned earlier, gel electrophoresis is probably the only method available for evaluating genetic divergence at the population and species levels without conducting time-consuming formal genetic experiments. The relatively low cost, ease of operation, and reproducibility make gel electrophoresis of allozymes preferable to other molecular techniques such as DNA–DNA hybridization and DNA sequencing. A recently published handbook describes numerous procedures and tips concerning allozyme electrophoresis (Richardson *et al*. 1986).

Allozyme electrophoresis is presently under-utilized in Coleoptera research and there is as yet relatively little information on the genetic diversity of this huge order of insects. The beetles, because of their great diversity, are ideal subjects for biosystematic research. Until the widespread advent of electrophoresis in the 1970s, the study of Coleoptera evolution and phylogeny was traditionally based on morphological criteria (Lawrence and Newton 1982). Allozyme electrophoresis and other biochemical and molecular techniques now also provide genetic and biochemical traits useful in interpreting phylogenetic relationships. These techniques should find widespread use in the biosystematic study of Coleoptera and it is thus hoped that this brief survey of the subject will stimulate further interest in the electrophoretic approach.

Acknowledgements

This work is supported by the Utah Agricultural Experiment Station, Utah State University, Logan, Utah 84322–4845. Approved as journal paper no. 3686.

References

Anderson, W. W., Berisford, C. W., and Kimmich, R. H. (1979). Genetic differences among five populations of the southern pine beetle. *Ann. ent. Soc. Am.*, **72**, 323–7.

Anderson, W. W., Berisford, C. W., Turnbow, R. H., and Brown, C. J. (1983). Genetic differences among populations of the black turpentine beetle, *Dendroctonus terebrans*, and an engraver beetle, *Ips calligraphus* (Coleoptera, Scolytidae) *Ann. ent. Soc. Am.*, **76**, 896–902.

Avise, J. C. and Selander, R. K. (1972). Evolutionary genetics of cave dwelling fishes of the genus *Astyanax*. *Evolution*, **26**, 1–19.

Ayala, F.J. (1978). Chemical genetics and evolution. In *Biochemistry of insects*, (ed. M. Rockstein), pp. 579–616. Academic Press, New York.

Baker, J.E. (1987). Electrophoretic analysis of amylase isozymes in geographical strains of *Sitophilus oryzae* (L.), *S. zeamais* Motsch., and *S. granarius* (L.) (Coleoptera, Curculionidae). *J. Stored Prod. Res*, **23**, 125–31.

Bancroft, H.R. and Jones, B.R. (1977). Genotypes of esterase II determined from frass of *Anthonomus grandis* Boh. (Coleoptera, Curculionidae). *Biochem Genet.*, **15**, 1175–80.

Bartlett, A.C. (1981). Isozyme polymorphisms in boll weevils and thurberia weevils from Arizona. *Ann. ent. Soc. Am.*, **74**, 359–62.

Bartlett, A.C., Randall, W.C., and May, J.E. (1983). Allozyme variation among populations of boll weevils in Arizona and Mexico. *Southwestern Ent.*, **8**, 118–30.

Beiras, M.J. and Petitpierre, E. (1981). Allozymic variability and genetic differentiation in three species of *Sitophilus* L. (Coleoptera, Curculionidae). *Egypt. J. Genet. Cytol.*, **10**, 95–104.

Berlocher, S.H. (1984). Insect molecular systematics. *A. Rev. Ent.* **29**, 403–33.

Biggers, C.J. and Bancroft, H.R. (1977). Esterases of laboratory-reared and field-collected cotton boll weevils, *Anthonomus grandis* Boh.: Polymorphism of esterases and formal genetics of esterase II. *Biochem. Genet.*, **15**, 227–33.

Blickenstaff, C.C. (1965). Partial intersterility of eastern and western U.S. strains of the alfalfa weevil. *Ann. ent. Soc. Am.*, **58**, 523–26.

Borror, D.J., De Long, D.M. and Triplehorn, C.A. (1981). *An introduction to the study of insects* (5th edn). Saunders College Publications, Philadelphia.

Bremner, T.A., Anderson, M.D., Pope, G.J., and Anderson, R.M. (1983). Genetic polymorphism of amylase in three species of *Tribolium* (Coleoptera, Tenebrionidae). *Comp. Biochem. Physiol.*, **74B**, 755–58.

Bush, G.L. and Kitto, G.B. (1978). *Application of genetics to insect systematics and analysis of species differences*, vol. 2, pp. 89–118; BARC Symposia, Biosystematics in Agriculture. Allanheld, Osmum & Co., Montclair.

Buth, D.G. (1984). The application of electrophoretic data in systematic studies. *A. Rev. Ecol. Syst.*, **15**, 501–22.

Cohen, E., Sverdlov, E., and Wool, D. (1977). Expression of esterases during ontogenesis of the flour beetle *Tribolium castaneum* (Tenebrionidae; Coleoptera). *Biochem. Genet.*, **15**, 253–64.

Crouau-Roy, B. (1986). Population studies on Pyrenean troglobitic beetles, local genetic differentiation and microgeographic variations in natural populations. *Biochem. Syst. Ecol.*, **14**, 521–26.

Crowson, R.A. (1981). *The biology of coleoptera.* Academic Press, London.

Dawson, P.S. and Jost, J. (1983). Linkage relationships among the malic enzyme, hexokinase-1, and red loci on the x-chromosome of *Tribolium confusum*. *Biochem. Genet.*, **21**, 661–65.

Dawson, P.S. and Hollingsworth, N.M. (1982). Sex-linkage of the glucose-6-phosphate dehydrogenase locus in the flour beetle, *Tribolium castaneum*. *Can. J. Genet. Cytol.*, **24**, 267–71.

Diehl, S.R. and Bush, G.L. (1984). An evolutionary and applied perspective of insect biotypes. *A. Rev. Ent.*, **29**, 471–504.

Frankel, J.S. and Duffield, R.M. (1984). Genetic variation in the cigarette beetle *Lasioderma serricorne* (Coleoptera, Anobiidae)- I. Esterase and water-soluble protein polymorphism. *Comp. Biochem. Physiol.*, **77B**, 337–40.

Giuseffi, S., Kane, T.C., and Duggleby, W.F. (1978). Genetic variability in the Kentucky cave beetle *Neaphaenops tellkampfii* (Coleoptera, Carabidae) *Evolution*, **32**, 679–81.

Graur, D. (1985). Gene diversity in Hymenoptera. *Evolution*, **39**, 190–9.

Higby, P.K. and Stock, M.W. (1982). Genetic relationships between two sibling species of bark beetle (Coleoptera, Scolytidae), Jeffrey pine beetle and mountain pine beetle, in northern California. *Ann. ent. Soc. Am.*, **75**, 668–74.

Hsiao, T.H. (1982). Geographic variation and host plant adaptation of the Colorado potato beetle. In *Proceedings of the 5th International Symposium on Insect-Plant Relationships* (ed. J.H. Visser, and A.K. Minks), pp. 315–24. Pudoc, Wageningen.

Hsiao, T.H. (1985). Ecophysiological and genetic aspects of geographic variations of the Colorado potato beetle. In *Proceeding of Symposium on the Colorado Potato Beetle, 17th International Congress of Entomology* (ed. D.N. Ferro, and R.H. Voss), pp. 63–77. Research Bulletin, 704, Massachusetts Agricultural Experiment Station, Amherst, Mass.

Hsiao, T.H. (1989). Geographic and genetic variation among alfalfa weevil strains. In *Evolution of insect pests: The pattern of variation* (ed. K.C. Kim). Wiley, New York. (In press.)

Hsiao, C. and Hsiao, T.H. (1984). Cytogenetic studies of alfalfa weevil, *Hypera postica*, strains (Coleoptera: Curculionidae). *Can. J. Genet. Cytol.*, **26**, 348–53.

Hsiao, T.H. and Hsiao, C. (1985*a*). Hybridization and cytoplasmic incompatibility among alfalfa weevil strains. *Entomologia exp. appl.*, **37**, 155–9.

Hsiao, C. and Hsiao, T.H. (1985*b*). Rickettsia as the cause of cytoplasmic incompatibility in the alfalfa weevil, *Hypera postica* (Gyllenhal). *J. Invert. Path.*, **45**, 244–6.

Hsiao, T.H. and Stutz J.M. (1985). Discrimination of alfalfa weevil strains by allozyme analysis. *Entomologia exp. appl.*, **37**, 113–21.

Huettel, M.D., Calkins, C.D., and Hill, A.J. (1976). Allozyme markers in the study of sperm precedence in the plum curculio, *Conotrachelus nenuphar*. *Ann. ent. Soc. Am.*, **69**, 465–8.

Jacobson, J.W. and Hsiao, T.H. (1983). Isozyme variation between geographic populations of the Colorado potato beetle, *Leptinotarsa decemlineata* (Coleoptera, Chrysomelidae). *Ann. ent. Soc. Am.*, **76**, 162–6.

Jones, B.R. and Bancroft, H.R. (1986). Distribution and probable physiological role of esterases in reproductive, digestive, and fat body tissues of the adult cotton boll weevil, *Anthonomus grandis* Boh. *Biochem. Genet.*, **24**, 499–508.

Jolivet, P., Petitpierre, E., and Hsiao, T.H. (ed.) (1988). *Biology of chrysomelidae*. Kluwer Academic Publishers, Dordrecht, Netherlands.

Kane, T.C. and Brunner, G.D. (1986). Geographic variation in the cave beetle *Neaphaenops tellkampfii* (Coleoptera, Carabidae). *Psyche*, **93**, 231–51.

Kuboki, M. (1978). Studies on the phylogenetical relationship in the *Henosepilachna-vigintioctomaculata* complex based on variation of isozymes (Coleoptera: Coccinellidae). *App. Ent. Zool.* **13**, 250–9.

Laing, C., Carmody, R., and Peck, S.B. (1976). Population genetics and evolutionary biology of the cave beetle *Ptomaphagus hirtus*. *Evolution*, **30**, 484–98.

Lanier, G.N and Wood, D.L. (1968). Controlled mating, karyology, morphology, and sex-ratio in the *Dendroctonus ponderosae* complex. *Ann. ent. Soc. Am.*, **61**, 517–26.

Lawrence, J.F. and Newton, A.F.Jr. (1982). Evolution and classification of beetles. *A. Rev. Ecol. Syst.*, **13**, 261–90.

Liebherr, J.K. (1983). Genetic basis for polymorophism in the ground beetle, *Agonum decorum* (Say) (Coleoptera, Carabidae). *Ann. ent. Soc. Am.*, **76**, 349–58.

Liebherr, J.K. (1986*a*). Cladistic analysis of North American Platynini and revision of the *Agonum extensicolle* species group (Coleoptera, Carabidae). *University of California Publications in Entomology*, **106**, 1–198.

Liebherr, J.K. (1986*b*). Comparison of genetic variation in two carabid beetles (Coleoptera) of differing vagility. *Ann. ent. Soc. Am.*, **79**, 424–33.

Liebherr, J.K. 1988. Gene flow in ground beetles (Coleoptera, Carabidae) of differing habitat preference and flight-wing development. *Evolution*, **42**, 129–37.

Lokki, J., Saura, A., Lankinen, P., and Suomalainen, E. (1976). Genetic polymorphism and evolution in parthenogenetic animals. V. Triploid *Adoxus obscurus* (Coleoptera, Chrysomelidae). *Genet. Res. Camb.*, **28**, 27–36.

McCauley, D.E. (1987). Population genetic consequences of local colonization, evidence from the milkweed beetle *Tetraopes tetraophthalmus*. *Florida Entomologist*, **70**, 21–30.

McCauley, D.E. and O'Donnell, R. (1984). The effect of multiple mating on genetic relatedness in larval aggregations of the imported willow leaf beetle (*Plagioderma versicolora*, Coleoptera, Chrysomelidae). *Behav. Ecol. Sociol.*, **15**, 287–91.

McCauley, D.E. and Eanes, W.F. (1987). Hierarchical population structure analysis of the milkweed beetle, *Tetraopes tetraophthalmus* (Forster). *Heredity*, **58**, 193–201.

McDonald, I.C., Krysan, J.L., and Johnson, O.A. (1982). Genetics of

Diabrotica (Coleoptera, Chrysomelidae), Inheritance of xanthine dehygrogenase, hexokinase, malate dehydrogenase, and esterase allozymes in two subspecies of *D. longicornis. Ann. ent. Soc. Am.*, **75**, 460–4.

McDonald, I. C., Krysan, J. L., and Johnson, O. A. (1985). Genetic variation within and among geographic populations of *Diabrotica barberi* (Coleoptera, Chrysomelidae). *Ann. ent. Soc. Am.*, **78**, 271–8.

Menken, S. B. J. and Ulenberg, S. A. (1987). Biochemical characters in agricultural entomology. *Agric. Zool. Rev.*, **2**, 305–60.

Namkoong, G., Roberds, J. H., Nunnally, L. B., and Thomas, H. A. (1979). Isozyme variations in populations of southern pine beetles. *Forest Sci.*, **25**, 197–203.

Petitpierre, E. (1988). Cytogenetics, cytotaxonomy and genetics of Chrysomelidae. In *Biology of chrysomelidae* (ed. P. Jolivet, E. Petitpierre, and T. H. Hsiao), Kluwer Academic Publishers, Dordrecht, Netherlands.

Phillips, T. W. and Lanier, G. N. (1985). Genetic divergence among populations of the white pine weevil, *Pissodes strobi* (Coleoptera, Curculionidae). *Ann. ent. Soc. Am.*, **78**, 744–50.

Piedrahita, O., Ellis, C. R., and Bogart, J. P. (1985). Electrophoretic identification of the larvae of *Diabrotica barberi* and *D. virgifera virgifera* (Coleoptera, Chrysomelidae). *Ann. ent. Soc. Am.*, **78**, 537–40.

Richardson, B. J., Baverstock, P. R. and Adams, M. (1986) *Allozyme electrophoresis, A handbook for animal systematics and population studies.* Academic Press, Sydney.

Riddle, R. A., Iverson, V., and Dawson, P. S. (1983). Dietary effects of fitness components at the GPM-1 locus of *Tribolum confusum. Genetics*, **103**, 65–73.

Ritzengruber, O. and Führer, E. (1986). Methods for analyzing isoenzymes of different populations in *Pityogenes-Chalcographus* Coleoptera Scolytidae I. Adaptation of methods enzyme polymorphism. *J. Appl. Ent.* **101**, 187–94.

Romero-Severson, J., Hogg, D. B., Kingsley, P. C., and Schwalbe, C. P. (1986). Age- and sex-dependent isozyme expression in a wisconsin population of adult alfalfa weevils, *Hypera postica* (Coleoptera, Curculionidae). *Ann. ent. Soc. Am.*, **79**, 364–8.

Sakanoue, S. and Fujiyama, S. (1987). Allozyme variation among geographic populations of *Chrysolina aurichalcea* (Coleoptera, Chrysomelidae). *Kontyu* (Tokyo), **55**, 437–49.

Samollow, P. B., Dawson, P. S., and Riddle, R. A. (1983). Sex-linked and autosomal inheritance patterns of homologous genes in two species of *Tribolium. Biochem. Genet.*, **21**, 167–76.

Sbordoni, V., Allegrucci, G., Caccone, A., Cesaroni, D., Sbordoni, M. C., and de Matthaeis, E. (1980). Preliminary report of the genetic variability in Troglobitic *bathysciinae Leptodirus-Hohenwarti* and 2 *Orostygia*-spp. Coleoptera Catopidae. *Fragment Ent.*, **15**, 327–6.

Schneider, K., Stubbe, A., Baldauf, F., and Tietze, F. (1984).

Electrophoretische Untersuchungen der Hamolymphe epigaisch lebender Coleopteren in unterschiedlich immissionsbelasteten Kiefernforsten. *Pedobiologia*, **26**, 107–16.

Sell, D.K., Armbrust, E.J., and Whitt, G.S. (1978). Genetic differences between eastern and western populations of the alfalfa weevil. *J. Hered.*, **69**, 37–50.

Smith. S.G. and Vikki, N. (1978). Coleoptera. In *Animal cytogenetics* (ed. B. John), Vol. 3, *Insecta* V. Gebruder Borntraeger, Berlin-Stuttgart.

Sokoloff, A. (1977). *The biology of tribolium*, Vol. 3. Oxford University Press, Oxford

Stock, M.W. and Amman, G.D. (1980). Genetic differentiation among mountain pine beetle populations *Dendroctonus ponderosae* from lodgepole pine in northeast Utah. *Ann. ent. Soc. Am.* **73**, 472–8.

Stock, M.W. and Guenther, J.D. (1979). Isozyme variation among mountain pine beetle *Dendroctonus ponderosae* populations in the pacific northwest. *Environ. Ent.*, **8**, 889–93.

Stock, M.W., Pitman, G.B., and Guenther, J.D. (1979). Genetic differences between douglas fir beetles from Idaho and coastal oregon. *Ann. ent. Soc. Am.*, **72**, 394–7.

Stock, M.W., Amman, G.D., and Higby, P.K. (1984). Genetic variation among mountain pine beetle (*Dendroctonus ponderosae*) (Coleoptera, Scolytidae) populations from seven western states. *Ann. ent. Soc. Am.*, **77**, 760–4.

Sturgeon, K.B. and Mitton, J.B. (1986). Allozyme and morphological differentiation of mountain pine beetles *Dendroctonus ponderosae* Hopkins (Coleoptera, Scolytidae) associated with host tree. *Evolution*, **40**, 290–302.

Suomalainen, E. and Saura, A. (1973). Genetic polymorphism and evolution in parthenogenetic animals. I. Polyploid Curculionidac. *Genetics*, **74**, 489–508.

Sweeney, B.W., Funk, D.H., and Vannote, R.L. (1987). Genetic variation in strean mayfly (Insecta: Ephemeroptera) populations of eastern North America. *Ann. ent. Soc. Am.*, **80**, 600–12.

Terranova, A.C. (1980). Inheritance patterns of aldehyde oxidase, glutamate-oxaloacetate transaminase and phosphoglucomustase allozymes in the boll weevil. *Ann. ent. Soc. Am.*, **73**, 653–57.

Terranova, A.C. (1981). Polyacrylamide gel electrophoresis of *Anthonomus grandis* Boheman proteins, profile of a standard boll weevil strain. *US Department of Agriculture/ARR-S-9*, pp. 1–48.

Terranova, A.C. (1982). Inheritance of esterases in *Anthonomus grandis grandis*. *Ann. Ent. Soc. Am.*, **75**, 261–5.

Terranova, A.C. and North, D.T. (1984*a*). Inheritance of alcohol and octanol dehydrogenases in the boll weevil. *J. Hered.*, **75**, 463–7.

Terranova, A.C. and North, D.T. (1984*b*). Genetics of glucose phosphate isomerase and glucose-6-phosphate dehydrogenase in *Anthonomous grandis* Boheman (Coleoptera, Curculionidae). *J. Agric. Ent.*, **1**, 380–5.

Terranova, A.C. and North, D.T. (1984*c*). Inheritance of isocitrate dehydrogenase in the boll weevil (Coleoptera, Curculionidae). *J. Agric. Ent.*, **1**, 397–404.

Terranova, A. C. and North, D. T. (1985). Inheritance of allozymes in the boll weevil (Coleoptera, Curculionidae). *Ann. ent. Soc. Am.*, **78**, 166–71.

Turanchik, E. J. and Kane, T. K. (1979). Ecological genetics of the cave beetle *Neaphaenops tellkampfii* (Coleoptera, Carabidae). *Oecologia* (Berlin), **44**, 63–7.

Wood, S. L. (1963). A revision of the bark beetle genus *Dendroctonus* Erichson (Coleoptera, Scolytidae). *Great Basin Naturalist*, **23**, 1–117.

Woods, P. E. and Guttman, S. I. (1987). Genetic variation in *Neodiprion* (Hymenoptera: Symphyta: Diprionidae) sawflies and a comment on low levels of genetic diversity within the Hymenoptera. *Ann. ent. Soc. Am.*, **80**, 590–9.

Wool, D. (1982). Critical examination of postulated cladistic relationships among species of flour beetles (Genus Tribolium, Tenebrionidae, Coleoptera). *Biochem. Genet.*, **20**, 333–49.

Wool, D. (1989). Electrophoretic variation in post-harvest agricultural pests, and its implications. This volume, pp. 341–62.

Wright, S. (1978). *Evolution and the genetics of populations.* Vol. 4. *Variability within and among natural populations.* University of Chicago Press, Chicago.

Yeh, F. C. H. and Scheinberg, E. (1974). Electrophoretic separation of enzymes of individual flour beetles, *Tribolium castaneum* on polyacrylamide gel. *Analyt. Biochem.*, **62**, 321–6.

9 Electrophoretic studies on geographic populations, host races, and sibling species of insect pests

STEPH B.J. MENKEN

Hugo de Vries-Laboratorium, University of Amsterdam, The Netherlands

Abstract

Allozyme analysis now plays a pivotal role in most areas of systematics and evolutionary biology. Allozymes are especially useful as characters for identification and discrimination of taxa at various levels of divergence and for description of patterns and processes such as population structure, breeding systems, gene flow, hybridization, and speciation. Many such studies deal with the controversy on the adaptive significance of the observed level of allozyme variation in natural populations.

In this contribution, lepidopterous and dipterous species are used to exemplify some of the above-mentioned aspects, with an emphasis on those encountered in pest insect research.

Accurate identification of specific pests is a prerequisite for progress in our knowledge about pest problems. This especially holds for insects, which are notoriously rich in sibling species. Electrophoresis repeatedly shows itself to be a superior technique in separating such sibling groups.

The study of the organization of genetic variation in space and time unravels the population substructuring and the amount of gene flow, information important for control programmes. Here we encounter the controversy on the relative importance of natural selection versus gene flow in the differentiation of populations. It also may help in finding the source area of an introduced species, and investigating the introductory event and the distribution thereafter.

Electrophoretic Studies on Agricultural Pests (ed. Hugh D. Loxdale and J. den Hollander), Systematics Association Special Volume No. 39, pp. 181–202. Clarendon Press, Oxford, 1989.

In the continuing lively debate concerning allopatric and non-allopatric modes of speciation, allozyme variability provides important information about the quantification of various evolutionary stages (host races, subspecies, semispecies, genuine species) and the dynamics of gene exchange and hybridization. In this respect, host-race formation in phytophagous insects is especially relevant to pest research.

Although electrophoresis has its shortcomings and the interpretation of enzyme banding patterns is often complex and difficult, both of which limit its application, it is nevertheless still at present the best technique available for studying population biology, speciation events, etc., in living organisms, including agricultural pests.

Introduction

Insects attack crops wherever these grow, either directly or indirectly, and reduce yields. Accurate identification is a prerequisite for any progress in our knowledge about specific pests. This especially holds for insects notoriously rich in sibling species, i.e. species that lack external morphological features to delineate species boundaries (Mayr 1976).

Molecular data and qualitative analyses are widely and increasingly used in systematics (Berlocher 1984*a*; Buth 1984; Menken and Ulenberg 1987). The zymogram technique (i.e. electrophoresis followed by a protein-staining procedure; Hunter and Markert 1957) can profitably be applied to various problems in pest-insect research [see Menken and Ulenberg (1987) for an extensive review of areas in which biochemical characters are used in agricultural entomology].

Nearly all animal and plant species harbour vast amounts of allozyme variation (Nevo *et al.* 1984; Graur 1985), thus providing an almost inexhaustible supply of characters for species identification and recognition of distinct populations. The genetic basis of allozyme variation is generally of a simple Mendelian nature and can either be readily determined or be reasonably inferred.

Moreover, in many aspects of population genetics and evolutionary and applied biology, enzyme variation analysis affords valuable information, e.g. on population structure, dispersal and gene flow, hybridization, modes of reproduction, and the speciation process (Oxford and Rollinson 1983). It also helps in discriminating between insecticide-resistant and insecticide-sensitive individuals (Stock and Robertson 1982; Oppenoorth 1985; Raftos and Hughes

1986), in monitoring genetic changes in laboratory strains (Singh 1984; Loukas *et al.* 1985; Leibenguth 1986; Menken and Ulenberg 1987), and in confirming the introduction of pest species into new areas (Harrison *et al.* 1983; Berlocher 1984*b*; Ross *et al.* 1985). Thus, electrophoresis is probably the most cost-effective technique available for studying the population biology of pest species.

This chapter centres on species identification, especially of sibling species, and the organization of genic variation in space and time, emphasizing aspects pertinent to the process of speciation. Before discussing these topics, however, some aspects of the sensitivity of electrophoresis will be outlined briefly.

Resolving power of electrophoresis

A major problem that plagues electrophoretic data is the fact that the majority of single amino-acid substitutions are undetected under standard electrophoretic conditions. The first demonstration of hidden variation preceded the introduction of electrophoresis in systematics and population genetics (Connell *et al.* 1962; Wright and MacIntyre 1965). The extent to which simple electrophoresis overlooks variability is not precisely known, but it is estimated at between 25 and 80 per cent (Henning and Yanofsky 1963; King and Ohta 1975; Johnson 1977*b*). Thus, many allozyme bands are effectively—populations of identically charged allozymes and are, therefore, designated more appropriately as electromorphs (King and Ohta 1975). These heterogeneous classes will separate when subject to a series of different electrophoretic conditions, so-called sequential electrophoresis (Johnson 1976; Singh *et al.* 1976; Johnson 1977*a*, *b*: McLellan 1984). From these and other investigations it appears that electrophoresis has a far higher resolution than has been assumed (Ramshaw *et al.* 1979; Fuerst and Ferrell 1980; Coyne 1982; Lewontin 1985; Watt 1985). The group of truly non-polar (neutral-to-neutral) amino-acid substitutions are not supposed to be amenable to electrophoretic separation, yet the recently developed technique of isoelectric focusing with immobilized pH gradients enables the separation of such 'electrophoretically silent' substitutions (Righetti and Gianazza 1987; Cossu and Righetti 1987, and references therein). For instance, two human foetal haemoglobin mutants with a difference between their isoelectric points of only 0.003 of a pH unit were successfully separated by Cossu and Righetti (1987). Moreover, there is increasing knowledge about the real determinants of electrophoretic mobility (Árnason and Chambers 1987, and references therein).

Screening under a variety of electrophoretic conditions generally adds little variation to nearly invariant loci, but at already highly polymorphic loci variability may increase dramatically (Coyne 1976; Singh *et al.* 1976; Johnson 1977*a*; Singh 1979, 1983; Menken 1987). As a consequence, cryptic variation not only leads to an overestimation of genic similarity between taxa but also to the common systematic problem that, if two individuals do not differ, it does not necessarily mean that they are indeed identical.

Since the first estimates of genic variation in man (Harris 1966) and *Drosophila pseudoobscura* (Lewontin and Hubby 1966), a huge amount of information on genetic polymorphisms in natural population has accumulated (see Nevo *et al.* 1984 for a recent review) from which the following generalizations can be made. Many species exhibit high levels of genetic variation: populations are polymorphic at 20–50 per cent of their enzyme loci and individuals are heterozygous at 5–25 per cent of their loci. However, there exists considerable heterogeneity in variation levels within and among taxonomic groups (Nevo *et al.* 1984; Graur 1985; Menken 1987). Plants and insects tend to harbour substantial amounts of allozyme variation, whereas vertebrates have relatively low levels. Hymenoptera, except for sawflies (Sheppard and Heydon 1986; Woods and Guttman 1987; see also Hung *et al.* 1986), exhibit levels of variation significantly lower than those of other insect groups (Berkelhamer 1983; Owen 1985). The variation levels of loci outside the strongly biased set of commonly screened loci is hardly known (Menken and Ulenberg 1987).

The unexpectedly large amount of allozyme variation found in natural populations has led to the neutralist–selectionist controversy over the mechanisms that maintain this variation. The controversy, which started as a clash of extreme viewpoints (e.g. Kimura 1968, 1982; Wills 1973), can still be characterized as 'remarkable for its temporal stability and its lack of resolution' (Endler 1977, p. 33). A variety of models have been proposed, but none seem to be fully satisfactory [reviewed by Ewens (1979); see also Mukherjee *et al.* (1987) and Singh and Rhomberg (1987) for different approaches to the controversy], because they lack power, make unrealistic assumptions, or demand data that are not (yet) available. The situation is further complicated by the fact that certain models produce predictions that are similar to those of opposing models (e.g. the SAS-CFF model; Gillespie 1982). However, identification of various classes of loci as based on major and minor polymorphisms 'elimininates the demand for a single simple theory that explains the aggregation of all observations' (Lewontin 1985, p. 89). It would appear well established, though, that a substantial amount of protein and nucleic-acid evolution is mutation-driven rather than selection-driven, although the exact amount remains to be determined. This being the case, so much the better, as the likelihood of convergence will be exceedingly low.

Sibling species and their diagnosis

Control programmes are often thwarted by unrecognized (sibling) species that show differences in their biology. Although reproductive isolation can be achieved by a few changes that do not result in morphological differences or genetic divergence in the set of commonly studied protein loci, most species still differ between one another in their allozyme patterns (Menken and Ulenberg 1987). Through the co-dominant expression of allozymes, reproductive isolation is readily indicated. If sympatric taxa are found to be fixed for different alleles at even a single locus, complete isolation and thus specific status can be assumed. A simple method for the calculation of the probability of correct diagnosis of an individual if allele frequencies slightly overlap is given in Ayala and Powell (1972) and Ayala (1983).

Since the first identification of sibling species by means of electrophoresis (Manwell and Baker 1963), the zymogram technique has proved to be superior in separating sibling species (Menken and Ulenberg 1987). The example of *Liriomyza* (Diptera, Agromyzidae) species, in which larvae, pupae, and adults can be discriminated through allozyme analysis (Menken and Ulenberg 1986) demonstrates an ideal diagnostic character: present in all members of both sexes of a species, absent from all other species, co-dominant, non-lethal, monomorphic, developmentally constant, and not under the influence of environmental factors.

Given the fact that electrophoresis requires only minute quantities of biological material, diagnosis is possible at an early stage of infestation, thus facilitating pest-control management (Menken 1989).

The status of allopatric taxa is not easily established. However, if biochemical divergence is even roughly proportional to time, then the level of protein differentiation between populations can be used to determine the proper taxonomic rank of a taxon. There is ample evidence that a molecular clock exists at the protein level (reviewed by

Wilson *et al.* 1977; Thorpe 1982). Data from a diverse collection of plant and animal species show that intraspecific diversity varies over a small range of genetic similarities (Thorpe 1983; Menken and Ulenberg 1987). Nearly all conspecific populations have Nei's genetic identity (Nei 1972) above 0.85 with the great majority above 0.95. The sibling species *Yponomeuta padellus* and *malinellus* (detailed below) form a notable exception to this rule. The critical value of 0.35 that distinguishes between species and genera (Thorpe 1982, 1983), however, is less useful (compare figure 2 in Menken and Ulenberg 1987).

Yponomeuta

Since the early 1970s the genus *Yponomeuta* (Lepidoptera, Yponomeutidae; the small ermine moths) has been studied as a model system for speciation in which insect–host-plant relationships are thought to play a key role (Wiebes 1976; Herrebout *et al.* 1976). Apart from this purely scientific interest, there is an important applied aspect because *Yponomeuta* species defoliate fruit trees, ornamentals and plants in natural stands. This is especially so with *Y. malinellus* feeding on apple and sometimes pear; *Y. padellus* on plum; and *Y. cagnagellus*, which infests spindle tree (*Euonymus* species). These taxa belong to the *padellus* complex, the taxonomic evaluations of which vary from five different species (Gershenson 1974) to one polytypic species (Friese 1960). Morphological differences among the taxa in the *padellus*-complex are minor or absent and therefore identification has hitherto been necessarily based on the food plant from which they were collected. The importance of correct recognition of siblings for optimal pest control can be well illustrated by *Yponomeuta*.

We now know that all the taxa are genuine biological species with diagnostic enzyme loci (Menken 1980; Arduino and Bullini 1985), slight differences in morphology (Povel 1986), and differing pheromone systems (Löfstedt *et al.* 1986; Hendrikse 1986; Löfstedt and Herrebout, pers. comm.). The sibling species *Y. malinellus* and *Y. padellus* appear to be almost identical genetically. The genetic identity (Nei 1972) between them amounts to either 0.98 (Menken 1982; based upon 51 genetic loci, none of which is diagnostic) or 0.92 (Arduino and Bullini 1985; calculated from a set of 32 loci, each of which broadly overlaps with the previous one, including two diagnostic loci). In addition, a cold-unstable larval alkaline phosphatase locus is likely to differentiate between the two species (Menken 1980).

Nevertheless, in the absence of diagnostic characters, sometimes the specific status of taxa can be determined with certainty if they occur sympatrically (Berlocher 1984*a*). For example, a deficiency of heterozygotes might indicate that two non-interbreeding populations were sampled and treated as an interbreeding one (Wahlund 1928). Furthermore, in the absence of selection, reduced (or zero) levels of gene flow can be inferred from statistically significant differences in allele frequencies at certain loci or the presence of rare alleles in one group only. Data in Table 9.1 illustrate the latter phenomenon. Allele distribution patterns for *Y. malinellus* and *Y. padellus* at the *Pgi* and *Pgm* loci are quite similar (geographic differentiation at all loci is low within each species, resulting in genetic identity estimates of 0.982 or higher; Table 9.1 and Menken 1981), probably reflecting the situation in their common ancestor, and are non-informative. The only exception is allele *Pgm-97*, which occurs exclusively in all *Y. padellus* populations. *Me* exhibits a slightly different pattern in that allele *98*, the second most common allele in *Y. padellus*, is rare or absent from *Y. malinellus*. At the *Hbdh*, *αGpdh*, and the sex-linked *Fudh* loci, alleles persist throughout the range of one species (sometimes they are absent owing to genetic drift or sampling error) and are virtually or completely absent from the sister species.

The patterns are highly constant in space and time and hold in sympatry (Table 9.1). Thus, the most likely explanation is that two non-interbreeding closely related species occur, although complex scenarios of various kinds and combinations of selection may be a (very unlikely) alternative. The occurrence of *αGpdh-95* in Beemster could indicate occasional gene flow (see also Arduino and Bullini 1985).

In Western Europe, small ermine moths can hardly be considered actual pests. The situation across the Altantic Ocean looks rather different. Since its introduction some 20 years ago, *Y. cagnagellus* has spread across the United States and parts of Canada (Hoebeke 1987). The introductory event was not accompanied by a bottleneck (Menken and Herrebout, in prep.; Menken and Ulenberg 1987). More threatening is the rapid advance of *Y. malinellus* since it was first reported in 1981 in British Columbia and Washington State. It defoliates trees in apple nurseries.

Population structure

Whether a species is genetically fragmented or coherent depends on the balance between gene flow, drift, and natural selection as well as

Table 9.1 Allozyme frequency distribution at six enzyme loci in populations of *Y. malinellus* and *Y. padellus*. All were sampled in the Netherlands, except for Kolham and Hartheim (West Germany), Gütting and Domat (Switzerland) and Helsinki (Finland). All *Y. malinellus* populations were sampled from apple, all *Y. padellus* from *Crataegus* except for Helsinki (*Sorbus aucuparia*), Domat II (*Prunus* sp.), and Beemster II (*Prunus cerasifera*). Domat I and II and Beemster I and II represent host races (Menken 1981).

		Pgi[a]						*Pgm*[b]							*Me*[c]	→	
Locality	Year	93	97	100	103.5	105	106.5	94	97	98.5	100	101.5	103	105	95	98	100
Y. malinellus																	
St. Agatha	1974	0.13		0.87												0.04	0.96
St. Agatha	1975	0.15		0.85													1.00
St. Agatha	1976	0.27		0.72		0.01				0.08	0.60	0.11	0.17	0.04		0.03	0.97
St. Agatha	1977	0.24		0.76						0.08	0.73	0.07	0.11	0.01			1.00
St. Agatha	1978	0.26	0.01	0.73						0.10	0.43	0.19	0.22	0.06		0.01	0.99
St. Agatha	1979	0.24		0.76						0.06	0.62	0.10	0.20	0.02			1.00
Kolham	1976	0.24		0.72		0.04				0.14	0.48	0.22	0.12	0.04	0.02	0.02	0.94
Gütting	1976	0.11		0.89						0.06	0.67	0.07	0.17	0.03		0.18	0.82
Beemster	1978	0.07		0.93						0.05	0.46	0.06	0.29	0.14			1.00
Lienden	1978	0.06		0.94							0.66	0.10	0.24				1.00
Y. padellus																	
Meyendel	1974															0.13	0.87
Meyendel	1975															0.23	0.77
Meyendel	1976	0.37		0.63					0.05	0.05	0.52	0.10	0.25	0.03		0.18	0.82
Meyendel	1977	0.28		0.72					0.11	0.04	0.45	0.05	0.34	0.01		0.29	0.70
Meyendel	1978	0.32		0.68					0.10	0.06	0.38	0.15	0.29	0.02		0.28	0.72
Hartheim	1977	0.29		0.71				0.02	0.13	0.08	0.44	0.05	0.26	0.02		0.22	0.78
Helsinki	1979	0.13		0.87					0.20	0.10	0.70					0.57	0.43
St. Agatha	1978	0.33	0.06	0.61					0.06	0.03	0.44	0.08	0.39			0.33	0.67
Domat I	1976	0.35		0.63	0.02			0.01	0.08	0.05	0.62	0.01	0.23			0.33	0.67
Domat II	1976	0.47		0.51			0.02	0.04	0.06	0.06	0.47	0.07	0.30			0.16	0.84
Beemster I	1978	0.36		0.64					0.04	0.07	0.42	0.22	0.24	0.01		0.33	0.67
Beemster II	1978	0.22		0.78					0.08	0.08	0.56	0.16	0.06	0.06		0.31	0.69

Table 9.1 *Continued*

	Hbdh[d]				*αGpdh*[e]					*Fudh*[f]			
102	98	100	102	104	93	95	97	100	104	95.5	97	98.5	100
						0.12		0.88			1.00		
						0.10		0.90			1.00		
						0.09		0.91			1.00		
		0.98		0.02		0.14		0.86			1.00		
		1.00				0.15		0.85			1.00		
		1.00				0.20		0.80			1.00		
0.02	0.02	0.98				0.14		0.86			1.00		
		1.00				0.13		0.87			1.00		
		1.00				0.12		0.88			1.00		
		1.00				0.60		0.40			1.00		
								1.00					
		0.75	0.25					1.00			0.92		0.08
	0.01	0.72	0.27					1.00			0.97		0.03
0.01	0.02	0.77	0.21					0.99	0.01		0.91		0.09
		0.73	0.27					1.00			0.91		0.09
		0.88	0.12				0.01	0.99			0.88		0.12
		0.88	0.12					1.00		0.02	0.90		0.08
		0.89	0.11					1.00			0.90		0.10
		0.80	0.20					1.00			0.85	0.10	0.05
		1.00						1.00		0.02	0.95		0.03
		0.80	0.20					1.00			0.95		0.05
	0.01	0.89	0.10		0.02	0.06		0.92			0.86		0.14

[a] *Pgi*, phosphoglucose isomerase. [b] *Pgm*, phosphoglucomutase. [c] *Me*, malic enzyme. [d] *Hbdh*, hydroxybutyrate dehydrogenase. [e] *αGpdh*, glycerol-3-phosphate dehydrogenase. [f] *Fudh*, fucose dehydrogenase.

on historical events. Unfortunately, there is no agreement about the relative evolutionary importance of gene flow (e.g. Mayr 1963; Ehrlich and Raven 1969; Stanley 1979; Cohan 1984; Lynch 1986). If gene flow is restricted between populations, geographic differentiation could be due to genetic drift and/or spatial variation in selection regimes. However, low-frequency alleles throughout a species' range indicate that gene flow among populations is likely to be substantial because stabilizing selection over temporally and spatially diverse habitats is not a likely alternative (Slatkin 1981; Daly and Gregg 1985; Pashley and Johnson 1986). Historical gene flow can also produce stable patterns of geographic variation (Liebherr 1988). It goes without saying that the amount of gene flow among populations has important consequences for pest management (reviewed in Roush and McKenzie 1987). This is so for both pest species and their parasites.

Population structure and associated gene flow can be investigated using two approaches, one ecological (direct measurement through mark-and-recapture methods) and one genetic (indirect measurement by analysis of patterns of gene frequency variability using theoretical models) (Coyne and Milstead 1987). The latter method is dealt with in this chapter.

Genetic identity or distance measures are too insensitive to reveal patterns of local differentiation, owing to the reduction of all information from mono- and polymorphic loci to a single figure. Monomorphic loci are fixed for the same allele over the entire range of a species. Often the most common allele for a given variable gene in one population is also the most common in all other populations of the species. Many loci may also show clinal variation (Menken and Ulenberg 1987; Singh and Rhomberg 1987). Individual polymorphic loci can be much more informative and data on gene frequencies are usually summarized by F-statistics as proposed by Wright (Wright 1978; Weir and Cockerham 1984; but see Nei 1987). F_{ST}, the among population variance in allele frequency, is best averaged over a set of loci, as single locus values have a large variance owing to genetic drift when the number of populations is small (Nei and Chakravarti 1977).

In general, there is a good agreement between the relative degree of population structuring (as measured by F_{ST}) of a species and its vagility. Active dispersers usually exhibit low F_{ST} values, highly sedentary species high ones (Eanes and Koehn 1978; McCauley and Eanes 1987 and references therein), unless strong selection is differentially working. Wright (1931) showed that, in the absence of selection, migration among populations prevents these populations from diverging unless $Nm < 1$ (Nm is the average number of migrants exchanged between populations). Theoretical and applied studies further demonstrated the homogenizing effect of the exchange of even a single individual per generation (Maruyama 1972; Nei and Feldman 1972; Frankel and Soulé 1981; Allendorf 1983).

F_{ST} values allow for estimates of Nm:

$$F_{ST} = \frac{1}{4Nm + 1}$$

(Wright 1931). For instance, F_{ST} averaged over loci, years and foodplants for *Y. padellus* equals 0.030 ± 0.012 (Menken 1981), similar to that for *Y. cagnagellus* (0.027 ± 0.013; Menken *et al.* 1980). Both species are known to be active fliers with good dispersal abilities. A similar value was obtained for *Y. rorellus* (0.030 ± 0.030) but this was based on the only locus among 75 that was reasonably polymorphic (Menken 1987). These three species exchange on the average some eight individuals per generation between their conspecific populations (Table 9.2). *Y. malinellus* has an intermediate rate of exchange (F_{ST} = 0.057 ± 0.032; Nm = 4.14). This species has a more local population structure with smaller population sizes after it was drastically cut back in numbers by pesticides and is mostly confined to abandoned orchards. All four of these species were sampled from more or less the same area. *Y. vigintipunctatus*, a species distantly related to the aforementioned species (Menken 1982), has a F_{ST} of 0.092 ± 0.023 (Menken, unpublished results) although over a somewhat smaller geographic area. *Y. vigintipunctatus* is the only western palearctic *Yponomeuta* species feeding on a forb (*Sedum telephium*). This foodplant is temporally and spatially less predictable than the trees and shrubs on which the other species rely and would therefore be expected to select for an insect with good dispersal ability

Table 9.2 Estimates of Nm based on fixation indices in five species of *Yponomeuta*.

Species	Mean no. populations	No. loci	No. years	Mean F_{ST}	Nm
Y. cagnagellus	6.7	4	3	0.027 ± 0.013	9.01
Y. malinellus	4	4	3	0.057 ± 0.032	4.14
Y. padellus	8	5	4	0.030 ± 0.012	8.08
Y. rorellus	3	1	3	0.030 ± 0.030	8.08
Y. vigintipunctatus	3	4	3	0.092 ± 0.023	2.47

and less localized population structure (Southwood 1962; Dingle 1972), somewhat contrary to the results (some 2–3 individuals migrate between populations; Table 9.2). Local populations of *Y. vigintipunctatus* regularly become extinct and are re-established, and this process augments the level of gene flow and hence prevents populations from differentiating by genetic drift (Slatkin 1977; Maruyama and Kimura 1980). However, in this process, populations often go through bottlenecks and the new population will rarely be completely representative of the source population. Owing to drift, there will be a general reduction in variability (depending on size of the bottleneck and the time the bottleneck persists) and a reduction in the average number of alleles per locus (depending on size alone; Nei *et al.* 1975; Chakraborty and Nei 1976). Moreover, population sizes in *Y. vigintipunctatus* are from one to several orders of magnitude smaller than those of *Y. padellus* and *Y. cagnagellus*, with a consequent greater probability of drift occurring. Smaller population sizes and bottleneck effects indeed lead to lower overall heterozygosity values for *Y. vigintipunctatus* (0.058 ± 0.007) versus 0.125 ± 0.020 for *Y. cagnagellus* and 0.127 ± 0.015 for *Y. padellus* (Menken 1982, 1987, and unpublished results). A similar pattern occurs for the average number of alleles per locus.

Recently, Slatkin (1985) developed a method to estimate *Nm* from the frequency of private alleles (i.e. alleles found in only one population). It appears that the logarithm of *Nm* shows an approximately linear relationship to the logarithm of $\bar{p}(1)$, the average frequency of private alleles. Furthermore, it was shown that the results depend mainly on *Nm* and the number of individuals analysed per population. The data in Table 9.3 were corrected for sample size as suggested by Slatkin (1985). The lower number of private alleles in *Y. padellus* compared to that of *Y. cagnagellus* most likely stems from the sampling of several adjacent populations of *Y. padellus* (Slatkin 1985).

Table 9.3 Estimates of *Nm* based on private alleles in five species of *Yponomeuta*.

Species	No. populations	No. loci	No. private alleles	$\bar{p}(1)$	*Nm*
Y. cagnagellus	6	32	23	0.028	7.12
Y. malinellus	4	32	12	0.042	3.42
Y. padellus	10	32	16	0.029	7.19
Y. rorellus	4	32	7	0.027	7.32
Y. vigintipunctatus	5	30	10	0.059	1.94

The estimates for *Nm* are highly consonant with those calculated from the F_{ST} values (Table 9.2). Thus, *Yponomeuta* species appear to exhibit moderate to high levels of gene flow. It should be remembered that these figures represent average values, leaving room for local differences (Slatkin 1985; Liebherr 1988; see also the section on host races below). For instance, the small Lienden population of *Y. malinellus* is either more isolated than the other populations or recently went through a bottleneck (Table 9.1), resulting in lower overall variability, absence of rare alleles (e.g. *Pgm-98.5* and *Pgm-105*), and reversed frequencies at the *αGpdh* locus. The aberrant pattern produces a disproportionally large contribution to F_{ST} (notice that in this case $\bar{p}(1)$ is not affected). Excluding the Lienden population brings the fixation index down to 0.039 ± 0.008, close to the values for *Y. padellus*, *Y. cagnagellus*, and *Y. rorellus*. Host-race formation, a special case of population structure, is dealt with in the next section.

Host races and speciation

Since Ernst Mayr's statement (Mayr 1963) that all that is known about the distribution patterns of related species and their geographic variation can be explained satisfactorily by the theory of speciation through geographic isolation and subsequent genetic divergence, many papers and books have been published advocating non-allopatric theories of speciation (Bush and Howard 1986, and references therein). A major argument against allopatric speciation as the one and only process is the observation that extremely 'speciose' groups such as phytophagous insects require the implausibly frequent occurrence of geographic barriers within species to account for the numbers of species actually observed (Strong *et al.* 1984).

Allozymes are frequently used to support evidence for various speciation models. It is thus well documented that in groups in which speciation takes place according to the allopatric model, increasing levels of evolutionary divergence exhibit decreasing levels of genetic identity (Ayala 1983).

Host races (defined as 'populations of a species partially reproductively isolated from other conspecific populations as a direct consequence of adaptation to a specific host'; Diehl and Bush 1984) play a pivotal role in discussions on mechanisms of evolution, sympatric speciation in particular. The well-documented shift of *Rhagoletis pomonella* from native *Crataegus* to introduced *Malus* and further to *Prunus* prompted Bush (1969) to formulate his theory of sympatric speciation through host-race formation. The chain of events leading to speciation in sympatry (Bush 1969, 1975) is supported by theoretical work (e.g. Maynard Smith 1966; Felsenstein 1981; Kondrashov 1986), although the stringent conditions may not often be met in nature.

Supposed examples of sympatric host or habitat races are manifold (references in Bush and Howard 1986). However, many turn out to be one of two other alternatives. Either genetic divergence between the populations feeding on two or more sympatric host plants cannot be demonstrated (which in view of the sensitivity of simple electrophoretic procedures and the small number of loci investigated is not identical to *absence* of divergence), or they are found to be completely isolated sibling species. The latter is, for example, the case in the treehopper *Enchenopa binotata* complex (Guttman *et al.* 1981; Wood 1987) and the lepidopterous species *Hyphantria cunea* (Jaenike and Selander 1980) and *Spodoptera frugiperda* (Pashley 1986; Pashley and Martin 1987).

In *Yponomeuta padellus*, the only western European small ermine moth species for which host-race formation (if any) is to be expected, host races still hold out. Statistically significant differences at particular allozyme loci were found among sympatric populations on various host plants (see the Domat and Beemster populations in Table 9.1; Menken 1981, 1982). Moreover, some rare alleles are consistently restricted to one host race. These differences could not be explained by sampling error or host-plant-specific selection against certain alleles at these loci or loci closely linked to and in linkage disequilibrium with them (Menken 1981). The pattern is most easily explained by a reduced level of gene exchange, i.e. by the occurrence of host races. For example, following the private allele method (Slatkin 1985), *Nm* equals 4.25 in the Malden population (Table 9.4). This figure for a sympatric situation is considerably lower than the average one found for 10 populations collected from all food plants and over a large geographic area (Table 9.3). This corroborates with earlier results based upon fixation indices (Menken 1981): geographic populations on *Crataegus* exchange more individuals than do sympatric host races.

So far no clear differences in morphology, larval development, pupal weight, parasite complexes, and pheromone system have been observed among the *Y. padellus* host races. The crucial assumption in sympatric speciation theory, however, that increased survival on a new host goes hand in hand with reduced survival on the old one has not been found in the few cases in which it was investigated (Via 1986; Futuyma and Philippi 1987).

Table 9.4 Allozyme frequencies at six polymorphic loci of *Y. padellus* on two sympatric food plants in Malden (Netherlands). Number of genomes sampled in parentheses.

	S. aucuparia (40)	*Crataegus* sp. (60)
Pgi		
87	0.025	—
93	0.425	0.167
100	0.550	0.833
103.5	—	0.050
Pgm		
97	0.025	—
98.5	0.100	0.067
100	0.450	0.417
101.5	0.050	—
103	0.375	0.483
105	—	0.033
Me		
98	0.450	0.317
100	0.550	0.683
Hbdh		
96	—	0.033
98	—	0.083
100	0.875	0.750
102	0.125	0.133
Mdh		
100	1.000	0.970
106	—	0.030
Fudh		
97	1.000	0.950
100	—	0.050

A comparable situation is described by Berlocher (this volume) for *Rhagoletis*.

In conclusion, enzyme electrophoresis is a highly versatile technique in population biological studies of insect pests, their parasites, and their predators. Its limitations are rather well defined and hence the technique is the best available.

Acknowledgements

I thank Wim Herrebout and Sandrine Ulenberg for their comments on an earlier draft of this manuscript, and Ria Touber for typing.

References

Allendorf, F. W. (1983). Isolation, gene flow, and genetic differentiation among populations. In *Genetics and conservation* (ed. C. M. Schonewald-Cox, S. M. Chambers, B. MacBryde, and W. L. Thomas), pp. 51–65. Benjamin/Cummings, London.

Arduino, P. and Bullini, L. (1985). Reproductive isolation and genetic divergence between the small ermine moths *Yponomeuta padellus* and *Y. malinellus* (Lepidoptera: Yponomeutidae). *Atti Acad. Naz. Lincei*, **18**, 33–61.

Árnason, E. and Chambers, G. K. (1987). Macromolecular interaction and the electrophoretic mobility of esterase-5 from *Drosophila pseudoobscura*. *Biochem. Genet.*, **25**, 287–307.

Ayala, F. J. (1983). Enzymes as taxonomic characters. In *Protein polymorphism: Adaptive and taxonomic significance* (ed. G. S. Oxford and D. Rollinson), pp. 3–26. Academic Press, London and New York.

Ayala, F. J. and Powell, J. R. (1972). Allozymes as diagnostic characters of sibling species of *Drosophila*. *Proc. natn. Acad. Sci. USA*, **69**, 1094–6.

Berkelhamer, R. C. (1983). Intraspecific genetic variation and haplodiploidy, eusociality, and polygyny in the Hymenoptera. *Evolution*, **37**, 540–5.

Berlocher, S. H. (1984*a*). Insect molecular systematics. *A. Rev. Ent.*, **29**, 403–33.

Berlocher, S. H. (1984*b*). Genetic changes coinciding with the colonization of California by the walnut husk fly, *Rhagoletis completa*. *Evolution*, **38**, 906–18.

Bush, G. L. (1969). Sympatric host race formation and speciation in frugivorous flies of the genus *Rhagoletis* (Diptera, Tephritidae). *Evolution*, **23**, 237–51.

Bush, G. L. (1975). Modes of speciation. *A. Rev. Ecol. Syst.*, **6**, 339–64.

Bush, G. L. and Howard, D. J (1986). Allopatric and non-allopatric speciation; assumptions and evidence. In *Evolutionary processes and theory* (ed. S. Karlin and E. Nevo), pp. 411–38. Academic Press, Orlando.

Buth, D. G. (1984). The application of electrophoretic data in systematic studies. *A. Rev. Ecol. Syst.*, **15**, 501–22.

Chakraborty, R. and Nei, M. (1976). Hidden genetic variability within electromorphs in finite populations. *Genetics*, **84**, 385–93.

Cohan, F. M. (1984). Can uniform selection retard random genetic divergence between isolated conspecific populations? *Evolution*, **38**, 495–504.

Connell, G. E., Dixon, G. H., and Smithies, O. (1962). Subdivision of the three common haptoglobin types based on 'hidden' differences. *Nature, Lond.*, **193**, 505–6.

Cossu, G. and Righetti, P. G. (1987). Resolution of G_γ and A_γ foetal haemoglobin tetramers in immobulized pH gradients. *J. Chromat.*, **398**, 211–6.

Coyne, J. A. (1976). Lack of genic similarity between two sibling species of *Drosophila* as revealed by varied techniques. *Genetics*, **84**, 593–607.

Coyne, J. A. (1982). Gel electrophoresis and cryptic protein variation. *Isozymes*, **6**, 1–32.

Coyne, J. A. and Milstead, B. (1987). Long-distance migration of *Drosophila*. 3. Dispersal of *D. melanogaster* alleles from the Maryland orchard. *Am. Nat.*, **130**, 70–82.

Daly, J. C. and Gregg, P. (1985). Genetic variation in *Heliothis* in Australia: species identification and gene flow in the two pest species *H. armigera* (Hübner) and *H. punctigera* Wallengren (Lepidoptera: Noctuidae). *Bull. ent. Res.*, **75** , 169–84.

Diehl, S. R. and Bush, G. L. (1984). An evolutionary and applied perspective of insect biotypes. *A. Rev. Entomol.*, **29**, 471–504.

Dingle, H. (1972). Migration strategies of insects. *Science*, **175**, 1327–35.

Eanes, W. F. and Koehn, R. K. (1978). An analysis of genetic structure in the monarch butterfly, *Danaus plexippus* L. *Evolution*, **32**, 784–97.

Ehrlich, P. R. and Raven, P. H. (1969). Differentiation of populations. *Science*, **165**, 1228–32.

Endler, J. A. (1977). *Geographic variation, speciation and clines.* Princeton University Press, Princeton, N.J.

Ewens, W. J. (1979). *Mathematical population genetics*. Springer Verlag, New York.

Felsenstein, J. (1981). Skepticism towards Santa Rosalia, or why are there so few kinds of animals? *Evolution*, **35**, 124–38.

Frankel, O. H. and Soulé, M. E. (1981). *Conservation and evolution.* Cambridge University Press, Cambridge.

Friese, G. (1960). Revision der paläarktischen Yponomeutidae unter besonderer Berücksichtigung der Genitalien (Lepidoptera). *Beitr. Ent.*, **10**, 1–131.

Fuerst, P. A. and Ferrell, R. E. (1980). The stepwise mutation model: An experimental evaluation utilizing hemoglobin variants. *Genetics*, **94**, 185–201.

Futuyma, D. J. and Philippi, T. E. (1987). Genetic variation and covariation in responses to host plants by *Alsophila pometaria* (Lepidoptera: Geometridae). *Evolution*, **41**, 269–79.

Gershenson, Z. S. (1974). Yponomeutidae, Argyresthiidae. *Fauna Ukraini*, **15**, 1–132.

Gillespie, J. H. (1982). A randomized SAS-CFF model of natural selection in a random environment. *Theor. Pop. Biol.*, **14**, 1–45.

Graur, D. (1985). Gene diversity in Hymenoptera. *Evolution*, **39**, 190–9.

Guttman, S.I., Wood, T.K., and Karlin, A.A. (1981). Genetic differentiation along host plant lines in the sympatric *Enchenopa binotata* Say complex (Homoptera: Membracidae). *Evolution*, **35**, 205–17.

Harris, H. (1966). Enzyme polymorphisms in man. *Proc. R. Soc. B*, **164**, 298–310.

Harrison, R.G., Wintermeyer, S.F., and Odell, T.M. (1983). Patterns of genetic variation within and among gypsy moth, *Lymantria dispar* (Lepidoptera: Lymantriidae), populations. *Ann. ent. Soc. Am.*, **76**, 652–6.

Hendrikse. A. (1986). Intra- and interspecific sex-pheromone communication in the genus *Yponomeuta*. *Physiol. Ent.*, **11**, 159–69.

Henning, V. and Yanofsky, C. (1963). An electrophoretic study of mutationally altered A proteins of the tryptophane synthetase of *Escherichia coli*. *J. molec. Biol.*, **6**, 16–21.

Herrebout, W.M., Kuijten, P.J., and Wiebes, J.T. (1976). Small ermine moths of the genus *Yponomeuta* and their host relationships (Lepidoptera, Yponomeutidae). *Symp. biol. Hung.*, **16**, 91–4.

Hoebeke, E.R. (1987). *Yponomeuta cagnagella* (Lepidoptera: Yponomeutidae): A palearctic ermine moth in the United States, with notes on its recognition, seasonal history, and habits. *Ann. ent. Soc. Am.*, **80**, 462–7.

Hung, A.C.F., Hedlund, R.C., and Day, W.H. (1986). High level of genetic heterozygosity in the hyperparasitic wasp, *Mesochorus nigripes*. *Experientia*, **42**, 1050–1.

Hunter, R. and Markert, C. (1957). Histochemical demonstration of enzymes separated by zone electrophoresis in starch gels. *Science*, **125**, 1294–5.

Jaenike J. and Selander, R.K. (1980). On the question of host races in the fall webworm, *Hyphantria cunea*. *Entomologia exp. appl.*, **27**, 31–7.

Johnson, G.B. (1976). Hidden alleles at the α-glycerophosphate dehydrogenase locus in *Colias* butterflies. *Genetics*, **83**, 149–67.

Johnson, G.B. (1977*a*). Assessing electrophoretic similarity: the problem of hidden heterogeneity. *A. Rev. Ecol. Syst.*, **8**, 309–28.

Johnson, G.B. (1977*b*). Characterization of electrophoretically cryptic variation in the alpine butterfly *Colias meadii*. *Biochem. Genet.*, **15**, 665–93.

Kimura, M. (1968). Evolutionary rate at the molecular level. *Nature, Lond.*, **217**, 624–6.

Kimura, M. (1982). The neutral theory as a basis for understanding the mechanism of evolution and variation at the molecular level. In *Molecular evolution, protein polymorphism and the neutral theory* (ed. M. Kimura), pp. 3–56. Japan Scientific Societies Press, Tokyo, and Springer-Verlag, Berlin, Heidelberg and New York.

King, J.L. and Ohta, T. (1975). Polyallelic mutational equilibria. *Genetics*, **79**, 681–91.

Kondrashov, A.S. (1986). Multilocus model of sympatric speciation. III.

Computer simulations. *Theor. Pop. Biol.*, **29**, 1–15.

Leibenguth, F. (1986). Genetics of the flour moth, *Ephestia kühniella. Agric. Zool. Rev.*, **1**, 39–72.

Lewontin, R.C. (1985). Population genetics. *A. Rev. Genet.*, **19**, 81–102.

Lewontin, R.C. and Hubby, J.L. (1966). A molecular approach to the study of genic heterozygosity in natural populations. II. Amount of variation and degree of heterozygosity in natural populations of *Drosophila pseudoobscura. Genetics*, **54**, 595–609.

Liebherr, J.K. (1988). Gene flow in ground beetles (Coleoptera: Carabidae) of differing habitat preference and flight-wing development. *Evolution*, **42**, 129–37.

Loukas, M., Economopoulos, A.P., Zouros, E., and Vergini, Y. (1985). Genetic changes in artificially reared colonies of the olive fruit fly (Diptera, Tephritidae). *Ann. Ent. Soc. Am.*, **78**, 159–65.

Löfstedt, C., Herrebout, W.M., and Du, J.-W. (1986). Evolution of the ermine morth pheromone tetradecyl acetate. *Nature, Lond.*, **323**, 621–3.

Lynch, M. (1986). Random drift, uniform selection, and the degree of population differentiation. *Evolution*, **40**, 640–3.

Manwell, C. and Baker, C.M.A. (1963). A sibling species of sea cucumber discovered by starch gel electrophoresis. *Comp. Biochem. Physiol.*, **10**, 39–53.

Maruyama, T. (1972). Distribution of gene frequencies in a geographically structured finite population. I. Distribution of neutral genes and of genes with small effect. *Ann. hum. Genet.*, **35**, 411–23.

Maruyama. T. and Kimura, M. (1980). Genetic variability and effective population size when local extinction and recolonization are frequent. *Proc. natn. Acad. Sci. USA*, **77**, 6710–4.

Maynard Smith, J. (1966). Sympatric speciation. *Am. Nat.*, **100**, 637–50.

Mayr, E. (1963). *Animal species and evolution.* Harvard University Press, Cambridge, Mass.

Mayr, E. (1976). *Evolution and the diversity of life.* Belknap Press, Cambridge and London.

McCauley, D.E. and Eanes, W.F. (1987). Hierarchical population structure analysis of the milkweed beetle, *Tetraopes tetraophthalmus* (Forster). *Heredity*, **58**, 193–201.

McLellan, T. (1984). Molecular charge and electrophoretic mobility in cetacean myoglobins of known sequence. *Biochem. Genet.*, **22**, 181–200.

Menken, S.B.J. (1980). Inheritance of allozymes in *Yponomeuta*. II. Interspecific crosses within the *padellus*-complex and reproductive isolation. *Proc. K. ned. Akad. Wet.*, **C83**, 425–31.

Menken, S.B.J. (1981). Host races and sympatric speciation in small ermine moths. *Entomologia. exp. appl.*, **30**, 280–92.

Menken, S.B.J. (1982). Biochemical genetics and systematics of small ermine moths (Lepidoptera, Yponomeutidae). *Z. zool. Syst., Evolut.-forsch.*, **20**, 131–43.

Menken, S.B.J. (1987). Is the extremely low heterozygosity level in

Yponomeuta rorellus caused by bottlenecks? *Evolution*, **41**, 630–7.

Menken, S.B.J. (1989). Identification of early immature stages by means of allozyme gel electrophoresis. In *Tortricoid pests, their biology. natural enemies and control. World Crop Pests*. Elsevier, Amsterdam. (In press.)

Menken, S.B.J. and Ulenberg, S.A. (1986). Allozymatic diagnosis of four economically important *Liriomyza* species (Diptera, Agromyzidae). *Ann. app. Biol.*, **109**, 41–7.

Menken, S.B.J. and Ulenberg, S.A. (1987). Biochemical characters in agricultural entomolgy. *Agric. Zool. Rev.*, **2**, 305–60.

Mukherjee, M., Skibinski, D.O.F., and Ward, R.D. (1987). A simulation study of the neutral evolution of heterozygosity and genetic distance. *Heredity*, **58**, 413–23.

Nei, M. (1972). Genetic distance between populations. *Am. Nat.*, **106**, 283–92.

Nei, M. (1987). Definition and estimation of fixation indices. *Evolution*, **40**, 643–5.

Nei, M. and Chakravarti, A. (1977). Drift variances of F_{ST} and G_{ST} statistics obtained from a finite number of isolated populations. *Theor. Pop. Biol.*, **11**, 307–25.

Nei, M. and Feldman, M.W. (1972). Identity of genes by descent within and between populations under mutation and migration pressures. *Theor. Pop. Biol.*, **3**, 460–5.

Nei, M., Maruyama, T., and Chakraborty, R. (1975). The bottleneck effect and genetic variability in populations. *Evolution*, **29**, 1–10.

Nevo, E., Beiles, A., and Ben-Shlomo, R. (1984). The evolutionary significance of genetic diversity: ecological, demographic and life history correlates. *Lect. Notes Biomath.*, **53**, 13–213.

Oxford, G.S. and Rollinson, D. (ed.)(1983). *Protein polymorphism: Adaptive and taxanomic significance.* Academic Press, London and New York.

Oppenoorth, F.J. (1985). Biochemistry and genetics of insecticide resistance. In *Comprehensive insect physiology, biochemistry and pharmacology*, Vol. 12. *Insect control* (ed. G.A. Kerkut and L.I. Gilbert), pp. 731–73. Pergamon Press Oxford.

Owen, R.E. (1985). Difficulties with the interpretation of patterns of genetic variation in the eusocial Hymenoptera. *Evolution*, **39**, 201–5.

Pashley, D.P. (1986). Host-associated genetic differentiation in fall armyworm (Lepidoptera: Noctuidae): A sibling species complex? *Ann. ent. Soc. Am.*, **79**, 898–904.

Pashley, D.P. and Johnson, S.J. (1986). Genetic population structure of migratory moths: the velvetbean caterpillar (Lepidoptera: Noctuidae). *Ann. ent. Soc. Am.*, **79**, 26–30.

Pashley, D.P. and Martin, J.A. (1987). Reproductive incompatibility between host strains of the fall armyworm (Lepidoptera: Noctuidae). *Ann. ent. Soc. Am.*, **80**, 731–3.

Povel, G.D.E. (1986). Pattern detection within the *Yponomeuta padellus*-complex of the European small ermine moths (Lepidoptera, Yponomeutidae). I. Biometric description and recognition of groups by numerical taxonomy. *Proc. K. ned. Akad. Wet.*, **C89**, 425–41.

Raftos, D.A. and Hughes, P.B. (1986). Genetic basis of a specific resistance to malathion in the Australian sheep blow fly, *Lucilia cuprina* (Diptera: Calliphoridae). *J. econ. Entomol.*, **79**, 553–7.

Ramshaw, J.A.M., Coyne, J.A., and Lewontin, R.C. (1979). The sensitivity of gel electrophoresis as a detector of genetic variation. *Genetics*, **93**, 1019–37.

Righetti, P.G. and Gianazza, E. (1987). Isoelectric focusing in immobilized pH gradients: theory and newer methodology. *Meth. biochem. Anal.*, **32**, 215–78.

Ross. K.G., Fletcher, D.J.C. and May, B. (1985). Enzyme polymorphisms in the fire ant, *Solenopsis invicta* (Hymenoptera: Formicidae). *Biochem. Syst. Ecol.*, **13**, 29–33.

Roush, R.T. and McKenzie, J.A. (1987). Ecological genetics of insecticide and acaricide resistance. *A. Rev. Ent.*, **32**, 361–80.

Sheppard, W.S. and Heydon, S.L. (1986). High levels of genetic variability in three male-haploid species (Hymenoptera: Argidae, Tenthredinidae). *Evolution*, **40**, 1350–3.

Singh, P. (1984). Insect diets: Historical developments, recent advances and future prospects. In *Advances and challenges in insect rearing* (ed. E.C. King and N.C. Leppla), pp. 32–44. Agricultural Research Series. USDA, Washington, DC.

Singh, R.S. (1979). Genic heterogeneity within electrophoretic 'alleles' and the pattern of variation among loci in *Drosophila pseudoobscura*. *Genetics*, **93**, 997–1018.

Singh, R.S. (1983). Genetic differentiation for allozymes and fitness characters between mainland and Bogotá populations of *Drosophila pseudoobscura*. *Can. J. Genet. Cytol.*, **25**, 590–604.

Singh, R.S. and Rhomberg, L.R. (1987). A comprehensive study of genic variation in natural populations of *Drosophila melanogaster*. II. Estimates of heterozygosity and patterns of geographic differentiation. *Genetics*, **117**, 255–71.

Singh, R.S., Lewontin, R.C., and Felton, A.A. (1976). Genetic heterogeneity within electrophoretic 'alleles' of xanthine dehydrogenase in *Drosophila pseudoobscura*. *Genetics*, **84**, 609–29.

Slatkin, M. (1977). Gene flow and genetic drift in a species subject to frequent local extinctions. *Theor. Pop. Biol.*, **12**, 253–62.

Slatkin, M. (1981). Estimating levels of gene flow in natural populations. *Genetics*, **99**, 323–35.

Slatkin, M. (1985). Rare alleles as indicators of gene flow. *Evolution*, **39**, 53–65.

Southwood, T.R.E. (1962). Migration of terrestrial arthropods in relation to habitat. *Biol. Rev.*, **37**, 171–214.

Stanley, S.M. (1979). *Macroevolution: Pattern and process*. Freeman, San Francisco.

Stock, M.W. and Robertson, J.L. (1982). Esterase polymorphism and response to insecticides during larval development of the western spruce budworm. *J. econ. Entomol.*, **75**, 183–7.

Strong, O.R., Lawton, J.H., and Southwood, T.R.E. (1984). *Insects on plants.* Harvard University Press, Cambridge, Mass.

Thorpe, J.P. (1982). The molecular clock hypothesis: biochemical evolution, genetic differentiation and systematics. *A. Rev. Ecol. Syst.*, **13**, 139–68.

Thorpe, J.P. (1983). Enzyme variation, genetic distance and evolutionary divergence in relation to levels of taxonomic separation. In *Protein polymorphism: Adaptive and taxonomic significance* (ed. G.S. Oxford and D. Rollinson), pp. 131–52. Academic Press, London and New York.

Via, S. (1986). Genetic covariance between oviposition preference and larval performance in an insect herbivore. *Evolution*, **40**, 778–85.

Wahlund, S. (1928). Zusammensetzung von Populationen und Korrelationserscheinungen vom Standpunkt der Vererbungslehre aus betrachtet. *Hereditas*, **11**, 65–106.

Watt, W.B. (1985). Allelic isozymes and the mechanistic study of evolution. *Isozymes*, **12**, 89–132.

Weir, B.S. and Cockerham, C.C. (1984). Estimating *F*-statistics for the analysis of population structure. *Evolution*, **38**, 1358–70.

Wiebes, J.T. (1976). The speciation process in the small ermine moths. *Neth. J. Zool.*, **26**, 440.

Wills, C. (1973). In defense of naive pan-selectionism. *Am. Nat.*, **107**, 23–34.

Wilson, A.C., Carlson, S.S., and White, T.J. (1977). Biochemical evolution. *A. Rev. Biochem.*, **46**, 573–639.

Wood, T.K. (1987). Host plant shifts and speciation in the *Enchenopa binotata* Say complex. In *Proc. 2nd Int. Workshop on Leafhoppers and Planthoppers of Economic Importance, Provo, Utah USA* (ed. M.R. Wilson and L.R. Nault), pp. 361–8. CAB International Institute of Entomology, London.

Woods, P.E. and Guttman, S.I. (1987). Genetic variation in *Neodiprion* (Hymenoptera: Symphyta: Diprionidae) sawflies and a comment on low levels of genetic diversity within the Hymenoptera. *Ann. Ent. Soc. Am.*, **80**, 590–9.

Wright, S. (1931). Evolution in Mendelian populations. *Genetics*, **16**, 97–159.

Wright, S. (1978). *Evolution and the genetics of populations.* Vol. 4. *Variability within and among natural populations.* University of Chicago Press, Chicago.

Wright, T.R.F. and MacIntyre, R. (1965). Heat stable and heat labile esterase-6 F enzymes in *Drosophila melanogaster* produced by different esterase-6 F alleles. *J. Elisha Mitchell scient. Soc.*, **81**, 17–9.

10 Study of the variation of aphid populations using enzymes and other traits

KLAUS WÖHRMANN and JÜRGEN TOMIUK

Biological Institute, Tübingen, Federal Republic of Germany

Abstract

The variability of different genetic traits — allozymes, insecticide resistance, body colour, reproductive rate, generation time and host specificity — were investigated within and between populations of the rose grain aphid *Macrosiphum rosae*. Polymorphism at some enzyme loci indicated balancing selection to be the mechanism responsible for maintaining observed variability in this species, whereas polymorphism in 16 other aphid species surveyed over a range of enzyme loci generally supported the neutral mutation hypothesis.

With the red or green forms of *M. rosae*, 'trade-offs' were observed with respect to different host plants, i.e. there were found to be interactions between the reproductive rates and insecticide resistance of clones and their host plants. Furthermore, the degree of endoparasitism by hymenopterous parasitoids was observed to depend on the colour morph and host plant. No correlation was observed between enzyme phenotypes and these particular quantitative characters, although genotypic heterogeneity in the European *M. rosae* populations studied appeared to depend on the life-cycle strategy employed—holocyclic (with sexual phase) or anholocyclic (asexual). Investigations on other aphid species also support the hypothesis that sexuality (combined with migration) governs the genetic structure of aphid populations.

Electrophoretic Studies on Agricultural Pests (ed. Hugh D. Loxdale and J. den Hollander), Systematics Association Special Volume No. 39, pp. 203–29. Clarendon Press, Oxford, 1989.

Genetic implications

As a rule, the individuals of a species population show variability in their characteristics. The extent of this variability, and its distinction from that of other populations of the same or different species, may give information about the population's history and its possible future evolution. The type and extent of this variation indicate factors that keep the variability stable or cause changes. Such knowledge is of purely theoretical interest — simply for the understanding of evolution, to find out where living organisms, including ourselves have come from and where possibly we are all going! Such knowledge is, however, also of applied interest in order to judge and to predict the response of the ecosystem to a man-made change. For example, an insect population under the selective pressure of an insecticide might be considered. In this instance, three questions arise: How much time does the population need to become resistant? What is the response if selection pressure is interrupted? What is the influence on neighbouring populations?

In assessing variation, it must be borne in mind that phenotypic variability is what is actually observed. Phenotypic variation is the result of the genotypic variation in interaction with the internal and external environment. Selection acts on the phenotype in total; the history of evolution, however, is described by changes in gene frequencies. At the end of each generation, the current geno-phenotypes and, therefore, the present phenotypic variation becomes extinct, whilst prior to this (by recombination) new geno-phenotypic variation is built-up. Evolution is based on genotypic variation only (Loeschcke and Wöhrmann 1984).

The phenotype is the result of the interaction between the genome and its internal and external environment. By reason of its complexity, it does not easily respond to genetic analysis. Using methods employed in quantitative genetics, attempts have been made to break down the phenotypic variation into its components. However, in the case of asexual reproduction as, for example, in aphids, genetic variances and covariance cannot be partitioned into additive and non-additive components (Falconer 1984; Futuyma and Phillipi 1987). On the other hand, by the use of aphid clones, parthenogenetically reproducing species offer the opportunity to estimate genotype–environment interations.

In aphids, investigations on variation can be performed on characteristics of different genetic complexity. For any particular case, neither the number of genes involved nor the contribution of a single gene can be determined exactly. Traits like reproductive rate, generation time, life span and others that are important for the understanding of, for example, population growth, belong to this category. From quantitative characteristics, traits can, however, be distinguished that seem to be determined by only one or a few genes, e.g. resistance against insecticides. In such cases, the effect of one 'major gene' becomes evident over and above the background of a series of other genes involved in the determination of the trait.

In view of these difficulties, there was understandable enthusiasm by aphid as well as other population geneticists following the introduction of electrophoretic techniques. By these techniques, enzyme variants (allozymes) could be observed and assigned to certain structural loci on the basis of the one-gene–one-enzyme hypothesis. The pattern on the gel corresponds both to the phenotype and the genotype. However, the demonstration of the existence of an allozyme does not provide information about its relative contribution to the fitness of an organism. In other words, the selective value of an allozyme cannot be estimated beyond doubt experimentally. Nevertheless, in the case of parthenogenetically reproducing aphid species and during the parthenogenetic phase of an otherwise holocyclic life cycle, enzyme loci are useful as markers for the whole genotype. In the meantime, the interest of population geneticists generally has shifted back to those quantitative characters that are primarily involved in life-cycle strategies, e.g. age structure and reproductive parameters.

In this chapter, we do not consider the type of polymorphism represented by the occurrence of different morphs during the life cycle of aphids. Despite the genetically determined differences between female and male aphids (XX and XO sex chromosomes respectively), all female morphs (apterae, alatae, gynoparae, and oviparae) should have the same genome. Differences between the morphs are due to different gene activities at different stages. Rather, we wish to concentrate on investigations of different trait groups, initially in our main object of study, the rose aphid, *Macrosiphum rosae* (L). The results will be compared and discussed with those of other aphid species. Lastly, an attempt will be made to draw more general conclusions with respect to the stability as well as the instability of aphid populations.

Biology of the rose aphid *Macrosiphum rosae*

The rose aphid, *Macrosiphum rosae*, is distributed world-wide. The aphid is holocyclic in regions with severe winters, although in temperate and warm climates, it is parthenogenetic all the year round. It is also heteroecious, its primary host being *Rosa* spp. and its secondary hosts, species of the families Valerianaceae and Dipsacaceae (Blackman and Eastop 1984).

Allozyme variation

Twenty-nine enzyme loci of the rose aphid were investigated; of these PGM, MDH, SDH, LAP, and EST* are polymorphic. The quaternary structure of PGM is monomeric, of MDH is dimeric, and of SDH is tetrameric (Tomiuk *et al.* 1979; Tomiuk and Wöhrmann 1983), whilst that of EST and LAP represent enzyme complexes in which the structure of each enzyme may be different (Wöhrmann *et al.* 1986). The genetics of the EST phenotypes are unknown. However, four classes of isoenzymes may be distinguished. Two different phenotypes are observed in the case of LAP. On the basis of these investigations, the average degree of heterozygosity over all loci investigated was found to be $\overline{H} = 0.032$ (Tomiuk and Wöhrmann 1980, 1983). Rhomberg *et al.* (1985) estimate a similar value $\overline{H} = 0.043$ for Canadian rose aphid populations. An average $\overline{H}$ of 0.01 was calculated by Tomiuk (1987) for 16 aphid species. In comparison with other species, this is a very low value. *Drosophila* spp. and other invertebrates are reported to have an $\overline{H}$ value of 0.12 (Ayala 1982).

During the last decade, data have been published on enzyme polymorphisms in many other aphid species (Chitty 1970; Beranek 1974; Baker 1977, 1978, 1979; May and Holbrook 1978; Wool *et al.* 1978; Takada 1979; Suomalainen *et al.* 1980; Rhomberg *et al.* 1985; Loxdale and Brookes 1987; Loxdale *et al.* 1985*a*, *b*; Wöhrmann *et al.* 1986; Hales and Lardner 1988; Hales *et al.* 1988). None of these will be discussed here. Suffice it to say that all of them confirmed the comparatively low degree of polymorphism in aphids. However, there is one remarkable exception in the case of the grain aphid, *Sitobion avenae*, which was polymorphic at 16 out of 25 loci and 5 out of 26 loci, respectively, in investigations performed over two successive years. The number of alleles has been reported to be up to eight per

* Full names of enzymes as abbreviated in text are PGM = phosphoglucomutase, EC 2.7.5.1; MDH = malate dehydrogenase, EC 1.1.1.37; SDH = sorbitol dehydrogenase, EC 1.1.1.14; LAP = leucine-amino peptidase, EC 3.4.11; EST = esterase, EC 3.1.1.

locus at some highly polymorphic ones, e.g. PEP-5 (Loxdale *et al.* 1985*b*).

We shall consider only those publications that are concerned with populations within large regions and/or over several years and consequently may give indications on the stability/instabilit of pop-latqoxs. ThesH qnclude investigations by Steiner *et al.* (1985) and Voegtlin *et al.* (1987) on *Rhopalosiphum maidis*, Singh and Rhomberg (1984) on *Aphis pomi*, Loxdale *et al.* (1985*a*) on *Sitobion avenae*, Takada (1986) on *Myzus persicae*, and Tomiuk and Wöhrmann (1982, 1984) and Wöhrmann and Tomiuk (1988) on *M. rosae.*

M. rosae was sampled in different climatic regions in Europe: Norway (mountains, coast), the United Kingdom (Scotland, Wales, England), Ireland, Denmark, West Germany, South Switzerland, North Italy, Hungary, Austria, Turkey, North Spain, Madeira, and the Canary Isles. The samples were investigated with respect to their variation in the enzymes MDH and PGM. Data from the population in Tübingen (West Germany) in the years 1977–1984 (Tomiuk and Wöhrmann 1984; Tomiuk *et al.* in press) and during the aphids' parthenogenetic phase within one life cycle have been gathered (Table 10.1) (Tomiuk and Wöhrmann 1981). From the data, the following may be concluded: the deviation from Hardy–Weinberg (H–W) expectation within a subpopulation (within a rose bed) is most significant in spring when aphids hatch from eggs. During the parthenogenetic phase, an approach to H–W expectations is observed because of migration between subpopulations (Wöhrmann and Tomiuk 1988).

With MDH, the ratio of homozygotes to heterozygotes remains constant, whereas the relation of the two homozygotes to each other may vary within one year and from year to year in populations from Tübingen. In 1977–1979, population density was correlated with homozygote genotype frequencies during the parthenogenetic phase (Tomiuk and Wöhrmann 1981).

This does not apply to the same extent to the PGM-locus (Fig. 10.1 below). Of the six possible genotypes, the m/m and m/f are the most common in Tübingen populations. The f/m genotype increases in frequency during the growth phase and decreases during the phase of population decline. The opposite is true for the m/m genotype.

The populations within one region, e.g. UK (Wales, England, Scotland) or middle Europe (Denmark, West Germany, Austria, etc.), and the Canary Isles, show great similarities in genotypic composition, i.e. the same genotypes occur in similar frequencies. Great differences can be observed, however, between regions with respect to the genotypes and their frequencies. To give examples of

Table 10.1 Frequencies of MDH collected genotypes and the s-allele in samples of N individuals from populations of *Macrosiphum rosae* collected in 1977, 1978, and 1979 on nineteen different sampling dates. χ^2-test on the ratio of homozygotes to heterotygotes.

Sampling dates	MDH-frequencies			N		Allele	
	s/s	s/f	f/f	s/s + f/f	s/f	s	
1977							
25.5.	0.25	0.50	0.25	14	14	0.50	
4.6.	0.28	0.59	0.13	39	57	0.58	$n = 4$
18.6.	0.32	0.42	0.26	22	16	0.53	$\chi^2 = 3.515$
4.7.	0.14	0.51	0.35	18	19	0.40	$p = 0.319$
1978							
7.6.	0.05	0.51	0.44	147	151	0.31	
15.6.	0.13	0.43	0.44	216	159	0.35	
20.6.	0.17	0.52	0.30	165	181	0.43	
27.6.	0.15	0.48	0.37	220	199	0.39	
6.7.	0.23	0.51	0.26	57	60	0.49	$n = 7$
12.7.	0.10	0.57	0.34	86	114	0.39	$\chi^2 = 14.907$
17.7.	0.07	0.44	0.49	49	39	0.29	$p = 0.021$
1979							
6.6.	0.36	0.43	0.21	43	32	0.48	
13.6.	0.32	0.52	0.16	77	84	0.58	
20.6.	0.31	0.48	0.21	97	90	0.55	
27.6.	0.37	0.49	0.14	97	94	0.52	
7.7.	0.28	0.52	0.20	89	97	0.54	
17.7.	0.37	0.44	0.23	71	55	0.59	$n = 8$
24.7.	0.34	0.40	0.27	115	76	0.54	$\chi^2 = 9.957$
6.8.	0.31	0.44	0.25	62	47	0.53	$p = 0.191$

Overall χ^2-test: $n = 19$, $\chi^2 = 31.593$, $p = 0.025$.

both findings, in Table 10.2 genotypic frequencies of the regions UK, North Norway, Switzerland, and Canary Isles are presented. (For further information, see Tomiuk and Wöhrmann 1984 and Wöhrmann and Tomiuk 1988). With parthenogenetic propagating species, the identity measure from Hedrick (1971)

$$I = \frac{\sum_j P_{xj} P_{yj}}{\frac{1}{2}\left(\sum_j P_{xj}^2 + \sum_j P_{yj}^2\right)} \tag{10.1}$$

is the most appropriate one for estimating the genetic relatedness of

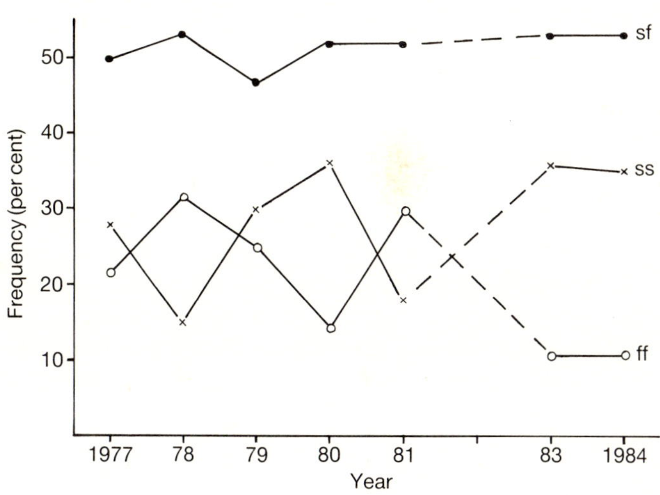

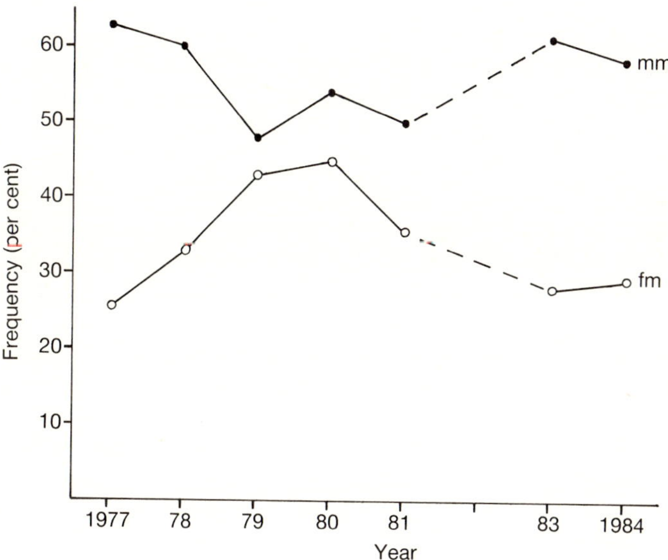

Fig. 10.1. Frequencies of (a) MDH and (b) PGM genotypes in *Macrosiphum rosae* populations in 1977–1984 at Tübingen. For PGM, only the most frequent genotypes (m/m, m/f) of the six possible are considered.

populations. In this formula, P_{xj} and P_{yj} are the frequencies of the jth genotype in the populations x and y. In the case of increasingly high identities between populations, I approaches 1.0; in the case of increasingly low identities, I approaches 0. It is apparent that the I values between populations on the Canary Isles are 1.0; within the region of Switzerland I is greater than 0.95; in the UK, I varies between 0.81 and 0.92; and in North Norway it ranges between 0.94 and 0.97 (Table 10.3). In contrast, the I values between regions indicate great differences (Table 10.3, e.g. Switzerland/North Norway 0.03–0.04). It is assumed that the high identities within regions are due to the high migration taking place during all life stages of *M. rosae*. Differences between regions could be due to selection or clinal drift, or both.

Comparison of the genotypic variability is possible using the expression:

$$V = \sum_{j=1}^{k} P_{xj}^2, \qquad (10.2)$$

where P_{xj} is the frequency of the jth genotype in the population with k genotypes (Hedrick 1971). In the case of *M. rosae*, 16 genotypes in total are possible for MDH and PGM. A value of $V = 1$ indicates no variability; decreasing values of V indicate increasing variability. On comparison (Table 10.4), it is found that populations from sites with severe winters (average temperature in January 0°C) show a high variability, whereas at more moderate sites (average temperature in January > 0°C) all possible V values are observed (Wöhrmann and Tomiuk 1988). This observation is understandable when sexuality i.e. holocycly enhances variability, whilst in the case of asexuality (parthenogenesis; anholocycly), fixation becomes more probable because of selection and/or random genetic drift (bottlenecks, founder effects).

This view is confirmed by studies on Canadian rose aphid populations (Rhomberg *et al.* 1985). There are large changes in genotype frequencies within one year, although differences are levelled out between years. The influence of the reproductive mode on the genotypic variability is also supported by investigations of Steiner *et al.* (1985) and Vocgtlin *et al.* (1987), who examined populations of *Rhopalosiphum maidis* in several southern states of the USA to explain the dispersal of populations in Illinois. *R. maidus* is anholocyclic. Survival through the winter is not possible in the north (i.e. Illinois) and the populations have to be re-founded each year. Nine enzyme loci out of a total of 21 are polymorphic and six were used in this

Table 10.2 Genotype frequencies at the MDH and PGM loci in populations of *Macrosiphum rosae* from different climatic regions. N = sample size.

Region Location			MDH			PGM					
	Year	N	f/s	s/s	f/f	f/m	f/s	m/s	f/f	m/m	s/s
Canary Isles											
La-Laguna	1984	136	0.00	0.00	1.00	0.00	0.00	0.00	0.00	1.00	0.00
Santa Cruz	1984	113	0.00	0.00	1.00	0.00	0.00	0.00	0.00	1.00	0.00
San Jose	1984	37	0.00	0.00	1.00	0.00	0.00	0.00	0.00	1.00	0.00
Switzerland											
Lago di Como	1985	25	0.80	0.00	0.20	0.08	0.00	0.00	0.00	0.92	0.00
Lago di Lugano	1985	25	0.96	0.00	0.04	0.00	0.00	0.00	0.00	1.00	0.00
Lago Maggiore	1981	180	0.97	0.00	0.03	0.00	0.00	0.00	0.00	1.00	0.00
UK											
England	1979	505	0.45	0.42	0.13	0.00	0.00	0.10	0.00	0.90	0.00
Wales	1979	191	0.49	0.36	0.15	0.01	0.00	0.24	0.00	0.74	0.01
Scotland	1979	212	0.46	0.37	0.17	0.12	0.02	0.00	0.00	0.86	0.00
North Norway											
Gjörvik	1981	94	0.10	0.00	0.90	0.34	0.15	0.01	0.34	0.11	0.05
Hamar	1981	188	0.09	0.00	0.91	0.24	0.21	0.06	0.34	0.10	0.06
Lillhamar	1981	158	0.17	0.00	0.83	0.40	0.15	0.02	0.29	0.05	0.09

Table 10.3 The probability of genetic identity (I) between populations of different regions (diagonal) and between regions (upper right half).

Region	Switzerland	UK	North Norway
Switzerland	1.00	0.63–0.79	0.03–0.04
UK		0.81–0.92	0.06–0.10
North Norway			0.94–0.97

Table 10.4 Genetic variability ($V = \sum_j P_{xj}^2$) for populations from different regions in Europe and average of the lowest daily temperature (T) in January at that site.

Region	T (°C)	V
North Norway	−9.1	0.20
Hungary	−4.9	0.26
South Norway	−2.0	0.17
Austria	−1.4	0.18
Germany	−1.0	0.18
Denmark	0.0	0.18
Switzerland	1.9	0.56–0.92
UK	3.5–6.0	0.29–0.33
Turkey	8.3	0.46
Spain	9.4	0.68
Madeira	15.0	0.37
Canary Isles	17.8	1.00

Table 10.5 Genetic variability at six polymorphic enzyme loci in 38 populations of North American *Rhopalosiphum maidis* (after Voegtlin *et al.* 1987).

Year	Number of populations	Loci $6 \times N$	Number of loci				
			Fixed in 1 hom	het	2 hom	Poly.	No analysis
1983	15	90	81	2	—	6	1
1984	10	60	51	3	—	4	2
1985	13	78	65	—	10	3	—
total	38	228	197	5	10	13	3

investigation. Sampling 38 populations in total, $6 \times 38 = 288$ 'loci' were considered (Table 10.5), of which 197 were fixed in one or the other allele, five showed only heterozygotes, 10 had both homozygotes, and only 13 exhibited all three possible genotypes. This corresponds to the populations of *M. rosae* in mild climates where no sexual phase is expected. In addition, it corresponds to the findings of Loxdale *et al.* (1985*a*) on *Sitobion avenae*, which showed spatial and temporal differences between populations that were mainly caused by a prevalence of homozygous allozyme variation. Loxdale *et al.* suspect *S. avenae* to be of predominantly anholocyclic population structure. Similar results were reported by Takada (1986) on *Myzus persicae*. Singh and Rhomberg (1984) found a more complex population structure in *Aphis pomi* (Table 10.6). Out of the expected 12 gene combinations between the EST-1 locus (three alleles) and the GOT locus (two alleles) they found only 4, and out of the expected 9 combinations between the EST-2 locus (two alleles) and the GOT locus (two alleles) they found only six combinations. Sixteen more loci were monomorphic. This can only be understood by assuming the existence of two 'subpopulations' that differ in their life cycles. The division results in one population (A) being monomorphic for 16 loci and polymorphic in EST and GOT, whilst the other population (B) is monomorphic at all 19 loci. The authors assume a holocyclic life cycle for subpopulation A and anholocyclic one for B. In subpopulation A, the yearly sexual phase evens out the possible changes in gene frequencies occurring during the parthenogenetic phase and leads to an overall stable situation. The assumption is supported by the following findings for type A:

Table 10.6 Number of genotypes marked by the enzymes GOT with EST-1, and EST-2, respectively, observed in a Canadian population of *Aphis pomi*. (After Singh and Rhomberg 1984.)

	EST-1				
	Null	s/s	s/f	f/f	Total
GOT s/s	0	438	26	136	600
s/f	0	1	0	0	1
f/f	581	1	0	0	582
Total	581	440	26	136	1 183
	EST-2				
	I	II	III		Total
GOT s/s	162	447	0		609
s/f	0	1	0		1
f/f	1	0	549		550
Total	163	448	549		1 160

(1) relatively constant genotype frequencies in every year and every orchard;
(2) the fact that the genotypic frequencies are close to Hardy-Weinberg proportions;
and the following for type B:
(3) remarkable variation in frequency from orchard to orchard;
(4) a great fluctuation between years;
(5) increase of genetic disequilibrium during the year;
(6) the fact that a migration to type B from east to west can be observed every year.

Variation in resistance to insecticides

In the past, the genetic basis of insecticide resistance has been demonstrated in many species and groups of insects. Cases of monogenic resistance as well as of 'oligogenic', i.e. involving a small number of genes, have been demonstrated. However, comparatively few experiments have been performed on resistance to insecticides in aphid species. In 1977, Devonshire demonstrated that increased detoxification of organophosphorus and carbamates is associated with an enhanced esterase activity (EST-4) in *Myzus persicae.* Sawicki *et al.* (1978) described variation of resistance within clones without giving on explanation of the possible mechanisms. This is of great interest because Blackman (1979), Suomalainen *et al.* (1980), and Tomiuk and Wöhrmann (1982) have all found that recombination of enzyme loci within aphid clones could be excluded with high probability. Gene duplications seem to be responsible for the variability of resistance (Blackman 1975). Devonshire and Sawicki (1979) and Lauritzen (1982) demonstrated the extent of such duplications in greenhouse populations under selection pressure of insecticides. Interruption of this pressure leads to a decrease in the frequency of resistant clones. Sawicki *et al.* (1980) suggested that highly resistant clones probably could only be maintained in glasshouses by continuous exposure to insecticides. Recently, ffrench-Constant *et al.* (1988) have modified an immunoassay for detection of EST-4 (E4) that enables the discrimination of different levels of resistance (S, susceptible; R_1, R_2, R_3 increasing resistance levels). They also found that *M. persicae* clones established from natural populations often spontaneously lose high levels of resistance in the absence of selection pressure but can subsequently regain higher levels in the presence of selection. This loss and recovery of selection of resistance has only occurred in translocated clones. The observation is explained by the possibility of regulation of gene amplification.

Variation in resistance is reported not only in greenhouse populations, but also in natural *M. persicae* populations in England (e.g. Sawicki *et al.* 1978), Switzerland (Büchi and Haeni 1984) and Japan (Takada 1986). The variation found in Japan applies not only to resistance to organophosphorus and carbamate insecticides, but also to pyrethroids. In contrast to the finding of Devonshire and Sawicki (1979), the resistance factors in Japanese populations remained constant in frequency after selection was interrupted. Presumably these different reactions may be understood assuming different genetic backgrounds — as shown by Wood and Bishop (1981) for the housefly, *Musca domestica*. The existence of resistance factors defined by RAE resistance-associated esterase; equivalent to E4 activity levels, e.g. RAE + 6, are found to be spread over large regions, but do not account for more than 15 per cent in each sample (Table 10.7) (Takada, 1986).

Besides these investigations on *M. persicae*, data on *Macrosiphum rosae* have been reported (Wöhrmann *et al.* 1987). In these investigations, the mortality of clones is determined after treatment with different insecticides (Pirimor: pirimicarb, Beosit: endosulfan, Metasystox: demeton-*S*-methylsulphoxide). Before treatment, *M. rosae* clones were cultivated on different hosts. It was shown that the observed susceptibility to the insecticides depended not only on the clones and the host, but also on the interaction of both factors (Table 10.8). The discrimination analysis of these data revealed the existence of groups of clones that are significantly different in their response to insecticides (Fig. 10.2). However, there were no significant correlations between mortality rates caused by different

Table 10.7 Frequencies of clones with RAE activity levels (+3, +4, +5, +6) in samples of *Myzus persicae* collected from tobacco at different localities in Japan (after Takada 1986). *N* = number of clones.

Locality	*N*	Percentage of RAE variants		
		(+3)+(+4)	(+5)+(+6)	Rest
Hokkaido	189	70	12	18
Yamagata	46	83	0	17
Miura	36	64	14	22
Kyoto	44	68	0	22
Kuraskiki	35	34	0	66
Kyushu	78	60	1	39
Ryukyu	84	88	10	2

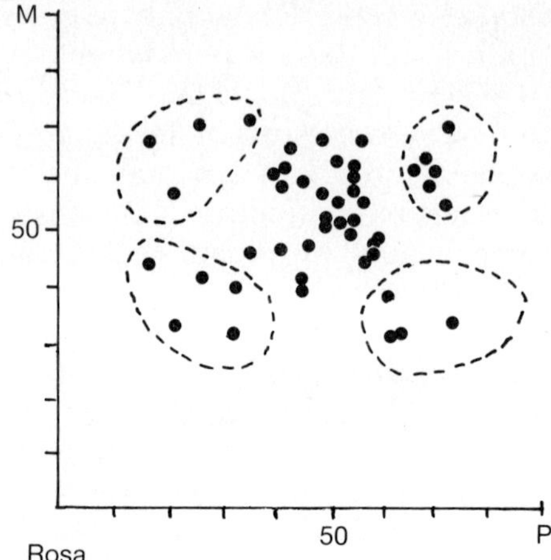

Fig. 10.2. The relationship between mortality rates of clones of *Macrosiphum rosae* grown on roses caused by Metasystox (M) and Pirimor (P). (After Wöhrmann *et al.* 1987.)

Table 10.8 Analysis of variance of the mortality rates of clones of *Macrosiphum rosae* treated with Pirimor, Beosit, and Metasystox. The clones were grown before treatment on *Rosa, Dipsacus fullonum*, and *D. sylvester*.

Source of variation	d.f.	Mean square sum	*F*
Host	2	1 571.21	13.49[a]
Insecticide	2	211 727.48	186.48[a]
Insecticide × host	4	747.25	6.42[a]
Clone	7	514.25	4.41[a]
Clone × host	14	354.19	3.04[a]
Clone × insecticide	14	626.19	5.37[a]
Clone × insecticide × host	28	510.36	4.38[a]
Error	168	116.52	

[a] Significant at the 5 per cent level

insecticides. This indicates a different genetic basis for the resistance to different insecticides. The same is true of the influence of the hosts on which the clones were propagated before treatment. *Rosa, Dipsacus fullonum* and *D. sylvester* were used. The influences of these hosts are not correlated (Fig. 10.3). However, a minor but non-significant correlation exists between the influence of *D. fullonum* and *D. sylvester*. An explanation of these findings might be the degree of

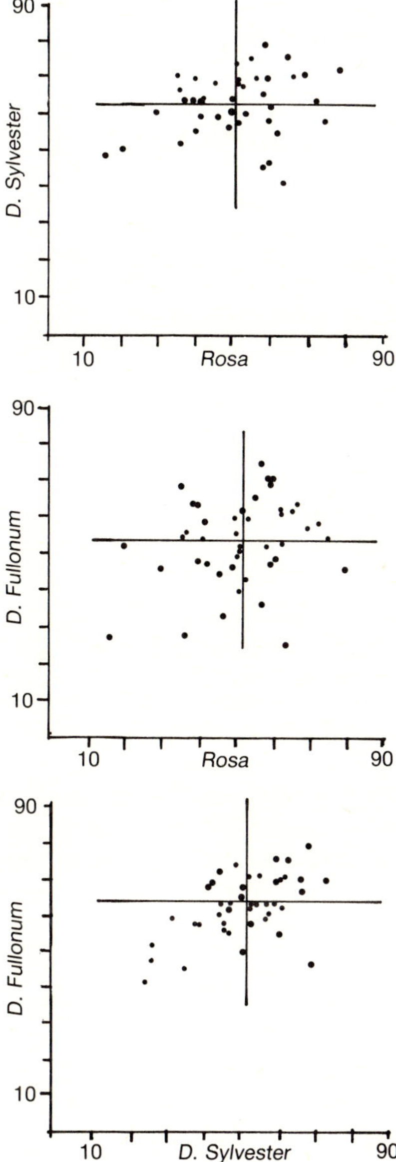

Fig. 10.3. Relationship of mortality rate of clones reared on different hosts (*Rosa*, *Dipsacus fullonum*, *D. sylvester*) before exposure to Pirimor. (After Wöhrmann *et al.* 1987.)

relatedness of both host species to each other. In the sample, studied, resistant and susceptible clones of *M. rosae* exhibited no differences in net reproductive rates and growth rates in the absence of selection pressure. This is in contrast to Eggers-Schumacher's (1983) findings on *M. persicae*. He showed that resistant clones have higher reproductive potential, but that susceptible clones have higher colonizing ability. He proposed that resistant clones were selected for high reproductive potential by intensive chemical control. Thus, in glasshouses resistant clones have an advantage, whereas, in unpredictable environments such as found in the field, susceptible clones are better adapted by their high colonizing ability on different host plants.

Variation in host specificity

Aphids are plant parasites. During the course of evolution they have developed, according to species, characteristic life cycles. Some spend all the year round on the same host (i.e they are monoecious), whilst others migrate from a primary host to a secondary host, returning to the primary host for mating and egg laying (i.e they are heteroecious). Some aphids feed on a wide range of secondary host species (polyphagy). In the past, such host specificities have been investigated several times (Dixon 1985*a*, *b*, 1987; Eastop 1973; Lees 1966).

To investigate aphid-plant interactions in *Macrosiphum rosae*, we sampled clones in and around Tübingen and determined the reproductive rate (m), generation time (T), and net reproductive rate (R_0). The reproductive rate (m) was estimated either by an exponential regression:

$$y = a\,\mathrm{e}^{mt}, \tag{10.3}$$

where y is the population size at time t and a is the initial density, or by the formula:

$$1 = \sum_{i=1}^{\infty} a_i\,\mathrm{e}^{-mi}, \tag{10.4}$$

where a_i is the number of larvae per age class i. The latter procedure estimates the actual reproduction per day and clone and considers the age structure of a population (cf. Tomiuk *et al.*, in press, for further details). The generation time (T) is the time between the birth and reproduction of the first larvae; the net reproductive rate (R_0) is the average number of larvae per individual. These parameters were estimated from clones reared in the laboratory under defined conditions on different hosts (i.e. *Rosa* spp., *Valeriana officinalis, Dipsacus fullonum* and *Dipsacus sylvester*). The investigation showed no significant differences between clones with respect to the generation time (T) and the net reproductive rate (R_0). However, there was a significant interaction between R_0 and host species. To obtain more detailed information, the clones were grouped according to their reproductive rates (m) relative to the mean of all clones cultivated on the same host species. Clones with rates higher than, or equal to, the mean are marked by a plus sign, clones with lower rates by a minus sign. The exact Tate-Clelland test (Lienert 1973) revealed that the sign combinations (clones on all four hosts) deviate significantly from a random distribution. It can be seen from Table 10.9 that none of the investigated clones are below average on all hosts. Only two clones, marked by an asterisk, were above average on all hosts. These two clones might indicate the existence of 'general purpose' genotypes. This analysis is, of

Table 10.9 The reproductive rates (m) of clones reared on four host-plant species. Clones with high or average reproduction on a particular host are marked by (+) and clones with low rates by (−).

Rosa	*D. fullonum*	*D. sylvester*	*V. officinalis*
0.18 +	0.18 −	0.22 +	0.15 +
0.18 +	0.16 −	0.22 +	0.10 −
0.17 +	0.17 −	0.23 +	0.10 −
0.17 +	0.18 −	0.22 +	0.10 −
0.17 +	0.21 +	0.21 −	0.08 −
0.16 +	0.19 +	0.20 −	0.14 +
0.17 +	0.22 +	0.21 −	0.10 −
0.15 +	0.22 +	0.22 +	0.11 −
0.17 +	0.18 −	0.22 +	0.12 +
0.18 +	0.19 +	0.23 +	0.12 + [a]
0.15 −	0.20 +	0.25 +	0.10 −
0.17 +	0.20 +	0.17 −	0.11 −
0.17 +	0.18 −	0.27 +	0.09 −
0.16 +	0.20 +	0.21 −	0.15 +
0.12 −	0.21 +	0.25 +	0.11 −
0.17 +	0.19 +	0.22 +	0.12 + [a]
0.17 +	0.20 +	0.19 −	0.16 +
0.14 −	0.19 +	0.22 +	0.14 +
0.14 −	0.20 +	0.22 +	0.16 +
$\overline{m} = 0.16$	$\overline{m} = 0.19$	$\overline{m} = 0.22$	$\overline{m} = 0.12$

[a] Above average on all hosts

course, a crude one and can both conceal as well as create apparent trends. For example, the two clones observed with an overall optimal function could be overvalued: the reproductive rate of one clone is actually higher than the mean only in the case of *Rosa* spp. and *D. sylvester*, whilst the other clone has above-average reproduction on *Rosa* spp. The existence of such 'general purpose' genotypes is in contrast to the results given by other investigators (McArthur 1972; Service and Lenski 1982). However, there is strong evidence for a close clone-host plant interaction (see, e.g., Iglisch 1968; Müller 1961, 1962; Shaposhnikov 1966, 1984; Moran 1981). In the *M. rosae* group, Holman (1972) described a species *M. knautia* that lives all the year round on *Knautia*, whilst Tomiuk *et al.* (unpublished) also found clones of *M. rosae* collected from *Knautia* that could not be propagated on roses, whereas all samples from roses could be grown on *Knautia*. The hypothesis of specialism can be tested in a linear regression model. Strong 'trade-offs' must result in significant negative correlations between traits related to host specificity. As may be seen from Table 10.10, we found predominantly negative correlations, which indicate the existence of trade-offs; but there is only one significant host combination, *Rosa/D. fullonum*, which shows that trade-offs do not generally exist, and appropriate plant 'contributions' must be chosen. However, in a further statistical procedure, the correlation coefficients were used to determine the number of factors (their genetic or physiological constitution is not defined!) that influence a series of quantitative traits. The factor analysis revealed that two factors mainly influence host preference (Table 10.11). One of these is responsible for growth on *Rosa* and *Valeriana officinalis*, the other for growth on *D. sylvester* and *D. fullonum*. There is a simple common effect of both factors: high fitness on one host signifies lower fitness on the other. The uniform tendency of both factors refers to the potential for a future possible evolution of trade-offs in rose aphids.

These results leads to the conclusion that apparently both strategies are employed by *M. rosae*: there are clones that become

Table 10.10 Estimates of the correlations of the reproductive rates (m) of *Macrosiphum rosae* on pairs of hosts.

	D. fullonum	*D. sylvester*	*V. officinalis*
Rosa	−0.47[a]	−0.27	−0.18
D. fullonum		−0.19	0.02
D. sylvester			−0.36

[a] Significant at the 5 per cent level

Table 10.11 Factor analysis of the data from Table 10.10. The reproductive rates on the four hosts are mainly influenced by two factors. The original factors are orthogonally rotated by about 45° and the unlinked factors (f1, f2) are represented.

	f1	f2
Rosa	- 0.70	0.05
D. fullonum	0.19	0.65
D. sylvester	0.11	- 0.60
V. officinalis	0.61	0.09

specialized for only one host within an otherwise heteroecious species (see Holman 1972; Tomiuk *et al.*, unpublished) and others that only develop potential host preferences. By means of host preferences, various subpopulations are formed during the course of the life cycle, thereby reducing, by spreading the risk, the possibility of a population crash. The usefulness of forming subpopulations in connection with predation by ladybird beetles was shown by Kareiva (1987). In discontinuous populations, the total population size was less reduced by predation than in continuous ones.

Trait combinations

It was shown in the previous section that character variation exists in *Macrosiphum rosae* that can be ascribed to a genetic basis. We now consider correlations between different traits. *M. rosae* is polymorphic for body colour. Red and green individuals are found in varying frequencies in the populations. There is no correlation between particular body colours and enzyme patterns (MDH, PGM). Both traits vary independently from each other (Tomiuk and Wöhrmann 1980; Tomiuk *et al.*, in press). Also, not host plant is settled preferentially by a given enzyme genotype.

In contrast to this finding, red morphs on roses exhibit a significantly higher reproductive rate than green ones (Table 10.12), whereas on the secondary hosts no such correlation could be shown. Theoretically, the higher reproductive rate of the red forms found on roses should lead to fixation of this colour. However, this expectation is in contradiction to our observations. The relationship of the red/green forms in the Tübingen population was observed to be constant over several years. This contradiction may be resolved by the observation that red forms showed a higher rate of endoparasitism

Table 10.12 Number of green and red clones with a mean reproductive rate above or equal to (+) and below (–) the mean value on *Rosa*.

Colour	(+)	(–)	χ^2
Green	38	42	
Red	26	14	7.53[a]

[a] Significant at the 5 per cent level

Table 10.13 The number of endoparasitized red and green forms of *Macrosiphum rosae* that could determined by electrophoretic methods in samples from natural populations in 1983 and 1984.

Colour	parasitized		
	No	Yes	χ^2
Green	1451	229	
Red	757	160	6.79[a]

[a] Significant at the 5 per cent level

by hymenopterous parasitoids than did the green ones (Table 10.13). Probably, the selection processes are antagonistic and therefore might cause a stabilized situation.

The importance of variation

The data presented allow an insight into the variation of different characters. Certainly these variations have a genetic basis. However, it is difficult to assess and to compare the extent of the variation at the protein level with that at the morphological and physiological level. Between populations and between species, the average degree of heterozygosity at enzyme loci $\overline{H}$ = average percentage of all heterozygotes at all loci investigated), and the degree of polymorphism (P = percentage of loci with more than one allele and the frequency of the most frequent allele not higher than 0.99) are usual measures of the genetic variability. In the case of aphid populations, both of them have to be interpreted with care, since P depends very much on the number of loci and the sample size, and $\overline{H}$ depends on the number of loci in addition to the breeding system. In a predominantly anholocyclic population in which only a few heterozygotes and mostly (different) homozygotes exists. $\overline{H}$ must give an underestimate of the existing genotypic variability. This was observed in populations of *Rhopalosiphum maidis* (Steiner *et al.* 1985; Voegtlin *et al.* 1987) and in *Sitobion avenae* (Loxdale *et al.* 1985*a*).

Estimates, as already mentioned, reveal a comparatively low enzyme variability in aphids. In general, however, investigations on numerous other species have shown a surprisingly large amount of variability that is not expected on the basis of selection theory. From theoretical considerations, it has been concluded that the large majority of mutants at the molecular level have negative, neutral, or at least quasineutral effect (Kimura, 1983, 1986*a, b*; Nei and Graur 1984). The disagreement between the neutralists and the selectionists that arose from this apparent discrepancy, will not be treated here in detail (see Dobzhansky 1970; Kimura 1983, 1986*a, b*; Nevo 1978; Ayala 1982, 1984; Nei and Graur 1984). However, suffice to say that, in aphids, the low variability at the protein level cannot be attributed uncritically to selection (Tomiuk 1987). This is also confirmed in a larger sample of hymenopteran species (Graur 1985). Nevertheless, strong indications were found for the selection of some traits, such as insecticide resistance, and the more complex traits, including host specificity, predation, and reproductive rates.

No correlations exist between variations in these characters and allozyme variation. This seems reasonable as, in general, the contribution of a single allozyme to the variation of a quantitative trait can be neglected. Only alleles coding for an essential protein might be an exception. The frequency change of an allozyme does not depend entirely on its own contribution to the trait, but rather on the selective values of the trait in total, and therefore, is more influenced by random genetic drift.

The data reported demonstrate the existence of genetic variation, which is believed to be the basis of any adaptability or future evolution. This is equally true for anholocyclic species e.g., *Rhopalosiphum maidis* (Steiner *et al.* 1985; Voegtlin *et al.* 1987), and for holocyclic species with varying proportions of clones exhibiting sexual and asexual propagation, e.g. *Sitobion fragariae* (Loxdale and Brookes 1987) or *Myzus persicae* (Takada 1986), and for the holocyclic species such as *Macrosiphum rosae* for which — at least in regions with severe winters — sexuality is assumed. Takada (1986) consolidated this point of view by looking at variation in a slightly different way. He described five esterase loci and expected, on the basis of the observed allozymes at these loci, 180 combinations, assuming independence between loci. If, in addition, body colours (yellow, green, red) are considered, theoretically 540 combinations are expected. In a sample of 1503 clones sampled at 26 sites in Japan from 23 host species, Takada found 76 of the possible combinations, including 26 'major types' and 49 'minor types'. The major types are defined on the basis of their frequencies and ubiquitous distribution. Such calculations could be extended *ad infinitum* if a greater number of traits is considered, whereby, however, free recombination is reduced by linkage with an increase of characters. At least, it may be shown that aphid species consist of a high number of genetically different clones.

A statement on the importance of this observed variability for the plasticity, maintenance and/or evolution of the species population would be speculation only. In any case, the enzymatic variability measured seems to be of minor significance in explaining the adaptability of aphid species, although they are useful markers for describing general mechanism(s). The total variability was/is great enough to facilitate the dispersal of species. The maintenance of this variability might be explained by particular population-biological factors. Dispersal in aphids is usually assisted by very high reproductive rates during the parthenogenetic phase, and by high migration ability. Large population sizes and splitting into subpopulations reduce the risk of genetic impoverishment and population collapse. Furthermore, local populations experience, during the year, several 'bottlenecks' caused by external factors, such climatic conditions, or by emigration to other host plants and the annual foundation of new populations in previously uncolonized regions. Because of such drift effects, even rare genotypes might also be successful, despite their otherwise relatively low reproductive rates.

Acknowledgement

This work was supported by several grants from the Deutsche Forschungsgemeinschaft. The manuscript was written during sabbatical leave of the senior author who was supported by a research fellowship from Macquarie University, N.S.W., Australia. We are grateful to Drs Dinah Hales, Hugh Loxdale, and Alan Devonshire for many helpful suggestions and for improving our English manuscript. We thank the *Canadian Journal of Genetics and Cytology* for permission to use the data in Table 10.6 of this chapter.

References

Ayala, F.J. (1982). Genetic variation in natural populations: Problem of electrophoretic cryptic alleles. *Proc. natn. Acad. Sci. USA*, **79**, 550–4.

Ayala, F.J. (1984). Molecular polymorphism: how much is there and why is there so much? *Devl. Genet.*, **4**, 379–91.

Baker, J.P. (1977). Changes in composition in populations of peach potato aphid, *Myzus persicae*, overwintering in Scotland in 1976/77. *Proc. 1977 Br. Crop Conf. Pest & Dis.*, pp. 255–61.

Baker, J.P. (1978). Electrophoretic studies on populations of *Myzus persicae* in Scotland from March to July 1976. *Ann. appl. Biol.*, **88**, 1–11.

Baker, J.P. (1979). Electrophoretic studies on populations of *Myzus persicae* in Scotland from October to December 1976. *Ann. appl. Biol.*, **91**, 159–64.

Beranek, A.P. (1974). Inherited changes in enzyme patterns within parthenogenetic clones of *Aphis fabae*. *J. Ent. (A)*, **48**, 141–7.

Blackman, R.L. (1975). A preliminary report of the genetics of resistance to organophosphates in *Myzus persicae*. *Proc. 8th Br. Insectic. & Fungic. Conf.*, pp. 79–85.

Blackman, R.L. (1979). Stability and variation in aphid lineages. *Biol. J. Linn. Soc., (Lond.)*, **11**, 259–77.

Blackman, R.L. and Eastop, V.F. (1985). *Aphids on the world's crops.* (*An identification guide*). Wiley, London.

Büchi, R. and Haeni, A. (1984). Das Resistenzspektrum in Blattlaus-populationen von *Myzus persicae* (Sulz) in der Schweiz in den Jahren 1977/78 und 1982. *Z. angew. Ent.* **98**, 239–46.

Chitty, D. (1970). Variation and population density. *Symp. Zool. Soc. Lond.*, **26**, 327–33.

Devonshire, A.L. (1977). The properties of a carboxyesterase from the peach-potato aphid, *Myzus persicae* (Sulz), and its role in conferring insecticide resistance. *Biochem. J.*, **167**, 675–83.

Devonshire, A.L. and Sawicki, R.M. (1979). Insecticide-resistant *Myzus persicae* as an example of the evolution by gene duplication. *Nature, Lond.*, **280**, 140–1.

Dixon, A.F.G. (1985*a*). Structure of aphid populations. *A. Rev. Ent.*, **30** 155–74.

Dixon, A.F.G. (1986*b*). *Aphid ecology*. Blackie, Glasgow and London.

Dixon, A.F.G. (1987). Aphid reproductive tactics. In *Population structure, genetics and taxonomy of aphids and Thysanoptera* (ed. J. Holman, J. Pelikan, A.F.G. Dixon, and L. Weisman), pp. 3–18. SPB Academic Publishing, The Hague.

Dobzhansky, T. (1970). *Genetics of the evolutionary process*. Columbia University Press, New York and London.

Eastop, V.F. (1973). Biotypes and aphids. *Bull. Ent. Soc. N.Z.*, **2**, 40–52.

Eggers-Schumacher, H.A. (1983). A comparison of the reproductive performance of insecticide-resistant and susceptible clones of *Myzus persicae*. *Entomologia exp. appl.*, **34**, 301–7.

Falconer, D.S. (1984). *Einführung in die quantitative Genetik*. UTB Ulmer, Stuttgart.

ffrench-Constant, R.H., Devonshire, A.L., and White, R.P. (1988). Spantaneous loss and reselection of resistance in extremely resistant *Myzus persicae* (Sulzer). *Pestic. Biochem. & Physio.*, **30**, 1–10.

Futuyma, D.F. and Phillipi, T.E. (1987). Genetic variation and covariation in response to host plants by *Alsophila pometaria* (Lepidoptera: Geometridae). *Evolution*, **41**, 269–79.

Graur, D. (1985). Gene diversity in Hymenoptera. *Evolution*, **39**, 190–9.

Hales, D.F., and Lardner, R.M. (1988). Genetic evidence for the occurrence of a new species of *Neophyllaphis takahashi* in Australia (Homoptera: Aphididae). *J. Australian ent. Soc.* (in press).

Hedrick, P.W. (1971). A new approach to measuring genetic similarity. *Evolution*, **25**, 276–80.

Holman, J. (1972). Description of *Macrosiphum knautiae* sp. with notes on the taxonomy of the *Macrosiphum rosae* group. (Homoptera: Aphididae). *Acta ent. bohemosl.*, **69**, 175–85.

Iglisch, I. (1968). Über die Entstehung der Rassen der 'schwarzen Blattläuse' (*Aphis fabae* Scop. und verwandte Arten), über ihre phytopathologische Bedeutung und über die Aussichten für erfolgversprechende Bekämpfungsmassnahmen (Homoptera: Aphididae). *Mittelungen aus der Biologischen Bundesanstalt für Land- und Forstwirtschaft Berlin Dahlem*. Paul Parey, Hamburg.

Kareiva, P. (1987). Habitat fragmentation and the stability of predator-prey interactions. *Nature (Lond.)*, **326**, 388–390.

Kimura, M. (1983). *The neutral theory of molecular evolution*. Cambridge University Press, Cambridge.

Kimura, M. (1986*a*). Evolutionary rates at the molecular level. *Nature, Lond.*, **217**, 624–6.

Kimura, M. (1986*b*). Genetic variability maintained in a finite population due to mutational production of neutral isoalleles. *Genet. Res.*, **11**, 247–69.

Lauritzen, M. (1982). Q-band and G-band identification of 2 chromosomal rearrangements in peach-potato aphids, *Myzus persicae*, resistant to insecticides. *Hereditas*, **97**, 95–102.

Lees, A.D. (1966). The control of polymorphism in aphids. *Adv. Insect Physiol.*, **3**, 207–77.

Lienert, G.A. (1973). *Verteilungsfreie Methoden in der Biostatistik*. Verlag Anton Hain, Meisenheim am Glan.

Loeschcke, V., and Wöhrmann, K. (1984). Populationsbiologie. *Nature, Rdsch., Stultg.*, **37**. 305–8.

Loxdale, H.D. and Brookes, C.P. (1987). Use of electrophoretic markers to study spatial and temporal genetic structure of populations of a holocyclic aphid species — *Sitobion fragariae* (Walker) (Hemiptera: Aphididae). In *Population structure, genetics and taxonomy of aphids and Thysanoptera*. (ed. J. Holman, J. Pelikan, A.F.G. Dixon, and L. Weismann), pp. 100–10. SPB Academic Publishing, The Hague.

Loxdale, H.D., Castañera, P., and Brookes, C.P. (1983). Electrophoretic study of enzymes from cereal aphid populations. I. Electrophoretic techniques and staining systems for characterising isozymes from six species of cereal aphids (Hemiptera: Aphididae). *Bull. ent. Res.*, **73**, 645–57.

Loxdale, H.D., Tarr, I.J., Weber, C.P., Brookes, C.P., Digby, P.G.N. and Castañera, P. (1985*a*). Electrophoretic study of enzymes from cereal aphid populations. III. Spatial and temporal genetic variations of *Sitobion avenae* (F.) (Hemiptera: Aphididae). *Bull. ent. Res.*, **75**, 121–41.

Loxdale, H.D., Rhodes, J.A., and Fox, J.S. (1985*b*). Electrophoretic study of enzymes from cereal aphid populations. IV. Detection of hidden genetic variation within populations of the grain aphid *Sitobion avenae* (F.) (Hemiptera: Aphididae). *Theo. appl. Genetic.*, **70**, 407–12.

May, B. and Holbrook, F.R. (1978). Absence of genetic variability in the green-peach aphid *Myzus persicae* (Hemiptera: Aphididae). *Ann. ent. Soc. Am.*, **71**, 809–12.

McArthur, R.H. (1972). *Geographical ecology*. Harper and Row, New York.

Moran, N. (1981). Intraspecific variability in herbivore performance and host quality: a study of *Ureleucon caligatum* (Homoptera: Aphididae) and Solidago hosts (Asteraceae). *Ecol. Ent.*, **6**, 301–6.

Müller, H.P. (1961). Stabilität und Veränderlichkeit der Färbung bei Blattlausen. *Arch. Naturgesch. Mecklenburgs*, **7**, 228–39.

Müller, H.P. (1962). Biotypen und Unterarten der 'Erbsenblattlaus' *Acyrthosiphum pisum* Harris. *Z. PflSchutz PflKrankh*, **69**, 129 36.

Nei, M. and Graur, D. (1984). Extent of protein polymorphism and the mutation theory. *Evol. Biol.*, **17**, 74–118.

Nevo, E. (1978). Genetic variation in natural populations: patterns and theory. *Theor. Pop. Biol.*, **13**, 121–77.

Rhomberg, L.R., Joseph, S., and Singh, R.S. (1985). Seasonal variation and clonal selection in cyclically parthenogenetic rose aphids (*Macrosiphum rosae*). *Can. J. Genet. & Cytol.*, **27**, 224–32.

Sawicki, R.M. Devonshire, A.L., Payne, R.W., and Petzing, S.M. (1980). Stability of insecticide resistance in the peach-potato aphid *Myzus persicae*. *Pestic. Sci.*, **11**, 33–42.

Sawicki, R.M., Devonshire, A.L., Rice, A., Moores, G.D., Petzing, S.M., and Cameron, A. (1978). The detection and distribution of organophosphorus and carbamate insecticide-resistant *Myzus persicae* (Sulz.) in Britain in 1976. *Pestic. Sci.*, **9**, 189–210.

Service, P.M. and Lenski, R.E. (1982). Aphid genotypes, plant phenotypes and genetic diversity: A demographic analysis of experimental data. *Evolution*, **36**, 1276–82.

Shaposhnikov, G.K. (1966). Origin and break down of reproductive isolation and criterion of species. *Ent. Rev.*, **45**, 1–8.

Shaposhnikov, G.K. (1984). Aphids and a step toward universal species concept. *Evol. Theory*. *7*, 1–39.

Singh, R.S. and Rhomberg, J. (1984). Allozyme variation, population structure and sibling species in *Aphis pomi*. *Can. J. Genet. & Cyto.*, **26**, 364–73.

Steiner, W.M., Voegtlin, D.J., and Irwin, M.E. (1985). Genetic differentiation and its bearing on migration in North American populations of the corn leaf aphid *Rhopalosiphum maidis* (Fitch). *Ann. ent. Soc. Am.*, **78**, 518–25.

Suomalainen, E., Saura, A., Lokki, J., and Terri, T. (1980). Genetic polymorphism and evolution in parthenogenetic animals. IX. Absence of variation within parthenogenetic aphid clones. *Theor. appl. Genet.*, **57**, 129–32.

Takada, H. (1979). Esterase variation in Japanese populations of *Myzus persicae* (Sulzer) (Homoptera: Aphididae) with special reference to resistance to organophosphorus insecticides. *App. Ent. & Zool.*, **14**, 245–55.

Takada, H. (1986). Genetic composition and insecticide resistance of Japanese populations of *Myzus persicae* (Sulzer) (Hom.: Aphididae). *Z. angew. Ent.*, **102**, 19–38.

Tomiuk, J. (1987). The neutral theory and enzyme poymorphism in populations of aphid species. In *Population structure, genetics and taxonomy of aphids and Thysanoptera*. (ed. J. Holman, J. Pelikan, A.F.G. Dixon, and L. Weismann), pp. 45–62. SPB Academic Publishing, The Hague.

Tomiuk, J. and Wöhrmann, K. (1980). Enzyme variability in populations of aphids. *Theor. appl. Genet.*, **57**, 125–7.

Tomiuk, J. and Wöhrmann, K. (1981). Changes in the genotype frequencies at the MDH-locus in populations of *Macrosiphum rosae* (L) (Hemiptera: Aphididae). *Biol. Zbl.*, **100**, 631–40.

Tomiuk, J. and Wöhrmann, K. (1982). Comments on the genetic stability of aphid clones. *Experientia*, **38**, 320–1.

Tomiuk, J. and Wöhrmann, K. (1983). Enzyme polymorphism and taxonomy of aphid species. *Z. zool. Syst. Evolutionsforsch.*, **21**, 266–74.

Tomiuk, J. and Wöhrmann, K. (1984). Genotypic variability in natural populations of *Macrosiphum rosae* (L) in Europe. *Biol. Zbl.*, **103**, 113–22.

Tomiuk, J., Wöhrmann, K., and Eggers-Schumacher, H.A. (1979). Enzyme patterns as a characteristic for the identification of aphids. *Z. angew. Ent.*, **88**, 440–6.

Tomiuk, J., Böhm, J., Stamp. J., and Wöhrmann, K. (1988). Variability of quantitative characters and enzyme loci in rose aphid populations. (In press.)

Voegtlin, D.J., Steiner, W.M., and Irwin, M.E. (1987). Searching for the source of annual spring migrants of *Rhopalosiphum maidis* (Hem.:

Aphididae) in North America). In *Population structure, genetics and taxonomy of aphids and Thysanoptera* (ed. J. Holman, J. Pelikan, A.F.G. Dixon, and L. Weismann), pp. 120–33. SPB Academic publishing, The Hague.

Wöhrmann, K. and Tomiuk, J. (1988). Life cycle strategies and genotypic variability in populations of aphids. *J. Genet.*, **67**, 43–52.

Wöhrmann, K., Tomiuk, J., and Weber, G. (1986). The search for hidden enzymatic variation in the aphid *Macrosiphum rosae* (L). *Theor. appl. Genet.*, **73**, 77–81.

Wöhrmann, K., Stamp, J., and Tomiuk, J. (1987). Interactions between insecticide resistance and net reproduction of rose aphid clones on different hosts. In *Population structure, genetics, and taxonomy of aphids and Thysanoptera*. (ed. J. Holman, J. Pelikan, A.F.G. Dixon, and L. Weismann), pp. 111–19. SPB Academic Publishing, The Hague.

Wood, R.J. and Bishop, J.A. (1981). Insecticide resistance: populations and evolutions. In *Genetic consequences of man-made change* (ed. J.A. Bishop and L.M. Cook), pp. 97–127. Academic Press, New York.

Wool, D., Bunting, S.M. and van Emden, H.F. (1978). Electrophoretic study of genetic variation in British *Myzus persicae* (Sulz) (Hemiptera: Aphididae). *Biochem. Genet.*, **16**, 987–1006.

11 Use of genetic markers (allozymes) to study the structure, overwintering and dynamics of pest aphid populations

HUGH D. LOXDALE and
CLIFF P. BROOKES

Entomology and Nematology Department, Rothamsted Experimental Station, Harpenden, Hertfordshire, UK

Abstract

Genetic markers can be used to gain further insight into the population biology of pest aphids, including the persistence and movement of given strains and the overwintering strategies of different populations. Electrophoretic markers, particularly allozymes, are of considerable value in this respect.

Although many aphid species have little biochemical polymorphism compared with other insects, some species do display significant allozyme variation (e.g. the grain aphid *Sitobion avenae*) or at least 'useful' polymorphism at one or a few enzyme loci (e.g. the blackberry-grain aphid *Sitobion fragariae* and the bird cherry-oat aphid *Rhopalosiphum padi*). Other species, e.g. the peach-potato aphid *Myzus persicae*, whilst virtually monomorphic at most enzyme loci studied, display genotypic variation of esterase activity related to resistance to insecticides, which is also very useful in population studies.

In this paper, the population genetics of the aphids *S. avenae*, *S. fragariae*, *R. padi*, and *M. persicae* are discussed on the basis of the use of enzyme markers separated electrophoretically from individual insects. We show how spatial and/or temporal allozyme frequency

Electrophoretic Studies on Agricultural Pests (ed. Hugh D. Loxdale and J. den Hollander), Systematics Association Special Volume No. 39, pp. 231–70. Clarendon Press, Oxford, 1989.

data for these species can be used to provide information on both long- and short-distance migration and the amount of population mixing. Additionally, we show how other aspects of aphid population biology/genetics may be deduced from such data. In the case of *S. fragariae* these include the proportions of populations overwintering sexually or asexually, the relative neutrality/selectivity of the marker genes used, and the discovery of apparently predominantly parthenogenetic 'strains' of this aphid.

With *M. persicae*, data are presented showing that populations in Britain appear to be structured, consisting of relatively few genotypes — as measured from the number of insecticide-resistant strains found in the field and their spatial frequencies.

Introduction

Aphids are major pests of many crops in Britain including cereals, potatoes, sugar beet, peas, beans, and hops, causing damage either by direct feeding or by transmission of viruses. It has been estimated recently (Tatchell 1989) that in terms of the value of yield lost, such damage currently runs to about £100 million per annum.

In order to forecast aphid outbreaks in this country, the Rothamsted Insect Survey (RIS) has since the 1960s sampled the aerial density of flying aphids using a nationwide network of 12.2 m suction traps (Macaulay *et al.* 1988; Woiwod *et al.* 1984, 1987; Taylor *et al.* 1981). The spatial and temporal data thus gathered are used to effect rational control strategies, especially the timing of insecticide sprays and also, to model the factors influencing aphid outbreaks (Woiwod *et al.* 1984). To these ends, it has been largely successful, with some studies deducing the source of migrating aphids e.g. the rose-grain aphid *Metopolophium dirhodum* (Walker) during its outbreak year in 1979 (Dewar *et al.* 1980) and the distance the damson-hop aphid *Phorodon humuli* Schrank travels (Taylor *et al.* 1979). Other studies have provided information on different aspects of population biology, e.g. the timing of the spring migration of certain cereal aphids (Walters and Dewar 1986), and of the peach-potato aphid *Myzus persicae* (Sulzer) (Bale *et al.* 1988), the spread of virus by aphids within and between potato crops (Harrington *et al.* 1986) and the flight behaviour of the winged morphs of the bird cherry-oat aphid *Rhopalosiphum padi* (L.) (Tatchell *et al.* 1988). However, the approach may be described as 'coarse-grained'. This is because for any given pest aphid species, suction trapping only shows their aerial presence and abundance at a particular site and time, whilst providing little direct information on the origin of individuals caught or of the population structure from which they originated.

To obtain further insight into the population biology and dynamics of pest aphids, especially the persistence and movement of particular strains, and overwintering strategies of populations, genetic markers can be studied. This approach, because it allows investigation of individuals from populations and hence, a more detailed examination of aphid dynamics, might be termed 'fine-grained'.

Since 1975, electrophoretic markers, especially allozymes, have been widely used throughout the world to study many aspects of aphid biology. This paper describes the use of such markers to study the population biology of several species of pest aphid in Britain and Europe, with particular reference to our own work at Rothamsted. Such markers also have considerable importance in taxonomic studies of aphids and parasitism by hymenopterous parasitoids, whilst certain enzymes (esterases) are known to be responsible for conferring resistance to insecticides. These latter aspects are described elsewhere in this volume (see Blackman *et al.*, Powell and Walton, and Devonshire, respectively).

Methodology

What are allozymes and why study them?

To study aphid populations using electrophoretic markers, individuals can be homogenized and run on a gel (usually conventional vertical-slab starch or polyacrylamide); the presence of one (or sometimes more) proteins can then be detected by staining, after which the banding patterns revealed are scored (cf. Sargeant and George 1975; Ferguson 1980; Richardson *et al.* 1986). Other techniques (e.g. isoelectric focusing, 2-D electrophoresis, etc.) may also be used for aphid studies and are detailed elsewhere (Sargent and George 1975; Furk 1979; Loxdale *et al.* 1985*b*; Bakker and Gommers, this volume; Robinson, this volume).

For most of these studies, enzymes/isoenzymes (Ferguson 1980; Richardson *et al.* 1986; Loxdale *et al.* 1983), revealed using specific enzyme-staining reaction mixtures, have been used in preference to general protein staining using Coomassie blue, since the bands can be attributed more readily to specific gene loci. If the genes on both homologous chromosomes of the individual aphid homogenized are the same, the band will be single [i.e. a homozygous (homomeric) banding pattern]. Alternatively, if the genes are different, i.e. if they

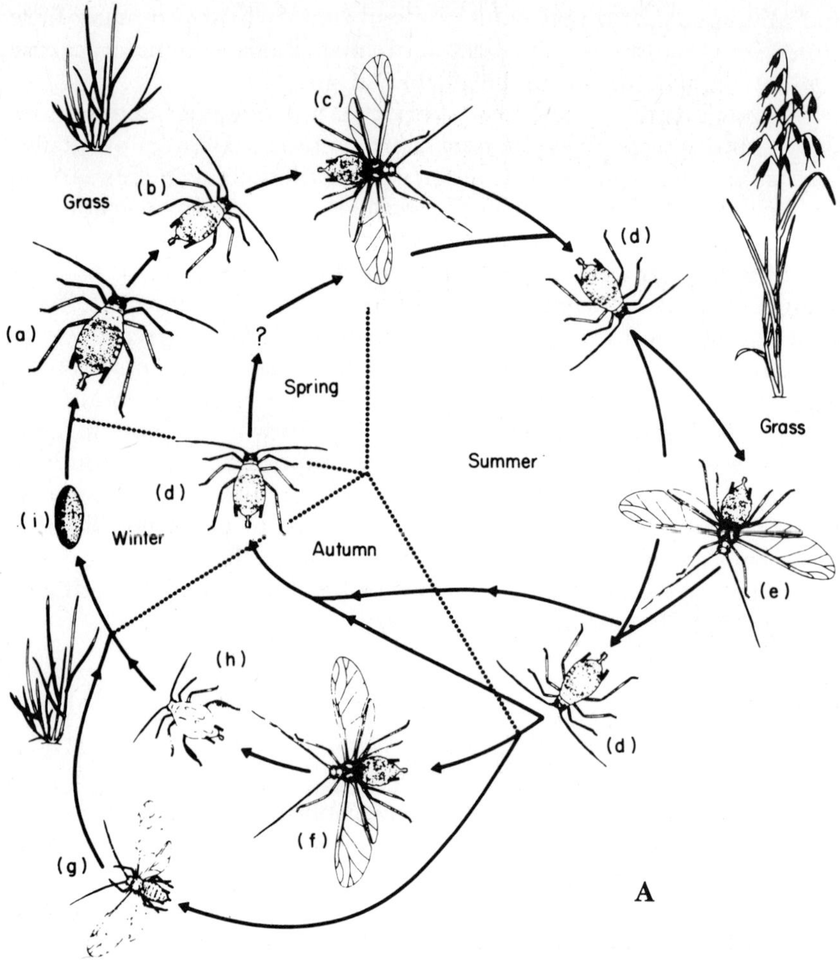

Fig. 11.1. Life cycles of two cereal aphid species, the grain aphid *Sitobion avenae* (A) and the bird cherry-oat aphid *Rhopalosiphum padi* (B). For both diagrams: a = fundatrix; b = apterous fundatrigenia; c = emigrant; d = apterous exule; f = gynopara; g = male; h = ovipara and i = egg. (From Carter *et al.* 1980.)

are alleles, this is most often expressed as a difference in electrophoretic mobility, the mobility variants detected being referred to as 'allozymes'. The banding pattern is then a heterozygous one. If the enzyme coded by the gene(s) in question has a polymeric structure (dimeric, trimeric, etc.), the heterozygous

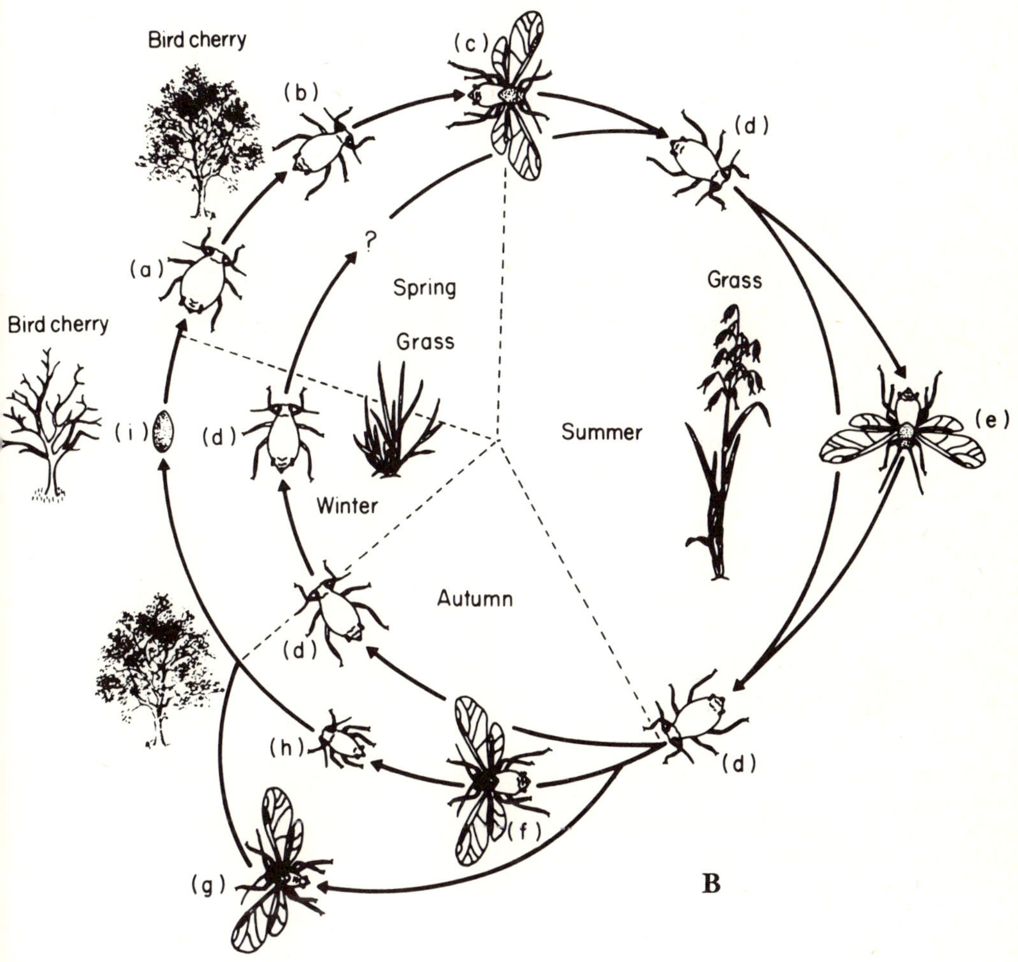

pattern will be increasingly complex, depending on the number of subunits present. In addition, some alleles code for inactive enzymes ('null' alleles), so that blank spaces appear on the gel, which may complicate interpretation of the banding pattern(s).

Such information obtained from individuals from different populations of the same species over a wide geographical area *or* from one population (or subpopulations) at particular sites over a period of time, enables spatial and temporal changes in gene and genotype frequency to be followed. Often such changes are obvious, but at other times sophisticated statistical approaches are necessary to examine the degree of genetic relatedness of populations in space or time. Conclusions can then be drawn concerning the population structure and genetic stability of individual populations, and the degree of between-population mixing (i.e. gene flow) occurring (cf. Daly, this volume).

In addition, by recording the proportion of heterozygotes to homozygotes at a particular locus within a sample, the goodness-of-fit of genotype frequencies to Hardy–Weinberg (H–W) expectations may be calculated using a chi-square test (see later). In the case of panmictic sexually-reproducing species and excluding the dual possibilities of pure chance and linkage disequilbrium owing to linked loci, genotype frequencies are most likely to conform to H–W expectations when migration, drift, mutation, and directional selection are negligible (Shorrocks 1978; Parkin 1979). Hence, this test provides information on the biological factors influencing populations. In addition, it gives an indication of overwintering strategies of species populations studied. For instance, a predominantly anholocyclic (parthenogenetic) species (e.g. as in Fig. 11.1A) is likely to show deviation of genotype frequencies from H–W expectations. Conversely, a population that has been through the sexual phase (holocycly) (Blackman 1974; Dixon 1985) (e.g. as in Fig. 11.1B), may show genotype frequencies in accordance with, or near to, H–W expectations.

Taxonomic separation of aphids using electrophoresis

This aspect of the use of electrophoresis is detailed by Blackman *et al.* (this volume).

In our work on the genetics of cereal aphids, we experienced difficulty in distinguishing between some individuals of the largely asexual grain aphid *Sitobion avenae* (F.) and the morphologically similar holocyclic species, the blackberry grain aphid *Sitobion fragariae* (Walker) (cf. Hille Ris Lambers 1939). However, the species *are* readily separated electrophoretically upon staining for PEP activity: *S. avenae* has three diagnostic isoenzymes, *S. fragariae* one (Loxdale *et al.* 1983; Loxdale and Brookes 1990*a,c*). Similarly, whilst studying aphids from blackberry (*Rubus fruticosus* agg.), we found the apterae, especially the nymphs, of the three main species infesting this host — *S. fragariae, Macrosiphum funestum* (Macchiati) and *Amphorphora rubi* (Kaltenbach) — difficult to separate morphologically, especially in frozen samples when antennae and/or siphunculi of individuals were often lost, although the species can readily be distinguished electrophoretically upon staining for GOT and EST activity (Loxdale and Brookes 1990*b*) (see Appendix for full names and EC numbers of enzyme abbreviations).

Sampling

The general theoretical and practical aspects of sampling populations for study of their structure using allozymes are described in detail by Richardson *et al.* (1986). Aspects of sampling pertinent to the study of aphids are described below.

1. Sample size and composition Having made sure of the taxonomic status of the aphid population(s) under study, it is important (1) to define what is meant by a population; (2) to sample sufficient of the genetic heterogeneity present, and (3) to collect a representative sample of the population.

As far as (1) and (2) are concerned, this often only becomes clear after allele frequencies have been determined. Such inferences drawn from our work on cereal aphids will be described later by reference to particular species populations. For a given site, a random sample must be taken from as many plants and over as wide an area as practicable. In most of our work, we have chosen an area of about 2–10 ha per site. For aphids infesting crops, samples could be collected throughout a large field. However, aphid populations are often not randomly distributed and may be concentrated in areas of shelter, such as in the lee of hedges and woods (Dean and Luuring 1970; Dean 1973). If aphid populations are small, aphids may have to be collected from such 'rich' areas and this could influence the assessment of allozyme heterogeneity, especially if this is marked between discrete colonies.

For aphids infesting bushes or trees, it is advisable initially to record all samples from individual plants so that any very local heterogeneity may be traced back to its source.

Individual, yet relatively small plants, e.g. wheat or sugar beet, may be heavily infested and might appear a good source of insect material to add to the total sample collected. However, it is possible that all individuals on one plant are derived from a single founder and that the colony is a clone. Addition of such clonal samples to the population sample, especially if they are large in relation to the total number of insects surveyed, will distort the true frequency of allelic variants present.

To demonstrate that colonies on single plants were derived from one or a few founders, we collected ears of cocksfoot grass *Dactylis glomerata* heavily infested with *S. fragariae.* All aphids from such ears were electrophoresed using polyacrylamide gels, whereafter these were stained for activity of the enzyme GOT, an enzyme known to have at least nine genotypes in this species. Six ears were sampled and three genotypes were found (SF, MF, and FF, where S, M, and F = slow, medium and fast alleles in terms of electrophoretic mobility). Of these, most ears were infested largely by one genotype (Table 11.1). Hence, most colonies were established by an original founder — usually one, but sometimes more (and of different genotype) per ear. Clearly, caution should be exercised over the collection of a few heavily infested plants.

Samples should be large enough to allow adequate statistical analysis. We have attempted to collect 100 aphids per site per enzyme locus to enable comparison of population allele frequency differences (see Richardson *et al.* 1986 for the rationale behind this).

2. *Spatial sampling* The number of sites chosen depends on the nature of the study. Spatial studies could involve examining local sites (say, within 50 km of each other) or more distant sites. Naturally, the more sites examined, the more chances there are of observing interpopulation differences. Also, the more distant samples are from one another, the greater the chance of populations being genetically distinct, either in terms of unique ('private') alleles or differences in common allele frequencies. Differences may be due to the restriction of gene flow and/or any locally differentiating selective forces. There is then little point sampling further if a population from a few sites (say, 4–6) is found to be rather similar in terms of alleles/allele frequencies over a wide geographical area. On the other hand, if substantial allelic heterogeneity is found even over a small geographical range, then intense sampling (possibly involving tens of sites) is warranted in order to attempt to delimit the extent of the variation. If data are collected for very distant sites, allele frequency

Table 11.1 GOT genotypes of *S. fragariae* collected from six separate heads of *D. glomerata* at Rothamsted in 1983; n = number of aphids tested. (From Loxdale and Brookes 1990*a*).

		Genotype		
Ear no.	n	S/F	M/F	F/F
1	35*	—	23	4
2	52	—	—	52
3	52	50	—	1
4	52	—	48	—
5	51	—	—	44
6	52	—	—	52

* 'Missing' individuals, e.g. n = 8, Ear no. 1, parasitized by parasitic hymenoptera and hence excluded from counts.

parameters (genetic distances, etc.) can be plotted directly against distance (Richardson *et al.* 1986).

3. Temporal sampling Wöhrmann and co-workers have been studying the population genetics of the rose aphid *Macrosiphum rosae* (L.) for over ten years (Wöhrmann 1984; Wöhrmann and Tomiuk, this volume). They have examined temporal changes in allele and genotype frequency both within the same year and between years. Since the host of this aphid, *Rosa* spp., remains nutritious throughout the growing season and the aphid reproduces parthenogenetically throughout this time, sampling is possible over quite a long period (*ca.* 3 months), although numbers of aphids actually only peak for about one month (June) (Wöhrmann 1984). With cereal aphids, problems in temporal sampling can arise in two ways. In the spring, colonies of some holocyclic species, e.g. *S. fragariae* and *R. padi* remain in large numbers for only a relatively short time (*ca.* 1 month) on the primary woody overwintering host Rosaceae before the alatae fly off to the secondary host Gramineae (grasses and cereals). Also, on the secondary host, numbers of aphids peak for a relatively short period [late June to mid-July in the case of *S. avenae* and *M. dirhodum* on wheat, barley, and oats (Dean and Luuring 1970)]: numbers decline rapidly as the crop senesces, making further sampling impossible at that exact location. With *S. fragariae*, we have managed, in some years and at some sites, to get three samples per growing season, one from the primary host and two from Gramineae prior to crop senescence, although this is exceptional (Loxdale and Brookes 1990*a*). Sampling annual crops is difficult owning to rotation; it may not be possible to sample from the same field for two or more consecutive years (Loxdale *et al.* 1985*a*). Adjacent or nearby fields can then be sampled, although clearly this is not wholly satisfactory, especially if it is known that the species is locally heterogeneous for genetic markers.

Clones

When starting work on a new aphid species, it is advisable to rear clones under constant-environment conditions. Insect material is then available to check the consistency of allozyme patterns within and between clones. Also, difficulties in species identification, or the presence of hymenopterous parasitoids (Castañera *et al.* 1983; Powell and Walton, this volume) and entomophthoralean fungi (Mardell *et al.*, in prep.) which may cause interference of the host allozyme pattern(s), can be resolved. Misidentified or infected individuals can then be discounted from allozyme frequency surveys. Lastly, should alternative alleles be present at a given enzyme locus in different clonal lineages, sexuales can be reared and crossed to determine the nature of the heterozygous enzyme banding pattern (Blackman and Devonshire 1978; Newton and Dixon 1988) and hence the polymeric structure of the enzyme in question (see later).

We have found that small plastic boxes 76 × 45 × 20 mm deep, as originally described by Blackman (1971), are ideal for rearing small numbers (fewer than about 30) of a clone on an excised leaf, e.g. potato. For cereal aphids, cereal seedlings can be grown individually in 6.5 cm diameter pots and then enclosed with a glass cylinder 15 cm high × 6.5 cm diameter and capped with fine-mesh Terylene netting. A collection of such cylinders can be maintained, as can the Blackman boxes, under constant-environment conditions.

Electrophoretic techniques and enzyme banding patterns

Sample storage

Most enzymes that we have investigated, with the exception of POD, retain their activities well when stored at −25 to −30°C for 3–6 months. Although the activity of some enzymes may not be lost, changes occur in their charge properties on prolonged deep-frozen storage so that 'pseudo' mobility variants appear on the gel. If this is suspected, it should be checked using freshly-killed material. We have detected such changes in Est-7 of *S. avenae* and ME of *Phorodon humuli*.

Sample preparation

We homogenize aphids in 15 per cent aqueous sucrose with or without 50 mM tris/HCl, pH 7.1 (depending on the stability of the enzyme examined), 0.5 per cent Triton-X100 (to solubilize membrane-bound enzymes, especially esterases), plus a few crystals of bromophenol blue (to act as a tracking dye during electrophoresis). Large numbers (>60) of individual aphids may be homogenized simultaneously using a multi-way homogenizer (Brookes and Loxdale 1985; ffrench-Constant and Devonshire 1987).

Electrophoresis and enzyme staining

Electrophoretic (vertical-slab polyacrylamide) and staining techniques for cereal aphids are detailed by Loxdale *et al.* (1983). A neutral staining system for pH-unspecific phosphatases found in some aphids (Loxdale *et al.* 1983) and a sensitive esterase staining system for improved staining of eserine-sensitive esterases, i.e. EST-1 and EST-2 such as found in the peach-potato aphid *Myzus persicae*, are detailed by Brookes and Loxdale (1987). Other electrophoretic techniques and staining systems suitable for aphids are referred to by May and Holbrook (1978), Rhomberg *et al.* (1985), Singh and Rhomberg (1984), Steiner *et al.* (1985), Suomalainen *et al.* (1980), Tomiuk and Wöhrmann (1983), and Wool *et al.* (1978), whilst other more general systems are detailed by Brewer and Sing (1970), Harris and Hopkinson (1977), Richardson *et al.* (1986), Shaw and Prasad (1970), and Thomson (1987).

Additional 'hidden' genetic variation has been revealed at certain polymorphic loci (e.g. PEP) in some aphid species (e.g. *S. avenae*) by changing electrophoretic running conditions, especially run time, percentage acrylamide (effectively thereby altering gel pore size), and pH (Loxdale *et al.* 1985*b*; Wool *et al.* 1978) (cf. also Menken and Ulenberg 1987). In other species (i.e. *M. rosae*), no additional variation has been detected using such 'serial' approaches (Wöhrmann *et al.* 1986).

Enzyme banding patterns

Interpretations of these in terms of submit number, presence of null alleles and number of likely loci involved are detailed for aphids by Brookes and Loxdale (1987), Loxdale *et al.* (1983; 1985*a*), May and Holbrook (1978), Rhomberg *et al.* (1985), Singh and Rhomberg (1984), Steiner *et al.* (1985), Suomalainen *et al.* (1980), Tomiuk and Wöhrmann (1983), and Wool *et al.* (1978), and for other insects and animals by Berlocher (1980), Ferguson (1980), Harris and Hopkinson (1977), and Richardson *et al.* (1986).

As in other organisms, crossing experiments are required to confirm the reality of putative heterozygotes in aphids. With aphids, this is difficult owing to the host alternation of many species and the time-scale of the life cycle. Production of sexuales and hence crossing can be speeded up under artificial conditions of light and temperature and has recently been performed in *S. avenae* (Hand 1982; Newton and Dixon 1988).

Statistical analysis

Various approaches to statistical analysis can be used, both single- and multilocus. These include chi-square testing of interpopulation allele frequencies (Workman and Niswander 1970), testing the congruence of genotype frequencies with Hardy–Weinberg expectations (including *F*-statistics; Wright 1978), and also the calculation of interpopulation genetic identity and distance coefficients (Nei, Rogers, etc.). Both allele frequencies *per se* (Powell *et al.* 1980) and genetic distance values can be subject to multivariate approaches to reveal additional information. Genetic identity and distance values can also be displayed as dendrograms, although many of those presented in the literature to show evolutionary relationships between geographical populations should be treated circumspectly, since the standard errors (rarely given) can overlap. Temporal data is required to establish the validity of such relations (see also Daly, this volume).

The statistics used in electrophoretic studies and various examples of their use are detailed by Ayala (1977), Ferguson (1980), Hartl (1980), Nei (1972, 1975, 1978), Oxford and Rollinson (1982), and most recently by Richardson *et al.* (1986).

Genetic variability in aphids

Aphids appear to be generally less variable genetically than other insects. Average (over all enzyme loci) percentage polymorphism (P) is less in aphids and average heterozygosity ($\bar{H}$) is below about 4.5 per cent compared to 7.3 per cent in most other insects (Loxdale *et al.* 1985*a*). However, British samples of both *S. avenae* and *S. fragariae* appear to be more polymorphic than the majority of aphids so far examined electrophoretically.

The reduced biochemical polymorphism in aphids compared to other insects undoubtably stems from their life cycle. This includes (1) total anholocycly (parthenogeneticity) or, the production of numerous asexual Spring and Summer generations on the secondary host; (2) annual population bottlenecks (Blackman 1981; Nei *et al.* 1975) eliminate rare alleles at given loci and may, over many generations, lead to fixation of the most common alleles, assuming elimination exceeds the production of new alleles by spontaneous mutation; and (3) founder effects (Mayr 1963; Parkin 1979). In some cases, selection for or against particular genes may also be responsible for the predominance of one or a few genotypes at a particular locus within a population, perhaps over a large geographical area.

The causes of invariability in aphids are discussed in more detail elsewhere (Loxdale *et al.* 1985*a*; Brookes and Loxdale 1987; Rhomberg *et al.* 1985; Steiner *et al.* 1985; ffrench-Constant *et al.* 1988; Dixon 1985).

Population structure, overwintering, and dynamics

The general lack of biochemical polymorphism in aphids restricts the type of study that can be performed. In reasonably polymorphic species, several or many enzyme loci can be examined, enabling estimates of interpopulation genetic distance to be calculated. In contrast, a species may have only one polymorphic locus, thereby necessitating a different approach.

Both multilocus and single-locus approaches have been followed in the study of aphid population biology. In order to discuss the available data and place it within a logical framework, we have categorized such work under these two headings.

Multilocus approach

Tomiuk and Wöhrmann (1980, 1983) used a multilocus approach to estimate the amount of biochemical polymorphism present in several aphid species. They also employed various statistical procedures to produce dendrograms of evolutionary relatedness of a wide range of aphid species (Tomiuk and Wöhrmann 1983). Lampel and Burgener (1987) and Eggers-Schumacher (1987) have applied similar approaches to aphid genera and families. In these studies and those described below, the enzymes used are a range of Group I and II enzymes (Gillespie and Kojima 1968) as conventionally examined in population studies using enzyme electrophoresis (e.g. esterase, malic enzyme, peptidase, glutamate-oxaloacetate transaminase, etc.).

1. Sitobion avenae Below the species level, Loxdale *et al.* (1985*a*) used a multilocus approach to examine the population genetics of the grain aphid *Sitobion avenae,* a major pest of wheat in Europe (Carter *et al.* 1980; Vickerman and Wratten 1979). This aphid, unlike the majority of aphid species examined to date, can be polymorphic at a large proportion of enzyme loci tested, i.e. P = 64 per cent (16/25 loci) in 1979, where polymorphism (P) is expressed as the probability that the most common allele at a given locus does not exceed a frequency of 95 per cent (Ayala and Valentine 1978). However, the following year this value had declined to 19 per cent (5/26 loci). The lower value falls within the range P = 0–42 per cent for a number of aphid species, pest and non-pest, cited by Tomiuk and Wöhrmann (1980, 1983). The average heterozygosity over all loci found for both years was *ca.* 2 per cent, half the maximum value cited by Tomiuk and Wöhrmann.

Samples of *S. avenae* were collected by hand in June–July 1979 and 1980 from grasses and cereals (mainly *Dactylis glomerata* and wheat) around Britain (Fig. 11.2) and reared on wheat seedlings under constant-environment conditions (Loxdale *et al.* 1983, 1985*a*). They were first tested for contamination by the sister species *S. fragariae*

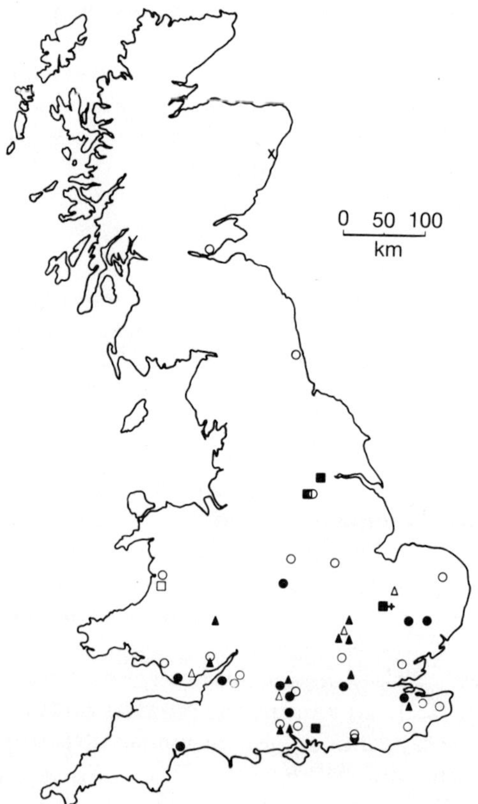

Fig. 11.2. Sampling sites for clones of *Sitobion avenae* collected from Gramineae in 1979 (white symbols) and 1980 (black symbols). (From Loxdale *et al.* 1985*a*; see that paper for further details.)

using the peptidase test. Thereafter, they were surveyed electrophoretically using 14 enzyme systems, including PEP.

Whilst *S. avenae* can be highly polymorphic, having numerous alleles at some loci (e.g. EST-7 = 7, PEP = 8), few of the alleles identified were in heterozygous condition; only at one locus (EST-1) were such putative genotypes commonly present. From this it may be deduced that the incidence of sexual crossing is infrequent. This supports other evidence that the species is largely asexual in Britain, especially the scarcity of overwintering eggs (Hand 1982) and of males in suction traps (G.M. Tatchell, pers. comm.).

Having identified the variable loci, five variable enzymes (EST, PEP, GOT, PHOS, and POD) representing 13 loci were used to calculate the degree of genetic relatedness between natural field populations on wheat sampled in June–July 1981 and 1982 at sites in Britain and Spain (Fig. 11.3). Approximately 200 aphids were surveyed per enzyme per site. At two British sites (Long Ashton and Norwich), populations were sampled temporally for each year.

At the majority of variable loci, most populations had statistically (chi-square) different allele frequencies from one another. Also, whilst 60 per cent of alleles were shared between both countries, 40 per cent were different (Loxdale *et al.* 1985*a*). Clearly, the populations in Britain and Spain do not form a continuum in terms of genetic similarity. The differences between sites both within and between countries show that populations comprise numerous clones in differing proportions. These differences could arise from random genetic drift, founder effects or locally differentiated selective forces (including perhaps crop variety) or a combination of factors.

With the one locus (EST-1) showing putative heterozygotes, genotype frequencies were, except for two sites in Spain, significantly different from Hardy–Weinberg expectations, again supporting the belief of the largely asexual reproduction of *S. avenae*.

However, although this species clearly showed spatial and temporal heterogeneity of allele frequencies, when the overall polymorphism detected was calculated as Nei's (1971, 1972) coefficient of average genetic distance ($\bar{D}$) and these values were plotted directly against geographical distance (or the square-root of distance so as to contract the x-axis), a 'scattergram' was produced consisting of two clouds of points—the one nearest the origin comprising intranational $\bar{D}$ values (within Britain and Spain), the further cloud comprising international values (between Britain and Spain) (Fig. 11.4). A regression could not be justified since the international samples do not form a geographical continuum.

The international difference in $\bar{D}$ is relatively small—about 0.025.

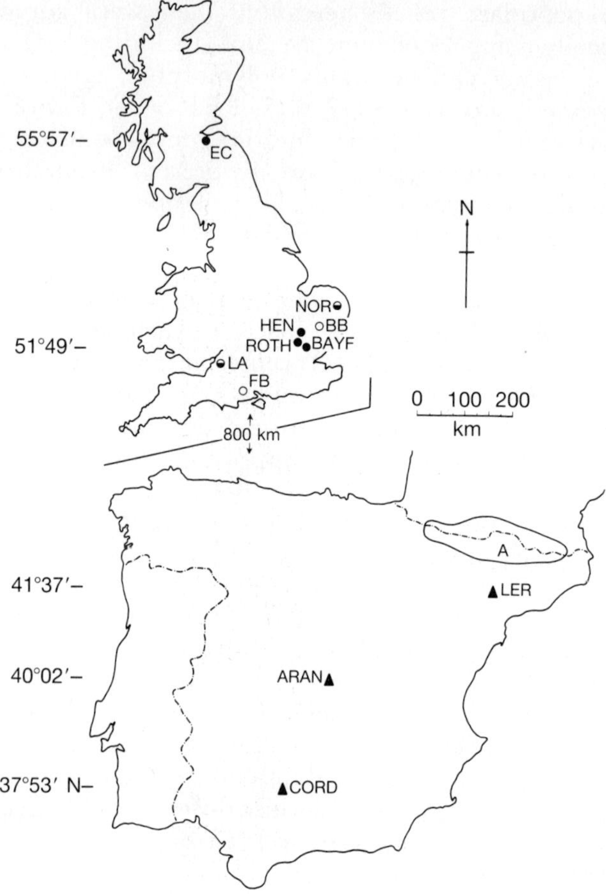

Fig. 11.3. Sampling sites for large *S. avenae* field populations in Britain in 1981 (open circles), 1982 (solid circles) and both years (part-solid circles) and in Spain in 1982 (solid triangles). Site name abbreviations: ARAN = Aranjuez, Castilla; BAYF = Bayfordbury, Hertfordshire; BB = Brooms Barn, Higham, Suffolk; CORD = Cordoba, Andalucia, EC = East Craigs, Mid-Lothian; FB = Fordingbridge, Hampshire; HEN = Henlow, Bedfordshire; LA = Long Ashton, Avon; LER = Lerida, Cataluña; NOR = Norwich, Norfolk; ROTH = Rothamsted, Hertfordshire. A = Pyrenees, maximum height above sea-level = *ca.* 3400 m. (From Loxdale *et al.* 1985*a*.)

Whilst larger than that found for geographical populations of some sexually reproducing insects examined (Loxdale *et al.* 1985*a*), this value is still at the geographical population level on the five-point scale of genetic differentiation proposed by Ayala and co-workers

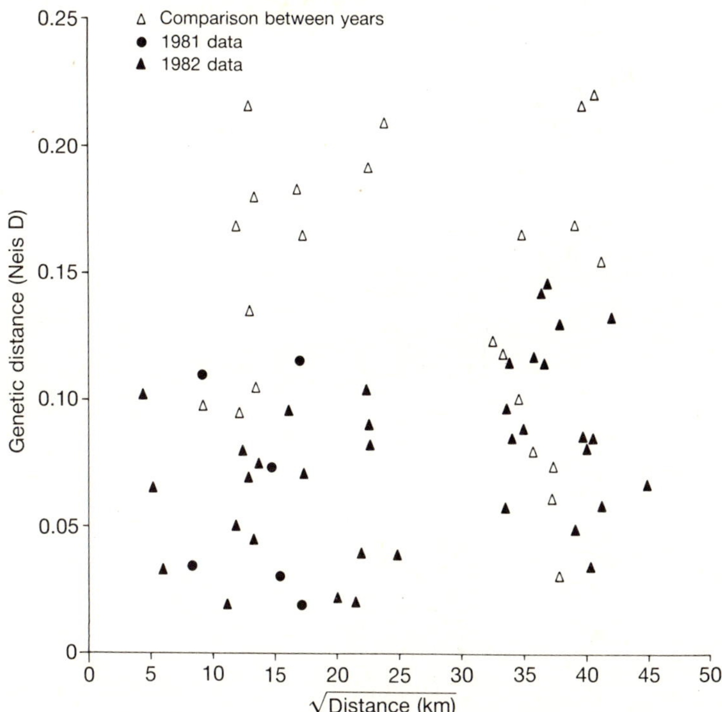

Fig. 11.4. Graph showing average genetic distance (Nei's D) values over 13 loci for *S. avenae* plotted against the square-root of geographical distance in kilometres. The group of points on the left represent intranational comparisons, those on the right represent international comparisons. Solid circles • within-year data for 1981; solid triangles ▲ within-year for 1982; open triangles △ between-year data. (From Loxdale and Brookes 1985*a*.)

from studies of the tropical American *Drosophila willistoni* 'species' complex (Avise 1977).

Using a multivariate approach (Procrustes' rotation; cf. Digby and Kempton 1986), the Spanish and British populations can be separated (Fig. 11.5). Whilst there is some international north–south distribution in $\bar{D}$ values, evidence for a latitudinal cline as such is not strong. From these data it may be inferred that gene flow exists between the *S. avenae* populations examined and that those in Britain and Spain, separated by geographical barriers, have diverged slightly, but probably not enough to form distinct races or biotypes. An alternative explanation could be that British and Spanish populations are in fact now totally isolated, but that not enough time

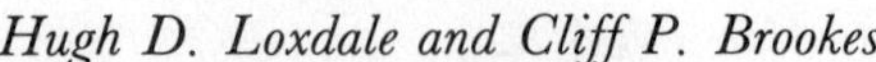

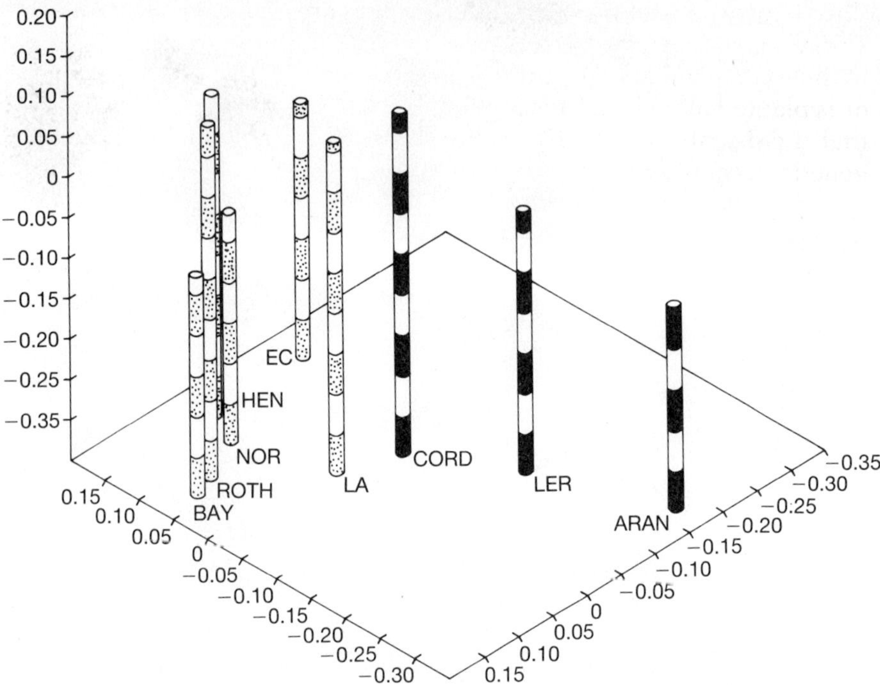

Fig. 11.5. Data of Fig. 11.4 (1982 data only) subjected to multivariate analysis (Procrustes' rotation) showing three-dimensional separation of British and Spanish sites in relation to square-root of geographical distance in kilometres. The points for the sites are at the tops of the vertical 'poles'. The distance between points represents the square-root of genetic distance; this figure recovers two-thirds of the genetic distance. Site name abbreviations as in Fig. 11.3. (From Loxdale *et al.* 1985*a*.)

has elapsed since they split to allow significant evolutionary divergence dependent upon mutation and other factors. This latter argument seems unlikely in the light of the known migratory abilities of *S. avenae*, the aphid having been captured at sea at least 160 km from the nearest land (Hardy and Cheng 1986).

2. *Rhopalosiphum padi* The multilocus approach was also used to examine the population biology of the bird cherry-oat aphid, *Rhopalosiphum padi* (L.) in Britain (Loxdale and Brookes 1988), a major vector of BYDV (barley yellow dwarf virus) in Europe (Plumb 1983). This species is largely holocyclic in Britain (Blackman and Eastop 1984), alternating between the primary host *Prunus padus*

(bird cherry), on which overwintering eggs are laid, and Gramineae. *P. padus* is native to and widespread in Britain, but is generally scarce in the south of the country, where it occurs mainly in parks or gardens or is planted along roadside verges. Hence, we considered it possible that the distribution of the primary host in Britain would influence genetic variability in *R. padi*.

R. padi was collected from *P. padus* in May and June 1985 and 1986 prior to their migrating to Gramineae. In 1985, 14 enzymes were tested from populations collected throughout Britain, but mainly in an area of *ca*. 30 km around Rothamsted (Fig. 11.6). Of the enzymes

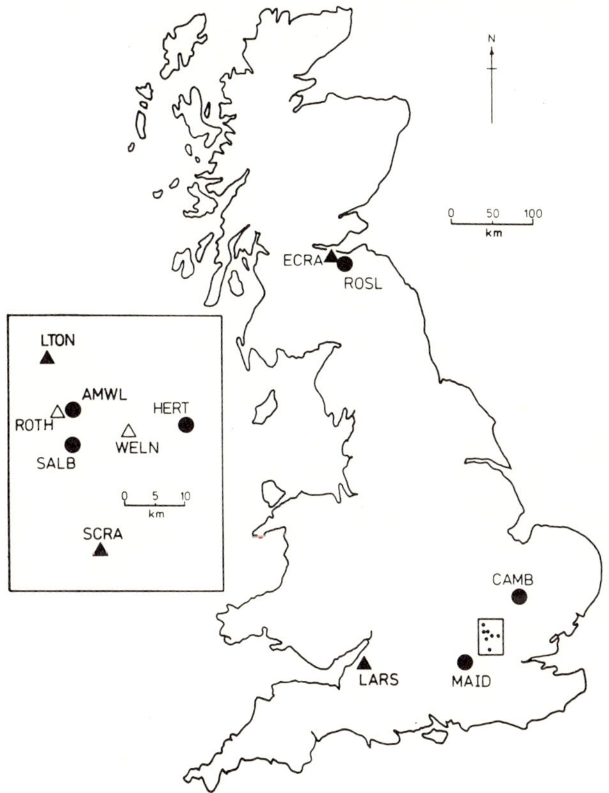

Fig. 11.6. Sampling sites for British *Rhopalosiphum padi* populations sampled from *Prunus padus* in 1985 and 1986. Site name abbreviations: AMWL = Amwell, HERT = Hertford; SALB = St. Albans, ROTH = Rothamsted estate and WELN = Welwyn Garden City, all Hertfordshire; CAMB = Cambridge; LARS = Long Ashton Research Station, Bristol, Avon; LTON = Luton, Bedfordshire; MAID = Maidenhead, Berkshire; SCRA = Scratchwood, Middlesex; ECRA = East Craigs and ROSL = Roslin, Mid-Lothian, Scotland. Populations sampled in 1985 (solid triangles), 1986 (open triangles) and both years (solid circles). (From Loxdale and Brookes 1988.)

surveyed, only two — GOT and SORDH — were polymorphic, i.e. 2/14 loci or 14 per cent using the 95 per cent frequency criterion stated earlier. These enzymes were surveyed at the sites shown (Fig. 11.6) the following year. Perhaps this general lack of variation is due to the scarcity of the primary host which acts as a population bottleneck and hence fixes most common alleles. The fact that GOT and SORDH have remained polymorphic may be due to some weak selective advantage of the alleles present at both loci.

The main findings from this study were that allele frequencies for the two enzymes GOT and SORDH were similar over the geographical range surveyed (Table 11.2), and that most populations conformed to H–W expectations. For both enzymes, the fast (F) allele was dominant in all populations examined ($f > 0.80$) (Table 11.2). The genetic homogeneity found suggests either that sufficient gene flow occurs between populations to alleviate the perturbing effects of drift, selection, etc., *or*, that some kind of balancing selection maintains the observed frequencies. We favour the former alternative because of the long-distance migratory ability of *R. padi* (Hardy and Cheng 1986; Tatchell *et al.* 1988).

The greater spatial allelic homogeneity of *R. padi* compared with *S. avenae* may arise because *R. padi* has to fly farther to find the widely dispersed primary host *P. padus*. Graminae on which *S. avenae* overwinters is generally abundant, but senescences in mid-summer. Such senescence causes perturbation of existing allele frequencies by various factors (genetic drift, etc.). It is known that males and gynoparae of *R. padi* are more abundant at a height of 12.2 m than at 1.5 m (Tatchell *et al.* 1988), a finding which may relate to the need of these winged morphs to find the tall, widely dispersed primary host.

3. Other studies A number of other studies have also concentrated on the use of several polymorphic enzyme loci to examine aphid population biology. For example, Tomiuk and Wöhrmann (1984) and Wöhrmann and Tomiuk (1988) used PGM and MDH to examine the population structure of *Macrosiphum rosae* in Europe. They found a latitudinal cline of allele frequencies for these two loci from Norway to Switzerland and have attempted to relate reproductive strategy to climatic conditions. They argue that continuous parthenogenesis (as found in countries with mild winters, such as Britain) results in low genotype variability caused by selection and/or random genetic drift. Conversely, the greater variability and population homogeneity of holocyclic populations found in countries with harsh winters (e.g. Norway), is attributed to recombination occurring during the sexual phase and to migration (Wöhrmann and Tomiuk 1988, and this volume).

Multilocus studies have also recently been performed on the apple aphid *Aphis pomi* (de Geer) in Canada by Singh and Rhomberg (1984), on Canadian populations of *M. rosae* (Rhomberg *et al.* 1985), on North American populations of the corn leaf aphid *Rhopalosiphum maidis* (Fitch) by Steiner *et al.* (1985) and Voegtlin *et al.* (1987) and, most recently, by Eggers-Schumacher and Sander (1988) on the damson hop aphid *Phorodon humuli* in West Germany.

In the case of *M. rosae* from Canada, sample genotype frequencies at some polymorphic loci show significant deviations from H–W expectations during the summertime asexual phase of reproduction (Rhomberg *et al.* 1985), which constrasts with the findings of Wöhrmann and Tomiuk (this volume) for similar cyclically parthenogenetic populations from Europe and to some extent with our own studies on *S. fragariae*. In addition, Rhomberg *et al.* (1985) found that at three loci in *M. rosae* (one a colour locus and the other two esterase loci, EST-2 and EST-4), dramatic fluctuations of genotype frequency occurred over the course of a season, which was thought to result from the domination of local infestations by a few particularly successful clones. They speculate that clonal competition followed by extensive migration purges much of the selectively neutral variation from populations, whilst the remaining polymorphic loci (which are mostly di- or tri-allelic), are subject to balancing selection at the gene or at closely linked loci, thereby preventing their disappearance. As with Wöhrmann and Tomiuk's study of *M. rosae* (1988; and this volume), and as we found in *S. avenae* (Loxdale *et al.* 1985*a*), parthenogenetic reproduction appears to lead to fluctuations of allelic stability, brought about by drift and selection. In cyclically parthenogenetic species, such fluctuations may lead to deviations of allele frequencies from H–W expectations during the summer asexual phase (Rhomberg *et al.* 1985). Both Wöhrmann and Tomiuk (this volume) and we (see p. 257) have also found some samples of *M. rosae* and *S. fragariae*, respectively, which show allele frequencies near to or in accordance with H–W expectations during the parthenogenetic phase of reproduction. Wöhrmann and Tomiuk suggest this as being due to migration between subpopulations.

Interestingly, Steiner *et al.*, (1985) working on *R. maidis*, found regional 'northern' and 'southern' North American populations of this asexual species that display fixed genetic difference. The authors conclude from this that the populations are clonally derived. They also conclude that there is very restricted gene flow between these populations, so that it is unlikely that northern (Illinois) populations, contrary to long-held belief, are founded from the southern regions sampled (i.e. in Oklahoma, Texas, and Louisiana).

Table 11.2 GOT and SORDH allele frequencies in British *R. padi* populations collected from *Prunus padus* in 1985 and 1986. (Data from Loxdale and Brookes 1988.)

Enzyme alleles		Sites											
		CAMB	HERT	SALB	AMWL	ROSL	ECRA	LTON	SCRA	MAID	WELN	ROTH	LARS
GOT													
1985													
	n =	104	104	104	104	102	104	104	104	104			
U (0.83)		—	—	0.043	—	0.005	—	0.005	0.01	—			
S (0.87)		—	—	—	—	0.015	—	—	—	—			
M (0.91)		0.005	—	0.010	—	0.005	0.005	0.024	0.01	—			
F (1.00)		0.995	1.00	0.947	1.00	0.907	0.995	0.971	0.98	1.00			
V (1.10)		—	—	—	—	0.068	—	—	—	—			
1986													
	n =	208	208	208	208	204					208	208	
U (0.83)		0.01	0.002	0.005	—	—					0.007	—	
S (0.87)		0.002	0.005	0.002	0.017	0.002					0.007	0.043	
M (0.91)		0.005	0.015	—	—	0.002					0.012	—	
F (1.00)		0.983	0.978	0.993	0.983	0.996					0.974	0.957	
V		—	—	—	—	—					—	—	

Table 11.2 *Continued*

Enzyme alleles		Sites											
		CAMB	HERT	SALB	AMWL	ROSL	ECRA	LTON	SCRA	MAID	WELN	ROTH	LARS
SORDH													
1985													
	$n =$	104	103	104	104	92		104	104	104			104
S (0.63)		0.091	0.034	0.067	0.010	0.005		0.038	—	—			0.005
F (1.00)		0.909	0.966	0.923	0.990	0.995		0.952	1.00	1.00			0.995
V (1.16)		—	—	0.010	—	—		0.010	—	—			—
1986													
	$n =$	384	152	185	205	181					208	196	
S (0.63)		0.133	0.191	0.022	0.076	0.072					0.058	0.018	
F (1.00)		0.867	0.809	0.978	0.924	0.928					0.942	0.982	
V (1.16)		—	—	—	—	—					—	—	

Site name abbreviations as in Fig. 11.6. n = sample size. U = Ultra slow; S = Slow, M = medium; F = Fast; and V = Very fast allele.

Single-locus approach

As well as the multilocus approach, much useful information relating to aphid population biology may be derived by studying one polymorphic locus.

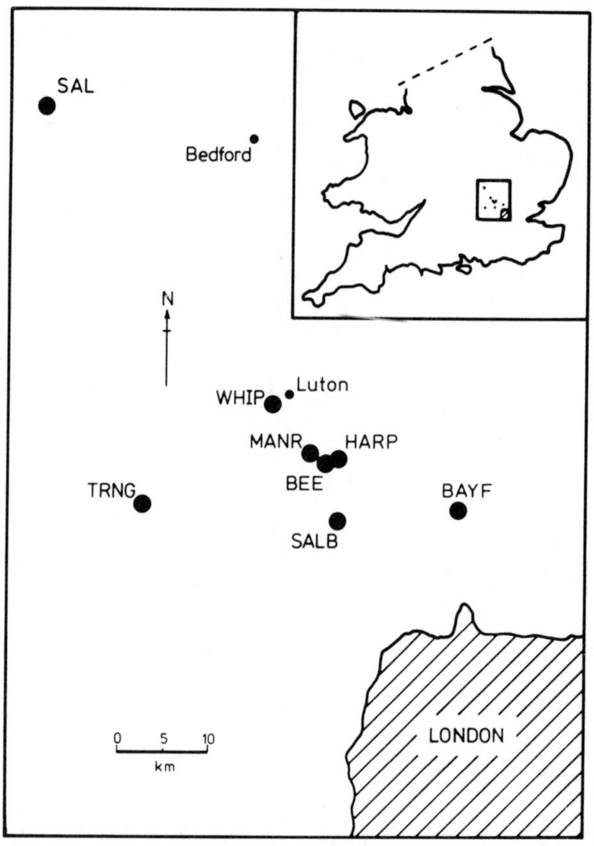

Fig. 11.7. Sampling sites for *Sitobion fragariae* around Rothamsted. Location of the area in relation to rest of south-east England shown inset. Insects sampled over period 1981–86, from both *R. fruticosus* and *D. glomerata*. Site name abbreviations = BAYF, Bayfordbury; BEE = Bee laboratory grounds, Rothamsted estate; HARP = Harpenden Common; TRNG = Tring Park; MANR = Manor house grounds, Rothamsted estate; SALB = St. Albans; WHIP = Whipsnade, all Hertfordshire; SAL = Salcey Forest, Northamptonshire. (From Loxdale and Brookes 1990*a*.)

1. Sitobion fragariae The blackberry-grain aphid, *Sitobion fragariae*, a minor pest of cereals in Europe (Vickerman and Wratten 1979), is thought to be predominantly holocyclic, alternating between the primary host *Rubus fruticosus* and Gramineae. We have studied local populations of this species within a 50 km radius around Rothamsted (Fig. 11.7) from 1981 to 1986. Aphids were collected from *R. fruticosus* in May–June and from grass (*D. glomerata*) in June–July.

We found initially that the species was polymorphic (7/14 loci; P = 50 per cent), although one locus, GOT, displayed particularly distinct and resolved allozymes and heterozygotes. Efforts were

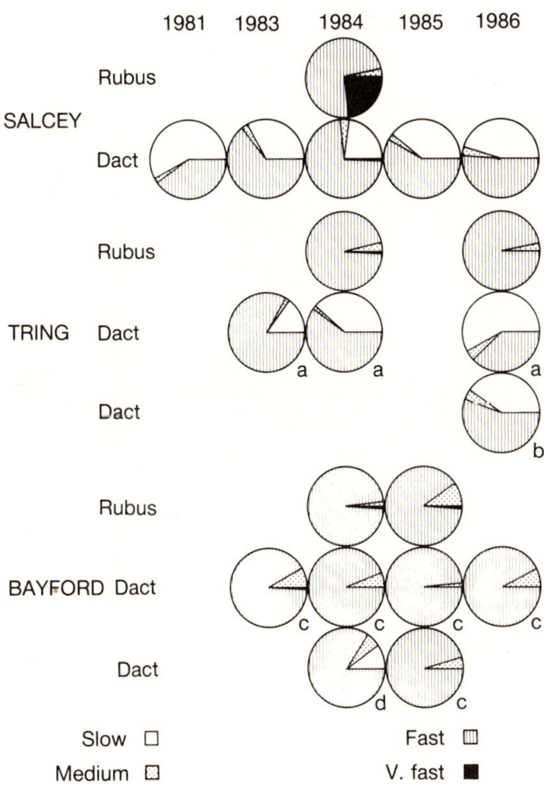

Fig. 11.8. GOT allele frequencies (represented as pie charts) for local populations of *S. fragariae* at eight sites in south-east England (SAL, TRNG, BAYF, HARP, BEE, MANR, SALB, WHIP; site abbreviations as in Fig. 11.7 legend). Years are shown horizontally at the top. Sampling sites and

therefore concentrated on this locus. By collecting at the same or nearby sites for four or five years, we have defined the spatial and temporal population structure of this species.

In all, six alleles have been found, represented by four homozygotes and seven heterozygotes. Of these, only three homozygotes were widespread within populations (SS, MM, FF), and three heterozygotes (SF, MF and FV), where S, M, F, and V = slow, medium, fast, and very fast in terms of their relative electrophoretic mobilities (*Rm*), i.e. *Rm* = 0.82, 0.91, 1.00, and 1.08, respectively, the F allele being the reference standard. The

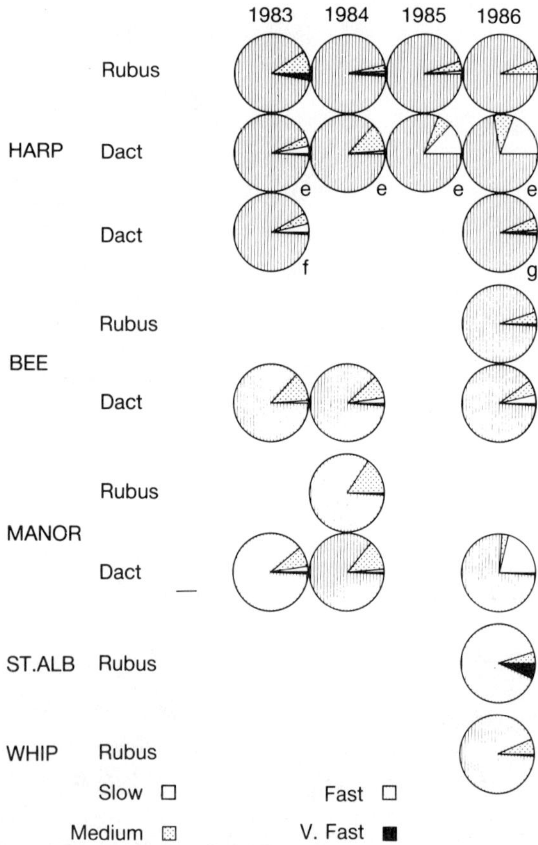

host [*Rubus fruticosus* or *Dactylis glomerata* are shown vertically at the left. At the foot is the graphic key for alleles—slow (S), medium (M), fast (F), and very fast (V)]. (From Loxdale and Brookes 1990*a*.)

absence of some expected heterozygotes, e.g. the SV genotype, out of the 15 calculated (i.e. $H = N(N - 1)/2$ where H = number of heterozygotes and N = number of alleles (6) at the locus), probably reflects failure to sample them due to their low frequency within populations, *or* because they are selectively disadvantageous and thus do not exist (cf. also Loxdale and Brookes, 1987; 1990*a*.).

In Fig. 11.8, the allele frequencies obtained in the study are presented as pie charts (the two rarest alleles are omitted owing to their scarcity in all populations examined). Each chart represents about 200–400 aphids surveyed. It is immediately clear that compared with *R. padi*, discussed earlier, this species shows considerable local differentiation of allele frequencies. The populations may be split into 'western' populations (i.e. SAL and TRNG) and 'eastern' ones (i.e. HARP, MANR, BEE, BAYF, etc.). The difference stems from the fact that whilst most sites are dominated by

Table 11.3 Deviations in GOT genotype frequencies from Hardy–Weinberg expectations in populations of *S. fragariae* collected from *R. fruticosus* and *D. glomerata* in 1981–86.

	R. fruticosus					*D. glomerata*				
Year Site	81	83	84	85	86	81	83	84	85	86
SAL			*			—	—	—	—	—
TRNG										
a(esc)			*		*		—	—		—
b(road)										—
BEE					*		*	*		*
MANR			**				—	—		—
HARP										
e(S)		*	*	*	*		*	*	—	—
g(N)							*			*
BAYF										
c(1)							*	*	*	*
c(2)									*	
d(pin)			*	*				*		
WHIP					*					
SALB					*					

*, $P = 0.05^{NS}$; **, $0.01 < P < 0.05$; (–), $P < 0.001$; NS = not significant.
(Data from Loxdale and Brookes 1990*a*). Site name abbreviations as in Fig. 11.7; a–g = subsites within main sampling site; esc = escarpment at Tring Park; S&N = south and north, two sites about 1 km apart on Harpenden Common; c(1) and c(2) = two sites about 0.5 km apart, Hook's Grove, Bayfordbury; d(pin) = Pinetum at Bayfordbury.

a high frequency of the F allele, western population samples taken from grass have a generally higher proportion of the S allele, whereas those from further east generally do not. Interestingly, allele frequencies of populations from *R. fruticosus* tend to be more homogeneous than those from grass. The frequency of the M allele is consistently low (about $f << 0.154$) at all sites and on both hosts. The S allele is conspicuously absent from populations infesting *R. fruticosus*. At some sites, populations have very similar GOT frequencies between primary and secondary hosts (e.g. BAYF, HARP, MANR), whereas at others, they are very different (e.g. SAL, TRNG). All population samples from *R. fruticosus* have genotype frequencies close to H–W expectations, as do some from grass, whereas those populations from grass bearing the S allele tend to deviate significantly from H–W expectations (Table 11.3).

The main inferences drawn from these findings are as follows.

1. The substantial allelic heterogeneity possibly reflects a lack of gene flow between populations. Certainly the populations at some sites show impressive temporal stability for the three main alleles S, M, and F over four consecutive years (Fig. 11.9).
2. Populations in which the allele frequencies for aphids on primary and secondary host are very different (i.e. SAL, Fig. 11.8) suggest that migration onto grass is not local in these instances, whereas, where it is similar on different hosts (i.e. BAYF, MANR; Fig. 11.8), migration is probably local.
3. That GOT genotype frequencies on grass are still at or near to H–W expectations is further evidence for a local migration of aphids between primary and secondary hosts. Additionally, this suggests little immigration in or out of the population(s), and also that the GOT alleles sampled are neutral or, at most, weakly selected.
4. S-bearing genotypes are a parthenogenetic strain(s) that do not go through the usual holocycle involving the principal primary host *R. fruticosus*. Since the S allele is often found in heterozygous combination, i.e. the SF genotype, presumably they once did so. An alternative hypothesis is that they go through the holocycle, but return to another primary host such as *Rosa* spp. and *Fragaria* spp. (Hille Ris Lambers 1939). We have never found *S. fragariae* on *Rosa* spp. in Hertfordshire.

2. Myzus persicae A few aphid species are resistant to insecticides, and if this is conferred by esterase activity, variation in this activity may be used to examine population structure and biology.

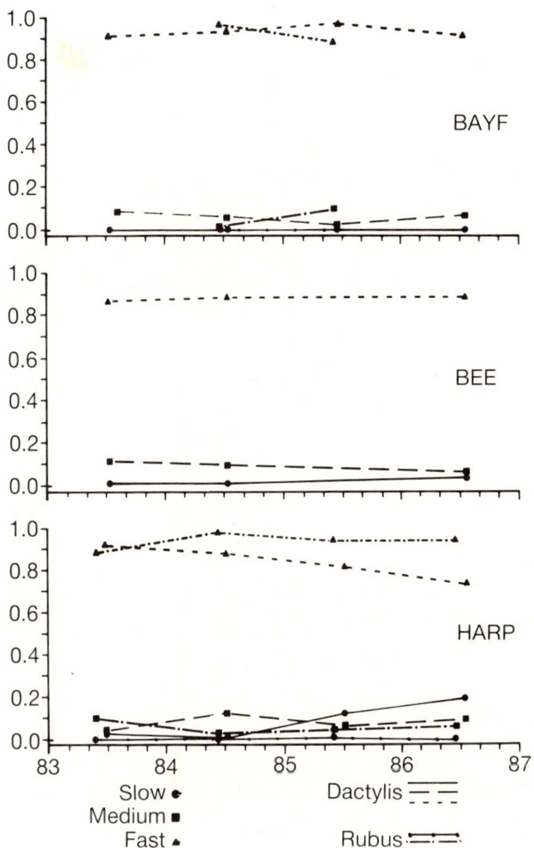

Fig. 11.9. Temporal stability of S, M, and F GOT allele frequencies (vertical axis) of *S. fragariae* on primary and secondary hosts at three sites (BAYF, BEE, HARP; site name abbreviations as in Fig. 11.7.) over four consecutive years, 1983–6 (horizontal axis). (From Loxdale and Brookes 1990*a*.)

A notable example is the peach-potato aphid *Myzus persicae* (Sulzer), perhaps the most economically important aphid in the world and vector of many plant virus diseases (Blackman and Eastop 1984). It has four resistance levels conferred by a single carboxylesterase, esterase-4 (EST-4, E-4). This enzyme hydrolyses the ester bonds of various insecticides (organophosphorus compounds, carbamates, and pyrethroids), thus conferring cross-resistance in this species.

It is thought that the four resistance genotypes (S, R1–R3) are the result of gene duplication of the EST-4 gene (Devonshire, this volume).

Because of the great economic importance of *M. persicae*, it was one of the first aphids to be examined electrophoretically, both by population geneticists as well as biochemists investigating insecticide resistance — not only the genetic/biochemical mechanism(s) *per se* (Devonshire 1977; Beranek and Oppenoorth 1977), but also the spread of resistance under field conditions (Devonshire *et al.* 1975).

As early as 1978, Wool and co-workers (Wool *et al.* 1978) investigated the genetic variation of 35 British clones of *M. persicae* using 11 enzymes representing 35 isoenzyme bands (35 putative loci). Of these, only two bands (EST-1 and EST-3) showed electrophoretic mobility variation in some clones. Thus, the species

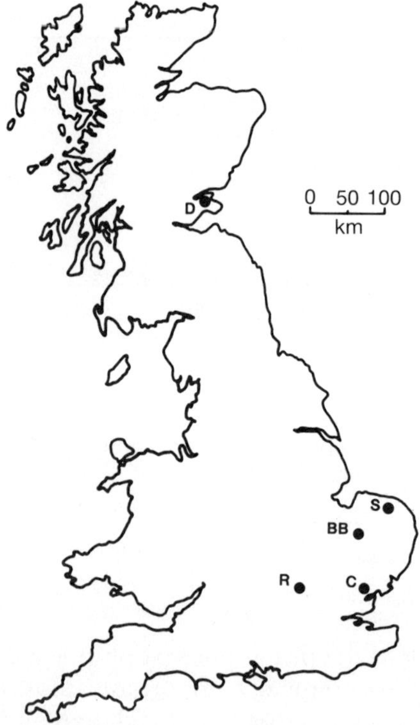

Fig. 11.10. Sampling sites for *Myzus persicae* clones in Britain in 1984. Abbreviations: D = Dairsie, Fife; S = Sall, Norfolk; BB = Brooms Barn, Suffolk; R = Rothamsted and C = Coggeshall, Essex. (From Brookes and Loxdale 1987.)

appeared to be largely invariant for enzymes markers—although it should be said that many of the clones used originated from greenhouses and so could have had a restricted variability. However, at the same time, May and Holbrook (1978) investigated the enzyme variation of 9 geographically-distant North American *M. persicae* field populations using 13 enzymes representing 19 loci and also showed the species to be (amazingly) invariant at all loci studied.

In 1984, in order to re-examine Wool *et al.*'s findings concerning British *M. persicae*, we (Brookes and Loxdale 1987) investigated five widely-separate field populations in Britain, but mainly from East Anglia (Fig. 11.10). Field populations were sampled at these various sites and clones were established on excised potato leaves under constant-environment conditions. These clones were then tested using eight enzyme systems representing 13 loci (Brookes and Loxdale 1987). As found previously by Wool *et al.* (1978) and May and Holbrook (1978), the majority of enzyme bands were completely invariant. However, EST-1/2 [a single locus with two alleles historically termed 1 and 2, 2 being the faster (Brookes and Loxdale 1987)] and EST-4 were variable, the latter showing only intensity differences related to the level of insecticide resistance noted above. Because EST-1/2 is not separated reliably on our polyacrylamide gel systems, being near the gel (cathodal) origin with allozymes of little mobility difference (Brookes and Loxdale 1987), we concentrated on studying EST-4 (Fig. 11.11).

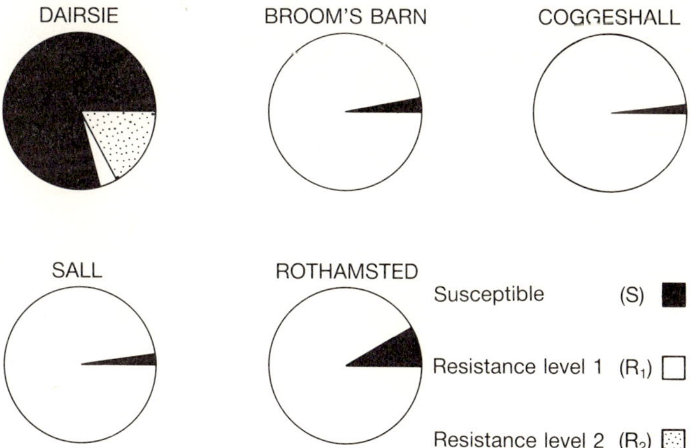

Fig. 11.11. Pie charts showing *M. persicae* insecticide resistance genotype frequencies at the five sampling sites listed in Fig. 11.10. (Modified from Brookes and Loxdale 1987.)

Clearly, the Scottish sample collected from potatoes (*Solanum tuberosum*) is very different from the English samples collected mainly from rape (*Brassica napus*), having a lower proportion of R1 (a resistant genotype with a 4-fold insecticide resistance compared to susceptible (S) individuals; Devonshire and Sawicki 1979) and a higher proportion of S individuals and R2 (a resistant genotype with a 16-fold resistance compared to S individuals). In contrast, the English population samples are dominated by R1 ($>>$ 80 per cent at all four sites), the remainder being S, whilst no R2 occurred. This pattern of genotype frequencies is very similar to that previously observed (Furk 1986). We concluded from these results that, in Britain, the R1 variant is heading for fixation in the south, probably as a result of the continued application of pesticides, while the R2 variant is common within populations on potatoes in southern Scotland because it is a major potato seed-growing region where the crop is heavily treated by pesticide sprays, thus possibly selecting for the higher-resistance variant there i.e. almost 80 per cent of potatoes in the area were treated at least twice by insecticides during the 1977 growing season (Steed *et al.* 1977; cf. also ffrench-Constant *et al.* 1988).

More recently, in 1985–86, ffrench-Constant and Devonshire (1988), using an immunoassay technique, have recorded *M. persicae* R2 variants on various *Brassica* species throughout England, including Humberside, Bedfordshire, Hertfordshire, Suffolk, and Hampshire, whilst another resistant variant, V8 intermediate in resistance between R1 and R2, was also found to be widespread and abundant. Hence, since our survey in 1984, variants of higher resistance now appear to be spreading rapidly throughout the country, surely a fact of considerable concern to farmers as well as the agrochemical industry.

We have no evidence of host-adapted races of susceptible and pesticide-resistant *M. persicae* on rape and sugar beet in Britain (cf. Brookes and Loxdale 1987), although as our own and previous results suggest, it is possible that there is a genetic relationship between the highly insecticide-resistant (R2) strains and adaptation to potatoes, as earlier proposed by Furk (1986). Weber (1985) provides some evidence to suggest host adaptation of *M. persicae* clones collected from several crops, including sugar beet and potatoes, in various parts of West Germany.

In Britain, as well as abroad (Germany and the USA), it seems that *M. persicae* field populations are highly structured, that is to say, comprise few genotypes. This limited variability does not appear to have arisen solely as a consequence of chemical selection for a small number of insecticide-resistant variants, since susceptible individuals collected from weeds (especially charlock, *Sinapis arvensis*) also show a lack of genetic variation over a range of enzyme loci unconnected with resistance (Brookes and Loxdale 1987). Presumably, such biochemical monomorphism is brought about by a combination of winter population bottlenecks, especially that of climate, affecting survival of parthenogenetically overwintering clones (Harrington and Cheng 1984), and scarcity of the primary host, peach *Prunus persica*. Over many years, both factors eliminate the rarer alleles at given loci, thereby fixing the most common ones (Brookes and Loxdale 1987). If there are indeed very few genotypes nationally and elsewhere, this makes it hard to believe in the existence of specific host adapted biotypes as proposed by Weber (1985) and others.

Conclusions

This chapter has shown how electrophoresis of enzymes may be used to study various aspects of aphid biology. These include persistence and dispersal of given strains (genotypes) and overwintering strategies. It may also be most successfully employed as a taxonomic tool, and as a method of assessing the presence of hymenopterous wasp parasitoids and entomophthoralean fungi, and to provide an estimate of the number of genotypes within a population — information of much value in host-adaptation studies. Lastly, it has proved invaluable in studying resistance to insecticides in *Myzus persicae*.

Acknowledgements

We are most grateful to Drs Brian Kerry and Mark Tatchell, Rothamsted, for their valuable comments concerning the manuscript of this paper. We thank Dr Nick Carter and Academic Press, New York, for permission to reproduce Figs. 11.1A and B, and C.A.B. International, Wallingford, Oxfordshire for permission to reproduce Figs. 11.2 to 11.11 and Tables 11.1, 11.2, and 11.3.

Appendix

Full name and Enzyme Commission (EC) classification numbers of enzymes cited as abbreviations in the text.

Abbreviation	*Full name*	*EC no.*
EST	Esterase	3.1.1.
GOT	Glutamate-oxaloacetate transaminase	2.6.1.1.
ME	Malic enzyme	1.1.1.40.
MDH	Malate dehydrogenase	1.1.1.37.
PEP	Leucine amino-peptidase	3.4.11.
PGM	Phosphoglucomutase	2.7.5.1.
POD	Peroxidase	1.11.1.7.
SORDH	Sorbitol dehydrogenase	1.1.1.14.

References

Avise, J.C. (1977). Genetic differentiation during speciation. In *Molecular evolution* (ed. F.J. Ayala), pp. 106–22. Sinauer Associates, Inc., Sunderland, Mass.

Ayala, F.J. (1977). *Molecular evolution.* Sinauer Associates, Inc., Sunderland, Mass.

Ayala, F.J. and Valentine, J.W. (1978). Genetic variation and resource stability in marine invertebrates. In *Marine organisms* (ed. B. Battaglia and J. Beardmore), pp. 23–51. Plenum, New York.

Bakker, J. and Gommers, F.J. (1989). Characterisation of populations (Pathotypes) of potato cyst nematodes. This volume, pp. 415–30.

Bale, J.S., Harrington, R., and Clough, M.S. (1988). Low temperature mortality of the peach-potato aphid *Myzus persicae. Ecol. Ent.*, **13**, 121–9.

Beranek, A.P. and Oppenoorth, F.J. (1977). Evidence that the elevated carboxylesterase (Esterase 2) in organophosphorus-resistant *Myzus persicae* (Sulz.) is identical with the organophosphate-hydrolysing enzyme. *Pestic. Biochem. Physiol.*, **7**, 16–20.

Berlocher, S.H. (1980). An electrophoretic key for distinguishing species of the genus *Rhagoletis* (Diptera: Tephritidae) as larvae, pupae, or adults. *Ann. ent. Soc. Am.*, **73**, 131–7.

Blackman, R.L. (1971). Variation in the photoperiodic response within natural populations of *Myzus persicae* (Sulz.). *Bull. ent. Res.*, **60**, 533–46.

Blackman, R.L. (1974). *Aphids.* Ginn and Company Ltd., London and Aylesbury.

Blackman, R.L. (1981). Species, sex and parthenogenesis in aphids. In *The evolving biosphere* (ed. P.L. Forey), pp. 75–85. Cambridge University Press, Cambridge.

Blackman, R.L. and Devonshire, A.L. (1978). Further studies on the genetics of the carboxylesterase regulatory system involved in resistance to organophosphorus insecticides in *Myzus persicae* (Sulzer). *Pestic. Sci.*, **9**, 517–21.

Blackman, R.L. and Eastop, V.F. (1984). *Aphids on the world's crops: An identification and information guide.* Wiley, Chichester.

Blackman, R.L., Brown, P.A., Furk, C., Seccombe, A.D., and Watson, Gillian W. (1989). Enzyme differences within species-groups containing pest aphids. This volume, pp. 271–95.

Brewer, G.J. and Sing, C.F. (1970). *An introduction to isozyme techniques.* Academic Press, New York and London.

Brookes, C.P. and Loxdale, H.D. (1985). A device for simultaneously homogenizing numbers of small insects for electrophoresis. *Bull. ent. Res.*, **75**, 377–8.

Brookes, C.P. and Loxdale, H.D. (1987). Survey of enzyme variation in British populations of *Myzus persicae* (Sulzer) (Hemiptera: Aphididae) on crops and weed hosts. *Bull. ent. Res.*, **77**, 83–9.

Carter, N., McLean, I.F.G. Watt, A.D., and Dixon, A.F.G. (1980). Cereal aphids: A case study and review. In *Applied biology,* Vol. V (ed. T.H. Coaker), pp. 271–348. Academic Press, London and New York.

Castañera, P., Loxdale, H.D., and Nowak, Kathy (1983). Electrophoretic study of enzymes from cereal aphid populations. II. Use of electrophoresis for identifying aphidiid parasitoids of the grain aphid *Sitobion avenae* (F.) (Hymenoptera: Aphidiidae; Hemiptera: Aphididae). *Bull. ent. Res.*, **73**, 659–65.

Daly, Joanne C. (1989). The use of electrophoretic data in a study of gene flow in the pest species *Heliothis armigera* (Hübner) and *H. punctigera* Wallengren (Lepidoptera: Noctuidae). This volume, pp. 115–41.

Dean, G.J.W. (1973). Aphid colonisation of spring cereals. *Ann. appl. Biol.,* **75**, 183–93.

Dean, G.J.W., and Luuring, B.B. (1970). Distribution of aphids in cereal crops. *Ann. appl. Biol.*, **66**, 485–96.

Dewar, A.M., Woiwod, I., and Choppin de Janvry, E. (1980). Aerial migrations of the rose-grain aphid, *Metopolophium dirhodum* (Wlk.), over Europe in 1979. *Pl. Path.*, **29**, 101–9.

Devonshire, A.L. (1977). The properties of a carboxylesterase from the peach-potato aphid, *Myzus persicae* (Sulz.), and its role in conferring insecticide resistance. *Biochem. J.*, **7**, 675–83.

Devonshire, A.L. (1989) The role of electrophoresis in the biochemical detection of insecticide resistance. This volume, pp. 363–74.

Devonshire, A.L. Needham, P.H., Rice, A.D. and Sawicki, R.M. (1975). Monitoring for resistance to organophoshorus insecticides in *Myzus persicae* from sugar beet. *Proc. 8th Br. Insecticide and Fungicide Conf.* pp. 21–4.

Digby, P.G.N. and Kempton, R.A. (1986). *Multivariate analysis of ecological communities.* Chapman and Hall, London.

Devonshire, A.L. and Sawicki, R.M. (1979). Insecticide-resistant *Myzus persicae* as an example of evolution by gene duplication. *Nature, Lond.*, **280**, 140–1.

Dixon, A.F.G. (1985). *Aphid ecology.* Blackie, Glasgow and London.

Eggers-Schumacher, H.A. (1987). Enzyme electrophoresis in biosystematics and taxonomy of aphids. In *Population structure, genetics and taxonomy of aphids and Thysanoptera* (ed. J. Holman, J. Pelikan, A. F. G. Dixon, and L. Weismann), pp. 63–70. SPB Academic Publishing, The Hague.

Eggers-Schumacher, H. A. and Sander, Evamarie (1988). Spatial and seasonal genetic (allozyme) variation within field populations of *Phorodon humuli* Schrank (Homoptera; Aphididae) on winter and summer hosts in southern Germany (FRG). *Entomologist*, **107**, 110–21.

Ferguson, A. (1980). *Biochemical systematics and evolution.* Blackie and Son Ltd, Glasgow.

ffrench-Constant, R. H. and Devonshire, A. L. (1987). A multiple homogenizer for rapid sample preparation in immunoassays and electrophoresis. *Biochem. Genet.*, **25**, 493–99.

ffrench-Constant, R. H. and Devonshire, A. L. (1988). Monitoring frequencies of insecticide resistance in *Myzus periscae* (Sulzer) (Hemiptera; Aphididae) in England during 1985–86 by immunoassay. *Bull. ent. Res.*, **78**, 163–71.

ffrench-Constant, R. H., Harrington, R., and Devonshire, A. L. (1988). Effect of repeated applications of insecticides to potatoes on numbers of *Myzus persicae* (Sulzer) (Hemiptera: Aphididae) and on the frequencies of insecticide-resistant variants. *Crop Prot.*, **7**, 55–61.

Furk, C. (1979). Field collections of *Aphis fabae* Scopoli *S. lat.* (Homoptera: Aphididae) studied by starch gel electrophoresis and isoelectric focusing. *Comp. Biochem. Physiol.*, **62B**, 225–30.

Furk, C. (1986). Incidence and distribution of insecticide-resistant strains of *Myzus persicae* (Sulzer) (Hemiptera: Aphididae) in England and Wales in 1980–84. *Bull. ent. Res.*, **76**, 53–8.

Gillespie, J. H. and Kojima, K. (1968). The degree of polymorphisms in enzymes involved in energy production compared to that of non-specific enzymes in two *Drosophila ananassae* populations. *Proc. natn. Acad. Sci. USA*, **61**, 582–5.

Hand, S. C. (1982). The overwintering and dispersal of cereal aphids. Ph.D. thesis, University of Southampton, UK.

Hardy, A. C. and Cheng, L. (1986). Studies on the distribution of insects by aerial currents. III. Insect drift over the sea. *Ecol. Ent.*, **11**, 283–90.

Harrington, R. and Cheng Xia-Nian (1984). Winter mortality, development and reproduction in a field population of *Myzus persicae* (Sulzer) (Hemiptera: Aphididae) in England. *Bull. ent. Res.*, **74**, 633–40.

Harrington, R., Katis, N., and Gibson, R. W. (1986). Field assessment of the relative importance of different aphid species in the transmission of potato virus Y. *Potato Res.*, **29**, 67–76.

Harris, H. and Hopkinson, D. A. (1977). *Handbook of enzyme electrophoresis in human genetics.* North-Holland, Amsterdam.

Hartl, D. L. (1980). *Principles of population genetics.* Sinauer Associates., Inc., Sunderland, Mass.

Hille Ris Lambers, D. (1939). Contributions to a monograph of the

Aphididae of Europe. II. The genera *Dactynotus* Rafinesque, 1818; *Staticobium* Mordvilko, 1914; *Macrosiphum* Passerini, 1860; *Masonaphis* Nov. Gen.; *Phalaris* Leach, 1826. *Temminckia*, **4**, 1–134.

Lampel, G. and Burgener, R. (1987). The genetic relationships between lachnid taxa as established by enzyme gel-electrophoresis. In *Population structure, genetics and taxonomy of aphids and Thysanoptera* (ed. J. Holman, J. Pelikan, A.F.G. Dixon, and L. Weismann), pp. 71–95. SPB Academic Publishing, The Hague.

Loxdale, H.D., Castañera, P., and Brookes, C.P. (1983). Electrophoretic study of enzymes from cereal aphid populations. I. Electrophoretic techniques and staining systems for characterising isoenzymes from six species of cereal aphids. *Bull. ent. Res.*, **73**, 645–57.

Loxdale, H.D., Tarr, I.J., Weber, C.P., Brookes, C.P., Digby, P.G.N., and Castañera, P. (1985*a*). Electrophoretic study of enzymes from cereal aphid populations. III. Spatial and temporal genetic variation of populations of *Sitobion avenae* (F.) (Hemiptera: Aphididae). *Bull. ent. Res.*, **75**, 121–41.

Loxdale, H.D., Rhodes J.A., and Fox, J.S. (1985*b*). Electrophoretic study of enzymes from cereal aphid populations. IV. Detection of hidden genetic variation within populations of the grain aphid *Sitobion avenae* (F.) (Hemiptera: Aphididae). *Theor. appl. Genet.*, **70**, 407–12.

Loxdale, H.D. and Brookes, C.P. (1987). Use of electrophoretic markers to study the spatial and temporal genetic structure of populations of a holocyclic aphid species—*Sitobion fragariae* (Walker) (Hemiptera: Aphididae). In *Population structure, genetics and taxonomy of aphids and Thysanoptera* (ed. J. Holman, J. Pelikan, A.F.G. Dixon, and L. Weismann), pp. 100–10. SPB Academic Publishing, The Hague.

Loxdale, H.D. and Brookes, C.P. (1988). Electrophoretic study of enzymes from cereal aphid populations. V. Spatial and temporal genetic similarity between holocyclic populations of the bird-cherry oat aphid *Rhopalosiphum padi* (L.) (Hemiptera: Aphididae) in Britain. *Bull. ent. Res.*, **78**, 241–9.

Loxdale, H.D. and Brookes, C.P. (1990*a*). Electrophoretic study of enzymes from cereal aphid populations. VI. Spatial and temporal allozyme variation at the GOT-locus in local populations of the blackberry-grain aphid *Sitobion fragariae* (Walker) (Hemiptera: Aphididae) in south-east England. *J. Anim. Ecol.*

Loxdale, H.D. and Brookes, C.P. (1990*b*). Separation of three *Rubus*-feeding species of aphid—*Sitobion fragariae* (Wlk.), *Macrosiphum funestum* (Macch.) and *Amphorophora rubi* (Kalt.) (Hemiptera: Aphididae)—by electrophoresis. *Ann. appl. Biol.*

Loxdale, H.D. and Brookes, C.P. (1990*c*). The cereal aphids *Sitobion avenae* (F.) and *Sitobion fragariae* (Wlk.) on wild Cocksfoot grass (*Dactylis glomerata*) in south-east England. *Bull. ent. Res.*

Macaulay, E.D.M., Tatchell, G.M., and Taylor, L.R. (1988). The Rothamsted Insect Survey '12-metre' suction trap. *Bull. ent. Res.*, **78**, 121–9.

Mardell, S.K., Wilding, N., Loxdale, H.D., and Brookes, C.P. Use of enzyme electrophoresis for the identification of *entomophthoralean* fungi infecting pest aphids (in prep).

May, B. and Holbrook, F.R. (1978). Absence of genetic variability in the Green-Peach Aphid *Myzus persicae* (Hemiptera: Aphididae). *Ann. ent. Soc. Am.*, **71**, 809–12.

Mayr, E. (1963). *Animal species and evolution.* Harvard University Press., Cambridge, Mass.

Menken, S.B. and Ulenberg, Sandrine A. (1987). Biochemical characters in agricultural zoology. *Agric. Zool. Rev.*, **2**, 305–60.

Nei, M. (1971). Interspecific gene differences and evolutionary time estimated from electrophoretic data on protein identity. *Am. Nat.*, **105**, 385–98.

Nei, M. (1972). Genetic distance between populations. *Am. Nat.*, **106**, 283–92.

Nei, M. (1975). *Molecular population genetics and evolution.* Elsevier, Amsterdam.

Nei, M. (1978). Estimation of average heterozygosity and genetic distance from a small number of individuals. *Genetics*, **89**, 583–90.

Nei, M., Maruyama, T., and Chakraborty, R. (1975). The bottleneck effect and genetic variability in populations. *Evolution,* **29**, 1–10.

Newton, Christine and Dixon, A.F.G. (1988). A preliminary study of variation and inheritance of life-history traits and the occurence of hybrid vigour in *Sitobion avenae* (F.) (Hemiptera: Aphididae). *Bull. ent. Res.*, **78**, 75–83.

Oxford, G.S. and Rollinson D. (1983) (eds) *Protein polymorphism: Adaptive and taxonomic significance.* Systematics Association Special Vol. 24. Academic Press, London.

Parkin, D.T. (1979). *An introduction to evolutionary genetics.* Edward Arnold, London.

Plumb, R.T. (1983). Barley yellow dwarf virus—a global problem. In *Plant virus epidemiology* (ed. R.T. Plumb and J.M. Thresh), pp. 185–98. Blackwell Scientific, Oxford.

Powell, J.R., Tabachnick, W.J., and Arnold, J. (1980). Genetics and the origin of a vector population: *Aedes aegypti*, a case study. *Science,* **208**, 1385–7.

Powell, W. and Walton, M.P. (1989). The use of electrophoresis in the study of hymenopteran parasitoids of agricultural pests. This volume, pp. 443–65.

Rhomberg, L.R., Joseph, S. and Singh, R.S. (1985). Seasonal variation and clonal selection in cyclically parthenogenetic rose aphids (*Macrosiphum rosae*). *Can. J. Genet. Cytol.*, **27**, 224–32.

Richardson, B.J., Baverstock, P.R., and Adams, M. (1986). *Allozyme Electrophoresis: A handbook for animal systematics and population studies.* Academic Press, Australia.

Robinson, M.P. (1989). Isoelectric focusing techniques for the identification of plant-parasitic nematodes. This volume, pp. 431–42.

Sargeant, J. R. and George, S. G. (1975). *Methods in zone electrophoresis*, 3rd edn. BDH Chemical Ltd, Poole, England.

Shaw, C.R. and Prasad, R. (1970). Starch gel electrophoresis-a compilation of recipes. *Biochem. Genet.*, **4**, 297-320.

Shorrocks, B. (1978). *The genesis of diversity.* Hodder and Stoughton, London.

Singh, R.S. and Rhomberg, L.R. (1984). Allozyme variation, population structure, and sibling species in *Aphis pomi. Can. J. Genet. Cytol.*, **26**, 364-73.

Steed, J. M., Sly, J. M. A., Tucker, G.G., and Cutler, J. R. (1977). *Arable Farm Crops—1977.* Survey Report No. 18, Pesticide Usage. Ministry of Agriculture, Fisheries & Food and Department of Agriculture and Fisheries for Scotland.

Steiner, W.W.M., Voegtlin, D.J., and Irwin, M.E. (1985). Genetic differentiation and its bearing on migration in North American populations of the corn leaf aphid *Rhopalosiphum maidis* (Fitch) (Homoptera: Aphididae). *Ann. ent. Soc. Am.*, **78**, 518-25.

Suomalainen, E., Saura, A., Lokki, J, and Teeri, T. (1980). Genetic polymorphism and evolution in parthenogenetic animals. Part 9: Absence of variation within parthenogenetic aphid clones. *Theor. appl. Genet.*, **57**, 129-32.

Tatchell, G. M. (1989). An estimate of the potential economic losses to some crops due to aphids in Britain. *Crop Prot.*, **8**, 25-9.

Tatchell., G. M., Plumb, R. T. and Carter, N. (1988). Migration of alate morphs of the bird cherry aphid (*Rhopalosiphum padi*) and implications for the epidemiology of barley yellow dwarf virus. *Ann. appl. Biol.*, **112**, 1-11.

Taylor L. R., Woiwod, I. P., and Taylor, R. A. J. (1979). The migratory ambit of the hop aphid and its significance in aphid population dynamics. *J. Anim. Ecol.*, **48**, 955-72.

Taylor, L.R., French, R.A., Woiwod, I.P., Dupuch, Maureen J., and Nicklen, Joan (1981). Synoptic monitoring for migrant insect pests in Great Britain and Western Europe. I. Establishing expected values for species content, population stability and phenology of aphids and moths. *Rothamsted Experimental Station, Report for 1980, Part 2*, pp. 41-104.

Thomson, G. (1987). Enzyme variation at morphological boundaries in *Maniola* and related genera (Lepidoptera: Nymphalidae: Satyrinae). Ph.D. thesis, University of Stirling, UK.

Tomiuk, J. and Wöhrmann, K. (1980). Enzyme variability in populations of aphids. *Theor. appl. Genet.*, **57**, 125-7.

Tomiuk J. and Wöhrmann, K. (1983). Enzyme polymorphism and taxonomy of aphid species. *Z. zool. Syst. Evolutionforsch.*, **21**, 266-74.

Tomiuk, J and Wöhrmann, K. (1984). Genotypic variability in natural Populations of *Macrosiphum rosae* (L.) in Europe. *Biol. Zbl.*, **103**, 113-22.

Vickerman, G. P. and Wratten, S. D. (1979). The biology and pest status of cereal aphids (Hemiptera: Aphididae) in Europe: a review. *Bull. ent. Res.*, **69**, 1-32.

Voegtlin, D.J., Steiner, W.W.M., and Irwin, M.E. (1987).Searching for the source of the annual spring migrants of *Rhopalosiphum maidis* (Homoptera: Aphididae) in North America. In *Population structure, genetics and taxonomy of aphids and Thysanoptera* (ed. J. Holman, J. Pelikan, A.F.G. Dixon, and L. Weismann, pp. 120-133. SPB Academic Publishing, The Hague.

Walters, K.F.A. and Dewar, A.M. (1986). Overwintering strategy and the timing of the spring migration of the cereal aphids *Sitobion avenae* and *Sitobion fragariae. J. appl. Ecol.*, **23**, 905-15.

Weber, G. (1985). Genetic variability in host plant adaptation of the green peach aphid, *Myzus persicae. Entomologia exp. appl.*, **38**, 49-56.

Wöhrmann, K. (1984) Population biology of the rose aphid, *Macrosiphum rosae.* In *Population biology and evolution* (ed. K. Wöhrmann and V. Loeschcke), pp. 208-16. Springer-Verlag, Berlin.

Wöhrmann, K., Tomiuk, J., and Weber, G. (1986). The search for hidden enzymatic variation in the aphid *Macrosiphum rosae* (L.). *Theor. appl. Genet.*, **73**, 77-81.

Wöhrmann, K. and Tomiuk, J. (1989). Study of the variation of aphid populations using enzymes and other traits. This volume, pp. 203-29.

Wöhrmann K. and Tomiuk, J. (1988). Life cycle strategies and genotypic variability in populations of aphids. *J. Genet.*, **67**, 43-52.

Woiwod, I.P., Tatchell, G.M., and Barrett, Angela M. (1984). A system for the rapid collection, analysis and dissemination of aphid-monitoring data from suction traps. *Crop Prot.*, **3**, 273-88.

Woiwod, I.P., Tatchell, G.M., Dupuch, Maureen J., Macaulay, E.D.M., Parker, Susan J., and Taylor, M.S. (1987). Rothamsted Insect Survey; Eighteenth Annual Summary; Suction Traps 1986. *Rothamsted Experimental Station, Report for 1986,* pp. 289-321.

Wool, D., Bunting, S., and van Emden, H.F. (1978). Electrophoretic study of genetic variation in British *Myzus persicae* (Sulz.) (Hemiptera: Aphididae). *Biochem. Genet.*, **16**, 987-1006.

Workman, P.L. and Niswander, J.D. (1970). Population studies on southwestern Indian tribes. II. Local genetic differentiation in the Papago. *Am. J. hum. Genet.*, **22**, 24-49.

Wright, S. (1978). *Evolution and the genetics of populations,* Vol. 4, *Variability within and among natural populations.* University of Chicago, Chicago.

12 Enzyme differences within species groups containing pest aphids

R.L. BLACKMAN[1], P.A. BROWN[1], C. FURK[2], A.D. SECCOMBE[1], and GILLIAN W. WATSON[1]

[1]*Department of Entomology, British Museum (Natural History), London UK*
[2]*Ministry of Agriculture, Fisheries and Food, ADAS, Harpenden Laboratory, Harpenden, Hertfordshire, UK*

Abstract

Many pest aphids belong to species groups or complexes, and are difficult to distinguish morphologically from other non-pest species within these groups. Pest species also show an exceptionally large range of intraspecific variation in morphology associated with season or host plant. The interpretation of such variation is difficult with our still very inadequate knowledge of the genetic structure of aphid populations.

Enzymes, particularly esterases, can show discrete differences in mobility between closely-related aphids, and help us not only to distinguish pest from non-pest species, but also to understand taxonomic relationships within such groups. This paper presents and discusses the findings of previously unpublished work on allozyme differences between closely-related species within the aphid genera *Myzus, Metopolophium, Macrosiphum,* and *Acyrthosiphon.*

Problems of pest aphid taxonomy

Many pest aphids are highly variable organisms, in which differences in host-plant-colonizing ability, life cycle, and morphology can be

Electrophoretic Studies on Agricultural Pests (ed. Hugh D. Loxdale and J. den Hollander), Systematics Association Special Volume No. 39, pp. 271–95. Clarendon Press, Oxford, 1989.

observed, raising questions about whether all the forms currently placed under a single name, such as *Aphis fabae* or *Myzus persicae*, can really be treated as single species. Such aphids usually have close relatives that look morphologically very much like them, but only feed on non-crop plants; hence, we need to be certain not only that these are in fact distinct species, but also to find reliable and convenient ways to distinguish them from the pest species.

Historically, aphid taxonomists recognized species by the association between a particular host plant and a particular morphology. As aphid life cycles were worked out, it became clear that aphid morphology changed greatly with the season, and often with the host plant. Experimental host-plant transfers became essential to clarifying taxonomic relationships. Some aphids previously described under different names from many different plant were subsequently recognized as single, polyphagous species, so that their currently valid names carry long lists of synonyms.

As the true extent of variation within species came to be appreciated, the morphological studies needed to discriminate between them became more complex. Aphids not only exist as discrete, recognizably distinct morphs — such as apterous and alate viviparous females, oviparous females, fundatrices, and so on — they also vary more or less continuously within morphs. Some of this variation can be ascribed to differences in general body size, and is then partly accommodated by using simple morphometric ratios to discriminate between species (Eastop 1985). There may, however, be more complex patterns of variation associated with aphid polymorphisms: for example, apterous females developing under certain conditions may show differing, lesser degrees of expression of characters that are fully expressed in the alate morph. Distinguishing between closely-related species on morphological grounds may then be rather difficult, requiring time-consuming multivariate techniques and complex discriminant functions that nevertheless may not be 100 per cent reliable for individual specimens (cf. Blackman 1987; Blackman and Paterson 1986).

Methods that circumvent these problems by focusing directly on gene products therefore offer considerable potential for discriminating between closely related aphid species. Standard techniques of electrophoresis using starch gels (Tomiuk *et al.* 1979; Tomiuk and Wöhrmann 1983; Eggers-Schumacher 1987), polyacrylamide gels (Loxdale *et al.* 1983), or cellulose acetate sheets (Easteel and Boussy 1987) are clearly the simplest way of looking at such gene products. Application of these techniques to studies of the genetic structure of populations within aphid species has been hampered by the apparent lack of variation in many of the enzyme systems most commonly investigated (Tomiuk 1987). However, for species separation, all that is normally required is one or two diagnostic enzyme loci (Ayala 1983). Diagnostic loci have been found in almost all cases in which electrophoretic comparisons have been made between closely-related (but recognizably distinct) aphid species, and in some cases have indicated the presence of sibling species.

Aphids have one important advantage over most other small invertebrates for such studies. They can be reared as clones, so that the electrophoretic method does not have to be destructive of the genotype. Thus, there is no theoretical limit to the numbers or the amounts of enzymes available for analysis, nor of the number of buffer systems and staining procedures that can be employed on a single genotype. Correlations between allozyme data and morphological or cytological characters can also be investigated on the same genotype, avoiding the problem of genetic heterogeneity within sampled populations.

This chapter will review instances of the use of electrophoretic methods to discriminate between aphid populations at or around the species level, paying particular attention to species groups involving pest aphids, and including previously unpublished information on several species complexes. For the most part, we shall be looking simply at differences at diagnostic enzyme loci, as in most cases insufficient data are available on too few loci to examine genetic relationships quantitatively using measures of genetic identity or distance (e.g. Nei 1972). Nor are we concerned here with the application of electrophoresis to problems of classification at the genus level and above (e.g. Lampel and Burgener 1987), which is a more controversial use of allozyme data (see Buth 1984).

Diagnostic loci for pest aphids

1. *Macrosiphum euphorbiae*

The potato aphid is a typical example of a polyphagous pest aphid that is liable to taxonomic confusion with several other morphologically similar, but more host-specific species. This species is an important pest, transmitting more than 50 plant viruses in 35 different crops throughout the world. The number of taxa recognized within the *M. euphorbiae* group in Europe has varied from 4 (Hille Ris Lambers 1939) to 10 (Meier 1961). *M. euphorbiae* itself is apparently a North American species, unknown in Britain before 1917, and the actual and potential confusion with its close relatives in North America is probably even greater than in Europe.

Table 12.1 Species of the *Macrosiphum euphorbiae* group in Britain, their host plants and life cycles.

Species	Main host plant(s)	Life cycle[a]
M. euphorbiae (Thomas)	Polyphagous	Virtually anholocyclic
M. tinctum (Walker) (= *M. epilobiellum* Theobald)	*Epilobium* spp.	Virtually anholocyclic
M. penfroense Stroyan	*Silene maritima*	Virtually anholocyclic
M. centranthi Theobald	*Centranthus ruber*, *Valeriana officinalis*	Unknown; probably mainly anholocyclic
M. hellebori Theobald and Walton	*Helleborus* spp.	Mainly anholocyclic
M. euphorbiellum Theobald (= *M. amygdaloides* Theobald)	*Euphorbia* spp.	Mainly anholocyclic
M. stellariae Theobald	*Stellaria holostea* and other Caryophyllaceae	Partly anholocyclic
M. melampyri Mordwilko	*Melampyrum pratense*	Holocyclic
M. cholodkovskyi Mordwilko	*Filipendula ulmaria*	Holocyclic
M. gei Koch	*Geum* and Umbelliferae	Holocyclic
M. daphinidis Borner	*Daphne mezereum*	Holocyclic

[a] Most of these species, except *M. euphorbiae*, are holocyclic in continental Europe

Watson (1982) applied multivariate techniques (mainly canonical variate analysis) to British populations of the *M. euphorbiae* group, on the basis of which she concluded that 11 taxa were present (Table 12.1). Esterases of ten of these taxa were examined, using starch gel electrophoresis, and 11 other enzyme systems were looked at in 6 of them. For esterases, each species was found to have a characteristic pattern of bands, fully confirming the conclusions reached by multivariate morphometrics. A typical gel with six species is represented in Fig. 12.1(a). The complexity of the banding patterns did not permit any genetic interpretation of the loci involved, but most species were polymorphic for at least one esterase locus, and *M. gei* was markedly polymorphic. *M. euphorbiae* itself, however, showed no clear polymorphisms at esterase loci in any of the 75 samples

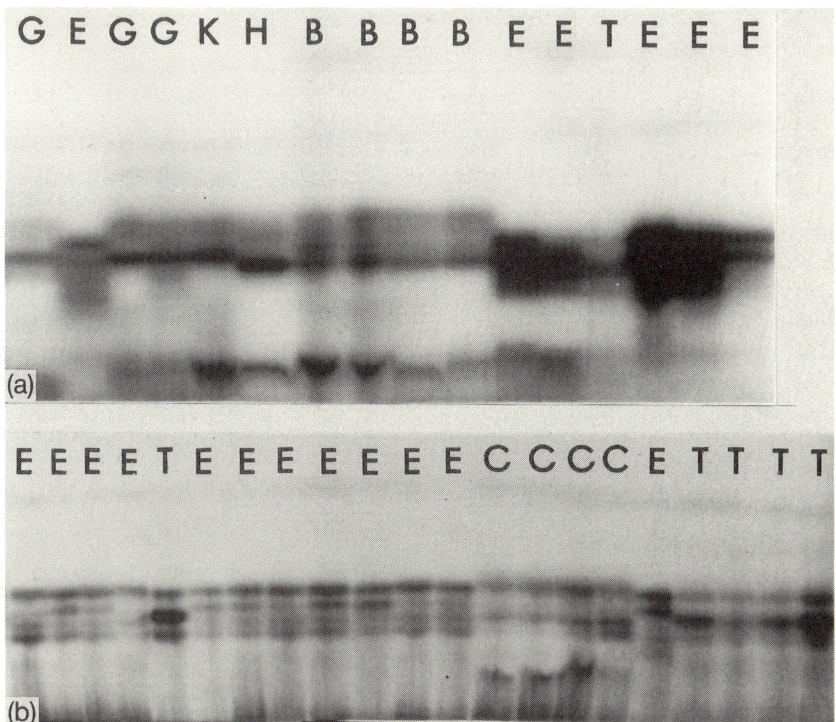

Fig. 12.1. Photographs of representative starch gels stained for esterases; species of the *Macrosiphum euphorbiae* group. (a) Samples of six species: E = *euphorbiae*, B = *euphorbiellum*, G = *gei*, H = *hellebori*, K = *cholodkovskyi*, T = *tinctum*. (b) 13 samples of *euphorbiae* (E), compared with four samples of *centranthi* (C), and five of *tinctum* (T).

examined, although the activity of individual allozymes varied greatly between samples (Fig. 12.1(b)).

The other enzyme systems examined provided at least one additional diagnostic locus for each species comparison (Table 12.2). *M. euphorbiae* and *M. tinctum* are the two species that are most difficult to separate morphologically and have most similar esterase patterns, yet they have different mobilities of 6-phosphogluconate dehydrogenase (6-PGD) and an isocitrate dehydrogenase locus (IDH-2). Contrary to the findings of Tomiuk and Wöhrmann (1983) in Germany, British populations of *M. euphorbiae* were found to be polymorphic at the IDH-1 locus. In North America, where *M. euphorbiae* is native, polymorphism has been found at a minimum of six loci, including IDH-1 and IDH-2 (Table 12.4; May and Holbrook 1978; Steiner *et al.* 1985*a*).

Table 12.2 Enzyme systems with diagnostic loci for six members of the *M. euphorbiae* group.

	tinctum	*hellebori*	*euphorbiellum*	*stellariae*	*gei*
euphorbiae	IDH 6-PGDH	HK IDH PGI	HK IDH PGI	HK IDH PGI	α-GPDH IDH PGI
tinctum		HK IDH PGI	HK IDH PGI	HK IDH PGI	α-GPDH IDH PGI
hellebori			HK	HK IDH	α-GPDH HK
euphorbiellum				HK IDH	α-GPDH HK
stellariae					α-GPDH HK IDH

Tomiuk and Wöhrmann (1983) examined up to 18 enzyme systems in three species of the *euphorbiae* group (*euphorbiae*, *gei*, and *hellebori*), and calculated genetic distance (D) values between them of 0.11–0.37. Eggers-Schumacher (1987) examined up to 12 enzyme systems in 8 species of the *euphorbiae* group (including one, *M. prenanthidis*, not included in Watson's study), and obtained values for the genetic distances between *euphorbiae*, *gei*, and *hellebori* comparable to those of Tomiuk and Wöhrmann. Both works also included data on *M. rosae* and *M. funestum*, two species which, in Watson's multivariate studies and according to most taxonomists, are placed outside the *euphorbiae* group. Again, the values of D they calculated between *rosae* and *funestum*, and between these two species and members of the *euphorbiae* group, agree fairly well. It is therefore instructive to compare phylogenetic trees produced from the two sets of data (Fig. 12.2(a,b)). The phylogeny produced by Tomiuk and Wöhrmann (Fig. 12.2(a)) accords with taxonomic relationships established on morphological grounds, the species of the *euphorbiae* group having a common ancestor not shared by *rosae* and *funestum.* A species of the genus *Sitobion*, which is sometimes considered as a subgenus of *Macrosiphum*, formed the nearest out-group. Tomiuk and Wöhrmann tried various methods of constructing phylogenetic trees, which for this group of taxa all produced similar results. The phylogeny obtained by Eggers-Schumacher, however, using another clustering technique, has *euphorbiae* less closely related to other members of the '*euphorbiae* group' than *rosae, funestum*, or another

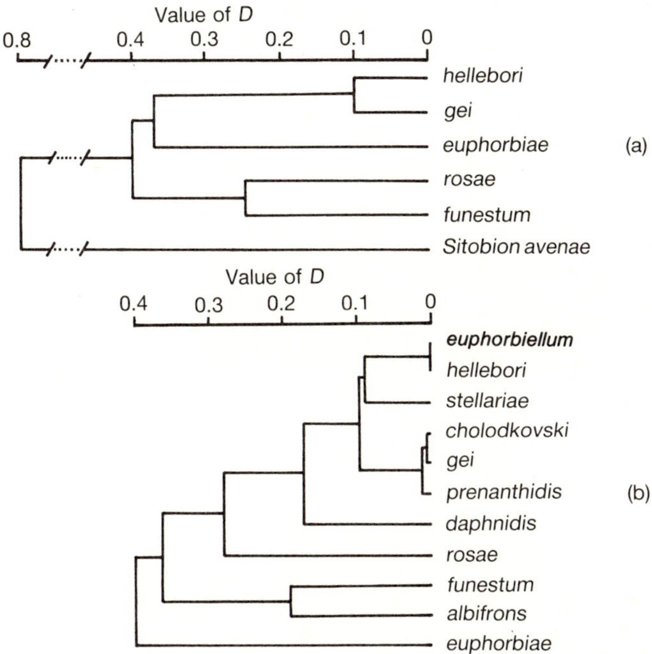

Fig. 12.2. Phylogenetic trees based on estimates of genetic distance (*D*) from allozyme data obtained by (a) Tomiuk and Wöhrmann (1980); (b) Eggers-Schumacher (1987).

species that is very different morphologically, *M. albifrons*. This illustrates the need for caution in constructing phylogenies on the basis of allozyme data alone, even at the intrageneric level of classification.

2. *Acyrthosiphon malvae*

A. malvae and its relatives colonize various plants in the families Geraniaceae and Rosaceae. Most species are of little economic importance, but some transmit viruses of cultivated strawberries, or infest zonal pelargoniums. As in the *M. euphorbiae* group, taxonomic confusion is rife. Eight taxa are recognized in the literature, most of which are usually given subspecific rank (e.g. Müller 1983). One of us (A.D.S.) has recently completed a morphometric and electrophoretic study identifying five new species, and demonstrating in particular that three separate taxa have been confused under the name of the strawberry aphid, *A. rogersii*.

Table 12.3 Enzyme systems with diagnostic loci for seven species of the *Acyrthosiphon malvae* group (letters apply to species listed in the legend to Fig. 12.3).

	C	D	E	F	H	I
A	HK	APH HK	HK	APH HK	HK MDH	HK
C		APH HK	HK	APH HK	HK MDH	
D			HK	APH HK	APH MDH	APH HK
E				HK	HK MDH	HK
F					APH HK MDH	APH
H						HK MDH

Attempts to use identical techniques of starch gel electrophoresis to those of Watson (1982) resulted in very poor resolution of bands, and polyacrylamide gels were therefore used. Differences in the relative quality of the results obtained with starch and polyacrylamide have been noticed for several different aphid genera, and are difficult to explain. Using polyacrylamide gels, 6 of the 16 enzyme systems investigated on nine taxa of the *A. malvae* group gave good staining and resolution. Esterase differences, despite the wealth of bands and the impossibility of genetic interpretation, again provided very good discrimination between the putative taxa recognized by multivariate morphometrics (Fig. 12.3; Table 12.3). A polymorphism of the fast-moving multimeric esterase of the species on *Geranium robertianum* was easy to recognize (Fig. 12.3,A), and not a source of confusion. The species that were most similar morphologically (B,C,G,H) had readily identifiable patterns of esterase mobility.

Of the other systems analysed, hexokinase was very polymorphic within species, but had species-specific patterns in all except two of the taxa examined. The newly recognized *Fragaria*-feeding species had two additional MDH bands in comparison with other species, and the new species distinguished on *Potentilla reptans* and *G. robertianum* had characteristic mobilities of APH-2.

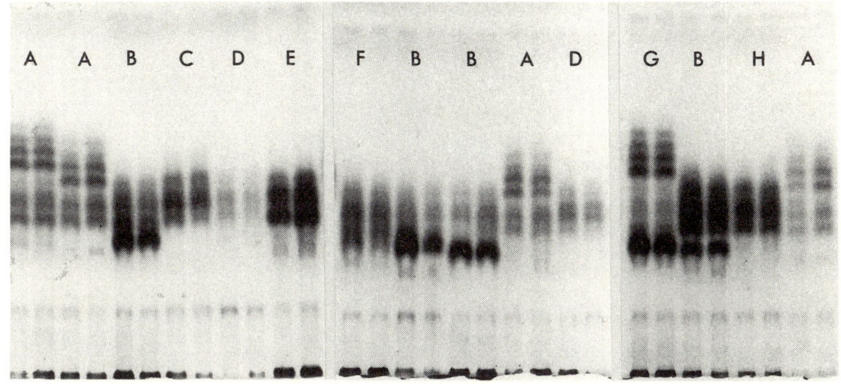

Fig. 12.3. Photographs of representative polyactylamide gels stained for esterases; species of the *Acyrthosiphon malvae* group. The pairs of samples consist of two individuals from a single clone. A = *Acyrthosiphon sp. nov.* from *Geranium robertianum*, B = *A. malvae* (Mosley), C = *A. geranii* (Kalt.), D = *A. poterii* Prior and Stroyan, E = *Acyrthosiphon sp. nov.* from *Geranium sanguineum*, F = *sp. nov.* from *Fragaria vesca*, G = *sp. nov.* from *Geranium palmatum*, H = *sp. nov.* from *Potentilla reptans*.

3. *Acyrthosiphon pisum*

The pea aphid is part of a complex of races or incipient species with different host-plant ranges and preferences. In Europe, several speciation events seem to be currently in progress, with parts of the original *pisum* genome in the process of being separated in aphids on *Ononis, Lotus, Sarothamnus*, and probably on *Pisum sativum* itself. The form introduced into North America and Australia is particularly destructive to alfalfa. In European populations, there is an MDH-1 polymorphism (Suomalainen *et al.* 1980), and at least in British populations there is also a polymorphism at the EST-3 locus (Fig. 12.4). The frequencies of alleles at this locus may vary between *pisum s. str.* and the forms on *Ononis* and *Sarothamnus*, respectively, but more samples are needed to confirm this. There is, however, a clear difference in mobility of glutamate-oxaloacetate transaminase (GOT) between *A. pisum* and the form (= *A. spartii*) on *Sarothamnus* (A. D. Seccombe, unpublished data). No allozyme data are available for populations outside Europe, although much work has been done in North America on the variability of *A. pisum* with respect to a variety of traits.

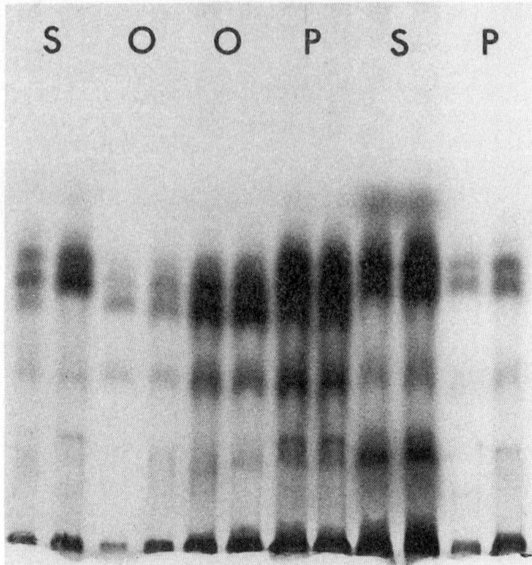

Fig. 12.4. Photograph of polyacrylamide gel stained for esterases; species of the *Acyrthosiphon pisum* group. Each pair of samples comprises two individuals of a single clone. O = *ononis* Koch, P = *pisum* Harris, S = *spartii* Koch.

4. *Myzus persicae*

The peach-potato aphid is well-known as a major crop pest throughout the world, and is probably the most polyphagous aphid species as well as the most important virus vector. In Europe, it has several close relatives with more specific host-plant associations. The problems of distinguishing alatae of *M. persicae* from those of *M. certus*, a common *Stellaria*- and *Viola*-feeding species in both Europe and North America, have long been recognized (Hille Ris Lambers 1959). Recently it has been shown that two other taxa, *M. antirrhinii* and *M. nicotianae*, can be separated from *M. persicae s. str.* using multivariate techniques (Blackman and Paterson 1986; Blackman 1987). However, morphological discrimination of these taxa is only possible using rather complex arithmetic functions involving several characters, and even then is not totally reliable for the identification of individual aphids.

Differences at esterase loci between *persicae* and *antirrhinii* have been noted by ffrench-Constant *et al.* (1988), and provide an easy method of distinguishing these taxa, especially as the esterase-4 activity of *M. persicae* is now routinely monitored because of its involvement with resistance to organophosphate and carbamate insecticides. *M. certus* also has a characteristic pattern of esterase bands, enabling it to be

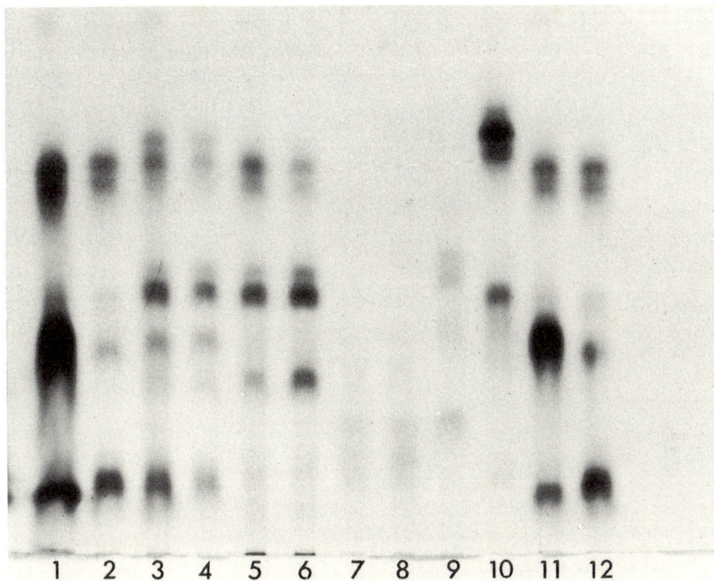

Fig. 12.5. Photograph of a polyacrylamide gel stained for esterases; clonal samples of *Myzus persicae* and its relatives. 1 and 11, *persicae* (insecticide resistant), with high EST-4 activity; 2 and 12, *persicae* (susceptible); 3 and 4, *antirrhinii*; 5 and 6, *certus*; 7 and 8, *cymbalariae*; 9, *ascalonicus*; 10, *ornatus*.

distinguished readily from these two species (Fig. 12.5). Other close relatives of *M. persicae* examined (including two not shown in Fig. 12.5: *M. myosotidis* on *Myosotis palustris* and *M. dianthicola* on *Dianthus caryophyllus*) all have characteristic esterases. The widely-distributed shallot aphid, *M. ascalonicus*, is often confused with the morphologically similar *M. cymbalariae*, and they often colonize the same plants (Brown 1983), yet they are easy to discriminate electrophoretically (Fig. 12.5).

No specific electrophoretic comparison has yet been made between *M. persicae* and the tobacco-feeding form *M. nicotianae*. However, May and Holbrook's (1978) survey of US populations of *persicae*, which amazingly failed to detect any variability at all at 19 enzyme loci, included specimens taken from tobacco in Maryland (population no. 2), and these were presumably *nicotianae*.

5. Metopolophium dirhodum and *M. festucae cerealium*

These cereal pests are relatively easy to tell apart, but each has close relatives on wild grasses that are virtually sibling species, and can only be distinguished morphologically by use of multiple discriminant functions (Stroyan 1982). For example, the rose-grain aphid *M. dirhodum*, typically migrating between its primary host *Rosa* spp. and numerous species of grasses and cereals, is liable to confusion with *M. fasciatum*, which lives only on *Arrhenatherum elatius* and *Bromus carinatus*. Similarly, *M. festucae cerealium*, which can be very damaging to winter-sown oats and barley, is not only very difficult to separate morphologically from *M. festucae s. str.* living commonly on meadow grasses, but may be confused with other species of more specialized food and habitat requirements such as *M. albidum* and *M. tenerum*. Probably there are other, as yet undiscovered, *Metopolophium* species

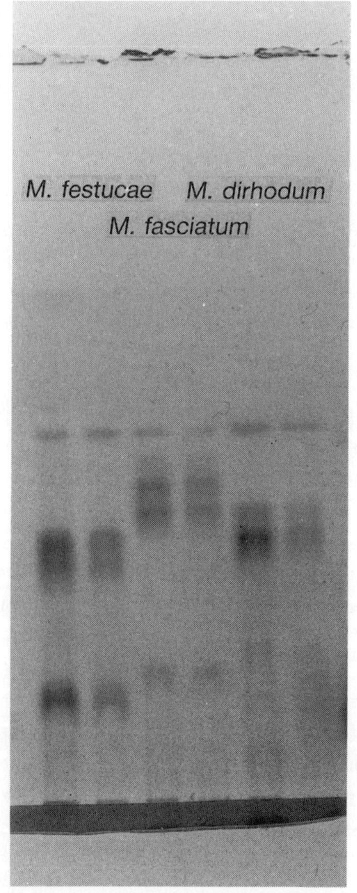

Fig. 12.6. Photograph of a polyacrylamide gel stained for esterases; samples of two clones of each of three *Metopolophium* species.

that will confuse the situation still further, especially in continental Europe where the group is in need of more detailed study.

One of us (C. F.) has examined the esterases of British populations of this group using polyacrylamide gel electrophoresis and isoelectric focusing. Differences in electrophoretic mobility of one or more esterases were found between *M. dirhodum, M. fasciatum, M. festucae, M. tenerum*, and *M. albidum* (Figs. 12.6, 12.7). Isoelectric focusing provided a much clearer separation, especially of *M. dirhodum* from *M. fasciatum* (Fig. 12.8), and of *M. festucae* from *M. tenerum*. The esterase system of *M. f. cerealium* was, however, not distinguishable from that of *M. festucae s. str.*

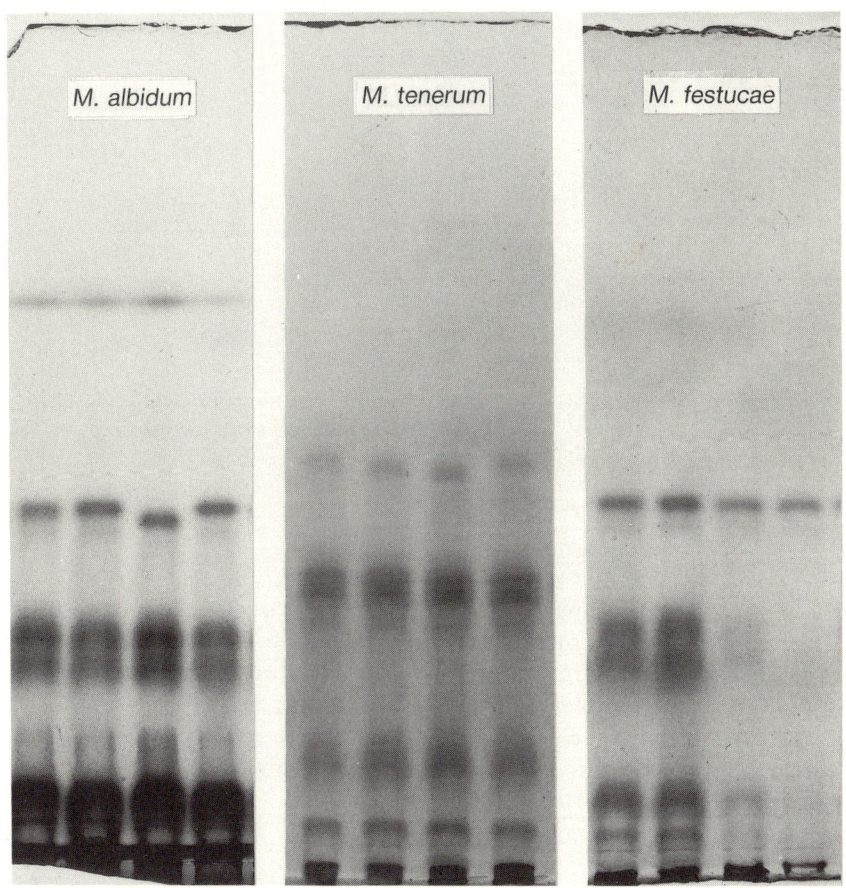

Fig. 12.7. Photograph of a polyacrylamide gel stained for esterases; samples of four clones of each of three *Metopolophium* species.

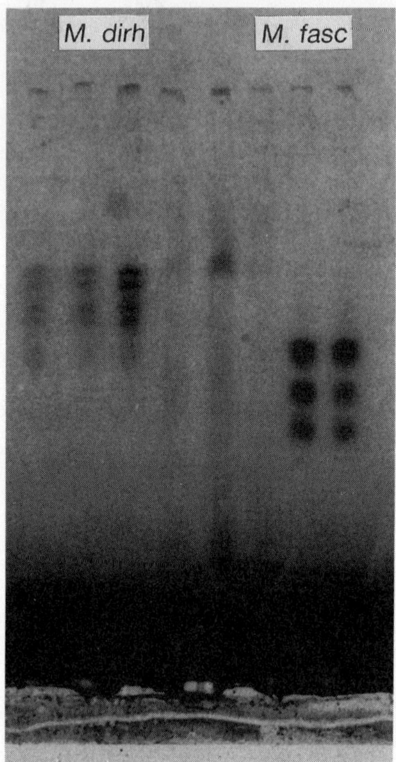

Fig. 12.8. Photograph of a polyacrylamide gel after isoelectric focusing, stained for esterases, showing different isoelectric points of esterases of *M. dirhodum* and *M. fasciatum*.

6. *Aphis fabae*

Even such a well known species as the black bean aphid presents a complex taxonomic problem, and in fact the only really reliable criterion for recognizing *A. fabae s. str.* is its ability to colonize broad bean (*Vicia faba*). Several other closely related species or subspecies share the same winter host (spindle, *Euonymus europea*) with *A. fabae*, and hybridization may sometimes occur (Müller 1982). Population studies of *A. fabae* on its winter host are used in Britain to predict infestation of field beans (Way *et al.* 1977), so it is important to clarify the taxonomy of the group as far as possible. *A cirsiiacanthoides* in particular is a common aphid on thistles and overwinters as eggs on *Philadelphus, Viburnum*, and *Euonymus*; on the latter host it is almost certainly still confused with *A. fabae* (Müller and Steiner 1986).

Furk (1979) collected samples of the black bean aphid group from *Tropaeolum*, *Calendula*, and *Dahlia*, none of which would transfer to *Vicia faba*, and compared their acid phosphatases, esterases, and total soluble proteins (the latter by isoelectric focusing) with those of 13 samples of *A. fabae s. str.*. The *Tropaeolum* aphids had a unique acid phosphatase pattern with additional bands that were not present in any of the other samples. This aphid was probably *A. mordwilkoi*, which is believed to colonize *Arctium lappa* and *Tropaeolum majus* in summer, and overwinter on *Viburnum opulus; A. barbarae*, described from *Arctium* and *Tropaeolum* in North America, may also be this aphid.

Isoelectric focusing revealed protein differences between *A. fabae s. str.* and the samples from *Calendula* and *Dahlia*; the *Calendula* aphids also showed an esterase difference. One of these samples may have been *A. cirsiiacanthoides*. Kiser (1979) found small differences between *A. fabae*, *A. cirsiiacanthoides* and *A. solanella* using thin-layer chromatography, and Odermatt (1981) found that *A. cirsiiacanthoides* had a diagnostic fast form of phosphoglucomutase.

However, the taxonomic significance of much of this variation is uncertain, since *A. fabae* itself also shows considerable variation between samples in the numbers and positions of certain bands. The *A. fabae* group clearly needs further detailed study in which allozyme data are correlated with the results of host-plant transference tests and morphometric analyses for numerous clonal samples from different sources.

7. *Aphis gossypii*

The melon or cotton aphid has a world-wide distribution on numerous plants, and pest populations probably comprise an indefinite number of permanently parthenogenetic lineages, some of which may have particular host plant associations. Until recently, the taxonomic situation with regard to *A. gossypii* in British glasshouses appeared to be relatively simple, with one form on cucumbers and another on chrysanthemums, only the latter having acquired resistance to insecticides. Furk *et al.* (1980) found that the esterases of these two forms could be distinguished, both by conventional starch gel electrophoresis and by isoelectric focusing.

One of us (C.F.) has shown recently that carbamate resistance is now present in populations on glasshouse cucumbers in England. The resistant aphids on cucumbers have the esterase pattern and morphology (V.F. Eastop, pers. comm.) of the chrysanthemum-colonizing form. In laboratory transference tests, the resistant cucumber aphids have so far shown reluctance to feed on *Chrysanthemum*. Nevertheless, it seems likely that the chrysanthemum-feeding form has acquired the ability to feed on cucumbers, rather than the cucumber form having evolved insecticide resistance.

An available name for the insecticide-resistant aphid is *Aphis parva* Theobald, which was described from *Chrysanthemum* in Egypt; however, the application of this name outside Britain, where there may be numerous other parthenogenetic lineages of the *A. gossypii* group, may be difficult.

8. *Aphis pomi* and *A. spiraecola*

There has in the past been considerable confusion, particularly in North America, between the green apple aphid, *Aphis pomi* de Geer, and the more polyphagous citrus or spiraea aphid, *A. spiraecola* Patch. Both species form very similar colonies on woody Rosaceae, although they can be reliably distinguished by several morphological characters when viewed under the microscope. Probably a recent example of this confusion is the report by Singh and Rhomberg (1984) of a consistent association of electromorphs for GOT and two esterase loci in populations sampled from apple in southern Ontario. They concluded that two 'differentiated, non-interbreeding subpopulations' were present. It seems probable that one of these was *A. pomi*, and that the other was *A. spiraecola*.

Steiner *et al*. (1985*a*) found differences at three loci (EST, MDH, ALD) between samples of *A. spiraecola* collected on *Rumex* and *Spiraea*, but more extensive sampling would be required to show whether there was any consistent association between electromorph and host plant.

9. *Other species complexes involving pest aphids*

There are a number of other species complexes that have been looked at electrophoretically, and in some cases the allozyme data either confirm the separation of previously suspected sibling species or suggest the presence of unsuspected ones.

The Fordini, which colonize the roots of grasses and cereals as their secondary hosts, are mostly very specific in their choice of *Pistacia* species as primary host plants. However, aphids identified as *Geoica utricularia* produce galls on both *P. atlantica* and *P. palaestina* in Israel. Koach and Wool (1977) found that all samples from *P. atlantica* had an additional esterase band when compared with those from *P. palaestina*, the constancy of this difference making it seem certain that two host-specific taxa were involved. The esterases of populations currently regarded as *G. utricularia*, on grasses and cereals in various parts of the world, have not been examined.

The willow-carrot aphid, *Cavariella aegopodii*, has a number of close relatives that share *Salix* species as primary hosts. One of these, *C. intermedia*, was for many years confused with *C. aegopodii*, and the apterous viviparae of the two species on *Salix* are difficult to separate morphologically (Hille Ris Lambers 1969), yet these species differ at at least five enzyme loci (Eggers-Schumacher 1987). The genus *Cavariella* appears to be a promising one for further electrophoretic study.

Detailed allozyme data might also contribute to the taxonomy of another genus, *Dysaphis*, in which morphologically similar species share primary hosts in the Rosaceae, some being pests of pome fruits. Eggers-Schumacher (1987) was able to find diagnostic loci for seven out of the eleven species he tested, and further electrophoretic work may help to show the extent of the natural hybridization thought to occur in this genus (Stroyan 1958).

The cherry blackfly, *Myzus cerasi* F., which has *Prunus cerasus* as primary host and various secondary hosts including *Galium* and *Veronica*, is one of a complex of species and/or subspecies with different host associations. Gruppe (1988) found populations on sweet cherry, *Prunus avium*, provisionally assigned to ssp. *pruniavium* Borner, to have a consistent esterase difference from *M. cerasi s. str.*

Without giving details of the enzyme systems involved, Eggers-Schumacher (1987) reported that the polyphagous aphid *Brachycaudus cardui* could be separated electrophoretically from closely related species living monophagously on *Myosotis* and *Malva*; that enzyme studies are potentially useful for sorting out the taxonomy and host plant relationships of the *Brachycaudus* subgenus *Appelia*; and that populations of *Pemphigus* spp. on roots of secondary host plants, which are almost impossible to identify to species by their morphology, can mostly be characterized using electrophoresis, and identified with the very different-looking aphids from galls on *Populus* spp. (the primary hosts). Concerning the last-named genus, Setzer (1980) obtained evidence of intergall migration in a North American *Pemphigus* species by analysing variation at an esterase locus.

Concluding points

The cases reviewed above illustrate the potential value of standard electrophoretic procedures for identifying aphid pest species and helping to clarify their taxonomic relationships. Generally speaking, as might be expected, the loci that are most commonly polymorphic within species also most often show differences between closely related species. However, in some cases, closely related taxa have fixed monomorphic enzymes of different mobilities, and sometimes systems such as esterases show such complex patterns of bands that the homologies between species are unclear.

In Table 12.4, we list the enzyme systems shown to vary in at least some populations of the main pest aphid species involved in taxonomically difficult species complexes. The scope for further work is apparent. For example, the studies of Loxdale and co-workers on *Sitobion avenae* and *S. fragariae* in Europe, where these two species are morphologically distinct, could provide a sound basis for investigations of the taxonomy of cereal-feeding *Sitobion* in other parts of the world such as Australasia, where a third species of intermediate morphology, *S. miscanthi*, confuses the issue.

Esterases alone can often provide the diagnostic loci needed for the characterization and identification of species, especially when electrophoretic data are supported by morphological and biological information. The well-known lack of variability in many other enzyme systems of aphids — particularly the Group I enzymes involved in central metabolic pathways — may not therefore be such a problem when electrophoresis is used for taxonomic purposes, as compared with population genetics. Nelson and Hedgecock (1980) suggested that trophic generalists — of which *Myzus persicae* and several other pest aphids seem to be prime examples — would tend to be more variable in their Group II enzyme systems: enzymes such as esterases that function in peripheral metabolism and process a variety of substrates (see Smith and Fujio 1982, for discussion of this topic). However, there are plenty of variable Group I loci now known, for example, in *Macrosiphum euphorbiae*, which is a very polyphagous species.

If more variation is needed, this may be obtained by use of isoelectric focusing (e.g. Furk *et al.* 1980), which also has the advantage that it is less sensitive to experimental and interpretational error than electrophoresis (Johnson 1973), or by serial electrophoresis in either one or two dimensions using different gel concentrations and/or run times (e.g. Loxdale *et al.* 1985*a*). Clonal aphid populations are particularly amenable to the latter approach. Serial electrophoresis has, however, rarely revealed any hidden variability at loci that are monomorphic using standard techniques (Lewontin 1985).

Finally, at the risk of stating the obvious, it seems necessary to

Table 12.4 Enzyme variability within pest aphid species and species complexes.

Pest species/ species complex	Variable systems or loci	Reference
Acyrthosiphon pisum	EST-3, GOT	This work
	MDH, PGI	Tomiuk and Wöhrmann (1980) and
	MDH-1	Suomalainen *et al.* (1980)
	EST, SDH	Weber and Wöhrmann, in Tomiuk and Wöhrmann (1984)
Acyrthosiphon malvae	APH-2, EST (several loci), HK, MDH	This work
Aphis gossypii	EST (several loci)	Furk *et al.* (1980)
Aphis fabae	APH, EST	Furk (1979)
	EST (several loci)	Beranek (1974), Beranek and Berry (1974)
	EST, HK	Tomiuk and Wöhrmann (1980)
	MDH, PGM	Odermatt (1981)
Aphis spiraecola	ALD, EST, HK, MDH, 6-PGDH	Steiner *et al.* (1985*a*)
Cavariella aegopodii	ADK, IDH-1, IDH-2, MDH, 6-PGDH, PGI	Eggers-Schumacher (1987)
Hyalopterus pruni	CA-1, CA-2, GPI	Spampinato *et al.* (1988)
Macrosiphum euphorbiae	EST (several loci), HK, IDH-1, α-GPDH, 6-PGDH, PGI	Watson (1982) and this work
	AAT, IDH-1, IDH-2, 6-PGDH	May and Holbrook (1978)
	ALD, EST-1, G-6-PDH, HK, 6-PGDH, PGM	Steiner *et al.* (1985*a*)

Metopolophium dirhodum	EST	This work
Metopolophium festucae	EST	This work
Myzus cerasi	EST	Gruppe (1988)
Myzus persicae	EST (several loci)	Brookes and Loxdale (1987), ffrench-Constant *et al.* (1988), and this work
Rhopalosiphum padi	PGM, MDH	Steiner *et al.* (1985*a*)
	EST, PHOS, GOT, ME, 6-PGDH, α-GPDH, SORDH, PEP, MDH	Loxdale and Brookes (1988)
Rhopalosiphum maidis	GOT, MDH, PEP	Steiner *et al.* (1985*a*)
	ALD, GOT, HK, IDH-1, MDH, PEP, PGI, PGM	Steiner *et al.* (1985*b*)
Schizaphis graminum	EST	Volkova and Titova (1983)
Sitobion avenae	6-PGDH, PEP (3 loci)	Loxdale *et al.* (1985*a*)
	AK, EST, α-GPDH, GOT, G-6-PDH, PEP, PGM, PHOS, POD	Loxdale *et al.* (1985*b*)
	EST (several loci)	Stribley *et al.* (1983)
Sitobion fragariae	GOT	Loxdale and Brookes (1987)

stress that the taxonomy of any group of organisms can only be improved if full consideration is given to *the cumulative total* of available knowledge about them. While electrophoretic methods may be very useful diagnostic tools for distinguishing between closely related aphids, it is most unwise to rely on enzyme differences alone for definitive answers about taxonomic and phylogenetic relationships within species groups containing pest aphids. A proper understanding can only be achieved by adopting a multidisciplinary approach, bringing together available information on host plants, life cycles, morphology, and karyotype, and taking all these into account when interpreting the results of an electrophoretic study.

References

Ayala, F.J. (1983). Enzymes as taxonomic characters. In *Protein polymorphism: adaptive and taxonomic significance* (ed. G.S. Oxford and D. Rollinson), Systematics Association Special Vol. 24, pp. 3–26. Academic Press, London.

Beranek, A.P. (1974). Esterase variation and organophosphate resistance in populations of *Aphis fabae* and *Myzus persicae*. *Entomologia exp. appl.*, **17**, 129–42.

Beranek, A.P. and Berry, R.J. (1974). Inherited changes in enzyme patterns within parthenogenetic clones of *Aphis fabae*. *J. Ent. (A),* **48**, 141–7.

Blackman, R.L. (1987). Morphological discrimination of a tobacco-feeding form from *Myzus persicae* (Sulzer) (Hemiptera: Aphididae), and a key to New World *Myzus (Nectarosiphon)* species. *Bull. ent. Res.*, **77**, 713–30.

Blackman, R.L. and Paterson, A.J.C. (1986). Separation of *Myzus (Nectarosiphon) antirrhinii* (Macchiati) from *Myzus (N.) persicae* (Sulzer) and related species in Europe. *Syst. Ent.*, **11**, 267–76.

Brookes, C.P. and Loxdale, H.D. (1987). Survey of enzyme variation in British populations of *Myzus persicae* (Sulzer) (Hemiptera: Aphididae) on crops and weed hosts. *Bull. ent. Res.*, **77**, 83–9.

Brown, P.A. (1983). A note on *Myzus (Sciamyzus) cymbalariae* Stroyan, with a description of the male. *J. nat. Hist.*, **17**, 875–80.

Buth, D.G. (1984) The application of electrophoretic data in systematic studies. *A. Rev. Ecol. Syst.*, **15**, 501–22.

Easteel, S. and Boussy, I.A. (1987). A sensitive and efficient isoenzyme technique for small arthropods and other invertebrates. *Bull. ent. Res.*, **77**, 407–15.

Eastop, V.F. (1985). The acquisition and processing of taxonomic data. In *Evolution and biosystematics of aphids* (ed. H. Szelegiewicz), Proceedings of the International Symposium at Jablonna, 1981, pp. 245–70. Ossolineum, Wroclaw, Poland.

Eggers-Schumacher, H.A. (1987). Enzyme electrophoresis in biosystematics and taxonomy of aphids. In *Population structure, genetics and taxonomy of aphids and Thysanoptera* (eds J. Holman, J. Pelikan, A.F.G. Dixon, and L. Weismann), pp. 63–70. SPB Academic Publishing, The Hague.

ffrench-Constant, R.H., Byrne, F.J., Stribley, Mary F., and Devonshire, A.L. (1988). Rapid identification of the recently recognised *Myzus antirrhinii* (Macchiati) (Hemiptera: Aphididae) by polyacrylamide gel electrophoresis. *Entomologist*, **107**, 20–3.

Furk, C. (1979). Field collections of *Aphis fabae* Scopoli *s. lat.* (Homoptera: Aphididae) studied by starch gel electrophoresis and isoelectric focusing. *Comp. Biochem. Physiol.*, **62**B, 225–30.

Furk, C., Powell, D.F., and Heyd, S. (1980). Pirimicarb resistance in the melon and cotton aphid, *Aphis gossypii* Glover. *Pl. Path.*, **29**, 191–6.

Gruppe, A. (1988). Elektrophoretische Untersuchungen zur Unterscheidung der Subspezies von *Myzus cerasi* F. (Hom., Aphididae). *J. appl. Ent.*, **105**, 460–5.

Hille Ris Lambers, D. (1939) Contributions to a monograph of Aphididae of Europe II. *Temminckia*, **4**, 1–134.

Hille Ris Lambers, D. (1959). *Myzus (Nectarosiphon) certus* (Wlk.) as a problem in studies on flights of *Myzus (Nectarosiphon) persicae* (Sulz.) (Homoptera: Aphididae). *Entom. Bericht.*, **19**, 17–19.

Hille Ris Lambers, D. (1969). Four new species of *Cavariella* del Guercio, 1911 (Homoptera, Aphididae). *Memoria della Societa entomologica Italiana* (Centenary Volume), **48**, 285–99.

Johnson, G.B. (1973). Enzyme polymorphism and biosystematics: the hypothesis of selective neutrality. *A. Rev. Ecol. Syst.*, **4**, 93–116.

Kiser, K. (1979). Dünnschichtchromatographie — als neue Methode zur Artdifferenzierung der 'Schwarzen Blattläuse': *Aphis fabae, A. solanella, A. cirsiiacanthoidis, A. sambuci* und *A. hederae* (Homoptera: Aphididae). *Z. angew. Ent.* **88**, 363–77.

Koach, J. and Wool, D. (1977). Geographic distribution and host specificity of gall-forming aphids (Homoptera, Fordinae) on *Pistacia* trees in Israel. *Marcellia*, 40, 207–16.

Lampel, G. and Burgener, R. (1987). The genetic relationships between lachnid taxa as established by enzyme-gel electrophoresis. In *Population structure, genetics and taxonomy of aphids and Thysanoptera* (ed. J. Holman, J. Pelikan, A.F.G. Dixon, and L. Weismann), pp. 71–95. SPB Academic Publishing, The Hague.

Lewontin, R.C. (1985). Population genetics. *A. Rev. Genet.*, **19**, 81–102.

Loxdale, H.D. and Brookes, C.P. (1987) Use of electrophoretic markers to study the spatial and temporal genetic structure of populations of a holocyclic aphid species — *Sitobion fragariae* (Walker) (Hemiptera: Aphididae). In *Population structure, genetics and taxonomy of aphids and Thysanoptera* (ed. J. Holman, J. Pelikan, A.F.G. Dixon, and L. Weismann) pp. 100–110. SPB Academic Publishing, The Hague.

Loxdale, H.D. and Brookes, C.P. (1988). Electrophoretic study of cereal aphid populations. V. Spatial and temporal genetic similarity between holocyclic populations of the bird cherry-oat aphid *Rhopalosiphum padi* (L.) (Hemiptera: Aphidae) in Britain. *Bull. ent. Res.*, **78**, 241–9.

Loxdale, H.D., Castañera, P. and Brookes, C.P. (1983). Electrophoretic study of enzymes from cereal aphid populations I. Electrophoretic techniques and staining systems for characterising isoenzymes from six species of cereal aphids (Hemiptera: Aphididae). *Bull. ent. Res.*, **73**, 645–57.

Loxdale, H.D., Rhodes, J.A., and Fox, J.S. (1985*a*). Electrophoretic study of enzymes from cereal aphid populations 4. Detection of hidden genetic variation within populations of the grain aphid *Sitobion avenae* (F.) (Hemiptera: Aphididae). *Theor. appl. Genet.*, **70**, 407–12.

Loxdale, H.D., Tarr, I.J., Weber, C.P., Brookes, C.P., Digby, P.G.N., and Castañera, P. (1985*b*). Electrophoretic study of enzymes from cereal aphid populations III. Spatial and temporal genetic variation of populations of *Sitobion avenae* (F.) (Hemiptera: Aphididae). *Bull. ent. Res.*, **75**, 121–41.

May. B. and Holbrook, F.R. (1978). Absence of genetic variability in the green peach aphid, *Myzus persicae* (Hemiptera: Aphididae). *Ann. ent. Soc. Am.*, **71**, 809–12.

Meier, W. (1961). Beiträge zur Kenntnis der grünstreifigen Kartoffelblattlaus, *Macrosiphum euphorbiae* Thomas 1870, und verwandter Arten (Hemipt., Aphid.). *Mitt. Schweiz. ent. Ges.*, **34**, 127–86.

Müller, F.P. (1982). Das Problem *Aphis fabae. Z. angew. Ent.*, **94**, 432–46.

Müller, F.P. (1983). Untersuchungen über Blattläuse der Gruppe *Acyrthosiphon pelargonii* im Freiland-Insektarium. *Z. angew. Zool.*, **70**, 351–67.

Müller, F.P. and Steiner, H. (1986). Morphologische Unterschiede und Variation der Geflügelten im Formenkreis *Aphis fabae* (Homoptera: Aphididae). *Beitr. Ent.*, **36**, 209–215.

Nei. M. (1972). Genetic distance between populations. *Am. Nat.*, **106**, 283–92.

Nelson, K. and Hedgecock, D. (1980). Enzyme polymorphism and adaptive strategy in the decapod Crustacea. *Am. Nat.* **116**, 238–80.

Odermatt, F. (1981). Die Stärke-Gel-Elektrophorese als Methode zur Artdifferenzierung der ‘Schwarzen Blattläuse’: *Aphis fabae* Scop. und verwandte Arten (Homoptera: Aphididae). *Z. angew. Ent.* **92**, 398–408.

Setzer, R. Woodrow (1980). Intergall migration in the aphid genus *Pemphigus. Ann. ent. Soc. Am.*, **73**, 327–31.

Singh, R.S. and Rhomberg, L. (1984). Allozyme variation, population structure and sibling species in *Aphis pomi. Can. J. Genet. Cytol.*, **26**, 364–73.

Smith, P.J. and Fujio, Y. (1982). Genetic variation in marine teleosts: high variability in habitat specialists and low variability in habitat generalists. *Mar. Biol.*, **69**, 7–20.

Spampinato, R., Ardvino, P., Barbagallo, S., and Bullini, L. (1988). Analisi genetica di *Hyalopterus pruni* e *H. amygdali* e dimostrazione dell'esistenza di una nuova specie (Homoptera, Aphidoidea). *Atti XV Congr. naz. ital. Ent., L'Aquila* 1988, 261–265.

Steiner, W.W.M., Voegtlin, D.J., Irwin, M.E., and Kampmeier, G. (1985*a*). Electrophoretic comparison of aphid species: detecting differences based on taxonomic status and host plant. *Comp. Biochem. Physiol.*, **81**B, 295–9.

Steiner, W.W.M., Voegtlin, D.J., and Irwin, M.E. (1985*b*). Genetic differentiation and its bearing on migration in North American populations of the corn leaf aphid, *Rhopalosiphum maidis* (Fitch) (Homoptera: Aphididae). *Ann. ent. Soc. Am.*, **78**, 518–25.

Stribley, M.F., Moores, G.D., Devonshire, A.L., and Sawicki, R.M. (1983). Application of the FAO-recommended method for detecting insecticide resistance in *Aphis fabae* Scopoli, *Sitobion avenae* (F.), *Metopolophium dirhodum* (Walker) and *Rhopalosiphum padi* (L.) (Hemiptera: Aphididae). *Bull. ent. Res.*, **73**, 107–15.

Stroyan, H.L.G. (1958). A contribution to the taxonomy of some British species of *Sappaphis* Matsumura 1918 (Homoptera, Aphidoidea) (corrected to *Dysaphis* Borner in footnote). *J. Linn. Soc. Lond. (Zool.)*, **43**, 644–713.

Stroyan, H.L.G. (1982). Revisionary notes on the genus *Metopolophium* Mordvilko 1914, with keys to European species and descriptions of two new taxa (Homoptera: Aphidoidea). *Zool. J. Linn. Soc.*, **75**, 91–140.

Suomalainen, E., Saura, A., Lokki, J., and Terri, T. (1980). Genetic polymorphism in parthenogenetic animals. Part 9. Absence of variation within parthenogenetic aphid clones. *Theor. appl. Genet.*, **57**, 129–32.

Tomiuk, J. (1987). The neutral theory and enzyme polymorphism in populations of aphid species. In *Population structure, genetics and taxonomy of aphids and Thysanoptera* (ed. J. Holman, J. Pelikan, A.F.G. Dixon, and L. Weismann), pp. 45–52. SPB Academic Publishing, The Hague.

Tomiuk, J. and Wöhrmann, K. (1980). Enzyme variability in populations of aphids. *Theor. appl. Genet.*, **57**, 125–7.

Tomiuk, J. and Wöhrmann, K. (1983). Enzyme polymorphism and taxonomy of aphid species. *Z. zool. Syst. Evolut.-forsch.*, **21**, 266–74.

Tomiuk, J. and Wöhrmann, K. (1984). Genotypic variability in natural populations of *Macrosiphum rosae* (L.) in Europe. *Biol. Zbl.*, **103**, 113–22.

Tomiuk, J., Wöhrmann, K., and Eggers-Schumacher, H.A. (1979). Enzyme patterns as a characteristic for the identification of aphids. *Z. angew. Ent.*, **88**, 440–6.

Volkova, R.I. and Titova, E.V. (1983). [Multiple molecular forms of esterases from *Aphis (Schizaphis) gramina*: inhibitory identification and stereospecificity.] (In Russian) *Biokhimiya*, **48**, 1634–42.

Watson, Gillian W. (1982). A biometric, electrophoretic and karyotypic analysis of British species of *Macrosiphum* (Homoptera: Aphididae). Ph.D. thesis, University of London, UK.

Way, M.J., Cammell, M.E., Alford, D.V., Gould, H.J., Graham, C.W., Lane, A., Light, W.I.St.G., Rayner, J.M., Heathcote, G.D., Fletcher, K.E., and Seal, K. (1977). Use of forecasting in chemical control of the black bean aphid, *Aphis fabae* Scop., on spring-sown field beans (*Vicia faba* L.). *Pl. Path.*, **26**, 1–7.

13 Electrophoretic studies on planthoppers and leafhoppers of agricultural importance

J. DEN HOLLANDER

School of Pure and Applied Biology, University of Wales, College of Cardiff, Cardiff, Wales, UK

Abstract

Planthoppers and leafhoppers belong to the Auchenorrhyncha (Homoptera). They include many important agricultural pest species that attack a variety of major crops, e.g. rice, corn, wheat, and barley. All have sucking mouthparts and cause damage by sucking the sap and cell contents of plants and also by transmission of phytopathogens. Their recognition as important pests is relatively recent and much basic work still needs to be done on the group.

There are many problems to be solved in the study of the biology of planthoppers and leafhoppers, which have involved, and will doubtless continue to involve for the foreseeable future, the use of electrophoresis. Examples include understanding of the nature of populations able to overcome resistant plant varieties, the identification of local populations and the extent of geographic variation in widespread species, the monitoring of local and large-scale migration, the development of insecticide resistance, and the resolution of species complexes. Although most of the elctrophoretic work in this group has been on the Brown Planthopper, *Nilaparvata lugens*, other examples are reviewed.

Electrophoretic Studies on Agricultural Pests (ed. Hugh D. Loxdale and J. den Hollander), Systematics Association Special Volume No. 39, pp. 297–315. Clarendon Press, Oxford, 1989.

Introduction

Planthoppers and leafhoppers are the common names for the superfamilies Fulgoroidea and Cicadeloidea of the Homoptera: Auchenorrhyncha. All Homoptera are plant feeders and have piercing and sucking mouthparts with which they suck the plant sap or cell contents. The entire life cycle is spent on the host plant, which is used for feeding, breeding, oviposition, and as a substrate for acoustic communication between the sexes (Claridge 1985). Plant and leafhoppers show a range of host specificity, from highly polyphagous to being strictly monophagous.

Many planthoppers and leafhoppers have been implicated as plant pests (Wilson and O'Brien 1987). Some of the damage they cause is mechanical — the feeding and oviposition sites cause physical damage to the host plant that may block the vascular bundles and allow pathogenic fungi and bacteria to enter passively (Williams 1957). Feeding also drains the plant of valuable nutrients. Feeding by large numbers may cause the plant to dessicate and wilt, a condition known as hopperburn (International Rice Research Insitute (IRRI), 1979), whilst sooty mould growing on the honeydew may reduce plant growth. Also, the saliva of some species is toxic (Raatikainen 1967, 1970). However, and perhaps most importantly, many hoppers are direct vectors of phytopathogens, transmitting viruses, mycoplasmas, spiroplasmas, and bacteria (Harris 1983; Conti 1985; Nielson 1985).

It is now recognized that the group contains many important pest species, but awareness of this is relatively recent. Before the turn of the century, plant and leafhoppers were considered only a minor nuisance; however, following the demonstration in the early 1900s of their ability to transmit plant diseases (Ou 1972; Conti 1985), more attention has been paid to the group. In more recent times, several species have come to prominence as major pests. This has been variously attributed to changes in farming practices, such as the use of irrigation leading to continuous cropping, the growing of vast areas under monocultures, the use of fertilizers and insecticides, and the growing of high-yielding crop varieties. Plant and leafhoppers attack many of the major food crops, including rice, corn, wheat, barley, and potatoes.

Electrophoresis — in a brief summary of how I use the term — is the movement of charged particles in a supporting medium under the influence of an electric field. Proteins and enzymes of different size or charge will move differing distances and can be visualized by selective staining (Smithies 1955; Shaw and Prasad

1970). A variety of supporting media have been used, e.g. starch, polyacrylamide, cellulose, agar, and a number of different techniques, e.g. horizontal or vertical slab, SDS, gradient, isoelectric focusing, and two-dimensional (2-D) electrophoresis. Starch gel electrophoresis, because of its ease and convenience, is perhaps the standard method. The technique is widely used in biology because it provides a rapid and routine way for measuring qualitative and quantitative differences at many protein loci. Large numbers of individuals can be screened in a short time, enabling the study of population genetics and evaluation of evolutionary problems.

Electrophoresis has been used to study variation within populations of the same species and geographic variation between different populations, to study and monitor migration, as a taxonomic tool to resolve species complexes, and in the study of insecticide resistance. I intend to review some of these aspects that have been studied in planthoppers and leafhoppers.

Despite the widespread use of electrophoresis and the importance of planthoppers and leafhoppers as agricultural pests, very few studies have, to date, used the technique on this group. Those that have been studied include the following

Pest species
Nilaparvata lugens: Kassai and Ogita (1965); Sogawa (1978); Claridge *et al.* (1983); Saxena and Mujer 1984; Saxena and Barrion (1985); Tranter (1983). *Nephotettix cincticeps*: Miyata and Saito (1979); Miyata *et al.* (1976). *Laodelphax striatellus*: Ozaki and Kassai (1970); Miyata *et al.* (1976).

Non-pest species
Alebra albostriella: Drosopolous and Loukas (1988). *Enchenopa binotata*: Guttman *et al.* (1981). *Philaenus spumarius*: Saura *et al.* (1973). *Ribautodelphax* species: den Bieman and Eggars-Schumacher (1987).

Variation within populations

The most-studied species is *Nilaparvata lugens* (Stal), the brown planthopper (BPH). This species causes direct feeding damage, resulting in hopperburn in extreme cases and, by transmission of the virus, grassy stunt and ragged stunt (Ling 1967, 1972; IRRI 1979). Before the 1960s, it was a relatively minor pest, having only sporadic outbreaks (Dyck and Thomas 1979). Since the development and release of high-yielding rice varieties in the mid-1960s, however, it has become one of the major pests of rice throughout Asia (Khush 1979; Claridge and den Hollander 1980). Plant resistance was earlier used to combat the pest, but it developed the ability to overcome this resistance (Khush 1979). Such a cycle was repeated several times. The populations of BPH able to attack different resistant varieties were called biotypes 1, 2, or 3, each possessing the ability to attack, respectively, cultivars with no known gene for resistance (TN-1), cultivars with BPH-1 and no genes for resistance (Mudgo and TN-1) and cultivars with BPH-2 and no genes for resistance (ASD-7 and TN-1) (Pathak and Kush 1979; Khush 1979).

Many suggestions have been made over the years about the possible nature of these biotypes: for example that they were members of a sibling species complex; that they represented host races and as such were transitionary stages that, given sufficient time, would lead to sympatric speciation (Saxena and Barrion 1985); that they represented a gene-for-gene relationship between plant resistance and insect virulence (IRRI 1979); they were polymorphic forms within a populations *or*, that they were selections from a continuously (polygenically) varying population (Claridge and den Hollander 1983). In the light of these possibilities Claridge and den Hollander have recently (1983) provided a critique of the biotype concept and its application to plant and leaf hoppers as well as to other insect pests of agriculture.

Because of the severe damage that brown planthoppers cause to rice crops, it has been considered highly desirable to resolve what, biologically, the biotypes represent. The aim was to develop a surveillance programme to identify biotypes in the field and to monitor the evolution of new biotypes. Several approaches have been used, including detailed study of BPH morphology (Sogawa 1981*a*; Saxena and Rueda 1982; Claridge *et al.* 1984), crossing and breeding experiments (Claridge and den Hollander 1980; Sogawa 1981*b*), studies of the morphology of their chromosomes (Liquido 1986; Saxena and Barrion 1982, 1983), and feeding ability on different rice varieties (Claridge and den Hollander 1982).

To identify biotypes of brown planthoppers, protein/enzyme electrophoresis was considered a possible method, since none of the above methods provided diagnostic differences. The assumption was that because the rice cultivars attacked differed in a single major gene for resistance (Lakshiminarayama and Khush 1977; Khush 1979), they probably differed in a major component, i.e. a secondary chemical or in the presence or absence of a particular

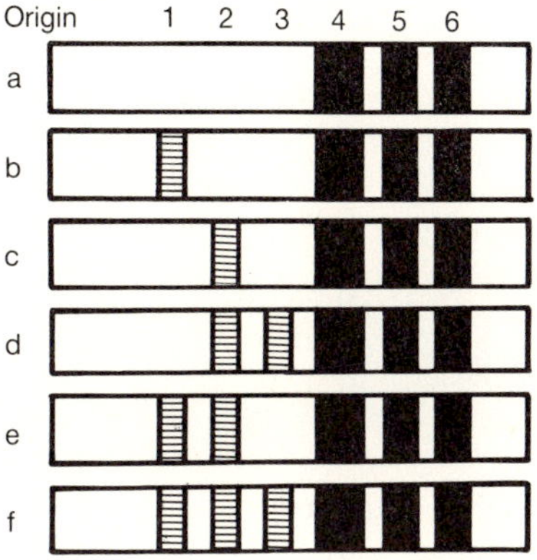

Electrophoretic phenotypes	Biotype 1 B	Biotype 1 M	Biotype 2 B	Biotype 2 M	Biotype 3 B	Biotype 3 M	wild M
a	9.4	10.2	12.3	1.9	8.5	13.0	1.9
b	1.9	0.9	0	0	0	0	0
c	54.8	68.5	63.2	79.3	70.8	63.0	89.8
d	0	0.9	11.3	9.4	0.9	0.9	0
e	29.2	16.7	11.3	7.5	18.9	23.1	8.3
f	0	0	1.9	1.9	0.9	0	0
g (no activity)	4.7	2.8	0	0	0	0	0

Fig. 13.1. Percentages of different electrophoretic variants of esterases in the three biotypes and a wild population of *N. lugens* B = brachypterous, M = macropterous. (After Sogawa 1978.)

amino acid. Perhaps the insect pest biotypes could be recognized by differences that correlated with this, such as detoxifying enzymes. Alternatively, if the biotypes represented sibling species, they might be recognizable by diagnostic protein/enzyme loci separated electrophoretically or by different gene (allozyme) frequencies.

Sogawa (1978) studied the esterase patterns of the BPH biotype cultures and a field population at IRRI using agar gel electrophoresis. He found seven phenotypes present (Fig. 13.1) and concluded that biotype 2 could be differentiated by the higher frequency of variant type D in the population. However, although differences such as these can be used to characterize populations, they cannot be used to characterize individuals (Pashley and Bush 1979) and are therefore no use for identifying biotypes in the field. Considerable differences in frequency occurred for some phenotypes between the different wing morphs of a biotype and between the biotypes and the field population.

Saxena and Mujer (1984) and Saxena and Barrion (1985) studied 18 enzyme systems of the biotypes of *N. lugens* and a population from

Table 13.1 Allele frequencies for three loci in biotypes 1,2, and 3 of the rice infesting *N. lugens* and a population (L) from *Leersia hexandra*. (After Saxena and Barrion 1985.)

Locus	Biotype	1	2	3
PGI	1	0.018	0.583	0.398
	2	0.012	0.904	0.084
	3	0.005	0.868	0.128
	L	0.047	0.668	0.285
IDH	1	0.037	0.963	—
	2	0.022	0.978	—
	3	0.003	0.997	—
	L	0.009	0.955	0.036
MDH	1	—	0.957	0.043
	2	—	0.998	0.001
	3	—	0.997	0.003
	L	—	0.984	0.016

the grass *Leersia hexandra* (Swartz) using starch gel electrophoresis. They found 11 gave activity and 5 were polymorphic. They presented data for three, PGI, IDH, and MDH (Table 13.1). All the polymorphisms were shared with the exception of the rare allele IDH, which occurred with a frequency of 0.036 in the Leersia population. Although the gene frequencies differed slightly between the biotype populations, no diagnostic differences suitable for biotype field identification were found. Populations of brown planthoppers on the weed grass *Leersia hexandra* are morphologically identical to populations from rice and are referred to as a primitive non-virulent biotype by Saxena and Barrion (1985). However, they are, in fact, probably a sibling species. This biotype does not survive on rice, and neither does the rice-feeding brown planthopper on *Leersia*. It also differs in terms of acoustic signals from the brown planthopper on rice and, when given a choice, no heterogametic matings occur between individuals from the two populations (Claridge *et al.* 1985).

I also conducted an investigation of BPH biotypes using starch gel electrophoresis. Of 18 enzymes studied 5 gave no activity, 7 were monomorphic and 6 were polymorphic, although no diagnostic loci or significant differences in allozyme frequencies were found. These results are not altogether surprising, as the biotypes were originally selected from a common field population by continuous rearing on different resistant cultivers. Crossing and acoustic analysis have yielded no differences in mate choice or mate recognition signals between the biotypes (Claridge and den Hollander 1980). Crossing between the biotypes have also indicated that virulence is inherited in a complex polygenic manner (den Hollander and Pathak 1981; Sogawa 1981*b*); thus there is unlikely to be a simple gene-for-gene relationship between the insect and host plant.

A note of caution must be introduced: even if differences had been found between the biotype cultures, the cultures were established from only a few individuals and have been maintained in culture for several years. Founder effects, bottlenecks, and selection for culture conditions could easily explain observed differences (Pashley and Proverbs 1981). A causal relationship between any differences and virulence would need to be demonstrated.

Biotypes have also been reported for the green leafhopper *Nephotettix virescens*, and the failure of electrophoresis as a means of identifying BPH biotypes may not hold for these.

Geographic variation

Many studies have used electrophoresis to study the variation between allopatric populations of a species (Powell 1975). Generally, if gene (allozyme) frequencies differ between populations, then this is evidence for low levels of gene flow between the populations, e.g. in *Philaenus spumarius* (Saura *et al.* 1973). If sufficient gene flow occurs, these differences will be obliterated and gene frequencies will be substantially the same over the entire range of the species. If gene frequencies are the same over a wide geographic area, this converse situation is generally taken as an indication that substantial gene flow has occurred or is occurring but it could also be due to selection pressures being the same over a wide area (den Boer 1978).

Many species of leafhopper and planthopper have been shown to migrate. The beet leafhopper *Circulifer tenellus* is a major pest in the western USA and attacks a wide variety of crops (including sugar beet) and it is also the principal vector of curly top virus. It migrates annually from the Sierras and Western desert into the agricultural lands of California (Taylor 1985). Similarly, *Empoasca fabae* is a pest of potato, soy bean, and legumes. It overwinters in and around the Gulf of Mexico and migrates north each summer. Other hopper species showing long-distance migration or seasonal expansion of their range include *Macrosteles fascifrons, Sogatella furcifera* (Kisimoto *et al.* 1982) and *Laodelphax striatellus*. A demonstration of the extent of possible migration in leafhoppers is the appearance of large numbers of

Balclutha pauxilla on Ascension Island in May 1976 (Ghauri 1983). The probable origin of these was south-west Africa around Angola. Migration in these small insects is usually not obvious from visual observation, but is inferred from the absence of overwintering stages, by the sudden synchronous appearance over a large area, and also by their presence in aeroplane net catches. Most plant and leafhoppers are weak fliers and rely on prevailing winds and frontal systems for migration. It is important to establish the extent and prevalence of migration in developing control strategies.

The nature and extent of migration in the brown planthopper has been studied in some detail in the northern subtropical and temporate regions, particularly in Japan and China (Kisimoto 1987). The brown planthopper has not been found overwintering in Japan, but appears suddenly and in large numbers in traps and in the field from May to June. Evidence from catches on weather ships stationed up to 500 km from land confirms that they can migrate large distances (Kisimoto, 1979). The migrations are associated with south-west winds, the insects originating from mainland China (Kisimoto 1987). This seasonal expansion of range has also been shown within China (Cheng *et al.* 1983). To what extent migration occurs in the tropics is not well documented.

In an effort to resolve this question, I have examined electrophoretically populations of brown planthoppers from several different geographic areas throughout their range. Populations were obtained from the localities shown in Table 13.2. In total, 19 enzyme systems were examined. Of these, 6 gave no activity or were too weak to score reliably, 7 were monomorphic and the same throughout the range of *N. lugens*, and 6 were variable. The variable enzymes included esterase, PGI, PGM, 6-PGDH, MDH, and AK. The last

Table 13.2 Details of localities from which populations of *N. lugens* were obtained for electrophoresis.

	Locality	Latitude	Longitude
Australia	Queensland, Mareeba	17.00S	145.28E
Fiji	Viti Levu, Suva	18.08S	178.25E
Solomon Islands	Guadalcanal, Honiara	9.28S	159.57E
Philippines	Luzon, Los Banos	14.10N	121.26E
Indonesia	Sumatra, Metro	5.08N	105.15E
Sri Lanka	Kandy	7.17N	80.40E
China	Sichuan, Chongqing	29.30N	106.35E
Japan	Honshu, Omagri	39.29N	140.29E

Table 13.3 Allele frequencies of PGI and PGM for *N. lugens* populations from different geographic localities.

Locality	PGI			PGM		
	1	2	3	1	2	3
Australia	—	—	1.00	0.03	0.02	0.95
Fiji	—	0.22	0.78	0.73	0.08	0.19
Solomon Islands	—	—	1.00	0.36	0.42	0.23
Philippines	—	0.95	0.50	0.55	0.33	0.11
Indonesia	0.08	0.92	—	0.57	0.30	0.13
Sri Lanka	—	—	1.00	0.84	0.15	0.01
China	—	0.10	0.90	0.95	0.05	—
Japan	—	—	1.00	0.95	0.05	—

three had only the occasional rare allele in some populations. Esterases were difficult to score reliably because the bands in some populations were extremely faint, but they showed clear differences between regions (I shall return to discuss these esterases later). PGI and PGM gave clear differences (Table 13.3).

For PGI, one allele (PGI-3) rare in the Philippines was absent in Indonesia but was fixed in the Australian, Solomon Islands, and Sri Lankan populations, whilst showing a high frequency in the Chinese samples examined (Table 13.3).

PGM also showed differences, the rarest alleles (PGM-3) in most populations being the most common in the Australian population.

Some reservations must, however, be made concerning this data:

(1) because cultures only were used for the Chinese and Japanese samples examined; and

(2) because of the nature of BPH multiplication in the field.

Populations occur as foci that are often started by single females. Nevertheless, these differences do suggest that large-scale migration is not a feature of the tropics. Migration seems to occur as a seasonal expansion and retreat at the margins of the species distribution. These conclusions are supported by other observations on BPH. Analysis of courtship songs shows significant differences to exist between Australian, Solomon Islands, and Philippine populations (Claridge *et al.* 1985). Virulence differences have also been found between various populations. Rice varieties resistant in the Philippines, where most of the resistant varieties have been developed, are not resistant in Sri Lanka and India. The universal susceptible, rice variety TN411 is highly resistant to Australian insects. Based on virulence differences, it appears that three major groups of BPH are distributed in different regions of Asia: one in all the countries of south-east Asia (Indonesia and Philippines), one in east Asia (Japan and Korea), and a third in south Asia (India and Sri Lanka). To this can be added a fourth in Australia (Claridge and den Hollander 1982).

Insecticide resistance studies

Electrophoresis is useful in insecticide resistance studies in the following ways.

(1) to monitor populations for resistant insects;
(2) to monitor the spread of resistance;
(3) to predict further development of resistance;
(4) to establish sound strategies for appropriate control measures to combat the build-up of resistance;
(5) to relate specific enzymes with insecticide resistance.

The acquisition of resistance is driven by selection through the use of insecticides (Voss 1983). With increasing use of insecticides to control leafhoppers and planthoppers, several species have developed resistance, e.g. *N. cincticeps* (Hama and Iwata 1971, 1978; Yamoto *et al.* 1977, 1978), *L. striatellus* (Ozaki and Kassai 1970), and *N. lugens* (Voss 1983; Hasui and Ozaki 1984). Resistance is usually caused by (1) target insensitivity towards the insecticide used or (2) enzymatic degradation of the insecticide(s). The enzymes involved are usually esterases and electrophoresis has been used to study the relationship between certain of these enzymes (bands) and resistance.

Insecticidal organophosphates inhibit acetylcholinesterase (AChE) and this may be combated in the insect by the development of insensitivity to such pesticides. This appears to be the mechanism in *N. cincticeps* (Hama 1980; Hama *et al.* 1980) and also in *N. lugens*. Chung and Sun (1981) showed that the AChE of a resistant strain of brown planthoppers was 16 times less sensitive than that of a susceptible strain. Organophosphate resistance in *N cincticeps, L. stiatellus*, and *N. lugens* is also characterized by high levels of aliesterases (= carboxylesterases). Increased level of aliesterase correlates with increased resistance in *N. cincticeps* (Ozaki and Koike 1965; Ozaki and Kassai 1970).

Tranter (1983) studied the mechanism of insecticide resistance in

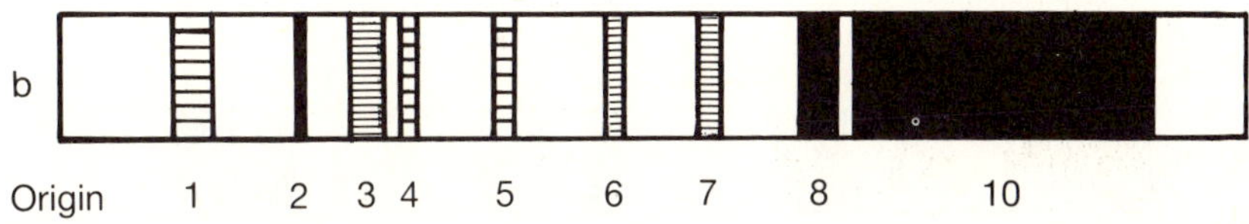

Fig. 13.2. Zymograms of esterases of susceptible and resistant *N. lugens* with 1-naphthyl acetate as a substrate. a = Queensland, Australia strain; b = Japanese carbofuran-resistant strain and the biotypes from Philippines. (After Tranter 1983.)

N. lugens. He compared populations from Japan, the Philippines, and Australia and found total esterase activity to vary about 10-fold between the populations and to be correlated with organophosphate resistance. Using polyacryamide gel electrophoresis, he resolved 11 esterases (E1–E11) (Fig. 13.2). All eleven occurred in the Australian (Queensland) insecticide-susceptible strain, whilst E1–E10 occurred in the resistant strains [Philippines (biotypes and a carbofuran-resistant strain), and Japan]. It was not possible to prove whether E9 occurred in the resistant strains because the activity of E10 was too great, which probably masked E9 activity. Tranter characterized the various esterase bands, using a variety of inhibitors, as 1 AChE, 2 arylesterases, 6 carboxylesterases, and 2 possible carboxylesterases.

Using polyacryamide disc electrophoresis, Hasui and Ozaki (1984) also studied esterase patterns of susceptible and resistant strains of *N. lugens*. While agreeing that resistance correlated with a markedly greater activity of certain esterase bands, the pattern of bands did not correspond with that described by Tranter (1983). These authors found two aliesterase bands (carboxylesterase of Tranter) with greater activity in resistant strains and eight bands of AChE activity, with both additional bands and increased levels of activity of bands in resistant strains (Fig. 13.3).

While surveying different populations for geographic variation in various enzymes, I also screened esterases. I was not able to resolve all the bands identified by Tranter (1983), but the geographic variation was interesting and may be related to insecticide usage and

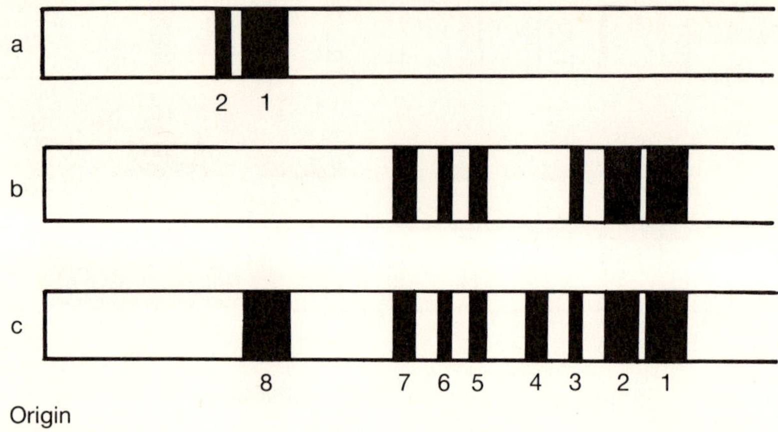

Fig. 13.3. Zymograms of *N. lugens*; a = aliesterase of susceptible and resistant strains; b = AChE of susceptible strain; c = AChE of fenitrothion resistant strain. (After Hasui and Ozaki 1984.)

evolution of resistance. Two to three bands were present in the Philippines population (Fig. 13.4). The band of greater activity probably corresponds to E10 of Tranter (1983). In the Japanese, Malaysian, and Sri Lankan populations, the same three bands were present, but E10 was polymorphic with very low or no activity in some individuals (Fig. 13.4).

While insecticide usage is probably greatest in Japan, *N. lugens* maybe less resistant there because it does not overwinter and a fresh invasion occurs each year, presumably from China. As insecticide usage in China is low, the reinvading insects are susceptible. Resistance has been shown to increase during the season in Japan, and in China it has also increased in recent years because of the increasing use of insecticides there (Kilin *et al.* 1981).

The BPH population from the Solomon Islands derives from an area where insecticide usage is high and this is reflected in high levels of esterase activity. However, the band of greatest acitivity is not E10 but probably E8 of Tranter (1983) (Fig. 13.4). The phenotype of this population is quite clearly different from that of the other populations. The Australian population is distinguishable by its very low esterase activity, making resolution of the bands difficult (Fig. 13.4). This population is also the most susceptible to insecticides. Insecticides have not been used on rice crops in this region, although there are plans to employ them in the future.

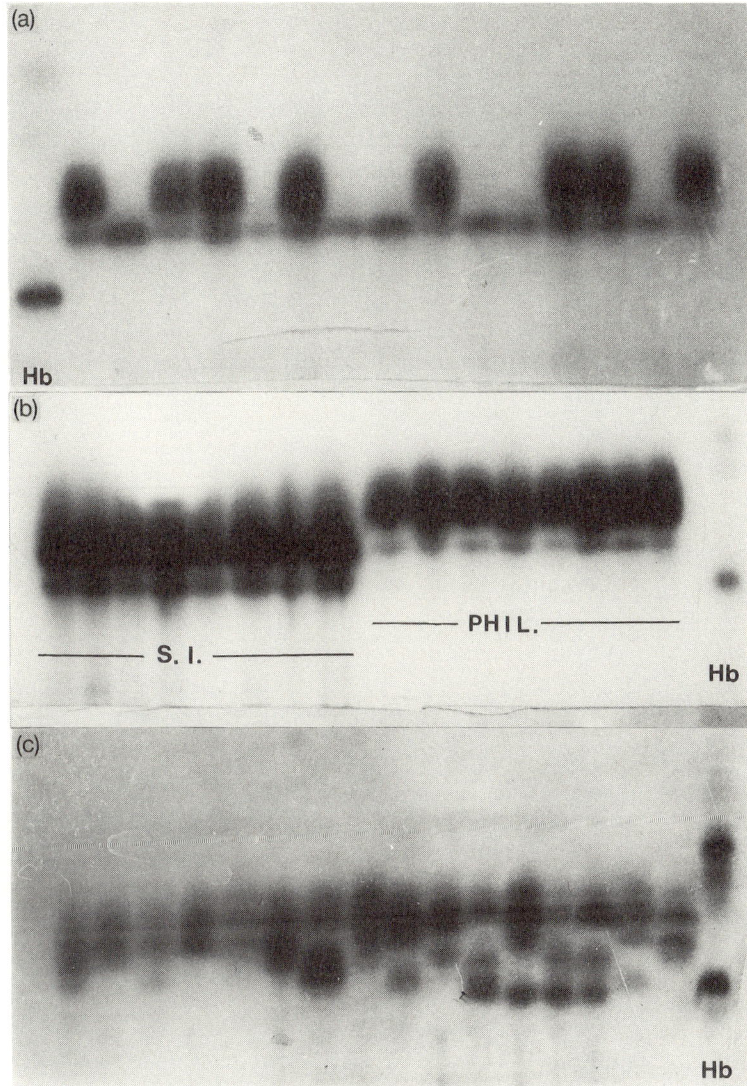

Fig. 13.4. Zymograms of esterases from (a) Japan, (b) Solomon Islands and Philippines, (c) Australia.

Conclusion

As shown above, gel electrophoresis has a wide variety of applications to plant and leafhopper research. However, despite the importance of those insects as agricultural pests, the technique has not been extensively used on the group. In particular, leafhopper and planthopper taxonomists traditionally use morphological characters especially of the male genitalia and apodemes, in species identification. But leafhoppers and plant hoppers are very rich in sibling species that classical morphology often fails to differentiate. Electrophoretic techniques could be profitably employed to help resolve many such species complexes.

Appendix

Full names and (EC) numbers (in parentheses) for enzymes cited in text: AChE, acetylcholinesterase (3.1.1.8); AK, adenylate kinase (2.7.4.3); carboxylesterase (3.1.1.1); IDH, isocitrate dehydrogenase (1.1.1.42); MDH, malate dehydrogenase (1.1.1.37); PGI, phosphoglucose isomerase (5.3.1.9); PGM, phosphoglucomutase (2.7.5.1); 6-PGDH, 6-phosphoglucose dehydrogenase (1.1.1.44).

References

Cheng, X.N., Chen, R.C., Hu, J.Z. and Zhu, S.S. (1983). Forecasting the migrations of brown planthopper (*Nilaparvata lugens* Stål) in China. In *1st International Workshop on Leafhoppers and Planthoppers of Economic Importance* (ed. W.J. Knight, N.C. Pant, T.S. Robertson, and M.R. Wilson), pp. 359–64. Commonwealth Institute of Entomology, London.

Chung, T.C. and Sun, C.N. (1981). Acetylcholinesterase sensitivity and carbamate resistance in brown planthopper. *Int. Rice Res. Newsl.*, **6**, 19–20.

Claridge, M.F. (1985). Acoustic signals in the Homoptera: Behaviour, taxonomy and evolution. *A. Rev. Ent.*, **30**, 297–317.

Claridge, M.F. and den Hollander, J. (1980). The 'biotypes' of the rice brown planthopper, *Nilaparvata lugens*. *Entomologia. exp. appl.*, **27**, 23–30.

Claridge, M.F. and den Hollander, J. (1982). Virulence to rice cultivars and selection for virulence in populations of the brown planthopper *Nilaparvata lugens*. *Entomologia. exp. appl.*, **32**, 213–22.

Claridge, M.F. and den Hollander, J. (1983). The biotype concept and its application to insect pests of agriculture. *Crop Prot.*, **2**, 85–95.

Claridge, M.F., den Hollander, J. and Haslam, D. (1984). The significance of morphometric and fecundity differences between the 'biotypes' of the brown planthopper, *Nilaparvata lugens*. *Entomologia exp. appl.*, **36**, 107–14.

Claridge, M.F., den Hollander, J., and Morgan, J.C. (1983). Variation within and between populations of the brown plant hopper, *Nilaparvata lugens* (Stål). In *1st International Workshop on Leafhoppers and Planthoppers of Economic Importance* (ed. W.J. Knight, N.C. Pant., T.S. Robertson, and M.R. Wilson), pp. 305–18. Commonwealth Institute of Entomology, London.

Claridge, M.F., den Hollander, J. and Morgan, J.C. (1985). Variation in courtship signals and hybridization between geographically definable populations of the Brown Planthopper, *Nilaparvata lugens* (Stål). *Biol. J. Linn. Soc.*, **24**, 35–49.

Conti, M. (1985). Transmission of plant viruses by leafhoppers and planthoppers. In *The planthoppers and leafhoppers* (ed. L.R. Nault and J.G. Rodriguez), pp. 289–307. Wiley, New York.

den Bieman, C.F.M. and Eggars-Schumacher.A. (1987). Allozyme polymorphism in planthoppers of the genus *Ribautodelphax* (Homoptera: Delphacidae), and the origin of the pseudogamous triploid forms. In *Biological and taxonomic differentiation in the Ribautodelphax collinus complex (Homoptera, Delphacidae)*, C.F.M. den Bieman, Ph. D. thesis, Wageningen, the Netherlands.

den Boer, M.H. (1978). Isoenzymes and migration in the African armyworm *Spodoptera exempta* (Lepidoptera: Noctuidae). *J. Zool.*, **185**, 539–53.

den Hollander, J. and Pathak, P.K. (1981). The genetics of the biotypes of the brown planthopper *Nilaparvata lugens*. *Entomologia exp. appl.*, **29**, 76–86.

Drosopolous, S. and Loukas, M. (1988). Genetic differentiation between coexisting color-types of the *Alebra albostriella* group (Homoptera:Cicadellidae). (In press).

Dyck, V.A. and Thomas, B. (1979). The brown planthopper. In *Brown planthopper: Threat to rice production in Asia,* pp. 3–17. International Rice Research Institute, Los Banos, Philippines.

Ghauri, M.S.K. (1983). A case of long distance dispersal of a leafhopper. In *1st International Workshop on Leafhoppers and Planthoppers of Economic Importance* (ed. W.J. Knight, N.C. Pant, T.S. Robertson, and M.R.Wilson), pp. 249–55. Commonwealth Institute of Entomology, London.

Guttman, S.I., Wood, T.K., and Karlin, A.A. (1981. Genetic differentiation along host plant lines in the sympatric *Enchenopa binotata* Say complex. *Evolution*, **35**, 205–17.

Hama, H. (1980). Studies on the mechanism of resistance to insecticides in the green rice leafhopper, *Nephotettix cincticeps* Uhler, with particular reference to reduced sensitivity of acetylcholinesterases. *Bull. natn. Inst. agric. Sci.*, **34**, 75–138.

Hama, H. and Iwata, T. (1971). Insensitive cholinesterase in the Nakagawara strain of the green rice leafhopper, *Nephotettix cincticeps* Uhler (Hemiptera:Cicadellidae), as a cause of resistance to carbamate insecticides. *Appl. Ent. Zool.*, **6**, 183–191.

Hama, H. and Iwata, T. (1978). Studies on the inheritance of carbamate resistance in the green rice leafhopper, *Nephotettix cincticeps* Uhler (Hemiptera:Cicadellidae). Relationships between insensitivity of acetylcholinesterase and cross-resistance to carbamate and organophophate insecticides. *Appl. Ent. Zool.*, **13**, 190–202.

Hama, H., Iwata, T., Miyata, T., and Saito, T. (1980). Some properties of acetylcholinesterase partially purified from susceptible and resistant green rice leafhoppers, *Nephotettix cincticeps* Uhler (Hemiptera: Deltocephalidae). *Appl. Ent. Zool.*, **15**, 249–61.

Harris, K.F. (1983). Auchenorrhynchous vectors of plant viruses; Virus-vector interactions and transmission mechanisms. In *1st International Workshop on Leafhoppers and Planthoppers of Economic Importance* (ed. W.J. Knight, N.C.Pant, T.S. Robertson, and M.R.Wilson), pp. 405–14. Commonwealth Institute of Entomology, London.

Hasui, H. and Ozaki, K. (1984). Electrophoretic esterase patterns in the brown planthopper, *Nilaparvata lugens* Stal (Hemiptera:Delphacidae) which developed resistance to insecticides. *Appl. Ent. Zool.*, **19**, 52–8.

International Rice Research Institute. (1979). Biotype studies: Genetics of BPH biotypes. *Annual Report for 1978,* p. 68.

Kassai, T. and Ogita, Z. (1965). Sudies on malathion resistance and esterase activity in green rice leafhoppers. *SABCO J.*, **1**, 130–40.

Khush, G.S. (1979). Genetics of and breeding for resistance to the brown planthopper. In *Brown planthopper: Threat to rice production in Asia,* pp. 321–32. International Rice Research Institute, Los Banos, Philippines.

Kilin, D., Nagata, T., and Matsuda, T. (1981). Development of carbamate resistance in the Brown planthopper, *Nilaparvata lugens* Stal (Homoptera:Delphacidae). *Appl. Ent. Zool.*, **16**, 1–6.

Kisimoto, R. (1979). Brown planthopper migration. In *Brown planthopper: Threat to rice production in Asia,* pp. 113–24. International Rice Research Institute, Los Banos, Philippines.

Kisimoto, R. (1987). Ecology of planthopper migration. In *2nd International Workshop on Leafhoppers and Planthoppers of Economic Importance* (ed. M.R. Wilson and L.R. Nault), pp. 41–54. Commonwealth Institute of Entomology, London.

Kisimoto, R., Hirao, J. Hirahara, Y., and Tanaka, A. (1982). Synchronization in migratory flight of planthoppers, *Nilaparvata lugens* Stal and *Sogatella furcifera* Horvath (Hemiptera:Delphacidae) in south western Japan. *Jap. J. appl. Ent. Zool.*, **26**, 112–18.

Lakshiminarayana, A. and Khush, G.S. (1977). New genes for resistance to the brown planthopper in rice. *Crop Sci.*, **17**, 96–100.

Ling, K.C. (1967). Transmission of rice viruses in Southeast Asia. In *The virus diseases of the rice plant.* Johns Hopkins University Press, Baltimore.

Ling, K.C. (1972). *Rice virus diseases*. International Rice Research Institute, Los Banos, Philippines.

Liquido, N.J. (1986). Meiosis in males of the brown planthopper, *Nilaparvata lugens*. *Entomologia. exp. appl.*, **41**, 61–7.

Miyata. T. and Saito, T. (1976). Mechanism of malathion resistance in the green rice leafhopper, *Nephotettix cincticeps* Uhler (Hemiptera: Deltocephalidae). *J. Pestic. Sci.*, **1**, 23–9.

Miyata, T, Honda, H., Saito, T., Ozaki, K., and Sasaki.Y. (1976). In vitro degradation of 14C-methyl malathion by organophosphate susceptible and resistant smaller brown planthopper, *Laodelphax striatellus* Fallen. *Botyu-Kagaku*, **41**, 10–15.

Nielson, M.W. (1985). Leafhopper systematics. In *The planthoppers and leafhoppers* (ed. L.R. Nault and J.G. Rodriuez), pp. 11–39. Wiley, New York.

Ou, S.H. (1972). *Rice diseases*. Commonwealth Mycological Institute, Association of Applied Biologists, Kew, Surrey, UK.

Ozaki, K. and Kassai, T. (1970). Biochemical genetics of malation resistance in the smaller brown planthopper, *Laodelphax striatellus* Fallen. *Entomologia exp. appl.*, **13**, 162–72.

Ozaki, K. and Koike, H. (1965). Naphyl acetate esterase in the rice green leafhopper, *Nephotettix cincticeps* Uhler, with special reference to resistant colony to the oganophosphorous insecticide. *Jap. J. appl. Ent. Zool.*, **9**, 53–9.

Pashley, D.P. and Bush, G.L. (1979). The use of allozymes in studying insect movement with special reference to the codling moth, *Laspayresia pomonella* (L.) (Olethreutidae). In *Movement of highly mobile insects: Concepts and methodology in research* (ed. R.L. Rabb and G.G. Kennedy). University Graphics, Raleigh, N.C.

Pashley, D.P. and Proverbs, M.D.. (1981). Quality control by electrophoretic monitoring in a laboratory colony of codling moths. *Ann. ent. Soc. Am.*, **74**, 20–3.

Pathak, M.D. and Khush, G.S. (1979). Studies on the varietal resistance to the brown planthopper at the International Rice Research Institute. In *Brown planthopper: Threat to rice production in Asia*, pp. 285–301. International Rice Research Institute, Los Banos, Philippines.

Powell, J.R. (1975). Protein variation in natural populations of animals. *Evol. Biol.*, **8**, 79–119.

Raatikainen, M. (1967). Bionomics, enemies and population dynamics of *Javasella pellucida* (F.) (Hom., Delphacidae). *Ann. Agric. Fenn.*, **6**, 1–149.

Raatikainen, M. (1970). Ecology and fluctuations in abundance of

Megadelphax sordidula (Stal) (Hom., Delphacidae). *Ann. Agric. Fenn.*, **9**, 315–24.

Saura, A., Halkka, O. and Lokki, J. (1973). Enzyme gene heterozygosity in small island populations of *Philaenus spumarius* (L.) (Homoptera). *Genetica*, **44**, 459–73.

Saxena, R.C. and Barrion, A.A. (1982). Cytogenetic variation in brown planthopper biotypes 1 and 2. *Int. Rice Res. Newsl.*, **7**, 5.

Saxena, R.C. and Barrion, A.A. (1983). Cytological variation among biotypes 1,2 and 3. *Int. Rice Res. Newsl.*, **8**, 14–15.

Saxena, R.C. and Barrion, A.A. (1985). Biotypes of the brown planthopper *Nilaparvata lugens* (Stal) and strategies in development of host plant resistance. *Ins. Sci. Appl.,* **6**, 271–89.

Saxena, R.C. and Mujer, C.V. (1984). Detection of enzyme poplymorphism among populations of brown planthopper (BPH) biotypes. *Int. Rice Res. Newsl.*, **9**, 18–19.

Saxena, R.C. and Rueda, L.M. (1982). Morphological variations among three biotypes of the brown planthopper *Nilaparvata lugens* in the Philippines. *Ins. Sci. Appl.*, **3**, 193–210.

Shaw, C.R. and Prasad, R. (1970). Starch gel electrophoresis of enzymes-a compilation of recipes. *Biochem. Genet.*, **4**, 297–320.

Smithies, O. (1965). Zone electrophoresis in starch gels: group variations in the serum protein of normal human adults. *Biochem. J.*, **61**, 629–41.

Sogawa, K. (1978). Electrophoretic variations in esterase among the biotypes of the brown planthopper. *Int. Rice Res. Newsl.*, **3**, 8–9.

Sogawa, K. (1981*a*). Biotypic variations in the brown planthopper *Nilaparvata lugens* (Homoptera: Delphacidae) at the I.R.R.I., the Philippines. *Appl. Ent. Zool.*, **16**, 129–37.

Sogawa, K. (1981*b*). Hybridization experiments on 3 biotypes of the brown planthopper *Nilaparvata lugens* (Homoptera: Delphacidae) at the I.R.R.I., the Philippines. *Appl. Ent. Zool.*, **16**, 193–99.

Taylor, R.A.J. (1985). Migratory behaviour in the Auchenorrhyncha. In *The planthoppers and leafhoppers* (ed. L.R. Nault and J.G. Rodriguez), pp. 259–88. Wiley, New York.

Tranter, B.C. (1983). Elucidation of insecticide resistance mechanisms in the brown planthopper *Nilaparvata lugens*. Ph. D. thesis, University of Reading, UK.

Voss, G. (1983). The biochemistry of insecticide resistant hopper races. In *1st International Workshop on Leafhoppers and Planthoppers of Economic Importance* (ed. W.J. Knight, N.C. Pant, T.S. Robertson, and M.R. Wilson), pp. 351–6. Commonwealth Institute of Entomology, London.

Williams, J.R. (1957). The sugar cane Delphacidae and their natural enemies in Mauritius. *Trans. R. Ent. Soc. Lond.*, **109**, 65–110.

Wilson, S.W., and O'Brien, L.B. (1987). A survey of planthopper pests of economically important plants (Homoptera: Fulgoroidea). In *2nd International Workshop on Leafhoppers and Planthoppers of Economic Importance* (ed. M.R. Wilson and L.R. Nault), pp. 343–60. Commonwealth Institute of Entomology, London.

Yamoto, I., Kyomura, N., and Takahashi, Y. (1977). Aryl N-propylcarbamates, a potent inhibitor of acetylcholinesterase from the resistant green rice leafhopper, *Nephotettix cincticeps*. *J. Pestic. Sci.*, **2**, 463–6.

Yamoto, I., Kyomura, N., and Takahashi, Y. (1978). Joint insecticidal effects of *N*-propyl and *N*-methyl-carbamates on the green rice leafhopper resistant to aryl *N*-methylcarbamates. *Rev. Plant Protect. Res.*, **11**, 79–91.

14 Speciation by hybridization in insects

LUCIANO BULLINI and GIUSEPPE NASCETTI

Department of Genetics and Molecular Biology, University of Rome "La Sapienza", Rome, Italy

Abstract

An increasing number of species of hybrid origin have been detected in recent years in different animal groups, mainly by the combined use of chromosome and isoenzyme studies. In this chapter, a number of cases are reviewed, including the stick-insects *Bacillus whitei*, *B. lynceorum*, *Leptynia hispanica* D, *Clonopsis gallica*, *Carausius morosus*, the grasshopper *Warramaba virgo*, some *Otiorrhynchus* weevils, spider beetles of the genus *Ptinus* and the planthopper *Muellerianella* - *2-fairmairei-brevipennis*.

The processes involved in the origin and evolution of hybrid species (hybridogens) are discussed. Two steps are required for a hybrid species to arise: (1) interspecific hybridizations, and (2) changes in the maturation divisions allowing the transmission of the hybrid genome to the next generation. The latter events are not caused by hybridization itself, and are undoubtedly rare; on the contrary, rather high rates of interspecific hybridization have frequently been detected in secondary contact zones, especially in habitats disturbed by man.

Insect hybrid species are all-female and reproduce in two different ways: thelytokous parthenogenesis or gynogenesis; the consequences of these modes of reproduction are examined.

Two phenonema are considered particularly relevant for the evolutionary success of insect hybridogens: (1) their reproductive potential, which is doubled with respect to that of their bisexual ancestors ('demographic advantage'); (2) their high level of

Electrophoretic Studies on Agricultural Pests (ed. Hugh D. Loxdale and J. den Hollander), Systematics Association Special Volume No. 39, pp. 317–39. Clarendon Press, Oxford, 1989.

heterozygosity, dramatically enhanced by interspecific hybridization ('heterotic advantage'). The short- and long-term effects of these phenomena are discussed. They account for the high competitiveness generally shown by hybrid species, which frequently overcome their bisexual ancestors, reducing the latter's ranges or even causing their extinction. Moreover, hybridogens often colonize wide areas unfavourable to their parental species and easily shift to new ecological niches. These features can explain the occurrence of hybrid species among agricultural pests and pests of medical importance.

Introduction

Speciation by hybridization in insects was, until recently, ruled out by most authors. Even for polyploid taxa, an autopolyploid origin was considered much more likely than an allopolyploid one. Isoenzyme and chromosome studies are dramatically changing this picture, since they provide definitive evidence that in many animal groups, including insects, various species arose by interspecific hybridization.

Insect hybrid species (hybridogens) reproduce either by thelytokous parthenogenesis or gynogenesis (cf. White 1978); they show a hybrid genome, combining the genes and chromosome sets of two (or more) sexual ancestors. For a number of parthenogens (e.g. the grasshopper *Warramaba virgo* and the stick insect *Bacillus whitei*), both the parental species have been identified and their hybrid origin has been genetically assessed. In other presumed hybrids (e.g. *Bacillus atticus*), only one of the bisexual parents has been detected; however, their hybrid origin appears to be supported by strong evidence, both at the isoenzyme and chromosome level. For other supposed hybrid stick insect species (e.g. *Clonopsis gallica* and *Carausius morosus*), no sexual ancestors have so far been identified, possibly because competition with their hybrid derivatives has made them excessively rare or perhaps extinct. An alternative hypothesis is that the supposed hybridogens arose as parthenogenetic derivatives of single sexual species (Suomalainen and Saura 1973; Lokki and Saura 1980). Accordingly, a random accumulation of allelic mutants and/or chromosome rearrangements in different individuals would be expected, giving rise to a chaotic array of biotypes scattered over the parthenogen's range. The finding in many cases of few alleles (often in the same heterozygous condition), and of the same heterozygous chromosome rearrangements throughout the species' range, makes a hybrid origin much more likely.

A model case is that of the parthenogenetic grasshopper *Warramaba virgo*, carefully studied by White and co-workers. Up to 1973, White considered this species to be a non-hybrid parthenogenetic derivative, differentiated from its presumed bisexual ancestor by several chromosome rearrangements. It was Hewitt (1975) who first suggested a hybrid origin for *W. virgo*. This hypothesis was confirmed by the discovery of both of the parental species of *W. virgo*, which combine their haploid genomes (White *et al.* 1977). Moreover, it was shown that the two population groups ('phylads') found in *W. virgo* arose by distinct hybridization events, involving two different races of one of the parental species (White 1980).

In this chapter, various insect species (especially stick insect species of hybrid origin) are examined, including several agricultural pests, e.g. some weevils of the genus *Otiorrhynchus* and spider beetles of the genus *Ptinus*. The hybrid species so far detected are apparently only a small fraction of those actually in existence, as indicated by their increasing rate of discovery.

Stick insects

Bacillus whitei

Bacillus whitei is a diploid all-female stick insect species found in southern Sicily (Fig. 14.1) that reproduces by parthenogenesis (Nascetti and Bullini 1982*a*). Its origin by interspecific hybridization was strongly suggested by isoenzyme and chromosome data and was confirmed when its sexual ancestors were recognized (Bullini 1983; Bullini and Nascetti 1986; Nascetti and Bullini 1982*a*, 1983). These are *B. rossius* and *B. grandii*. The latter is a relict species recently discovered in southern Sicily that consists of a few small populations, and the former is widespread in the Mediterranean region.

B. rossius and *B. grandii* show distinct alleles at many of the enzyme loci tested; their genetic distance (Nei's *D*) is therefore high: average $D = 1.14$ in Sicily. The alleles of *B. rossius* and *B. grandii* are combined in *B. whitei*, which shows fixed heterozygosity at the loci differentiated in its parental species (Table 14.1 and Fig. 14.2). The observed mean heterozygosity per locus in *B. whitei* is therefore very high, i.e. 0.68, whereas is less than 0.05 in *B. grandii* and *B. rossius* populations from Sicily.

The karyotypes of *B. rossius* and *B. grandii* appear well-differentiated, both in chromosome number and shape; their haploid sets can be easily identified in the karyotype of *B. whitei* (Nascetti *et al.* 1985).

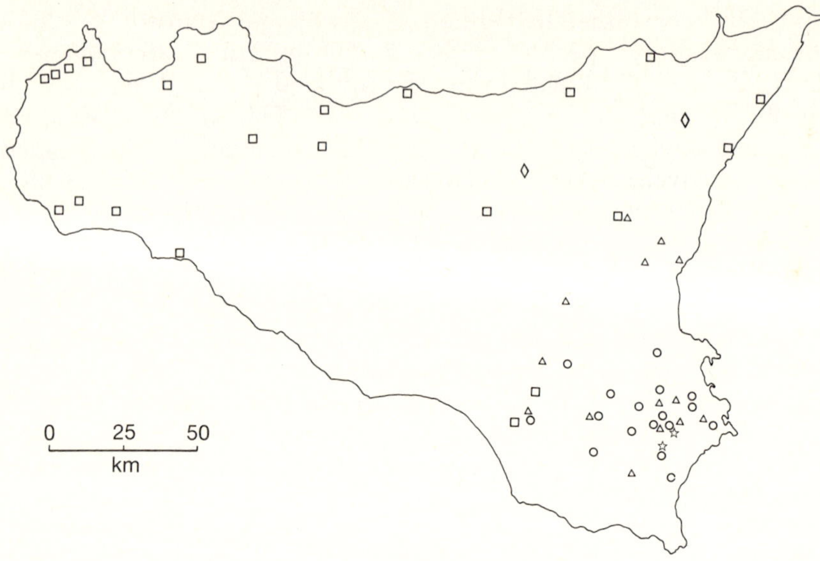

Fig. 14.1. Sicily, showing the collecting sites of *Bacillus rossius* (squares), *B. grandii* (stars), *B. whitei* (circles), *B. atticus* (diamonds), and *B. lynceorum* (triangles).

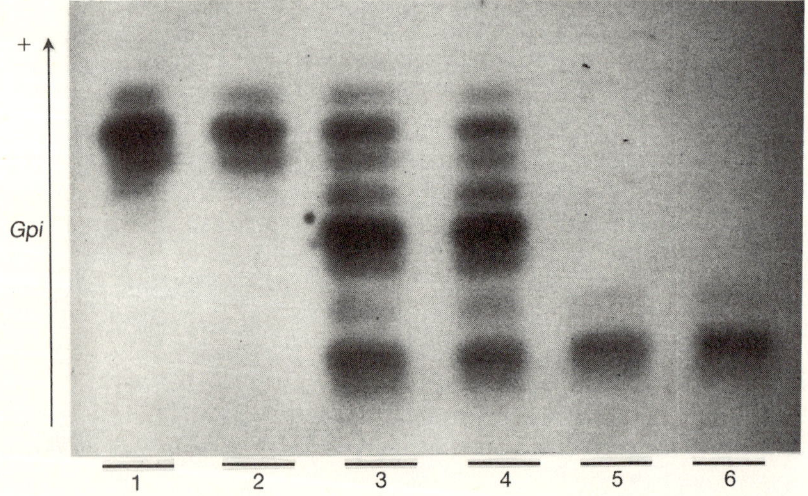

Fig. 14.2. Zymograms of glucose phosphate isomerase (GPI) in *Bacillus rossius* (1,2), *B. whitei* (3,4), and *B. grandii* (5,6) from southern Sicily. The genotypes are as follows: 1,2 = $Gpi^{100/100}$; 3,4 = $Gpi^{73/100}$; 5,6 = $Gpi^{73/73}$.

Table 14.1 Genotype frequencies at 16 enzyme loci in population samples of *Bacillus rossius* from Granieri (Catania), *B. grandii* from Noto (Siracusa), and *B. whitei* from Floridia (Siracusa). The loci α-*Gpdh, Ldh, Got-2, Hk-2, Adk* and *Ald* are monomorphic for the same allele in the samples of the three species from the study area.

Loci	Genotypes	*B. rossius*	*B. grandii*	*B. whitei*
Sdh	100/100	1.00	—	—
	100/115	—	—	1.00
	115/115	—	1.00	—
Hbdh	82/82	—	1.00	—
	82/100	—	—	1.00
	100/100	1.00	—	—
Mdh-1	88/88	1.00	—	—
	88/95	—	—	1.00
	95/95	—	0.61	—
	95/108	—	0.34	—
	108/108	—	0.05	—
Mdh-2	100/100	1.00	—	—
	100/105	—	—	1.00
	105/105	—	1.00	—
Idh-1	90/90	—	0.02	—
	90/100	—	0.12	—
	100/100	1.00	0.86	1.00
Idh-2	92/92	—	1.00	—
	92/100	—	—	1.00
	100/100	1.00	—	—
6Pgdh	100/100	1.00	—	—
	100/106	—	—	1.00
	106/106	—	0.96	—
	106/120	—	0.04	—
G3pdh	100/100	1.00	—	—
	100/115	—	—	1.00
	115/115	—	1.00	—
Sod-1	82/82	—	1.00	—
	82/100	—	—	1.00
	100/100	1.00	—	—
Got-1	92/92	—	1.00	—
	92/100	—	—	1.00
	100/100	1.00	—	—
Hk-1	100/100	1.00	—	—
	100/107	—	—	1.00
	107/107	—	1.00	—
Pgm	100/100	1.00	—	—
	100/105	—	—	1.00
	105/105	—	1.00	—

Table 14.1 *Continued*

Loci	Genotypes	*B. rossius*	*B. grandii*	*B. whitei*
Ca-1	100/100	1.00	—	—
	100/105	—	—	1.00
	105/105	—	1.00	—
Ca-2	95/110	—	0.20	—
	100/100	1.00	—	—
	100/110	—	—	1.00
	110/110	—	0.80	—
Mpi	100/100	1.00	—	—
	100/110	—	—	1.00
	110/110	—	1.00	—
Gpi	73/73	—	1.00	—
	73/100	—	—	1.00
	100/100	1.00	—	—

The offspring of *B. whitei* are genetically identical with their mother; the cytological mechanism of parthenogenesis, which is not yet fully understood, possibly involves a chromosome extra-duplication at the first meiotic division, or a fusion of pronucleus with the first polar body (Bullini 1985; Bullini and Nascetti 1986). This mechanism well explains the genetic homogeneity of *B. whitei*, which has a single clone widespread throughout its range; the genotypes of this clone at different enzyme loci are shown in Table 14.1. Only a few other clones were found, geographically confined to restricted areas; they differ from each other at single loci (e.g. for $Mdh\text{-}2^{100/108}$ instead of $Mdh\text{-}2^{100/105}$). These clones presumably originated from the most common one by gene mutation (Bullini and Nascetti 1986). Some interclonal chromosome variation was also found, bringing the chromosome number from the normal 35 to 36 or 37 (Scali and Marescalchi 1987).

Many 'synthetic' *B. whitei* individuals of both sexes were obtained in the laboratory by crossing *B. rossius* and *B. grandii*: hybrid females produced eggs, although these did not reach development; males produced either no sperm or abnormal sperm (Nascetti *et al.* 1985).

The Bacillus atticus *complex*

Another parthenogenetic stick insect species whose hybrid origin was revealed by isoenzyme and chromosome evidence is *Bacillus atticus* (Bullini 1982; Goday *et al.* 1982; Nascetti and Bullini 1983). This taxon was divided into a number of species or subspecies (their taxonomic rank is obviously only conventional), either diploid or triploid, including *B. atticus atticus* from Greece (diploid); *B. atticus mülleri* from Dalamtia (diploid); *B. atticus caprai* from southern Italy and Sicily (diploid); *B. creticus* from Crete (diploid); *B. diplocarius* (diploid) and *B. carius* (triploid) from south-western Turkey; *B. rhodius* from Rhodes (triploid); *B. cyprius* from Cyprus (diploid) (Gasperi *et al.* 1986; Marescalchi *et al.* 1985; Nascetti and Bullini 1982*b*; Scali and Marescalchi 1987).

All these parthenogenetic forms appear to be related to *B. grandii*, both at the isoenzyme and at the chromosome level; this species represents one of the sexual ancestors of the *B. atticus* complex [all the alleles found in *B. grandii* are present in *B. atticus* (Nascetti and Bullini 1988)]. The other parental species has not yet been identified and may be extinct; evolution of the *B. atticus* complex from parthenogenetic forms of *B. grandii* involving summation of mutations appears to be ruled out by isoenzyme data (Nascetti and Bullini 1988). These data show that in *B. atticus* from Greece, Yugoslavia, and Italy (the data from the other regions are still lacking) only one extra allele exists in addition to those found in *B. grandii* at several loci (i.e. *Sdh, Hbdh, Mdh-1, Sod-1, Got-1*).

In *B. atticus*, several dozen biotypes were detected; the most common ones are shown in Table 14.2. These biotypes show a high proportion of heterozygous loci (from 0.20 to 0.40). The highest variation was found in Greece, with several biotypes coexisting in the same locality, whereas Italian and Yugoslavian samples are remarkably homogeneous (Nascetti and Bullini 1988). The high number of biotypes found in *B. atticus* seems to be due both to multiple hybridization events, and to the cytological mechanism of parthenogenesis. The latter involves the fusion of the pronucleus with a polar body. This mechanism takes place either in the first or in the second meiotic division, maintaining respectively all, or only a part, of the initial heterozygosity. Analysis of genetic data suggests that the first mechanism is more common in populations from Yugoslavia, Italy, and Sicily, whilst the second occurs more frequently in those from Greece. When *B. atticus* is crossed with *B. rossius* or *B. grandii* males, a normal meiosis occurs. The offspring, often sterile, are diploid and show Mendelian segregation (Bullini and Nascetti 1987; Nascetti and Bullini 1982*b*; and our unpublished data).

Table 14.2 Genetic structure of 21 biotypes (A to Z) observed in samples of *Bacillus atticus* from Greece (A–N), Yugoslavia (O–P) and Italy (Q–Z).

		Greece												Yugoslavia		Italy						
Loci	Genotypes	A	B	C	D	E	F	G	H	I	L	M	N	O	P	Q	R	S	T	U	V	Z
Sdh	110/110			*			*	*				*										
	110/115	*	*		*	*			*	*	*		*	*	*	*	*	*	*	*	*	*
Hbdh	65/65											*				*	*	*	*	*	*	*
	65/82	*	*	*	*	*	*	*	*	*	*		*									
	82/82													*	*							
Mdh-1	95/95								*	*	*	*				*	*	*	*	*	*	*
	95/108	*		*	*	*	*	*					*	*	*							
	108/108		*																			
Idh-1	90/90					*			*	*	*	*		*	*					*		*
	90/93	*			*								*									
	90/100		*	*			*									*	*	*	*		*	
	93/93							*														
6Pgdh	106/106		*	*					*	*	*			*	*	*	*	*	*	*	*	*
	112/112	*			*	*	*	*				*	*									
Sod-1	70/82			*			*	*	*	*	*	*				*	*	*	*	*	*	*
	82/82	*	*		*	*							*	*	*							

Got-1	92/92								*	*	*											
	92/104	*	*		*	*	*						*	*		*	*					
	104/104			*				*				*			*	*			*	*	*	*
Pgm	95/95							*					*									
	95/100	*																				
	95/105						*															
	100/100		*							*												
	100/105			*	*				*			*				*	*	*	*	*	*	*
	105/105					*					*			*	*							
Ṁpi	105/105													*	*							
	105/110			*			*					*				*	*	*	*	*	*	*
	110/110	*	*		*	*		*	*	*	*		*									
Gpi	73/73	*											*									
	73/80													*	*							
	73/83				*	*	*															
	73/95		*						*	*	*											
	83/83																*			*		
	83/87															*		*	*			*
	85/95											*										
	87/87																				*	
	90/100							*														
	95/95			*																		

Bacillus lynceorum

A third hybrid stick insect species is *B. lynceorum*. This triploid parthenogenetic species, endemic to Sicily, combines two *B. grandii*-like genomes and one from *B. rossius*, as indicated by isoenzyme and chromosome evidence (Bullini *et al.* 1984; Bullini and Nascetti 1986). Various biotypes were detected in *B. lynceorum*, nine of which are shown in Table 14.3. The alleles present in these biotypes are the same found in *B. rossius* and *B. atticus*. In previous papers (Bullini *et al.* 1984; Bullini and Nascetti 1986) it was suggested that *B. lynceorum* arose from crosses between *B. whitei* and *B. grandii* (or a *B. grandii*-like species). Futher isoenzyme data on *B. atticus, B. whitei* and *B. lynceorum* seem to be more in agreement with the hypothesis

Table 14.3 Genetic structure of nine biotypes (A to I) observed in samples of *Bacillus lynceorum* from Sicily.

Loci	Genotypes	A	B	C	D	E	F	G	H	I
Sdh	100/115/115	*	*	*	*	*	*	*	*	*
Hbdh	65/82/100	*	*	*	*	*	*	*	*	*
Mdh-1	88/95/95	*	*	*	*	*	*	*		*
	95/95/95								*	
Mdh-2	100/105/105		*	*	*	*	*	*	*	*
	105/105/105	*								
Idh-1	90/100/100			*		*	*	*	*	*
	100/100/100	*	*		*					
Idh-2	92/92/100	*	*	*	*	*	*	*	*	*
6Pgdh	100/106/106	*	*	*	*	*	*	*	*	*
G3pdh	100/115/115	*	*	*	*	*	*	*	*	*
Sod-1	82/82/100	*	*	*	*	*	*	*	*	*
Got-1	92/92/100							*	*	
	92/100/104	*	*	*						*
	100/104/104				*	*	*			
Hk-1	100/107/107	*	*	*	*	*	*	*	*	*
Pgm	100/105/105	*	*	*	*	*	*	*	*	*
Ca-1	100/105/110	*	*	*	*	*	*	*	*	*
Ca-2	90/110/110						*			
	100/110/110	*	*	*	*	*		*	*	*
Mpi	100/110/110	*	*	*	*	*	*	*	*	*
Gpi	73/83/100		*	*	*	*	*	*	*	
	73/87/100									*
	83/83/100	*								

that *B. lynceorum* arose in Sicily by a cross between *B. grandii* and a diploid F_1 hybrid, *B. atticus* × *B. rossius*, which exceptionally produced unreduced (i.e. diploid) eggs. These F_1 hybrids, generally sterile, are frequently observed in the field, especially where the chief food plants of *B. atticus* (lentisc, *Pistacia lentiscus*) and of *B. rossius* (bramble, *Rubus fruticosus* agg.) intermingle. These hybridization zones are strictly limited to the areas inhabited by sexual populations of *B. rossius* and parthenogenetic *B. atticus*, such as south-eastern Italy and Sicily. In the western Mediterranean countries, *B. rossius* is often bisexual, but *B. atticus* is not present; in contrast, in the eastern Mediterranean, both *B. atticus* and *B. rossius* are widespread, but the latter is parthenogenetic.

Clonopsis gallica

Clonopsis gallica is a parthenogenetic species, presumably triploid with $3n = 54\text{–}57$ (Bullini and Bianchi Bullini 1971), and is widespread in southern Europe and northwestern Africa. In the latter region, it is often found together with the sexual *Cl. algerica*, wrongly considered by various authors as a geographic race of *Cl. gallica.*

Cl. gallica shows a highly heterozygous karyotype and a widespread heterozygosity at a number of enzyme loci, often in fixed condition (Bullini and Bianchi Bullini 1971; Nascetti and Bullini 1983; and our unpublished data). This suggests a hybrid origin for this species (Bullini 1983, 1985); however, its presumed parental species are still unknown. They cannot be *Cl. algerica* or the recently-described *Cl. maroccana*, as is clearly shown by isoenzyme data (Nascetti and Bullini 1983; Bullini and Nascetti 1987).

The Leptynia hispanica *complex*

Other parthenogenetic stick insects of hybrid origin are found in the *Leptynia hispanica* complex, which inhabits southern France and the Iberian peninsula (Nascetti *et al.* 1983). *L. hispanica* was previously considered to be a triploid parthenogen, derived from the diploid sexual species *Leptynia attenuata* (Cappe de Baillon and de Vichet 1940). Electrophoretic and chromosome evidence, however, showed that *L. hispanica* is in fact a species complex including both sexual and parthenogenetic species, the latter being either triploid or tetraploid (Nascetti *et al.* 1983, and our unpublished data). The parthenogens presumably originated from bisexual members of the *L. hispanica* complex, mainly through hybridization, and not from *L. attenuata.* The latter, also including a number of species, is quite differentiated from the *L. hispanica* complex, in morphology, chromosomes, and isoenzymes (Bullini and Nascetti 1986, 1987).

One of the supposed hybrid species in the *L. hispanica* complex is *L. hispanica* D. This parthenogen shows a fixed heterozygosity at a number of loci (Table 14.4). *L. hispanica* A can reasonably be assumed to be one of its parental species (Nascetti *et al.* 1983). The second one has not yet been discovered. Also, the hypothesis that *L. hispanica* D is a parthenogenetic derivative of *L. hispanica* A appears to disagree with the genetic data. Many of the alleles found in *L.*

Table 14.4 Genotype frequencies at nine enzyme loci in population samples of the sexual stick insect species *Leptynia hispanica* A from Benisa (Alicante, Spain) and of the parthenogenetic *L. hispanica* D from Boniches (Cuenca, Spain). In the latter, the doses of alleles are not indicated. The loci *α-Gpdh, Ldh, Mdh-1, Mdh-2, G3pdh, Sod, Hk-1,* and *Mpi* are monomorphic for the same allele in the two populations.

Loci	Genotypes	*L. hispanica* A	*L. hispanica* D
Sdh	85/100	—	1.00
	100/100	1.00	—
Idh-1	95/95	1.00	—
	95/105	—	1.00
6Pgdh	77/84	0.11	—
	77/91	0.09	—
	84/84	0.23	—
	84/91	0.40	—
	84/100	—	1.00
	91/91	0.17	—
Got-1	108/108	0.04	—
	108/116	0.32	1.00
	116/116	0.64	—
Got-2	85/85	1.00	—
	85/95	—	1.00
Hk-2	96/106	0.07	—
	96/112	0.05	—
	106/106	0.31	—
	106/112	0.43	—
	112/112	0.14	1.00
Adk	93/100	—	1.00
	100/100	1.00	—
Pgm	115/115	0.03	—
	115/120	0.28	—
	120/120	0.69	1.00
Gpi	97/105	—	1.00
	105/105	1.00	—

hispanica D were not found in *L. hispanica* A; if they were mutants, a much higher genetic heterogeneity would be expected in *L. hispanica* D. The hybrid hypothesis again correlates well with the observed data.

Carausius morosus

Carausius morosus is a parthenogenetic species introduced into Europe from the native Palmi Hills (southern India) at the end of the last century; since that time, it has been bred in many laboratories throughout the world. The karyotype of this species is presumed to be triploid, with 64 chromosomes (Pijnacker and Harbott 1980). Oocytes undergo a chromosome doubling during prophase of the first meiotic division, followed by a normal meiosis that restores the somatic number of chromosomes (Koch *et al.* 1972). The offspring are genetically identical with their mother (Pijnacker 1966).

C. morosus is highly heterozygous both at the chromosome and gene levels; about two-thirds of the loci studied were found to be fixed in heterozygous condition, as shown in Table 14.5. Hybrid origin appears to be probable; sexual *Carausius* species closely related to *C. morosus* live in the area where the latter was first found, and their males seem to be able to mate with non-conspecific females (Pantel 1917). Unfortunately, no isoenzyme work has been carried out so far on these bisexual species, which are also poorly studied at the chromosome level.

Table 14.5 Electrophoretic phenotypes at 13 enzyme loci in *Carausius morosus*.

Loci	Phenotypes	Loci	Phenotypes
α-Gpdh	100/110	*Got-1*	100
Hbdh	100/110	*Got-2*	100/105
Mdh-1	100/115	*Hk-1*	100
Mdh-2	100	*Adk*	100/108
Idh-1	100	*Pgm*	100
6Pgdh	100/108	*Mpi*	100/120
G3pdh	100/110		

Grasshoppers

Warramaba virgo

Warramaba virgo is a parthenogenetic wingless grasshopper belonging to the family Eumastaciidae. This all-female species inhabits

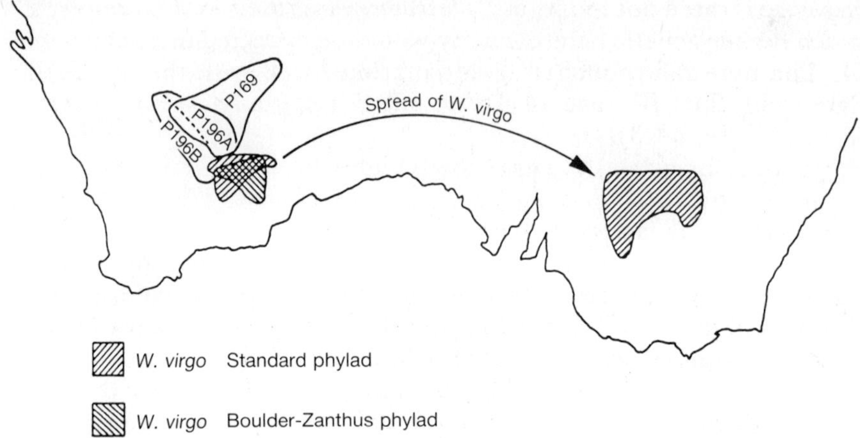

Fig. 14.3. Southern Australia, showing the areas inhabited by the two phylads of *Warramaba virgo* (Standard and Boulder–Zanthis), together with the approximate ranges of their parental species P169 and P196 (with the two races, A and B). From its origin in western Australia the Standard phylad is believed to have migrated across the continent to eastern Australia and then undergone extinction in the central part of its range. The Boulder–Zanthus phylad never crossed the continent. (Modified from White 1980.)

two areas of the arid zone of Australia, one in the east, the other in the west, separated by a gap of 177 km (Fig. 14.3). It feeds on several species of shrubs belonging to the genus *Acacia* and lays its eggs in sandy soil. *W. virgo* arose by hybridization between two related bisexual species, as yet unnamed and provisionally indicated as P169 and P196 (Hewitt *et al.*, 1975; White *et al.* 1977). The genetic distance between these parental species, calculated using Nei's formula on the basis of 38 gene loci, is $D = 0.38$ (White, pers. comm.).

W. virgo consists of several clones, at least 27, which fall into two groups: the Standard phylad (including 25 clones) and the Boulder–Zanthus phylad (with 2 clones). The two phylads differ in their external morphology, in many cytogenetic features, and at the molecular level — both in enzyme loci and DNA and RNA structure (Webb *et al.* 1978; White 1980; White and Contreras 1981, 1982; White *et al.* 1982). They originated in south-western Australia from two independent hybridization events occurring at different times and involving, as indicated by genetic data, two distinct races of the species P196. The most ancient is the Standard phylad, originating from the cross between P169 and the 'A' race of P196. This phylad migrated across the continent to New South Wales and then underwent extinction in the central part of its range. More recent is the Boulder–Zanthus phylad, originating from the cross between P169 and the 'B' race of P196, inhabiting only south-western Australia (Fig. 14.3).

The two phylads of *W. virgo* show a highly heterozygous karyotype (15 chromosomes in most clones), which can be related directly to the karyotype of the two bisexual parental species on the basis of chromosome morphology and C-banding patterns (Fig. 14.4). The cytological mechanism of parthenogenesis involves a premeiotic doubling of the chromosome number, followed by synapsis between the sister chromosomes, formation of 15 bivalents, and two meiotic

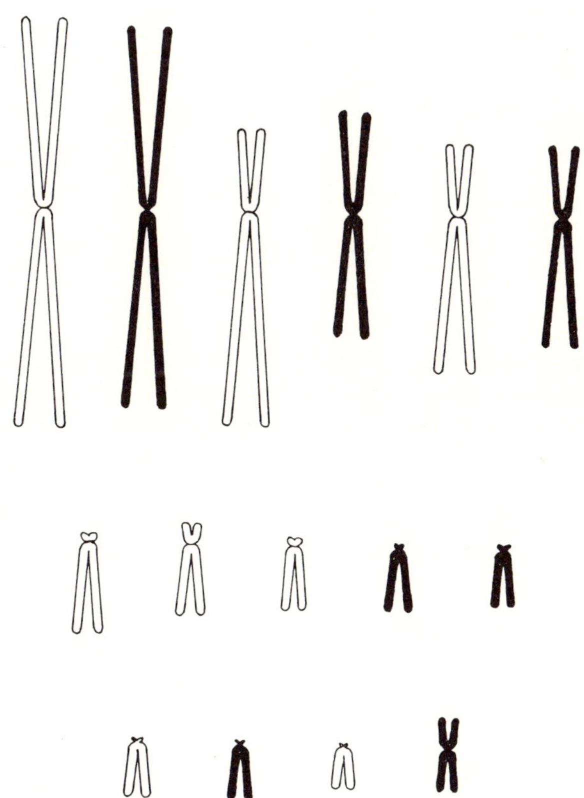

Fig. 14.4. The 'Standard' karyotype of *Warramaba virgo*. Chromosomes derived from species P169 in outline, those derived from P196 in black. (Modified from White 1980.)

divisions restoring the diploid condition. This mechanism preserves the heterozygosity from one generation to the next, and produces offspring genetically identical to their mother (White 1966, 1980).

White and co-workers obtained 'synthetic' *W. virgo* by crossing the two parental species (White *et al.* 1977; White and Contreras 1978). Synthetic male individuals were totally sterile, whilst synthetic females occasionally produced a few eggs that developed parthenogenetically into embryos with the same karyotype of their mothers. When *W. virgo* is crossed with males of either P169 or P196, triploid offspring are obtained. Triploid males are sterile, and females have a much reduced fecundity. The latter can reproduce by parthenogenesis, showing the same premeiotic doubling mechanism as *W. virgo* (White 1980).

The absence of overlapping areas between the present range of *W. virgo* and that of its sexual ancestors seems to indicate that competition has been taking place between hybrid and non-hybrid species. The successful spread of *W. virgo* in wide areas unfavourable to its parental species appears to be related mainly to the heterotic advantage involved in its hybrid origin.

Saga pedo

A hybrid origin was considered by Matthey (1948) in discussing the evolution of the parthenogenetic long-horned grasshopper, *Saga pedo* (Tettigonoidea, Sagidae). This all-female species is found in central and southern Europe, from Spain to Russia, while the range of its sexual relatives includes mainly the regions around the eastern Mediterranean (i.e. the Balkans, Syria, Lebanon, Palestine, Israel, etc.).

S. pedo is apparently tetraploid ($4n = 68$), with no meiotic reduction of the chromosomes (Matthey 1941, 1948). The chromosome number in males of the sexual species studied so far is $2n = 31$ $(30 + x)$ in *S. gracilipes* (Matthey 1946) and *S. cappadocica* (Matthey 1950), and $2n = 33$ $(32 + x)$ in *S. epiphiggera* (Matthey 1946). At the karyotype level, none of these species appears to be one of the sexual ancestors of *S. pedo*. However, many other sexual *Saga* species exist that have not yet been investigated cytologically. Moreover, neither *S. pedo* nor any of its sexual relatives have until now been analysed at the isoenzyme level. The data so far available do not, therefore, allow us to draw any conclusion on the hybrid origin of *S. pedo*.

Weevils

Many Curculionid species are polyploid and reproduce by (supposedly apomictic) parthenogenesis (Suomalainen and Saura 1973; Suomalainen *et al*. 1976). A number of these polyploid weevils, from central and northern Europe, were studied electrophoretically (Saura *et al*. 1976*a, b*; Lokki *et al*. 1976; Lokki and Saura 1980). Their hybrid origin was considered by these authors as highly improbable, since the differences found between populations could in general be explained by single mutations, and few alleles alien to the bisexual race were found. Accordingly, these authors favoured an origin of these forms as parthenogenetic derivatives from a sexual species. In our opinion, many of the data reported (e.g. widespread fixed heterozygosity for the same alleles at a number of loci) agree well with the hybrid hypothesis.

Recent studies on populations of the *pupillatus* and *cribricollis* groups from Italy and northern Africa show a similar picture. The polyploid parthenogens *Otiorrhynchus pupillatus* and *O. pasubianus* have fixed heterozygosity at most of the loci analysed. They seem to be derived from crosses between *O. cyclophthalmus* and two species of the *O. pupillatus* complex, not yet detected or maybe extinct (Magnano *et al*. 1983, and our unpublished data). Again for *O. cribricollis*, a parthenogenetic triploid species, evidence was found of a hybrid origin, involving the sexual parental species *O. reticollis* and a new taxon recently discovered by Magnano in Tunisia (Bullini 1985, and our unpublished data).

Gynogenetic insects of hybrid origin

A triploid gynogenetic species was discovered by Drosopoulos (1976) in the planthopper genus *Muellerianella* (Homoptera, Delphacidae). This all-female taxon presumably arose through hybridization between the originally south-European species *M. fairmairei* and the more northern species, *M. brevipennis* (Drosopoulos 1977, 1978; Booij 1981, 1982). This hybridogen appears to have two genomes of *M. fairmairei* and one of *M. brevipennis*; accordingly, it was named *M. -2-fairmairei-brevipennis*. This latter hybrid species depends on the sperm of either *M. fairmairei, M. brevipennis*, or *M. extrusa* to activate egg development, but is more frequently associated with the first one, sharing the same host-plant species, the grass *Holcus lanatus*. Its range approximately coincides with the area of sympatry between the two parental species. The frequency of gynogens relative to the sexual species providing sperm differs considerably geographically, but appears temporally stable at most sites. It tends to increase with altitude, latitude, and wetness of climate, and to decrease with the length of the growing season. At a more local scale, the gynogens appear to be favoured in cool, wet, and disturbed habitats, such as meadows, whereas the sexual species prevails in warmer, moist, and less-disturbed habitats, such as pastures (Booij and Guldemond 1984).

M.-2-fairmairei-brevipennis was successfully produced in the laboratory by crossing *M. fairmairei* females × *M. brevipennis* males, then backcrossing F_1 females with *M. fairmairei* males. A few progeny were obtained, all of which were triploid females, from unreduced diploid eggs. When crossed with *M. fairmairei*, these females produced gynogenetically numerous offspring, very similar to the wild gynogens (Drosopoulos 1978).

Another triploid gynogenetic insect is the spider beetle, *Ptinus mobilis* (Moore *et al.* 1956; Sanderson and Jacobs 1957; Sanderson 1960). This all-female species produces unreduced triploid eggs. Apparently, the oocyte nucleus undergoes an endomitotic doubling of chromosome number after the last oogonial division, thereafter followed by a single maturation division, with formation of 27 'pseudobivalents'. However, the egg does not develop beyond the metaphase unless it is activated by sperm of the related diploid sexual species *P. claviceps*. A hybrid origin for *P. mobilis* has not been demonstrated, but appears likely.

Discussion and conclusions

Two main problems are involved in the evolution of hybrid species: their origin, and their adaptive significance. The origin of an animal hybridogen requires (1) an interspecific hybridization, and (2) changes in the maturation divisions allowing the transmission of the hybrid genome to the next generation. The latter (premeiotic doubling of the chromosomes, fusion of the pronucleus with a polar body, etc.) are apparently not caused by hybridization itself, and their occurrence is undoubtedly rare, as indicated by the frequent sterility of 'synthetic' hybridogens. High rates of interspecific hybridization are, therefore, required. They readily occur when two previously allopatric species come into contact, without having developed effective premating barriers, or when the latter break down, e.g. owing to habitat disturbance (Remington 1968). Such an example is that of the sexual ancestors of *B. whitei* — *B. rossius* and *B. grandii* — which presumably came into contact when the barriers between the Iblea region (inhabited by *B. grandii*) and the rest of Sicily (inhabited by *B. rossius*) broke down; these two species still interbreed easily, apparently not having yet evolved effective antihybridization mechanisms.

The evolutionary success of animal hybrid species seems to involve two distinct phenomena: the so-called 'demographic' advantage, and the 'heterotic' advantage. The former, shared with the other unisexual species, consists in the doubling of the intrinsic rate of increase with respect to that of their sexual ancestors. As to heterosis, interspecific hybridization dramatically enhances heterozygosity within a single generation, especially when the parental species are well-differentiated genetically. These selective advantages may well explain the high competitiveness generally shown by insect hybrid species, which frequently out-compete their sexual ancestors, reducing the latter's ranges or even causing their extinction (e.g. *B. atticus* versus its parental species). Moreover, hybridogens often successfully colonize wide areas not inhabited by their parental species (as in the case of *W. virgo*), and easily shift to new ecological niches. These factors are probably the main reasons for the occurrence of hybrid pest species of agricultural and medical importance.

When the bisexual ancestors are unknown, it is often difficult to distinguish between hybrid species and non-hybrid parthenogens showing high levels of heterozygosity, produced by mutation both at the gene and chromosome level. It should now be possible to solve such puzzling cases by combined use of the new techniques made available by cytogenetics, biochemical genetics, and molecular biology.

References

Booij, C.J.H. (1981). Biosystematics of the *Muellerianella* complex (Homoptera, Delphacidae), taxonomy, morphology and distribution. *Neth. J. Zool.*, **31**, 572–95.

Booij, C.J.H. (1982). Biosystematics of the *Muellerianella* complex (Homoptera, Delphacidae): host plants, habitats and phenology. *Ecol. Ent.*, **7**, 9–18.

Booij, C.J.H. and Guldemond, J.A. (1984). Distributional and ecological differentiation between asexual gynogenetic planthoppers and related sexual species of the genus *Muellerianella* (Homoptera, Delphacidae). *Evolution*, **38**, 163–75.

Bullini, L. (1982). Genetic, ecological, and ethological aspects of the speciation processes. In *Mechanisms of speciation* (ed. C. Barigozzi), pp. 241–64. Alan R. Liss Inc., New York.

Bullini, L. (1983). Taxonomic and evolutionary inference from electrophoretic studies on various animal groups. In *Adaptation and taxonomic significance of protein variation* (ed. D. Rollinson and G.S. Oxford), pp. 179–92, Academic Press, London.

Bullini, L. (1985). Speciation by hybridization in animals. *Boll. Zool.*, **52**, 121–37.

Bullini, L. and Bianchi Bullini, A.P. (1971). Ricerche sulla riproduzione e sul corredo cromosomico del fasmide *Clonopsis gallica* (Cheleutoptera, Bacillidae). *Acc. Naz. Lincei, Rend. Cl. Sc. fis. mat. nat., Ser. VIII*, **51**, 563–9.

Bullini, L. and Nascetti, G. (1986). Speciation by hybridization in stick insects (Insecta, Phasmatodea). *Proceedings of the 1st International Symposium on Stick Insects, Siena* (ed. M.Mazzini and V.Scali), pp. 179–94. University of Siena Publications.

Bullini, L. (1987). Genetic and taxonomic studies on the genus *Clonopsis*, with the description of a new species (Phasmatodea, Bacillidae). *Boll. Ist. Ent. Univ. Bologna*, **4**, 325–53.

Bullini, L., Nascetti, G., and Bianchi Bullini, A.P. (1984). A new stick-insect of hybrid origin: *Bacillus lynceorum* n.sp. (Cheleutoptera, Bacillidae). *Acc. Naz. Lincei, Rend. Cl. Sc. fis. mat. nat., Ser. VIII*, **75**, 169–76.

Cappe de Baillon, P. and de Vichet, G. (1940). La parthénogenèse des espéces du genre *Leptynia* Pant. (Orthopt. Phasmidae). *Bull. biol. Fr. Belg.*, **74**, 43–87.

Drosopoulos, S. (1976). Triploid pseudogamous biotype of the leafhopper *Muellerianella fairmairei. Nature, Lond.*, **263**, 499–500.

Drosopoulos, S. (1977). Biosystematics studies on *the Muellerianella* complex (Delphacidae, Homoptera, Auchenorryncha). *Meded.LandbHoogesch., Wageningen*, **77**, 1–33.

Drosopoulos, S. (1978). Laboratory synthesis of a pseudogamous triploid 'species' of the genus *Muellerianella* (Homoptera, Delphacidae). *Evolution*, **32**, 916–20.

Gasperi, G., Malacrida, A., Mantovani, B., and Scali, V. (1986). Preliminary data on electrophoretic multilocus analysis of eastern taxa of the genus Bacillus (Phasmatodea, Bacillidae). *Proceedings of the 1st International Symposium on Stick Insects, Siena* (ed. M.Mazzini and V.Scali), pp. 195–201, University of Siena Publications.

Goday, C., Bianchi Bullini, A.P., Nascetti, G., and Bullini, L. (1982). Chromosome studies on *Bacillus atticus, B. rossius*, and their hybrids (Cheleutoptera, Bacillidae). *Acc. Naz. Lincei, Rend. Cl. Sc. fis. mat. nat., Ser. VIII*, **71**, 126–33.

Hewitt, G.M. (1975). A new hypothesis for the origin of the parthenogenetic grasshopper *Moraba virgo. Heredity*, **34**, 117–36.

Koch, P., Pijnacker, L.P., and Kreke, J. (1972). DNA reduplication during meiotic prophase in the oocyte of *Carausius morosus* Br. *Chromosoma*, **36**, 313–21.

Lokki, J., and Saura, A. (1980). Polyploidy in insect evolution. In *Polyploidy: Biological relevance* (ed. W.H. Lewis), pp. 277–312. Plenum Press, New York.

Lokki, J., Suomalainen, E., Saura, A., and Lankinen, P. (1976). Genetic polymorphism and evolution in parthenogenetic animals. VI. Diploid and triploid *Polydrosus mollis* (Coleoptera: Curculionidae). *Hereditas*, **82**, 209–16.

Magnano, L., Nascetti, G., Cianchi, R., and Bullini, L. (1983). Primo contributo alla tassonomia biochimica dei gruppi *pupillatus* e *cribricollis* (genere *Otiorrhynchus*). *Atti XIII Congr. naz. ital. Entomol., Sestriere* (1983), pp. 535–38.

Marescalchi, O., Corni, M.G., and Scali, V. (1985). I cromosomi di tre specie di *Bacillus* (Insecta Phasmatodea) del Mediterraneo orientale. *Atti XIV Congr. naz. ital. Ent., Palermo, Erice, Bagheria* (1985), 201–7.

Matthey, R. (1941). La cytologie de la parthénogenèse chez *Saga pedo* Pall. *Revue suisse Zool.*, **48**, 523–4.

Matthey, R. (1946). Démonstration du caractère géographique de la parthénogenèse de *Saga pedo* Pallas et de sa polyploidie, par comparaison avec les espèces bisexuées *S. ephipiggera* Fisch. et *S. gracilipes* Uvar. *Experientia*, **2**, 1–3.

Matthey, R. (1948). Données nouvelles sur les chromosomes des Tettigonides et la parthénogenèse de *Saga pedo* Pallas. *Revue suisse Zool.*, **55**, 45–56.

Matthey, R. (1950). Les chromosomes de *Saga cappodocica* Werner (Orthoptera–Tettigoniidae). *Neunter Jb. schweiz. Ges. f. Vererbungsforsch.*, **25**, 44–6.

Moore, B.P., Woodroffe, G.E., and Sanderson, A.R., (1956). Polymorphism and parthenogenesis in a Ptinid beetle. *Nature, Lond*, **177**, 847–8.

Nascetti, G. and Bullini, L. (1982*a*). *Bacillus grandii* n.sp. and *B. whitei* n.sp.: two new stick insects from Sicily (Cheleutoptera, Bacillidae). *Boll. Ist. Ent. Univ. Bologna*, **36**, 245–58

Nascetti, G. and Bullini, L. (1982*b*). A new phasmid from Italy: *Bacillus atticus caprai* (n. subsp.) (Cheleutoptera, Bacillidae). *Fragm. Ent.*, **16**, 143–59.

Nascetti, G. and Bullini, L. (1983). Differenziamento genetico e speciazione in fasmidi dei generi *Bacillus* e *Clonopsis* (Cheleutoptera, Bacillidae). *Atti XII Congr. naz. ital. Entomol., Roma* (1980) Vol. II, pp. 215–23.

Nascetti, G. and Bullini, L. (1988). Struttura genetica di *Bacillus atticus* (Phasmatodea, Bacillidae). *Atti XV Congr. naz. ital. Entomol., L'Aquila (1988)*, pp. 247–53.

Nascetti, G., Bianchi Bullini, A.P., and Bullini, L. (1983). Speciazione per ibridazione nei fasmidi del bacino del Mediterraneo (Cheleutoptera: Bacillidae). *Atti XIII Congr. naz. ital. Entomol., Sestriere* (1983), pp. 475–8.

Nascetti, G., Bianchi, A.P., and Bullini, L. (1985). Speciation by hybridization in the stick-insects *Bacillus whitei* and *B. lynceorum* (Cheleutoptera: Bacillidae). *Atti Ass. genet. Ital.*, **31**, 135–6.

Pantel, P.J. (1917). Description de *Carausius* nouveaux (Orth. Phasmidae) et notes sur les *Carausius* de l'Inde meridionale. *Ann.Soc. ent. Fr.*, **86**, 267–306.

Pijnacker, L. P. (1966). The maturation divisions of the parthenogenetic stick insect *Carausius morosus* Br. (Orthoptera, Phasmidae). *Chromosoma*, **19**, 99–112.

Pijnacker, L. P. and Harbott, J. (1980). Structural heterozygosity and aneuploidy in the parthenogenetic stick insect *Carausius morosus* Br. (Phasmatodea: Phasmatidae). *Chromosoma*, **76**, 165–74.

Remington, C. L. (1968). Suture-zones of hybrid interaction between recently joined biotas. *Evol. Biol.*, **2**, 321–428.

Sanderson, A. R. (1960). The cytology of a diploid spider beetle, *Ptinus clavipes* Panzer and its triploid gynogenetic form *mobilis* Moore. *Proc. R. Soc. Edin.*, **67**, 333–50.

Sanderson, A. R. and Jacob, J. (1957). Artificial activation of the egg in a gynogenetic spider beetle. *Nature, Lond.*, **179**, 1300.

Saura, A., Lokki, J., Lankinen, P., and Suomalainen, E. (1976*a*). Genetic polymorphism and evolution in parthenogenetic animals. III. Tetraploid *Otiorrhynchus scaber* (Coleoptera: Curculionidae). *Hereditas*, **82**, 79–100.

Saura, A., Lokki, J., Lankinen, P., and Suomalainen, E. (1976b). Genetic polymorphism and evolution in parthenogenetic animals. IV. Triploid *Otiorrhynchus salicis* (Coleoptera: Curculionidae). *Ent. Scand.*, **7**, 1–6.

Scali, V. and Marescalchi, O. (1987). The evolution of the genus *Bacillus* (Phasmatodea, Phasmatidae) from a karyological point of view. In *Evolutionary biology of Orthopteroid insects* (ed. B. Baccetti), pp. 541–9, Ellis Horwood, Chichester, UK.

Suomalainen, E. and Saura, A. (1973). Genetic polymorphism and evolution in parthenogenetic animals. I. Polyploid Curculionidae. *Genetics*, **74**, 489–508.

Suomalainen, E., Saura, A., and Lokki, J. (1976). Evolution of parthenogenetic insects. *Evol. Biol.*, **9**, 209–57.

Webb, G. C., White, M. J. D., Contreras, N., and Cheney, J. (1978). Cytogenetics of the parthenogenetic grasshopper *Warramaba* (formerly *Moraba*) *virgo* and its bisexual relatives. IV. Chromosome banding studies. *Chromosoma*, **67**, 309–39.

White, M. J. D. (1966). Further studies on the cytology and distribution of the Australian parthenogenetic grasshopper *Moraba virgo*. *Revue suisse Zool.*, **73**, 2383–98.

White, M. J. D. (1978). *Modes of speciation* (ed. C. I. Davern). W. H. Freeman, San Francisco.

White, M. J. D. (1980). The genetic system of the parthenogenetic grasshopper *Warramaba virgo*. In *Insect cytogenetics*: Tenth Symposium of the Royal Entomological Society of London (ed. M. Ashburner and G. M. Hewitt), pp. 119–31. Blackwell, Oxford.

White, M. J. D. and Contreras, N. (1978). Cytogenetics of the parthenogenetic grasshopper *Warramaba* (formerly Moraba) virgo and its bisexual relatives. III. Meiosis of male 'synthetic virgo' individuals. *Chromosoma*, **67**, 55–61.

White, M.J.D. and Contreras, N. (1981). Chromosome architecture of the parthenogenetic grasshopper *Warramaba virgo* and its bisexual ancestors. *Chromosome Today*, **7**, 165–75.

White, M.J.D. and Contreras, N. (1982). Cytogenetics of the parthenogenetic grasshopper *Warramaba virgo* and its bisexual relatives. VIII. Karyotypes and C-banding patterns in the clones of *W. virgo*. *Cytogenet. cell. Genet.*, **34**, 168–77.

White, M.J.D., Contreras, N., Cheney, J., and Webb, G.C. (1977). Cytogenetics of the parthenogenetic grasshopper *Warramaba* (formerly *Moraba*) *virgo* and its bisexual relatives. II. Hybridization studies. *Chromosoma*, **61**, 127–48.

White, M.J.D., Dennis, E.S., Honeycutt, R.L., Contreras, N., and Peacock, W.J. (1982). Cytogenetics of the parthenogenetic grasshopper *Warramaba virgo* and its bisexual relatives. IX. The ribosomal RNA cistrons. *Chromosoma*, **85**, 181–99.

15 Electrophoretic variation in post-harvest agricultural pests, and its implications

DAVID WOOL

Department of Zoology, Tel Aviv University, Tel Aviv, Israel

Abstract

Among organisms that attack stored commodities, insects have a prominent role, in particular beetles (Coleoptera) and moths (Lepidoptera). These insects have adapted themselves to an ecosystem created by man. The original environment (and food) in which their evolution took place can often only be guessed at. It is very likely that the domesticated species today are genetically unlike their ancestors.

Three processes may shape the genetic structure of these pest populations. These include: (1) founder effects — since new populations are established by small numbers of immigrants, genetic differences among populations can be expected; (2) migration between warehouses, countries, and even continents, which tends to reduce the genetic differences among populations; and (3) 'domestication', including all genetic changes that follow the transition from wild to human habitats (mainly from inbreeding and selection).

The extent of genetic variation in pest populations may have important implications for their control. The widespread resistance to pesticides in many stored-products pests demonstrates that they do possess genetic variation. Electrophoretic studies provided estimates of genetic variation in many organisms, but very few stored-products pests have to date been investigated electrophoretically. An extensive literature search for information of stored-products beetles and moths (representing 14 genera), yielded 34 papers and abstracts dealing with only 12 species (belonging to 6 genera). The majority of these papers

Electrophoretic Studies on Agricultural Pests (ed. Hugh D. Loxdale and J. den Hollander), Systematics Association Special Volume No. 39, pp. 341–62. Clarendon Press, Oxford, 1989.

are limited to the study of single enzyme systems, focusing either on the inheritance patterns of isoenzymes or on the physiological or biochemical properties of the enzymes. Population genetic data relevant to the questions of evolution and domestication are available for only one genus, *Tribolium*. These data were re-analysed to provide estimates of genetic diversity among and within strains of two *Tribolium* species. The results showed that 14–16 per cent of total diversity is among strains and that the remaining 84–86 per cent is allocated to variation among individuals within strains.

These results are compared with experimental data on differentiation of *Tribolium* populations under extreme conditions, and with electrophoretic data on island populations of other organisms. The final conclusion is that migration seems to be a stronger factor in determining genetic variation in these post-harvest pests than inbreeding and founder effects.

The ability to infest new commodities in the future, and to overcome future control measures, depends on this variation. It is very important to estimate the extent of genetic variation in more populations and species of such pests.

Introduction

Many organisms, particularly insects, thrive on products stored by man. Most of the insects attacking stored agricultural commodities belong to two groups: beetles (Order Coleoptera) and moths (Order Lepidoptera).

Pooling these diverse organisms under one heading (post-harvest pests) for the present discussion is justified for two reasons. The first is, of course, that they all live in human grain stores and other warehouses and that they pose a threat to human economy, on global as well as local levels. Considerable efforts are spent annually on the application of pesticides, fumigants, pheromone traps, and other methods to control these pests: costs have continued to increase because of the development of insecticide resistance in many major pests consequent upon continued spraying (e.g. in *Tribolium*: Dyte and Blackman 1970; in *Ephestia cautella, Plodia interpunctella*, and related moths: Attia *et al*. 1979; Attia 1981; Zettler 1982). The second reason is that all these organisms are 'domesticated' species. Their original habitat can only be guessed at (the recent discovery of *Tribolium brevicornis* in a natural habitat in California provides interesting data on this point (cf. Sokoloff *et al*. 1983)). Their association with human habitats is as old as man's food gathering activities (stored-product pests have been discovered in Egyptian tombs 5000 years old (Levinson and Levinson 1985)), although, on the evolutionary time scale, the domestication of stored-products insects has taken but a brief moment.

Domestication is familiar to most people as the deliberate process of selecting strains of plants and animals from wild ancestors that possess traits beneficial to man. The domestication of the dog, cow, horse, and laboratory rat and mouse are obvious examples; similarly, in the plant world, the domestication of wheat, corn, rice, tomatoes, and many other crops. Domestication, however, has often happened without active participation of man and even against his wishes, as a result of ecological niches being created or destroyed by human activities. For example, the house mouse, house sparrow, housefly, and cockroach — as well as post-harvest pests — are now human commensals, i.e. they share our food and lodgings, much to our distress. Once the new niche became available, domestication went on by natural processes.

Whether it occurs naturally, or by deliberate human efforts, domestication involves genetic changes in the populations. In the latter case, these are mainly due to the artificial selection of desirable traits, but, as plant and animal breeders know all too well, the consequent inbreeding (owing to the small numbers of selected parents) is a powerful modifying factor that often causes genetic changes in selected populations against the wishes of the breeder. In the case of domesticated pest populations, the processes of both natural selection and inbreeding must have been at work.

Present-day post-harvest pest insect populations are often founded by small numbers of individuals — either those brought into the warehouse with the stored commodity or those that find their own way into it. 'Founder effects' (Mayr 1963) may cause genetic differences among populations in different storage facilities. This process will be enhanced by the inevitable inbreeding, and possibly by selection on different foods and in different localities. On the other hand, migration of insects between warehouses, brought about by flying or crawling adults, or by shipping commodities in human domestic and world trade, will tend to make the different populations more similar to each other and reduce genetic differentiation.

We do not know what the level of genetic variation was in these pests before domestication, but we do have some evidence that domestication causes considerable genetic changes in these insects. For example, Waterhouse and Nowosielski-Selpowron (1964) discovered that freshly collected stocks of the flour beetle *Cathartus quadricollis* could survive and develop in a considerably wider range of temperatures and humidities than the laboratory stock (16 years in culture). Toro and Lopez-Fanjul (1979) found that mean pupal weights in six strains of *Tribolium castaneum*, collected from granaries in Spain, were significantly lower than in two laboratory strains (10 years in culture). In the first four generations after transfer to the laboratory, mean pupal weights did not change, although the phenotypic and genetic variances of the character increased, perhaps owing to the relaxation of natural selection under the optimal laboratory conditions. More recently, Mulder and Sokoloff (1982) compared one feral and two domesticated strains of *Tribolium brevicornis* when cultured in the laboratory on the same flour medium. Although their results are based on very small samples, it is interesting that the feral strain laid considerably fewer eggs, and survival to adulthood was lower than in the domesticated strains. The field collected beetles were twice as variable in weight as the domesticated strains, but this variation was not genetic. The feral population also showed higher egg-cannibalism and dispersal tendency than the laboratory strains. In my own recent experience, strains of the almond moth (*Ephestia cautella*) that were highly resistant to malathion when collected in the USA in the 1970s have lost this property when tested after 10–15 years of laboratory rearing (Wool and Brower, unpublished).

It would be useful to measure present levels of genetic variation in stored-products pests, because this level determines their ability to respond to future control measures (like new pesticides) and to adapt to new kinds of human commodities that may become available in the future. It is interesting to estimate the relative impact of inbreeding and migration on the genetic structure of these pest populations.

The widespread use of electrophoresis in population genetics since 1966 has enabled the accumulation of data on genetic variation in hundreds of animal and plant species. The most recent survey (Nevo *et al.* 1984) lists electrophoretic information on 1111 species, among them 166 species of insects. Only one stored-products pest is listed (*Ephestia kuehniella*), whereas 34 are species of *Drosophila*. It is interesting to speculate why post-harvest pests were not studied more intensively using electrophoresis, a method that has become very fashionable in the last 20 years. Possibly, researches working with these insects have been unaware of the potential value of estimates of genetic variation in these populations.

This chapter is divided into two parts. The first is a review of the literature on electrophoretically detectable proteins in stored-products pests. The second is a new analysis of data on two species, *Tribolium castaneum* and *T. confusum*, from my own laboratory. I shall use these data to estimate the relative importance of inbreeding and migration in shaping the genetics of post-harvest pest populations.

Part 1: Literature review

Literature search

The earliest papers on electrophoretic variation in populations were published around 1966 (cf. Hubby and Lewontin 1966; Lewontin and Hubby 1966). In the present study, data were gathered by a computer search of the BIOSIS data file, in which most biological journals have been catalogued since 1970. The search asked for any publication that included, either in the title or in the abstract, at least one of the following terms: electrophoresis, electrophoretic, isozymes, allozymes, isoenzymes, in connection with any one of 16 insect genera (Table 15.1; three of the names are synonymous, so actually only 14 genera were included). I also searched the computer-listed papers and my own files for earlier references. Finally, I wrote

Table 15.1 List of the insects investigated. Species for which at least one references was found are listed. (–) indicates no data for the genus.

Order	Family	Genus	Species
Coleoptera	Tenebrionidae	*Alphitobius*	–
		Tenebrio	*T. molitor* L.
		Tribolium	*T. castaneum* (Herbst)
			T. confusum Du Val
			T. brevicornis Le Conte
			T. destructor Uyttenb.
	Dermestidae	*Attagenus*	*A. megatoma* (Fabr.)
		Trogoderma	–
	Cucujidae	*Oryzaephilus*	–
		Cryptolestes	–
	Curculionidae	*Sitophilus*	*S. granarius* (L.)
			S. oryzae (L.)
			S. zeamais (Motsch.)
	Bostrychidae	*Rhyzopertha*	–
	Bruchidae	*Callosobruchus*	–
	Anobiidae	*Lasioderma*	*L. serricorne* (Fabr.)
Lepidoptera	Gelechiidae	*Sitotroga*	–
	Pyralidae	*Plodia*	*P. interpunctella* (Hubner)
		Anagasta (= Ephestia)	*A. kuehniella* (Zeller)
		Ephestia (= Cadra)	*E. cautella* (Walker)

to 15 colleagues who worked on stored-products pests, asking them for references I might have missed (six of these colleagues replied and I am grateful for their help).

The effort yielded disappointingly little information (Table 15.2). A total of 34 publications was collected, very unequally distributed among six genera. In the other eight, no information on any use of electrophoresis was published. Within the six genera, most of the papers deal with two species of beetles — both of the genus *Tribolium* and one species of moth, *Ephestia* (= *Anagasta*) *kuehniella*. Some details on the available data for each species are given below. Enzyme names are abbreviated to save space. The full names are listed in the heading for Table 15.3.

Results

Tenebrio molitor Only one enzyme system, amylase, was studied. Four isoenzymes were detected electrophoretically in homogenates of larval midguts (pooled from four larvae) (Applebaum *et al.* 1961). This was a physiological investigation and no genetic information is available.

Tribolium castaneum Several electrophoretic studies were performed on this species (Table 15.3), although some of them are published only as preliminary notes. Lahr and Costantino (1973) used disc-gel electrophoresis and a general protein stain on pooled homogenates of several individuals of two strains. They reported the presence of many bands, and some differences between strains. Yeh and Scheinberg (1972) reported a two-allele genetic polymorphism at the ADH locus in a strain of this species, of which they analysed 2500 beetles. However, only an abstract of this work was published. Two years later, the same authors (1974) reported good results with four enzyme systems, again using the disc-gel apparatus. In adult homogenates, they discovered three isoenzymes of ADH, five of LDH, three of IDH and two of G6PD. In my laboratory, using acrylamide slab gels and homogenates of adults, we failed to obtain any staining for these systems (Sverdlov and Wool 1978). We did, however, get good staining with EST and ACPH. We did not succeed in working out the genetics of the EST isoenzymes, but discovered a sex-related, two-allele locus in ACPH. We described the ontogenetic patterns of the major EST isoenzymes, which are active in larvae and adults but almost disappear in the pupa (Sverdlov *et al.* 1976; Cohen *et al.* 1977; Sverdlov and Wool 1978).

Table 15.2 Distribution of papers and abstracts on electrophoresis in post-harvest pests (in alphabetical order).

Genus	Number of species studies	Number of publications	Main purpose of research			
			Population variation	Genetics	Physiology	Other
Alphitobius	—	—	—	—	—	—
Anagasta	See *Ephestia*					
Attagenus	1	1	—	—	—	—
Cadra	See *Ephestia*					
Callosobruchus	—	—	—	—	—	—
Cryptolestes	—	—	—	—	—	—
Ephestia	1	12	—	9	—	3
Lasioderma	—	—	—	—	—	—
Oryzaephilus	—	—	—	—	—	—
Plodia	1	4	—	2	—	2
Rhyzopertha	—	—	—	—	—	—
Sitophilus	3	4	1	—	3	—
Sitotroga	—	—	—	—	—	—
Tenebrio	1	1	—	—	1	—
Tribolium	4	11	2	4	1	4
Trogoderma	—	—	—	—	—	—
Total	12	34	3	15	5	9

Table 15.3 Summary of electrophoretic information on 12 species of stored-products insects.

Organism	Enzyme system analysed for: Physiology, biochemistry	Development pattern	Genetics of isoenzymes	Population genetics	Other (preliminary)	Reference
Tenebrio molitor	AMY					Applebaum *et al.* (1961)
Tribolium castaneum					Proteins	Lahr and Costantino (1973)
					IDH,ADH,LDH,G6PD	Yeh and Scheinberg (1974)
					EST	Sverdlov, Wool and Cohen (1976)
					EST,ACPH	Wool and Sverdlov (1976)
		EST				Cohen, Sverdlov and Wool (1977)
				ADH		Yeh and Scheinberg (1972)
				EST,ACPH, AMY,MDH, AO		Wool (1982)
			PGM	PGM		Riddle, Iverson and Dawson (1983)
			ME,HK			Samollow, Dawson and Riddle (1983)
			G6PD			Dawson and Hollingsworth (1982)
Tribolium confusum					EST	Sverdlov, Wool and Cohen (1976)
				EST,ACPH, AMY,MDH, AO		Wool (1982)
			ME,HK			Samollow, Dawson and Riddle (1983)
			ME,HK			Dawson and Jost (1983)
			G6PD			Dawson and Hollinsworth (1982)
Tribolium brevicornis				EST,ACPH, AMY,MDH, AO		Wool (1982)

Tribolium destructor				EST,ACPH, AMY,MDH, AO		Wool (1982)
Attagenus megatoma	Peptidases					Baker (1982*a*)
Sitophilus granarius				AMY		Baker (1987*a*)
	AMY					Baker (1982*b*)
	Proteinases					Baker (1982*c*)
Sitophilus oryzae	Proteinases					Baker (1982*c*)
				AMY		Baker (1987*a*)
	AMY					Baker (1987*b*)
Sitophilus zeamais				AMY		Baker (1987*a*)
	AMY					Baker (1982*b*)
	Proteinases					Baker (1982*c*)
Lasioderma serricorne	EST, protein					Frankel and Duffield (1984)
Plodia interpunctella			Catalase			Lampropoulou and Gelti-Douka (1977*a*,*b*)
					EST	Beeman and Schmidt (1982)
					EST	Zettler (1974)
Ephestia (= *Angasta*)		Proteins				Imberski and Gertson (1974)
Kuehniella			EST,PGM			Jelnes (1971)
			EST			Leibenguth (1973*a*)
					EST	Leibenguth (1973*b*)
			EST			Beregovoy and Riemann (1983)
			EST,ADH			Leibenguth and Schafer (1984)
	ADH		ADH			Leibenguth and Kutz (1985)

Abbreviations: AMY = amylase, ADH = alcohol dehydrogenase, LDH = lactate dehydrogenase, MDH = malate dehydrogenase, ACPH = acid phosphatase, EST = esterase, IDH = isocitrate dehydrogenase, G6PD = glucose 6-phosphate dehydrogenase, PGM = phosphoglucomutase, HK = hexokinase, ME = malic enzyme.

Genetic work on *T. castaneum* isoenzymes was performed by Dawson and his associates. A one-locus, two-allele system of PGM, and one additional monomorphic isoenzyme were found in pupae. The frequencies of the PGM genotypes were similar in populations feeding on different diets (Riddle *et al.* 1983). Dawson and Hollingsworth (1982) discovered a polymorphic, sex-linked G6PD locus in pupae in three laboratory populations, with genotype frequencies at Hardy–Weinberg equilibrium. Two other polymorphic systems were studied in pupae: the first locus (ME) contained two alleles; the second (HK-1) contained three. There were two other presumptive polymorphic HK loci, but their genetics were not worked out (Samollow *et al.* 1983).

Tribolium confusum There is far less information on electrophoresis in this species than in *T. castaneum*, and fewer strains have been used. In our preliminary work (Sverdlov *et al.* 1976), we found one EST isoenzyme active in the pupa but not in larvae or adults, although most other isoenzymes were inactive in the pupa. Dawson and Hollinsworth (1982) report that all *T. confusum* populations they tested were monomorphic at a G6PD locus, in contrast to *T. castaneum*. Samollow *et al.* (1983) found, in a laboratory strain of *T. confusum*, an ME locus with three alleles and a HK locus segregating for three alleles. Some inbred lines of this species were fixed homozygotes of one or the other allele. Dawson and Jost (1983) studied the HK and ME loci further and showed that they are located on the X-chromosome. They also mention some polymorphic laboratory populations that show Hardy–Weinberg equilibrium frequencies at these loci.

The only comparative population genetic study of *Tribolium* is my own (Wool 1982). The frequencies of isoenzyme systems — EST, AMY, AO, ACPH, and MDH — were compared in 19 strains of *T. castaneum*, 7 of *T. confusum*, and one each of *T. brevicornis* and *T. destructor*. These data were re-analysed for this review and will be discussed in part 2.

Attagenus megatoma Using isoelectrofocusing techniques, Baker (1982*a*) discovered eight isoenzymes of exo- and endopeptidases in larval homogenates and studied their physiological properties. No other electrophoretic work was published on this species.

Sitophilus granarius In this species, amylases and proteinases are located in the larval midgut (Baker *et al.* 1984). A single amylase isoenzyme was detected (Baker 1982*b*). In a later study, one of six geographical strains carried an additional isoenzyme (Baker 1987*a*). Proteinases and their physiological properties were also studied (Baker 1982*c*), and several isoenzymes were found. In all these physiological studies, pooled homogenates of 20 midguts were used, and there is no data on the frequencies of these variants.

Sitophilus oryzae has two amylase isoenzymes. Their physiological properties were studied by Baker (1987*b*). Geographic variation among strains from different countries was reported: two strains (from Japan and Tanzania) lacked the most frequent α-AMY isoenzyme that was present in all the other 12 strains. The pattern of proteinases was not the same in different strains. As in the previous species, pools of 20 guts were used, and data on allele frequencies are not available.

Sitophilus zeamais Amylase isoenzymes in 17 geographical strains were studied. An early examination of one strain revealed two isoenzymes (Baker 1982*b*). However, in a larger study (Baker 1987*a*) four major amylase patterns were found among 17 strains, unrelated to the geographical origin of the strains. Again, pools of 20 larvae midguts were used, and there is no information on isoenzyme frequencies. The physiological properties of some proteinase isoenzymes were studied by Baker (1982*c*).

Lasioderma serricorne In this species, disc-gel electrophoresis revealed two esterase loci and one water-soluble protein. Some physiological work was carried out to characterize the esterase isoenzymes (Frankel and Duffield 1984).

Plodia interpunctella The inheritance of catalase isoenzymes was studied by Lampropoulou and Gelti-Douka (1977*a, b*). The three observed phenotypes appear to be controlled by one sex-linked locus. Some data on developmental patterns of the isoenzymes is presented. Two other papers deal with esterases in malathion-susceptible and resistant strains (Zettler 1974; Beeman and Schmidt 1982). These investigations used pooled homogenates of several larvae per sample and the main purpose was to study enzyme activity, although the electrophoretic patterns are described. It is not clear whether the differences in pattern are at all related to insecticide resistance.

Ephestia (= Anagasta) kuehniella. The genetics of some isoenzymes in this species were reported in two early abstracts (Imberski 1972; Eicher and Caspari 1968; and see Caspari and Gottlieb 1975).

Imberski (1971) studied the developmental pattern of ADH during metamorphosis, and Imberski and Gertson (1974) worked on general protein bands in relation to larval age. Jelnes (1971) deals with the inheritance of two linked EST loci, carrying two and three alleles, respectively, and a PGM locus with two alleles, in a 'wild'-caught strain. Beregovoy and Riemann (1983) confirmed the observations on the two-allele esterase locus. They reported that the genotype frequencies in a sample of moths agreed with Hardy–Weinberg expectations. Leibenguth (1973*a, b*) studied the genetics and the tissue-specific pattern of the three-allele EST system (esterase-2). Leibenguth and Kutz (1985) describe the genetics and physiology of an ADH locus with two alleles (the description does not agree with that of Imberski 1972), and Leibenguth and Schäfer (1984) crossed laboratory strains homozygous for EST and ADH alleles and studied the effects of hybridization on fitness. They found some effect on F_1 offspring but concluded that it was *not* due to heterosis at the marker loci. A recent summary of these results is included in Leibenguth (1986).

As a summary of the entire literature survey, I offer the following points.

1. Variable electrophoretic patterns were found in all the investigated species.
2. In several systems, Mendelian inheritance of isoenzymes was confirmed by appropriate crosses.
3. In a few cases, samples from laboratory populations contained the genotypes in equilibrium (i.e. Hardy–Weinberg) proportions.
4. Almost no electrophoretic information is available on 'wild' (warehouse) strains (!)
5. Apart from my own work (Wool 1982), nothing has so far been published on the levels of genetic variation within and among strains of any post-harvest pest species.

Part 2: Genetic variation within and among populations

I shall use my 1982 study on genetic variation in *Tribolium* strains and species as a model for stored-products pests in general.

The study included 19 laboratory strains of *T. castaneum* (CS), 7 of *T. confusum* (CF), and one each of *T. brevicornis* (BR) and *T. destructor* (DE).

Five enzyme systems were studied: EST, ACPH, MDH, AO, and AMY. About half of 61 isoenzymes detected were reliable and repeatable in terms of staining and resolution. (Wool 1982). The numbers of adults assayed individually per enzyme system were 200–600 in CS, 100–380 in CF, 20–90 in BR, and 7–30 in DE.

The objective of the study was to investigate the genetic similarity of strains and species to each other, and to deduce from this similarity the cladistic relationships among them — for comparison with the traditional phylogeny suggested by Hinton (1984). From the frequencies of the 61 isoenzymes in the 28 strains, I calculated measures of genetic similarity and genetic distance (Nei 1972) and several of the statistical measures of genetic association conventionally used in numerical taxonomy (cf. Sokal and Sneath 1963; Sneath and Sokal 1973).

The results showed that the four species were clearly separable electrophoretically. Each species had some isoenzymes that were not shared with other species (Table 15.4). Strains within species were similar to each other (see Table 3 and figures 3 and 4 in Wool 1982). CS and CF were rather dissimilar genetically. This conclusion remained unchanged when, instead of the electrophoretic data, I compared the occurrence of similar morphological mutants in the two species using the known genetic maps of *Tribolium* (Sokoloff 1966, 1977).

One problem with using electrophoretic markers (or, for that matter, genetic markers of any type) in comparisons amongst species is that the assumption must be made that two isoenzymes, in two different species migrating to the same distance on a gel, are the products of the same gene. Owing to reproductive isolation between species (in practice and in principle!) this hypothesis cannot be tested. This difficulty does not arise in comparisons among individuals or strains within species.

Table 15.4 Percentage of all comparable isoenzymes that were present in one species and absent in the other (number of comparable bands—those present in at least one of the species—are given in parentheses). Data from Wool (1982).

	CS	CF	BR	DE
CS	—			
CF	42.0(50)	—		
BR	47.8(46)	53.1(49)	—	
DE	82.6(46)	82.0(50)	75.0(40)	—

I have re-analysed the 1982 data matrix, excluding one strain (CS MSG 345), of which less than 10 individuals were available. Here I am concerned with the diversity among strains as a measure of genetic differentiation among populations in CS and CF.

Materials and methods

Isoenzyme frequencies

The original data matrix includes isoenzyme frequencies for each strain: 31 isoenzymes is CS (18 strains) and 30 (7 strains) in CF. BR, and DE, with one strain per species, were not used in the present analysis.

Very few isoenzymes have been analysed genetically in *Tribolium* (see Table 15.3). Assigning the 61 isoenzymes to loci and alleles in the conventional manner seemed highly speculative and inappropriate. I therefore made the assumption — in 1982 and in the present analysis — that each isoenzyme is a product of a different gene that carries two alleles: active and inactive ('null'). Individuals showing the isoenzyme may be homozygous at the locus (i.e. active/active) or heterozygous (active/null). To calculate allele frequencies, I assumed Hardy–Weinberg equilibrium at each locus. Then the frequency (q) of the *inactive* allele is the square-root of Q, the frequency of 'null' individuals who do not show the isoenzyme, and the frequency of the active allele is $p = 1 - q$. This procedure inflates the number of loci, but should not distort the average diversity among populations.

Calculation of diversity

Diversity among strains was calculated using Lewontin's algorithm, based on the Shannon diversity index $H = - \Sigma P_i \log_2 P_i$, and expressed as a percentage (Lewontin 1972). Diversity was calculated for each isoenzyme (locus) separately and then averaged over all isoenzymes. Diversity *within* strains was calculated as unity minus the diversity among strains.

Results

Distribution of isoenzyme frequencies in strains of Tribolium

The distribution of allele frequencies (the frequency of the active allele, P_i) are illustrated in Fig. 15.1. The active allele was present in

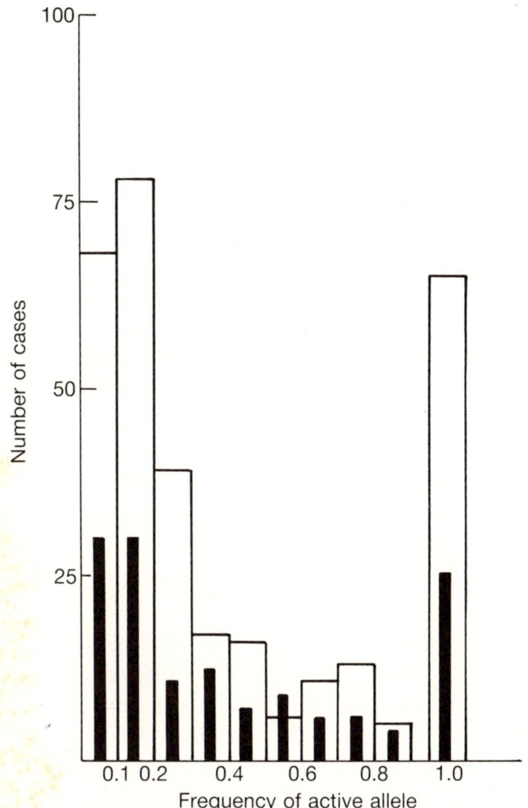

Fig. 15.1. Distribution of the frequency of the active allele (all loci combined) in 18 laboratory strains of *Tribolium castaneum* (open bars) and in 7 strains of *T. confusum* (solid bars). Re-analysed data of Wool (1982).

318 (60 per cent) of the total possible cases (18 strains × 31 alleles = 558) in CS, and in 117 (55.6 per cent) of 210 cases (7 strains × 30 isoenzymes) in CF.

The figure illustrates that the majority of isoenzymes were present at either low or high frequencies in both species. Intermediate frequencies were rather uncommon. In CS, an average of three isoenzymes were fixed per strain, and in CF about four. The distributions are similar to the form expected of selectively neutral alleles affected by random genetic drift (U-shaped). The frequency of fixed alleles is, of course, underestimated: *inactive* alleles could not be scored if fixed in all strains, and the cases in which an allele was fixed in *some* (but not all) strains could not be identified with certainy

Table 15.5 Diversity among and within strains of *T. castaneum* (CS) and *T. confusum* (CF) (n = 29 diversity estimates per species)

Percentage diversity	CS	CF
Among strains	14.3 ± 2.40	11.9 ± 1.72
Within strains	85.7	88.1
Range of estimates (among strains) from different isoenzymes	0.04–51.66	1.14–33.33

because weak isoenzymes could be overlooked. If we assign all cases of absence of the active allele to fixation of the inactive one, we get a frequency of about 40 per cent in both species.

It is important to point out that all these strains are laboratory-reared, and their genetic origin as separate gene pools dates back no longer than 50 years. During this time, however, they suffered numerous bottlenecks, considerable inbreeding, and often selection in order to maintain them in homozygous condition for specific genetic markers.

Diversity among strains was calculated separately for each isoenzyme, and the ranges of these estimates were wide. The mean diversity among strains was similar in the two species, with rather large standard errors (Table 15.5) ($P > 0.05$ by t-test). About 12–14 per cent of the gene diversity was found among strains within species, which leaves 86–88 per cent of the variation among individuals within strains. The distribution of among-strain diversity values (for the two species combined) is illustrated in Fig. 15.2. The distribution is very skewed with most alleles showing low diversity values.

In order to evaluate the magnitude of the among-strain diversities, estimates of differentiation in other groups of organisms are needed. In a recent paper (Wool 1987), I described the process of differentiation of artificial 'island' populations of *Tribolium*, under founder effects, extreme bottlenecks, and inbreeding, and compared it with reports on real island populations of *Drosophila*, lizards, and rodents. The *Tribolium* data were based on frequencies of a morphological marker (*black*). The other frequencies were electrophoretic alleles calculated in the conventional way by the authors of the eight respective source papers (Table 2 in Wool 1987).

The experimental islands were maintained for 11 generations in isolation, with inbreeding and extremely small bottlenecks (one pair per population and generation). (Such bottlenecks are rarely used when *Tribolium* strains are to be maintained in the laboratory for

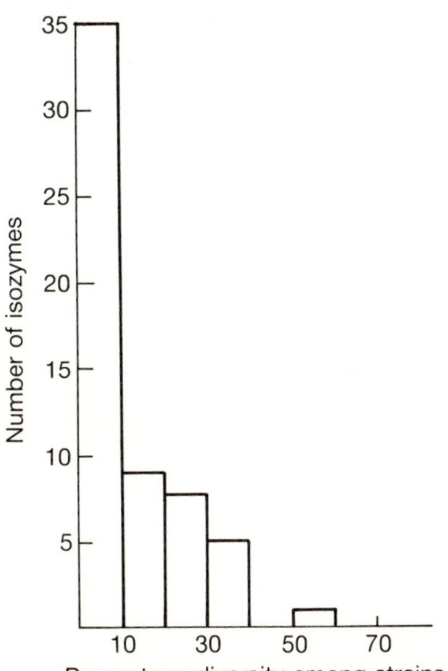

Fig. 15.2. Distribution of among-strain diversity values, estimated by different isoenzyme markers, in 18 strains of *T. castaneum* and 7 strains of *T. confusum* (combined). Re-analysed data of Wool (1982).

extended periods, owing to their deleterious effects on fitness; see Wool and Sverdlov 1976 and Wool and Bergerson 1986). Diversity among the island populations, which were identical initially, increased rapidly and reached 60 per cent. When migration was allowed between islands, diversity decreased as expected and stabilized around 35 per cent. Comparisons with real island data showed that only two data sets (lizards from the Galapagos Islands) were more highly differentiated than the *Tribolium* 'islands'. *Drosophila* island populations (four data sets) gave values of less than 20 per cent diversity among islands for 65 out of 66 marker alleles, and the shape of the distributions was very similar to the distribution of *Tribolium* strains in the present study (Fig. 15.2).

The implication of this is that after 50 years or less of laboratory culture (probably 100–500 generations), *Tribolium* strains in the present study are as differentiated from each other as real island populations of *Drosophila*. In comparison with the data on lizards and mice, I interpret the *Drosophila* data to mean that these small, mobile organisms are travelling (or are transported) among islands rather frequently. The similarity of the diversity distribution with that of *Tribolium* strains may mean that, in these strains, genetic differentiation is not very strong (despite obvious differences among strains that are homozygous for some morphological markers!). This situation is analogous to the morphological differentiation of chimpanzee and man despite the similarity in electrophoretic alleles and DNA.

We do not know how different 'natural' warehouse populations of *Tribolium* are from each other, but based on the present results, my prediction would be that the diversity among such populations should be even less pronounced than that amongst the laboratory strains. I would therefore conclude that migration and intermixing of *Tribolium* populations would be (or has been) a more important factor in shaping their genetics than have inbreeding and founder effects.

It remains to be discovered how representative these data are of 'natural' *Tribolium* populations, and of other stored products pests. I am convinced that valuable information would be accumulated if the effort were made to obtain the data.

This group of organisms, both from an economic and from an evolutionary point of view, deserves more attention from population geneticists than it has received so far.

Note added in proof

Two papers on population genetics of stored-products pests have come to my attention since the review was first prepared. The first is on three species of *Sitophilus* (Bairas and Petitpierre 1981). Population genetic work was carried out on three strains each of *S. granarius* and *S. oryzae*, and two of *S. zeamais*. The strains were obtained from different countries in Europe and more than 100 individuals were analysed per strain. AMY, α-GPDH, MDH, LAP, and ADH were monomorphic within species (but the isoenzyme patterns differed between species). EST was the only polymorphic system, and allele frequencies were not the same in different strains. One EST locus (EST-3) was analysed genetically and Mendelian inheritance of the isoenzymes was confirmed.

The second paper deals with the relationship of ME genotypes and fitness in an experimental population of *Tribolium confusum* (Zirkle and Dawson 1988). Survival rate, developmental time, fecundity and rate of egg cannibalism of different genotypes at the ME locus were investigated and some genotype-environment interactions were detected.

References

Applebaum, S.W., Jankovic, M., and Birk, Y. (1961). Studies on the midgut amylase activity of *Tenebrio molitor* (L.) larvae. *J. Insect Physiol.*, **7**, 100–108.

Attia, F.I. (1981). Insecticide resistance in pyralid moths of grain and stored products. *Gen. appl. Genet.*, **13**, 3–8.

Attia, F.I., Shipp, E., and Shanahan, G. (1979). Survey of insecticide resistance in *Plodia interpunctella* (Hübner), *Ephestia cautella* (Walker) and *E. kuehniella* Zeller (Lepidoptera, Pyralidae) in New South Wales. *J. Aust. ent. Soc.*, **18**, 67–70.

Baker, J.E. (1982*a*). Synthesis of endopeptidases and exopeptidases in actively feeding larvae of the black carpet beetle. *J. Georgia. ent. Soc.*, **17**, 221–8.

Baker, J.E. (1982*b*). Properties of amylases from midguts of larvae of *Sitophilus zeamais* and *Sitophilus granarius*. *Insect Biochem.*, **13**, 421–8.

Baker, J.E. (1982*c*). Digestive proteinases of *Sitophilus* weevils (Coleoptera, Curculionidae) and their response to inhibitors from wheat and corn flour. *Can. J. Genet. Cytol.*, **60**, 3206–14.

Baker, J.E. (1987*a*). Electrophoretic analysis of amylase isozymes in geographical strains of *Sitophilus oryzae* (L.), *S. zeamais* Motsch., and *S. granarius* (L.) (Coleoptera, Curculionidae). *J. Stored Prod. Res.*, **23**, 125–31.

Baker, J.E. (1987*b*). Purification of isoamylases from rice weevil, *Sitophilus oryzae* (L.) (Coleoptera, Curculionidae) by high-performance liquid chromatography and their interaction with partially-purified amylase inhibitors from wheat. *Insect Biochem.*, **17**, 37–44.

Baker, J.E., Woo, S.M., and Byrd, R.V. (1984). Ultrastructural features of the gut of *Sitophilus granarius* (L.) (Coleoptera, Curculionidae) with notes on distribution of proteinases and amylases in crop and midgut. *Can. J. Genet. Cytol.*, **62**, 1251–9.

Bairas, M.J. and Petitpierre, E. (1981). Allozyme variability and genetic differentiation in three species of *Sitophilus* L. (Coleoptera, Curculionidae). *Egypt. J. Genet. Cytol.*, **10**, 95–104.

Beeman, R.W. and Schmidt, B.A. (1982). Biochemical and genetic aspects of malathion-specific resistance in the Indian meal moth, (Lepidoptera, Pyralidae). *J. econ. Ent.*, **75**, 945–9.

Beregovoy, V., and Reimann, J.G. (1983). Polymorphism and inheritance patterns in laboratory colonies of two pyralid moth species. *J. Hered.*, **74**, 117–18.

Caspari, E.W. and Gottlieb, F.J. (1975). The Mediterranean meal moth, *Ephestia kuehniella*. In *Handbook of genetics*, vol. III (ed. R.C. King), pp. 125–47. Plenum Press, New York.

Cohen, E., Sverdlov, E., and Wool, D. (1977). Expression of esterases during ontogenesis of the flour beetle, *Tribolium castaneum* (Tenebrionidae, Coleoptera). *Biochem. Genet.*, **15**, 253–64.

Dawson, P.S. and Hollingsworth, N.M. (1982). Sex-linkage of the glucose-6-phosphate dehydrogenase locus in the flour beetle, *Tribolium castaneum*. *Can. J. Genet. Cytol.*, **24**, 267–71.

Dawson, P.S. and Jost, J. (1983). Linkage relationships among the Malic enzyme, Hexokinase-1, and red loci on the X-chromosome of *Tribolium confusum*. *Biochem. Genet.*, **21**, 661–5.

Dyte, C.E. and Blackman, D.G. (1970). The spread of insecticide resistance in *Tribolium castaneum* (Herbst) (Coleoptera, Tenebrionidae). *J. Stored Prod. Res.*, **6**, 225–61.

Eicher, A.M. and Caspari, E.W. (1968). Genetic control of the disappearance of a protein in *Ephestia*. *Genetics*, **60**, 175 (abstract only).

Frankel, J.S. and Duffield, R.M. (1984). Genetic variation in the cigarette beetle, *Lasioderma serricorne* (Coleoptera, Anobiidae). I. Esterase and water-soluble protein polymorphism. *Comp. Biochem. Physiol.*, **77**B, 337–40.

Hinton, H.E. (1948). A synopsis of the genus *Tribolium* Macleay, with some remarks on the evolution of its species groups (Coleoptera, Tenebrionidae). *Bull. ent. Res.*, **39**, 1–55.

Hubby, J.L. and Lewontin, R.C. (1966). A molecular approach to the study of genic heterozygosity in natural populations. I. The number of alleles at different loci in *Drosophila pseudoobscura*. *Genetics*, **54**, 577–94.

Imberski, R.B. (1971). Alcohol dehydrogenase activity during metamorphosis of *Ephestia kuehniella*. *Experientia*, **28**, 26.

Imberski, R.B. (1972). Genetic control of alcohol dehydrogenase in *Ephestia kuehniella*. *Genetics*, **71**, S27 (abstract only).

Imberski, R.B. and Gertson, P.N. (1974). Changes in basic proteins during metamorphosis of the moth, *Ephestia kuehniella*. *Insect Biochem.*, **4**, 341–4.

Jelnes, J.E. (1971). The genetics of three isozyme systems in *Ephestia kuehniella Z*. *Hereditas*, **69**, 138–40.

Lahr, J.R. and Costantino, R.F. (1973). Disc gel electrophoresis of the fatty-acid sensitive mutant of *T. castaneum*. *Tribolium Inf. Bull.*, **16**, 79–89.

Lampropoulou, M.S. and Gelti-Douka, H. (1977*a*). Catalase in *Plodia interpunctella*. I. Genetic control and enzymatic patterns during development. *Genetica*, **47**, 49–52.

Lampropoulou, M.S. and Gelti-Douka, H. (1977*b*). Catalase in *Plodia interpunctella*. II. Biochemical variation associated with an X–Y translocation. *Genetica*, **47**, 53–8.

Leibenguth, F. (1973*a*). Esterase-2 in *Ephestia kühniella*. I. Genetics and characterization. *Biochem. Genet.*, **10**, 219–29.

Leibenguth, F. (1973*b*). Esterase-2 in *Ephestia kuehniella*. II. Tissue-specific patterns. *Biochem. Genet.*, **10**, 231–42.

Leibenguth, F. (1986). Genetics of the flour moth, *Ephestia kühniella*. *Agric. Zool. Rev.*, **1**, 39–72.

Leibenguth, F. and Schäfer, M. (1984). Heterosis in the flour moth, *Ephestia kühniella*. *Genetica*, **64**, 209–19.

Leibenguth, F. and Kutz, U. (1985). Alcohol dehydrogenase polymorphism in the flour moth, *Ephestia kühniella*. *Comp. Biochem. Physiol.*, **82**B, 455–9.

Lewontin, R.C. and Hubby, J.L. (1966). A molecular approach to the study of genic heterozygosity in natural populations. II. Amount of variation and degree of heterozygosity in natural populations of *Drosophila pseudoobscura*. *Genetics*, **54**, 595–609.

Lewontin, R.C. (1972). The apportionment of human diversity. *Evol. Biol.*, **6**, 381–98.

Levinson, H.Z. and Levinson, A.R. (1985). Storage and insect species of stored grain and tombs in ancient Egypt. *Z. angew. Ent.*, **100**, 321–39.

Mayr, E. (1963). *Animal species and evolution.* Harvard University Press, Cambridge, Mass.

Mulder, G. and Sokoloff, A. (1982). Observations on populations of *Tribolium brevicornis* LeConte (Coleoptera, Tenebrionidae). III. Preliminary comparisons between feral and domesticated populations. *J. adv. Zool.*, **3**, 33–45.

Nei, M. (1972). Genetic distance between populations. *Am. Nat.*, **106**, 283–92.

Nevo, E., Beiles, A., and Ben Shlomo, R. (1984). The evolutionary significance of genetic diversity: Ecological, demographic and life history correlates. *Evolutionary Dynamics of Genetic Diversity,* Lecture Notes in Biomathematics, **53**, 13–213.

Riddle, R.A., Iverson, V., and Dawson, P.S. (1983). Dietary effects on fitness components at the PGM-1 locus of *Tribolium castaneum*. *Genetics*, **103**, 65–73.

Samollow, P.B., Dawson, P.S., and Riddle, R.A. (1983). Sex-linked and autosomal inheritance patterns of homologous genes in two species of *Tribolium*. *Biochem. Genet.*, **21**, 167–76.

Sokal, R.R. and Sneath, P.H.A. (1963). *Numerical taxonomy.* Freeman, San Francisco.

Sneath, P.H.A. and Sokal, R.R. (1973). *Numerical taxonomy.* Freeman, San Francisco.

Sokoloff, A. (1966). *The genetics of Tribolium and related species.* Academic Press, London and New York.

Sokoloff, A. (1977). *The biology of Tribolium*, Vol. III. Oxford University Press, Oxford.

Sokoloff, A., Wilson, R., and Mulder, G. (1983). Observations of populations of *Tribolium brevicornis* Le Conte (Coleoptera, Tenebrionidae). II. The habitat niche of a local population in southern California. *Pan-Pacif. Ent.*, **58**, 177–83.

Sverdlov, E., Wool, D., and Cohen, E. (1976). Esterase isozymes of some *Tribolium* strains. *Tribolium Inf. Bull.*, **19**, 120–25.

Sverdlov, E. and Wool, D. (1978). Isozymes of *Tribolium*. *Tribolium Inf. Bull.* **21**, 155–60.

Toro, M.A. and Lopez-Fanjul, C. (1979). Genetic changes of pupa weight in *Tribolium castaneum* under domestication. *Experimentia*, **35**, 1296–7.

Waterhouse, F. L. and Nowosielski-Slepowron, B. J. A. (1964). Differences between identically-reared laboratory and field stocks of the flour beetle, *Cathartus quadricollis* Guer. *Proc. 12th Int. Congress Ent.*, **9b**, 664–5.

Wool, D. (1982). Critical examination of postulated cladistic relationships among species of flour beetles (genus *Tribolium*, Tenebrionidae, Coleoptera). *Biochem. Genet.*, **20**, 333–49.

Wool, D. (1987). Differentiation of island populations: a laboratory model. *Am. Nat.*, **129**, 188–202.

Wool, D. and Sverdlov, E. (1976). Sib-mating populations in an unpredictable environment: effects on components of fitness. *Evolution*, **30**, 119–29.

Wool, D. and Bergerson, O. (1986). Random environmental variation and inbreeding: effects on pure-strain and hybrid populations of flour beetles (*Tribolium*). *Can. J. Genet. Cytol.*, **28**, 889–98.

Yeh, F. C. H. and Scheinberg, E. (1972). Maintenance of alcohol dehydrogenase polymorphism in an experimental population of *Tribolium castaneum*. *Genetics*, **71**, S70 (abstract only).

Yeh, F. C. H. and Scheinberg, E. (1974). Electrophoretic separation of enzymes of individual flour beetles, *Tribolium castaneum*, on polyacrylamide gel. *Analyt. Biochem.*, **62**, 321–6.

Zettler, J. L. (1974). Esterases in a malathion-susceptible and a malathion-resistant strain of *Plodia interpunctella* (Lepidoptera, Phyticidae). *J. Georgia ent. Soc.*, **9**, 207–13.

Zettler, J. L. (1982). Insecticide resistance in selected stored-products insects infesting peanuts in the southeastern United States. *J. econ. Ent.*, **75**, 359–62.

Zirkle, D. F. and Dawson, P. S. (1988). An experimental analysis of the relationships between fitness components and Malic Enzyme genotypes in *Tribolium confusum*. *Biochem. Genet.*, **26**, 277–86.

16 The role of electrophoresis in the biochemical detection of insecticide resistance

A.L. DEVONSHIRE

Insecticides and Fungicides Department, Rothamsted Experimental Station, Harpenden, Hertfordshire, UK

Abstract

The biochemical factors responsible for insecticide resistance are described, and the role of electrophoresis for characterizing biochemical mechanisms in individual insects is set against a background of the variety of techniques available, from straightforward spot tests to more precise immunological or enzyme-kinetic methods. The clear advantage of biochemical methods is their ability to monitor accurately the changes in resistance gene frequency resulting from insecticide selection, unlike conventional bioassays that measure the average response of a heterogeneous population. This more precise information should greatly improve the development and appraisal of resistance-management strategies.

The majority of insecticides currently used are esters, and cleavage of the ester bond is a common detoxication mechanism. This is difficult to assay in individual insects with insecticides as substrates, but model substrates such as naphthyl esters are widely used. Because of the extensive esterase polymorphism in insects, electrophoresis plays a major part in resolving and identifying the enzyme involved. However, a prerequisite for exploitation of this or any such *in vitro* technique is supporting biochemical evidence implicating the enzyme measured as a cause of resistance, as illustrated by studies of resistance in the peach-potato aphid (*Myzus persicae*). The recent cloning of a cDNA sequence corresponding to esterase-4 (E4) the enzyme responsible for resistance in this species, also illustrates the new

Electrophoretic Studies on Agricultural Pests (ed. Hugh D. Loxdale and J. den Hollander), Systematics Association Special Volume No. 39, pp. 363–74. Clarendon Press, Oxford, 1989.

opportunities for exploiting DNA probes to study resistance genes directly.

Key considerations for any resistance monitoring technique are the abilities to analyse large numbers of insects and to provide both qualitative and quantitative data representative of the population sampled. The advantages and disadvantages of the various techniques are discussed, and a multiple homogenizer to help achieve high throughput is described.

Biochemical basis of resistance

Insects develop resistance as a result of changes either in the factors that control the amount of insecticide reaching the critical target (the nervous system in the case of most current commercial insecticides) or in the nature of the target itself (Oppenoorth 1985).

Thus, unidentified modifications of the cuticle can slow the rate at which chemicals enter the insect; at the other end of the poisoning process, the target may be altered so that, whilst retaining its normal physiological function, its affinity for insecticides is decreased. This latter form of resistance affects the performance of the pyrethroids (and DDT) through unidentified changes in the sodium channel or its immediate environment, and the organophosphorus (OP) and carbamate insecticides as a consequence of mutation of the structural gene for acetylcholinesterase.

Between penetrating the cuticle and binding to its target, a large proportion of the insecticide dose can be detoxified by three main groups of enzymes, any of which can be more active in resistant insects as a result of either quantitative or qualitative changes. Monooxygenases (mixed-function oxidases) are membrane-bound enzymes with a broad substrate specificity that render lipophilic chemicals more polar, and hence more readily excreted — either directly or after conjugation of sugars, etc., to the hydroxyl group they have introduced. Glutathione transferases catalyse the alkylation of the thiol group of glutathione, using both aromatic and aliphatic compounds as donors. This is a common metabolic route for organophosphorus insecticides, which leads to ester bond cleavage, and hence detoxication. The third group of enzymes, the esterases, effect the hydrolytic cleavage of the ester bonds present in most insecticides, although, in some circumstances, esterases can also act by merely sequestering a large proportion of the toxic dose of an insecticide (Devonshire and Moores 1982), as discussed below.

Identification of metabolic mechanisms involved in a particular instance of resistance generally requires mass homogenates of homogeneous insect strains, coupled with the use of radiotracers to identify metabolites. Similarly, homogeneous strains are needed for the kinetic studies necessary for detailed characterization of target insensitivity, although the *identification* of modified acetylcholinesterase forms, even when heterozygous, is possible in single insects (Moores *et al.* 1988). Direct techniques for identifying metabolic resistance mechanisms have not been applicable to single insects because the techniques lack sensitivity when insecticides are used as substrates. The analysis of single insects is clearly essential if biochemical monitoring is to be used for studying the population genetics of insecticide resistance.

Biochemical monitoring

Model substrates have been used extensively in place of insecticides for assaying all three groups of degrading enzymes. These substrates give readily assayed products, usually based on spectrophotometric determination, that allow enzyme activity to be measured in single insects. However, since these enzymes all exist in multiple forms (Clark and Dauterman 1982; Devonshire 1977; Stanton *et al.* 1978), the activity detected by the model substrate chosen need not be due to the same enzyme responsible for insecticide detoxication (Clark *et al.* 1986). This multiplicity of enzymes is apparent, without the need for separation techniques, when they are assayed with different model substrates. Thus, the two aromatic donors commonly used to assay glutathione transferase activity (chloro-2, 4-dinitrobenzene, and 3, 4-dichloronitrobenzene) can give quite different relative activities between insect strains (Clark and Dauterman 1982). Similarly, whilst 1-naphthyl acetate reveals massive differences in esterase (E4) activity of crude homogenates between insecticide-susceptible and resistant peach-potato aphids (*Myzus persicae*), this is barely apparent with 1-naphthyl butyrate because the other esterases common to all strains hydrolyse this substrate very effectively, so masking the increased activity of the esterase (E4) responsible for resistance.

However, using an appropriate model substrate, differences in activities of these enzymes are commonly identified between susceptible and resistant insects. Whilst their activity is usually greater in resistant strains, a negative correlation is occasionally found, notably with esterases; in both houseflies (van Asperen and Oppenoorth 1959) and blowflies (Hughes and Raftos 1985), this has been interpreted as being due to the loss of an enzyme's ability to hydrolyse the model substrate when it has mutated to a form that can hydrolyse insecticides.

With few exceptions, such as *M. persicae* (Devonshire and Moores 1982), the use of model substrates for diagnosing resistance mechanisms is based on circumstantial evidence, i.e. a correlation between enzyme activity and resistance. If these techniques are to acquire greater credibility and play a more important role in resistance monitoring, this often tenuous relationship needs biochemical validation in insect pests to which the test is applied.

Since changes in enzymes associated with resistance can be hidden by the 'background' of irrelevant enzymes able to metabolize the same model substrate, electrophoresis has been used to separate the enzyme forms. However, for the following reasons, it has only found significant use for studying esterases. The membranous nature of monooxygenases prevents their analysis in this way because they lose activity when solubilized, although they can be detected on SDS gels by their haem-associated peroxidase activity (Thomas *et al.* 1976). Similarly, glutathione transferases can be detected on electrophoresis gels by negative staining (i.e. depletion of glutathione), but this method is not used routinely (Ricci *et al.* 1984). Finally, although some of the qualitative changes in acetylcholinesterase are potentially identifiable by electrophoresis, this is complicated by the hydrophobic nature of this membranous enzyme, which aggregates to give multiple forms when solubilized (Steele and Manackjee 1979), and inhibition kinetic analysis remains the most appropriate method for identifying this resistance mechanism (Moores *et al.* 1988).

The role of electrophoresis

Electrophoresis separates charged molecules; it is the subsequent identification of particular molecules by appropriate methods that detects resistance mechanisms. This requires the *critical* use of model substrates as discussed above, and more recently of nucleic-acid hybridization techniques. These approaches are illustrated here by our work on the esterases of *M. persicae*, which shows how electrophoresis can reveal both qualitative and quantitative changes in these enzymes and their corresponding genes.

M. persicae esterases

The increased production of one of two very closely related esterases (Fig. 16.1), E4 or FE4, is responsible for the broad cross-resistance of

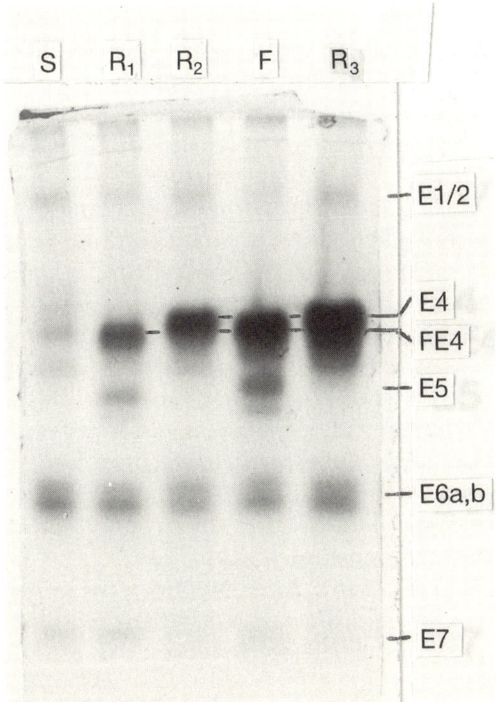

Fig. 16.1. Polyacrylamide gel electrophoresis of homogenates of individual aphids, stained for esterase activity with 1-naphthyl acetate (see Devonshire and Moores (1982) for details of electrophoresis and staining conditions). Clone US1L (S) is the susceptible reference, and clones 405D (R_1), T1V (R_2), Ferrara (F) and 794J (R_3) are progressively more resistant to a wide range of organophosphorus, carbamate, and pyrethroid insecticides. S, R_1, and F clones have a normal karyotype, whereas R_2 and R_3 clones have an A1,3 translocation. R_1 and F clones, in common with other non-translocated aphids, overproduce FE4 and also have the E5 band, whilst translocated aphids (R_2, R_3, and others) overproduce E4 only.

M. persicae to many organophosphorus, carbamate, and pyrethroid insecticides (Devonshire and Moores 1982). The chromosome complement (karyotype) of the aphid appears to determine which of these two esterase forms is overproduced. Thus, very resistant aphids with an autosomal translocation (A1, 3; Blackman *et al.* 1978) contain E4, wherease non-translocated aphids overproduce an electrophoretically faster (on this buffer system) version of the enzyme, FE4 (and also contain a further esterase, E5, not involved in resistance). Whilst important for understanding the evolution of resistance, the distinctiveness of E4 and FE4 is of little toxicological significance

(Devonshire *et al.* 1983). Both are typical B-esterases that were first identified by their ability to hydrolyse 1-naphthyl acetate, and both are readily inhibited by organophosphorus and carbamate insecticides, but re-activate slowly. The inhibition step detoxifies the insecticide by sequestration, and the re-activation constitutes turnover, or hydrolysis, of the insecticide.

Seven levels of esterase production have been identified in a series of aphid clones that each show a corresponding level of resistance (Devonshire and Sawicki 1979). Clonal cultures of all seven types are readily identified by their characteristic *mean* total esterase activity with 1-naphthyl acetate. However, since only one esterase changes in resistant aphids, against a background of several other enzymes hydrolysing this substrate, the unambiguous classification of an unknown individual aphid, especially when only slightly resistant (R_1), is not possible by total esterase determination. Electrophoresis has therefore been used extensively to resolve the esterases and so provide a more specific assay for the relevant enzyme. This can identify even slightly resistant aphids, but relies on the subjective assessment of esterase band intensity because gel scanning is not practicable on the large scale necessary for field studies. However, this disadvantage is compensated by several other advantages of electrophoresis.

1. Careful identification of *M. persicae* is not necessary, since even very similar species such as *Myzus antirrhinii* and *Myzus certus* give distinct esterase band patterns (ffrench-Constant *et al.* 1988*a*).
2. Aphids parasitized by hymenopterous parasitoids are readily identified from the additional (often superimposed) parasitoid bands, whilst the E4/FE4 band retains its characteristic activity.
3. The close association between FE4 and E5 provides a readily-scored qualitative character to aid classification; the presence or absence of E5 also gives a strong indication of the aphid's karyotype.
4. The activity of E4/FE4 can be judged against the overall 'background' of other esterases to compensate partially for the size of the aphid.
5. Qualitative changes in other esterases, in particular E1/E2, can be identified as a further aid to studying the population structure (Baker 1979).

Some clones of the most resistant (R_3) type spontaneously lose resistance as a consequence of their switching off of E4 production (Sawicki *et al.* 1980; ffrench-Constant *et al.* 1988*b*). This appears within each generation as a variegation effect, with most clonal progeny retaining the characteristic E4 level of their parent, but with 1–5 per cent showing complete 'reversion' to the susceptible E4 level. Such revertants, unlike normal susceptible aphids, retain the A1,3 translocation of their parent and can be partially reselected for increased esterase activity at the R_1 level. However, this activity is due to E4 rather than FE4, as identified electrophoretically (ffrench-Constant *et al.* 1988*b*).

Alternatives to enzyme electrophoresis

Immunoassay

Although electrophoresis has played a major role since the mid-1970s in studying resistance in *M. persicae*, it has recently been eclipsed by an immunoassay for E4/FE4 that serves the same purpose, namely, the isolation of the relevant enzyme, so that it can be assayed without the background contribution of other esterases (Devonshire *et al.* 1986). This enables the *quantitative* analysis of many more insects (more than ten times the throughput of electrophoresis), but at the expense of sacrificing all five advantages of electrophoresis outlined above (species identification is important, since susceptible *M. persicae* give readings close to blank incubations). The immunoassay has been used to monitor the occurrence of resistant variants in the field (ffrench-Constant and Devonshire 1988), to study the consequences of spraying different insecticides on mixed populations of susceptible and resistant aphids (ffrench-Constant *et al.* 1987, 1988*c*), and to examine the intraclonal loss of resistance (ffrench-Constant *et al.* 1988*b*).

A major factor contributing to the high capacity of the immunoassay was the development of a multiple homogenizer that allows 96 individual insects to be homogenized simultaneously in the wells of a microtitration plate (ffrench-Constant and Devonshire 1987). It provides very effective homogenization of insects ranging in size from whiteflies (*ca.* 20 μg) to houseflies (ca. 20 mg), with the further advantage that all liquid transfers are done rapidly with 8-channel pipettes.

The efficiency of this homogenizer has also been exploited for electrophoresis by adapting gel combs to give wells at 4.5-mm spacing, so that multi-tip samplers can be used to load samples on to the gel (ffrench-Constant *et al.* 1988*a*).

DNA hybridization

The use of biochemical markers in combination with bioassays can thus give a clearer insight into resistance in heterogeneous insect populations. In addition to studying primary gene products biochemically, the lure of molecular biology to identify genotype directly is obvious (Brown and Brogdon 1987), but in view of the effort involved, it must offer a clear advantage over more straightforward biochemical assays. The major obstacle to such an approach is the need to clone the appropriate genes, which in turn requires a thorough understanding of the underlying biochemistry. It is therefore perhaps not surprising that, at the time of writing, only two resistance genes have been cloned, both of which encode esterases and are highly amplified (Mouchès *et al.* 1986; Field *et al.* 1988). However, with the growth of work in this area, there is optimism that a wider range of resistance genes will become more amenable to cloning, especially by exploiting heterologous probes.

Once a cloned gene is available, electrophoresis will continue to play a major role, at least in the early stages. This is again illustrated by our work on the E4 esterase gene of *M. persicae*, which demonstrates an unanticipated advantage of DNA probing over straightforward esterase analysis.

Details of cloning this gene, and its use to confirm gene amplification, have been given previously (Field *et al.* 1988). Briefly, it entailed the preparation of a cDNA library, screening by differential hybridization, and final confirmation that a cloned sequence corresponded to E4 by hybrid arrested translation combined with immunoprecipitation. The radiolabelled probe prepared from this cloned sequence was used to identify esterase-related sequences on a Southern blot of DNA from susceptible and resistant aphids after agarose gel electrophoresis (Fig. 16.2). Amplification of the esterase gene is apparent, involving either a 4 kb or an 8 kb fragment according to the aphids' karyotype and associated esterase form — E4 or FE4. These qualitative patterns were consistent through several other aphid clones of known karyotype, and their origins have been discussed (Field *et al.* 1988).

Although these results are with DNA isolated from mass homogenates of aphid clones, the quantitative differences between the susceptible and most resistant variants are readily seen in dot blots prepared from homogenates of single aphids. The development of microtechniques for purifying and analysing DNA, for example in DNA 'fingerprinting' of forensic samples (Higuchi *et al.* 1988),

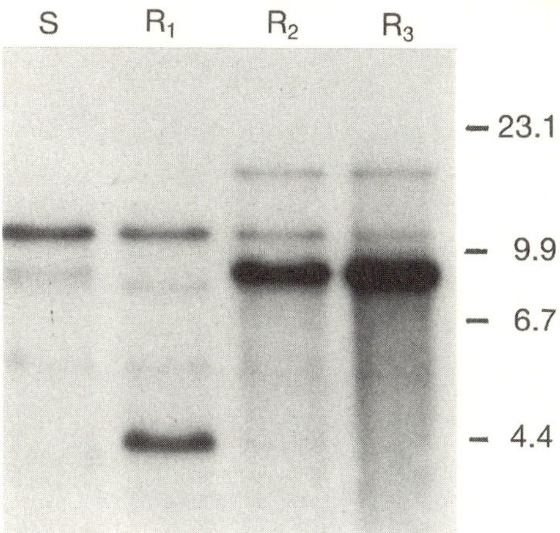

Fig. 16.2. Autoradiograph showing esterase-related restriction fragments of aphid DNA. Details of aphid clones are given in Fig. 16.1. DNA (10 μg) from each clone was digested with *Eco*RI, electrophoresed on an agarose gel and Southern blotted on to a nylon filter, and was then hybridized with the ^{32}P-labelled E4 cDNA probe. A 4 kb fragment is amplified in R_1 aphids (also, to a greater extent in F aphids; data not shown), whereas the translocated clones R_2 and R_3 have an amplified 8 kb fragment. The 10 kb and 15 kb fragments probably correspond to terminal non-amplified regions together with their respective flanking sequences according to whether the gene is associated with a translocated (heterozygous) or non-translocated karyotype. (Reprinted by permission from *Biochem. J.* **251**, 309–12, copyright © 1988, The Biochemical Society, London.)

should make possible the electrophoretic analysis of restriction fragments of single insects.

So far, apart from the more dramatic differences in electrophoretic pattern of DNA according to karyotype, DNA probing would appear to offer no advantage over esterase electrophoresis as a means of studying *M. persicae* populations. However, recent work (Field *et al.* 1989*a*) has shown that when very resistant aphids lose resistance, their susceptible offspring retain the E4 gene amplification, demonstrating that loss of enzyme involves transcriptional control. This has important practical implications since, unlike susceptible aphids of normal karyotype, such revertants retain both their translocated karyotype and a potential for re-selection (ffrench-Constant *et al.* 1988*b*); their identification in field populations is therefore important. This is now possible by a combination of

DNA probing and enzyme immunoassay, both of which can performed on the homogenates of individual insects (Field *et al.* 1989*b*).

Such techniques have so far relied on the use of ^{32}P-labelled probes, but improvements in non-radioactive labelling, notably the biotin–streptavidin system, are likely to make this approach more accessible to non-specialist laboratories.

Conclusions

Electrophoresis will continue to have a role in the biochemical monitoring of insecticide resistance, although this is likely to be directed predominantly at esterases either known or suspected to be involved in insecticide degradation. However, other high-throughput techniques, notably immunoassays and DNA probing, will become more important. All should benefit from advances in rapid sample handling, especially homogenization, associated with the use of 96-well microplate technology. Finally, the use of these *in vitro* techniques must involve thorough biochemical validation if their credibility is to be established and maintained.

Acknowledgements

I thank Dr R.M. Sawicki, L.M. Field, and G.D. Moores for commenting on the manuscript.

References

Asperen, K. van and Oppenoorth, F.J. (1959). Organophosphate resistance and esterase activity in houseflies. *Entomologia exp. appl.*, **2**, 48–57.

Baker, J. P. (1979). Electrophoretic studies on populations of *Myzus persicae* in Scotland from October to December, 1976. *Ann. appl. Biol.*, **91**, 159–64.

Blackman, R. L., Takada, H., and Kawakami, K. (1978). Chromosomal rearrangement involved in insecticide resistance of *Myzus persicae*. *Nature, Lond.*, **271**, 450–2.

Brown, T. M. and Brogdon, W. G. (1987). Improved detection of insecticide resistance through conventional and molecular techniques. *A. Rev. Ent.*, **32**, 145–62.

Clark, A. G. and Dauterman, W. C. (1982). The characterization by affinity chromatography of glutathione *S*-transferases from different strains of house fly. *Pestic. Biochem. Physiol.*, **17**, 307–14.

Clark, A.G., Shamaan, N.A., Sinclair, M.D., and Dauterman, W.C. (1986). Insecticide metabolism by multiple glutathione *S*-transferases in two strains of the house fly, *Musca domestica* (L). *Pestic. Biochem. Physiol.*, **25**, 169–75.

Devonshire, A.L. (1977). The properties of a carboxylesterase from the peach-potato aphid *Myzus persicae* (Sulz.), and its role in conferring insecticide resistance. *Biochem. J.*, **167**, 675–83.

Devonshire, A.L. and Moores, G.D. (1982). A carboxylesterase with broad substrate specificity causes organophosphorus, carbamate and pyrethroid resistance in peach-potato aphids (*Myzus persicae*). *Pestic. Biochem. Physiol.*, **18**, 235–46.

Devonshire, A.L. and Sawicki, R.M. (1979). Insecticide-resistant *Myzus persicae* as an example of evolution by gene duplication. *Nature, Lond.*, **280**, 140–1.

Devonshire, A.L., Moores, G.D., and Chiang, C. (1983). The biochemistry of insecticide resistance in the peach-potato aphid (*Myzus persicae*). *Pestic. Chem. (Proceedings of the 5th International Congress of Pesticide Chemistry)*, **3**, 191–6.

Devonshire, A.L., Moores, G.D., and ffrench-Constant, R.H. (1986). Detection of insecticide resistance by immunological estimation of carboxylesterase activity in *Myzus persicae* (Sulzer) and cross reaction of the antiserum with *Phorodon humuli* (Schrank) (Hemiptera: Aphididae). *Bull. ent. Res.*, **76**, 97–107.

ffrench-Constant, R.H. and Devonshire, A.L. (1987). A multiple homogeniser for rapid sample preparation in immunoassays and electrophoresis. *Biochem. Genet.*, **25**, 493–9.

ffrench-Constant, R.H., Devonshire, A.L., and Clark, S.J. (1987). Differential rate of selection for resistance by carbamate, organophosphorus and combined pyrethroid and organophosphorus insecticides in *Myzus persicae* (Sulzer) (Hemiptera: Aphididae). *Bull. ent. Res.*, **77**, 227–38.

ffrench-Constant, R.H., Byrne, F.J., Stribley, M.F., and Devonshire, A.L. (1988*a*). Rapid identification of the recently recognized *Myzus antirrhinii* (Macchiati) by polyacrylamide gel electrophoresis. *Entomologist*, **107**, 20–3.

ffrench-Constant, R.H., Devonshire A.L., and White, R.P. (1988*b*). Spontaneous loss and reselection of resistance in extremely resistant *Myzus persicae* (Sulzer). *Pestic. Biochem. Physiol.*, **30**, 1–10.

ffrench-Constant, R.H., Clark, S.J., and Devonshire, A.L. (1988*c*). Effect of decline of insecticide residues on selection for insecticide resistance in *Myzus persicae* (Sulzer) (Hemiptera: Aphididae). *Bull. ent. Res.*, **78**, 19–29.

Field, L.M., Devonshire, A.L., and Forde, B.G. (1988). Molecular evidence that insecticide resistance in peach-potato aphids (*Myzus persicae*, Sulz.) results from amplification of an esterase gene. *Biochem. J.*, **251**, 309–12.

Field, L. M., Devonshire, A. L., ffrench-Constant, R. H., and Forde, B. G. (1989*a*). Changes in DNA methylation are associated with loss of insecticide resistance in the peach potato aphid *Myzus persicae* (Sulz.). *FEBS Lett.*, **243**, 323–7.

Field, L. M. Devonshire, A. L., ffrench-Constant, R. H., and Forde., B. G. (1989*b*). The combined use of immunoassay and a DNA diagnostic technique to identify insecticide-resistant genotypes in the Peach-Potato Aphid, *Myzus persicae* (Sulz.). *Pestic. Biochem. Physiol.* (in press).

Higuchi, R., von Beroldingen, C. H., Sensabaugh, G. F., and Erlich, H. A. (1988). DNA typing from single hairs. *Nature, Lond.*, **332**, 543–6.

Hughes, P. B. and Raftos, D. A. (1985). Genetics of an esterase associated with resistance to organophosphorus insecticides in the sheep blowfly, *Lucilia cuprina* (Wiedemann) (Diptera: Calliphoridae). *Bull. ent. Res.*, **75**, 535–44.

Moores, G. D., Devonshire, A. L., and Denholm, I. (1988). A microtitre plate assay for characterising insensitive acetylcholinesterase genotypes of insecticide resistant insects. *Bull. ent. Res.*, **78**, 537–44.

Mouchès, C., Pasteur, N., Bergé, J. B., Hyrien, O., Raymond, M., de Saint Vincent, B. R., de Silvestri, M., and Georghiou, G. P. (1986). Amplification of an esterase gene is responsible for insecticide resistance in a California *Culex* mosquito. *Science*, **233**, 778–9.

Oppenoorth, F. J. (1985). Biochemistry and genetics of insecticide resistance. In *Comprehensive insect physiology, biochemistry and pharmacology*, Vol. 12, *Insect control* (ed. G. S. Kerkut and L. J. Gilbert), pp. 731–73. Pergamon Press, Oxford.

Ricci, G., Bello, M. L., Caccuri, A. M., Galilazzo, F., and Federici, G. (1984). Detection of glutathione transferase activity on polyacrylamide gels. *Analyt. Biochem.*, **143**, 226–30.

Sawicki, R. M., Devonshire, A. L., Payne, R. W., and Petzing, S. M. (1980). Stability of insecticide resistance in the peach-potato aphid, *Myzus persicae* (Sulzer). *Pestic. Sci.*, **11**, 33–42.

Stanton, R. H., Plapp, F. W. Jr., White, R. A., and Agosin, M. (1978). Induction of multiple cycochrome P-450 species in housefly microsomes—SDS-gel electrophoresis studies. *Comp. Biochem. Physiol.*, **61B**, 297–305.

Steele, R. W., and Maneckjee, A. (1979). Toxicological significance of acetylcholinesterase of the housefly thorax. *Pestic. Biochem. Physiol.*, **10**, 322–32.

Thomas, P. E., Ryan, D., and Levin, W. (1976). An improved staining procedure for the detection of the peroxidase activity of cytochrome P-450 on sodium dodecyl sulfate polyacrylamide gels. *Analyt. Biochem.*, **75**, 168–76.

17 Genetic fingerprinting of house sparrows, *Passer domesticus*

JON H. WETTON and DAVID T. PARKIN

Department of Genetics, School of Medicine, Queen's Medical Centre, Nottingham, UK

Abstract

There have been comparatively few studies of avian pest species using electrophoresis for the analysis of genetic structure. However, since these pest species are usually to be found in close association with man, their population dynamics and behaviour are frequently well known. This has made them attractive to evolutionary biologists, not directly because of their pest status, but because they represent a well-quantified group of species whose genetics can consequently be better understood.

Here, we show that although the number of polymorphic enzyme loci is relatively high within those wild bird populations so far examined, especially house sparrows (*Passer domesticus*), enzyme electrophoresis is of little use for the recognition of individuals within an avian population. In contrast, the analysis of hypervariable minisatellite DNA ('genetic fingerprinting') reveals extensive variation that allows unique recognition of individuals and the unequivocal assignment of parentage.

Introduction

Electrophoresis of soluble proteins was introduced to population genetics in the mid-1960s and rapidly gained popularity. It provided a quick and relatively inexpensive means of assaying many gene loci in large numbers of individuals. Over the following 20 years, it resulted in a deluge of research reports documenting levels of heterozygosity and polymorphism in species as diverse as wheat and

Electrophoretic Studies on Agricultural Pests (ed. Hugh D. Loxdale and J. den Hollander), Systematics Association Special Volume No. 39, pp. 375–91. Clarendon Press, Oxford, 1989.

wheat-bulb fly (Nevo 1978). The statistical analysis of these data sets progressed equally spectacularly (Nei 1975), so that it soon became possible to compare the genetic structure of populations and higher taxa with a facility that had never before been realized.

For some problems, the technique proved almost ideal. Population geneticists were freed from the constraints of having to sieze any fortuitous variation for study, and could select subjects for analysis for their intrinsic interest, or technical convenience, safe in the knowledge that electrophoresis would reveal polymorphisms that could be used to assess genetic structure.

There were difficulties, of course. Some species, such as species of *Drosophila*, revealed remarkably high levels of variation, whereas others appeared to be less variable. However, many of these differences could be explained in terms of population biology, and the research flourished.

Ornithologists were rather slow to climb aboard this particular electrophoretic bandwagon, and their tardiness was very unfortunate. Birds have been studied extensively, so that a great deal of information exists concerning their taxonomy, evolution, ecology, behaviour, and physiology. Thus, avian geneticists were able to apply their skills to biological problems, confident that any difficulties in the interpretation of their results could be overcome by reference to the vast library of ornithological literature. However, because there has been a great deal of interest in the demographic study of wild birds, one particular limitation of enzyme electrophoresis struck avian geneticists very forcibly. This is that the technique cannot easily be applied to genetic studies within individual populations.

Birds have the extremely valuable attribute of putting their eggs into a nest, and in many cases of caring for the young for a period of days, or even weeks. It is thus relatively easy to measure a series of components of fitness, such as clutch size, hatching success, growth rate, or fledging success. These parameters had been measured for populations and species, and it seemed appropriate to marry this methodology with studies of polymorphic enzymes to determine the fitnesses of the various genotypes.

Problems arose almost at once. For example, an enzyme locus (isocitrate dehydrogenase, IDH) is polymorphic in European populations of the house sparrow, *Passer domesticus* (Cole and Parkin 1981). Two alleles (A and B) segregate in a population in the east midlands of England that we have been studying for several years. When we compare parents with 10-day-old nestlings in their nests, we find an apparent distortion in genotype ratios (Table 17.1). There is clear evidence of a departure from the expected 1 : 1 segregation in

Table 17.1 The genotypes of 566 progeny scored for an isocitrate locus in a natural population of house sparrows. The parental genotypes are shown at the left-hand side of the table. Progeny whose genotypes are incompatible with their 'parents' are shown in **bold** type.

Male genotype	Female genotype	Progeny genotypes AA	AB	BB
AA	AA	119	**8**	0
	AB	44	47	**4**
	BB	0	12	2
AB	AA	65	39	0
	AB	38	66	28
	BB	0	16	18
BB	AA	0	31	0
	AB	**3**	14	12
	BB	0	0	0

families where the male is genetically AB ($P = 0.05$). This is not apparent in the reverse matings. We have found similar anomalies at other enzyme loci and, in all cases, heterozygous males appear to transmit the common allele more frequently than would be expected. This phenomenon is due to the birds' behaviour, rather than any effect associated with the loci themselves. The apparent distributions are explicable by a significant amount of extra-pair fertilizations in the population. If a female copulates with a male other than her (heterozygous) mate, she is more likely to be fertilized by a sperm carrying the commoner allele. Conclusive proof of extra-pair copulation is provided in Table 17.1, in which 17 nestlings (3 per cent) have genotypes that are incompatible with either one or specifically the male adult. To study the fitness of particular genotypes, it is important initially to exclude all cases of non-parentage.

Enzyme data such as these ought to be an efficient way of searching for evidence of extra-pair copulations in wild birds. They reveal extensive polymorphism, and it should be possible to recognize each individual more or less uniquely, and thereby pick out mismatching offspring in groups of nestlings. This approach has indeed been tried by a series of researchers, but the success rate is relatively low, because of the inefficiency of the loci under analysis. The house sparrow, for instance, is one of the more polymorphic birds when monitored by electrophoresis (Evans 1987). Fifteen enzyme loci have been identified that show variation is most populations, although several of these reveal only the occasional rare variant. The allele frequencies recorded at 11 of these in a large sample of 290 birds from Northampton, England, are shown in Table 17.2. For the purpose of this analysis, the identity of the individual alleles is not relevant. The number of different genotypes at each locus is also shown in Table 17.2, and multiplying these together suggests a staggering total of 47 239 200 possible genotypes! However, many of these are vanishingly rare, and only a tiny fraction of the total has ever been recorded.

We can calculate the probability that two individuals have the same genotype at every locus as follows. Assuming that the population is in Hardy–Weinberg equilibrium, then we can estimate the frequency of every possible genotype from the polynomial expansion:

$$p^2, 2pq, q^2, 2pr, 2qr \ldots$$

The probability of two separate individuals having the same genotype is given by the sum of the square of each of these probabilities, and these are also shown in Table 17.2. Despite the astonishing potential for variation, relatively little is realized, and the probability of two random individuals having exactly the same enzymic profile is 0.0023, or about 1 in 500. With populations of house sparrows that number only a few thousand (Parkin and Cole 1984), individual recognition is very unlikely.

In fact, the position is worse than this, for several of the loci included in Table 17.2 can only be recognized from liver tissue. If the analysis is restricted to those enzymes that can be recognized from blood samples, the probability rises to 0.033. Demographic analysis then becomes very tedious: it is only necessary for the male and the extra-pair male to share one allele at each locus for potential extra-pair fertilization to be unrecognizable. Since the commonest allele at almost all of these loci has a frequency that exceeds 50 per cent, this occurs quite frequently, and complex statistical techniques are required to allow for families that cannot be used to detect mismatching progeny (Chakraborty *et al.* 1988). Using these we find (Wetton, in prep.) that approximately 20 per cent of mismatches are detectable using IDH, and only about 50 per cent when all seven loci are included. This suggests that about 15 per cent of nestling house sparrows are the result of extra-pair fertilizations.

It has been recognized for some time that a technique is required that is much more sensitive to intrapopulation variability, one that allows the unique recognition of individuals and also allows the determination of relatedness. Genetic fingerprinting is such a technique.

Table 17.2 Allele frequencies at 11 polymorphic protein loci in a sample of house sparrows taken from a population near Northampton, England. The penultimate column shows the number of possible genotypes at that locus, and the final column (**bold** type) shows the probability that two random individuals have the same genotype at that locus.

Adenosine deaminase	0.4948	0.4536	0.0412	0.0052	0.0052	15	**0.3069**
Di-peptidase	0.9620	0.0380				3	**0.8618**
Tri-peptidase	0.9541	0.0245	0.0102	0.0102		10	**0.8316**
Esterase	0.6543	0.3191	0.0266			6	**0.3695**
Isocitrate DH A	0.9032	0.0914	0.0054			6	**0.6929**
Isocitrate DH C	0.7500	0.2500				3	**0.4609**
Sorbitol DH	0.6735	0.3061	0.0204			6	**0.3854**
6-Phosphogluconate DH	0.9897	0.0103				3	**0.9598**
Phosphoglucomutase 1	0.8298	0.1702				3	**0.5548**
Phosphoglucomutase 2	0.8814	0.1134	0.0052			6	**0.6437**
Aconitase (sex-linked)	0.8658	0.1342				3	**0.6717**

Genetic fingerprinting

This new and sensitive technique was first reported by Jeffreys *et al.* (1985*a*). They discovered a class of hypervariable minisatellite DNA that occurs scattered throughout the human genome. It consists of a series of tandemly-repeated DNA sequences in which the repeated units are all variations of a common 'core'. The great variability associated with the minisatellites results from differences in the number of repeats at particular sites in individual chromosome.

The hybridization probes that we have used are designated 33.6 and 33.15 (Jeffreys *et al.* 1985*a*). They consist of slight variations of the core sequence, tandemly repeated 18 and 29 times, respectively. They can be used to detect any fragments of minisatellite DNA with which they share some homology.

The DNA from an individual to be tested is extracted from a suitable tissue (usually blood in birds, because avian erythrocytes are nucleated) and purified (see Wetton *et al.* 1987 for technical details). It is then digested with an appropriate restriction enzyme, usually HaeIII. This enzyme cleaves DNA at GGCC sites that occur frequently through the genome (approximately every 256 base pairs). There is no recognition site for HaeIII within the minisatellite sequence, so that the cleaved DNA comprises a large number of very small fragments, with a proportion of larger fragments that include, more or less intact, the minisatellite sequences.

The fragments of DNA are separated by electrophoresis through an agarose gel, under conditions that partition by molecular weight rather than charge. Smaller fragments are allowed to migrate through the gel into the buffer at the other end. Larger fragments are retained in the gel, and then transferred directly onto a nylon membrane for ease of handling. The DNA is dissociated into its component strands, and then treated to make it bind firmly to the nylon membrane. The nylon is flooded with radioactively labelled probe, which hybridizes with the minisatellite sequences bound to the membrane. Presence of minisatellite fragments is detected by autoradiography.

Results

Figure 17.1 shows a typical autoradiograph from a group of house sparrows probed with 33.6. These birds comprise two adults and 17 progeny that they reared over three breeding seasons. We have measured the position of 30 bands in this blot, and present

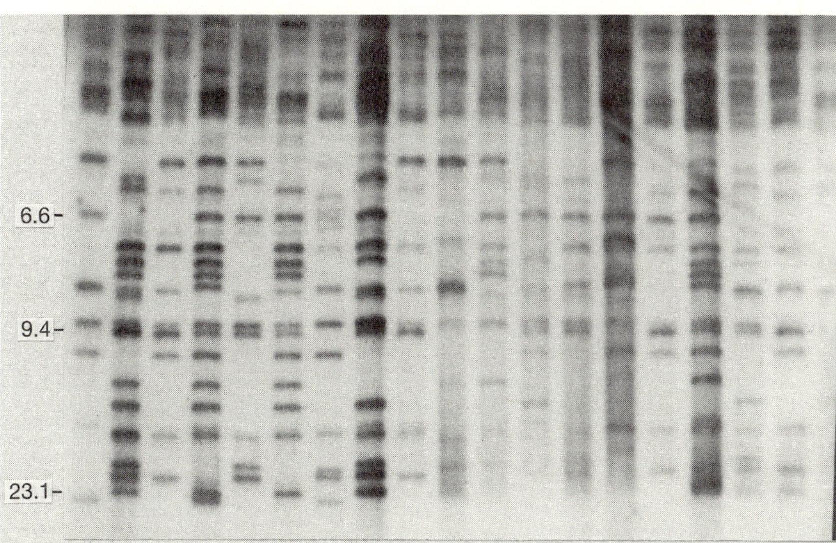

Fig. 17.1. Autoradiograph of a family of house sparrows having the genetic fingerprint pattern revealed with probe 33.6. The vertical tracks consist of fragments of sizes in kilobases shown on the margin. F and M are the two parents; 1 to 17 are their progeny. See text for further details.

these as Table 17.3. The membrane was re-probed with 33.15, and the results of this are also shown. Lastly, we also scored seven polymorphic protein loci in the 19 birds, and present the genotypes as the last seven lines of Table 17.3. We shall discuss the enzyme data in more detail later, but note for the present that the esterase genotype of progeny 5 is incompatible with that of the male, as is the transferrin genotype of progeny 12.

Returning to Fig. 17.1, we can quantify the resemblances between individuals using a 'coefficient of resemblance', *F*, previously developed for the analysis of restriction-fragment-length polymorphisms (Lansman *et al.* 1981). This statistic is given as

$$F = \frac{2N_{AB}}{N_A + N_B},$$

where N_A and N_B are the number of bands in the tracks being compared, and (N_{AB}) is the number of these that are present in both. Thus, the adult male and female have 16 and 7 bands respectively, of which 1 is shared, giving a value of $F = 0.087$. This is merely

Table 17.3 Diagrammatic representation of the autoradiographs produced when two adult house sparrows and their 17 progeny were analysed for their genetic fingerprint. The first 30 rows show the presence (+) or absence (−) of bands using probe 33.6. The next 14 show the same for probe 33.15. The final 7 rows give the genotype for 7 polymorphic protein loci; genotypes shown in **bold** type show data that genetically mismatch with those of the 'parents'.

	F	M	1	2	3	4	5	6	7	8	9	10	11	12	13	14	15	16	17
1	+	−	+	+	+	−	−	−	+	+	+	−	−	+	−	+	−	−	−
2	−	−	−	−	−	−	+	−	−	−	−	−	−	−	−	−	−	−	−
3	−	+	−	−	+	−	−	+	−	+	+	+	+	−	−	−	+	−	+
4	−	+	+	+	−	+	−	+	+	−	−	−	−	−	+	+	−	+	+
5	+	−	−	+	+	+	−	+	−	−	+	+	+	+	+	+	−	−	−
6	−	−	−	−	−	−	+	−	−	−	−	−	−	−	−	−	−	−	−
7	−	+	+	+	+	+	−	+	−	−	−	+	−	−	−	−	+	+	+
8	−	+	+	+	−	+	−	+	+	−	+	−	+	+	+	+	+	+	−
9	−	−	−	−	−	−	+	−	−	−	−	−	−	−	−	−	−	−	−
10	−	−	−	−	−	−	−	−	−	−	−	+	−	−	−	−	−	−	−
11	−	+	−	+	−	+	−	+	−	−	+	−	−	−	−	+	+	−	+
12	−	+	−	+	−	+	−	−	−	−	+	−	−	−	−	+	−	−	−
13	+	−	−	+	−	−	+	−	−	+	−	+	+	−	−	+	+	+	+
14	−	+	+	−	−	+	−	+	+	+	−	+	+	−	−	−	+	−	+
15	−	+	−	−	+	−	−	+	−	−	−	−	−	−	−	−	−	−	−
16	+	+	+	+	+	+	+	+	+	+	+	+	+	+	+	+	+	+	+
17	−	+	+	+	+	+	−	+	+	−	−	+	+	−	+	+	+	+	+
18	+	−	+	+	−	+	+	−	−	−	−	+	−	+	+	+	−	+	−
19	−	+	−	+	−	+	−	−	−	+	+	−	−	−	−	+	−	−	−
20	−	+	−	+	−	+	−	+	−	−	−	+	−	−	−	−	+	−	+
21	+	−	−	+	−	+	+	+	+	−	−	+	−	−	−	+	−	−	−
22	−	−	−	−	−	−	−	−	−	−	−	−	−	+	−	−	−	−	−
23	−	−	−	−	−	−	−	−	−	−	−	−	−	+	−	−	−	−	−
24	−	+	+	+	+	+	+	+	+	−	+	+	+	−	+	−	+	+	−

25	–	–	–	–	–	–	–	–	–	+	–	–	–	–	–	–	–	–	–
26	–	+	–	–	+	–	–	+	–	–	+	+	+	–	–	–	+	–	+
27	–	–	–	–	–	–	+	–	–	+	–	–	–	–	–	–	–	–	–
28	–	+	+	–	+	–	+	+	+	–	–	+	+	+	+	+	+	+	+
29	–	+	–	+	–	+	–	+	–	–	–	–	–	–	–	+	–	–	–
30	+	–	–	+	–	–	+	–	–	–	–	–	–	–	–	–	+	+	–
1	–	–	–	–	–	–	+	–	–	–	–	–	–	–	–	–	–	–	–
2	–	+	–	+	–	–	–	+	–	–	+	–	+	–	+	–	–	+	+
3	–	–	–	–	–	–	+	–	–	–	–	–	–	–	–	–	–	–	–
4	–	–	–	–	–	–	–	–	–	+	–	–	–	–	–	–	–	–	–
5	+	–	+	–	–	+	–	–	–	+	–	+	–	–	–	+	–	+	–
6	–	+	+	+	–	–	–	–	–	–	+	–	+	–	–	+	–	–	+
7	–	+	–	–	–	+	–	–	–	+	–	+	–	–	+	–	–	+	–
8	–	–	+	–	–	–	+	–	–	–	–	–	–	–	–	–	–	–	–
9	–	+	–	–	+	+	–	–	+	–	–	+	–	–	–	+	+	–	–
10	–	–	–	–	–	–	–	–	–	+	–	–	–	–	–	–	–	–	–
11	+	–	–	–	–	–	–	–	+	+	–	–	–	–	+	+	–	+	–
12	–	+	+	+	–	–	–	–	+	–	–	–	–	–	–	–	–	–	–
13	+	–	+	+	+	+	+	+	–	–	+	+	+	+	–	–	+	–	+
14	+	–	–	+	+	+	+	+	–	+	+	–	+	+	–	+	–	+	+
PepD2	BB	BC	BC	BB	BB	BB	BB	BC	BB	BB	BB	BB	BC	BB	BC	BB	BB	BB	BB
PepD3	BB	BB	BB	BB	BB	BB	BB	BB	BB	BB	BB	BB	BB	BB	BB	BB	BB	BB	BB
Pep-T	DD	BD	DD	DD	DD	DD	DD	DD	BD	DD	DD	DD	DD	DD	DD	DD	DD	DD	BD
Idh-C	AA	AB	AB	AA	AA	AA	AA	AA	AB	AA	AA	AA	AA	AB	AA	AB	AA	AB	AA
Trf	AB	BB	BB	BB	BB	AB	BB	BB	**	BB	AB	AB	BB	AA	BB	BB	BB	AB	AB
Est	AB	BB	AB	BB	BB	BB	AA	BB	BB	BB	BB	AB	BB	AB	BB	BB	BB	**	**
6 Pgd	BB	BB	BB	BB	BB	BB	BB	BB	BB	BB	BB	BB	BB	BB	BB	BB	BB	BB	BB

Table 17.4 Coefficients of identity (*F*) between the two adults in Table 17.3 and the 17 progeny that they reared during three breeding seasons, based upon the genetic fingerprints produced with probe 33.6, and illustrated in Fig. 17.1.

	P1	P2	P3	P4	P5	P6	P7	P8	P9
Female	0.353	0.583	0.353	0.364	0.556	0.261	0.375	0.400	0.353
Male	0.615	0.667	0.615	0.774	0.222	0.875	0.560	0.333	0.615

	P10	P11	P12	P13	P14	P15	P16	P17
Female	0.476	0.353	0.533	0.400	0.571	0.300	0.471	0.222
Male	0.600	0.615	0.250	0.500	0.600	0.769	0.538	0.741

a measure of phenotypic resemblance: a moment's thought reveals that F can vary between zero when no bands are shared to unity when the two tracks are identical.

Table 17.4 shows the values of F for both parents compared with each of the 17 progeny based upon the data from probe 33.6. Considering the female first, the similarities range from 0.22 to 0.58 with a mean of 0.407 (SD = 0.108, see Table 17.5). These data are too few to test formally for normality, but they seem reasonably homogeneous. Such is not the case for the male, however. The coefficients are more variable, ranging from 0.22 to 0.88 (mean = 0.581, SD = 0.177), and three of them are distinct 'outliers' with values below 0.35 while the remainder all exceed 0.5. These three birds are progeny numbers 5, 8, and 12 and, interestingly, two of them (5 and 12) were found to mismatch genetically with the male parent from their enzyme genotypes. Could it be that these three individuals are all unrelated to the male, having been produced by the female indulging in extra-pair copulation? We have several strands of evidence that suggest this to be the case.

First, as we have already shown, the mean coefficient of similarity between the male and the 14 progeny is 0.648 (SD = 0.104) which differs significantly ($P < 0.05$) from the mean of 0.268 (SD = 0.058) with the other three.

Secondly, we find equally convincing differences when we compare the progeny with each other, rather than with the adults. The 14 progeny resemble each other with a mean of 0.598 (SD = 0.117), whereas 5, 8, and 12 resemble the rest with F = 0.375 (SD = 0.104). Thus, the resemblance between these three and the remainder is substantially less than the latter among themselves.

Table 17.5 Mean and SD of coefficients of identity (*F*) between various groups of birds in the pedigree shown in Table 17.3. Calculations were performed using the banding patterns derived with both probes 33.6 and 33.15.

	F	SD	*N*
Female c.w. all 17 progeny	0.438	0.094	17
Female c.w. 14 matching progeny	0.420	0.092	14
Male c.w. 14 matching progeny	0.593	0.076	14
Female c.w. 3 mismatching progeny	0.523	0.046	3
Male c.w. 3 mismatching progeny	0.214	0.064	3
14 matching progeny compared among themselves	0.557	0.106	91
Three mismatch progeny compared among themselves	0.300	0.073	3
Three mismatch progeny c.w. 14 matching progeny	0.369	0.073	42

Thirdly, and most convincingly, there is clear evidence of mismatches from the banding patterns in Fig. 17.1. It has been shown by Jeffreys *et al.* (1985*a, b*), Wetton *et al.* (1987), and Burke and Bruford (1987) that the pattern of inheritance of minisatellite bands is relatively straightforward. Apart from the occasional mutation, every band present in an offspring is also found in one or other (or both) of the parents. Conversely, the more bands an individual possesses that are absent from both parents, the less likely it is to be their offspring. Careful matching of the offspring bands with those of the putative parents will reveal which (if either) of the adults is a parent, and which is not. In Fig. 17.1 and Table 17.3, it can be seen that there is extensive band mismatching between the adults and each of progeny 5, 8, and 12. Closer inspection reveals that several bands in each offspring match with the female, but very few with the male. These results confirm that the three progeny are indeed the consequences of extra-pair copulation (each in a different year). Furthermore, if the probability of fertilization is directly related to the frequency of copulation, it implies that 17 per cent of the matings in which the female participated were promiscuous.

It seems unlikely that the same male was involved in these three extra-pair copulations. We cannot presently assign probabilities to this, since we do yet know the detailed frequencies of the various bands in the wild population (see Chakraborty *et al.* 1988; Brookfield, in prep). However, there are two further strands of evidence that suggest this to be the case. First, there is little homogeneity among the paternal bands in the three extra-pair progeny. it is possible to reconstruct the fingerprint pattern of missing parents from those of the progeny and the other adult (Wellbourn *et al.*, in prep). We do not have enough extensive families for this in the present case, but rather few bands co-occur among the paternal components of 5, 8, and 12. Secondly, the similarity coefficient of the three mismatching progeny and the remainder of the brood (0.375; Table 17.5) is close to that of the mismatches among themselves (0.294; SD = 0.038). These three are half-siblings to the remainder of the family, since they share a mother. The similarity of the coefficients suggests that they also are half-siblings among themselves. If the same male were the father of these three, they would be full siblings with a coefficient closer to 0.5, so that we can deduce that a different male was involved in each mating.

Using these banding patterns to infer relatedness implies that each band is independent of the rest: in genetic terms, that they are not allelic and are unlinked. Certainly, among the 14 full siblings and across both autoradiographs, there is a high degree of polymorphism, for only one band is present is every individual. Thus, apart from bnd 15 in 33.6, both parents must have been heterozygous at each minisatellite locus that we can detect. There is one instance in which a pair of bands that are present in an adult are found exclusively as alternatives among the 14 offspring. Bands 6 and 8 in autoradiograph 33.15 both occur in the female, and apparently segregate among the 14 genuine offspring (and, as might be expected, among the three mismatches as well). So, apart from this single example, it seems that each locus is heterozygous and that the alternative alleles are too small to be identified by our detection methods. We therefore have 43 loci, only 1 of which (band 15, autoradiograph 33.6) may be homozygous — a level of heterozygosity of over 95 per cent.

We can also examine the linkage of the fragments by a comprehensive analysis of the banding combinations of the parents and their 14 progeny. Bands representing minisatellite loci that are closely linked will be inherited together more frequently than expected under random assortment. Hence, allowing for the single example of allelism detected by probe 33.15, we can compare the segregation of pairs of bands present in one adult but lacking from the other. For example, bands 1 and 5 in autoradiograph 33.15 are both present in the female, but absent from the male. If they occurred at closely linked sites, these bands should co-occur in, or be absent from, the 14 progeny more often than by chance. We have analysed this using lod scores (Cavalli-Sforza and Bodmer 1971), and find no evidence of linkage for this pair of bands.

A comprehensive analysis of the co-segregation of all the bands in Table 17.3 from both autoradiographs reveals 319 instances in which we can look for evidence of linkage. In 12 of these, there is evidence that θ (the linkage estimator) departs significantly from 0.5. However, with such an extensive array of data, it is wrong to use the conventional $P = 0.05$ level of significance. A modified value is given by Sidak (in Sokal and Rohlf 1981) as

$$P' = 1 - (1 - P')^{1/N}$$

where P is the conventional level of significance and N is the number of analyses. In this case, $P' = 0.00016$. Applying this value of P' to the data reveals that only two have probabilities that are statistically significant. Bands 3 and 26, and 12 and 19 detected by probe 33.6 give evidence of linkage: in both cases the bands occur in the male and are either both present or both absent in all of the progeny. Extending this analysis to the enzyme data reveals no evidence of linkage between a minisatellite band and any enzyme allele.

Discussion

It has been shown elsewhere (Wetton *et al.* 1987; Burke and Bruford 1987) that minisatellites of the class reported by Jeffreys *et al.* (1985*a*) are as variable among birds as in humankind. The core sequence appears to be sufficiently conserved that a probe developed from human sources will hybridize freely with avian DNA and produce fragment patterns ('genetic fingerprints') that are equally variable. The bands are inherited in a Mendelian fashion, with only two instances of linkage in the pedigree reported here. The lack of linkage between bands detectable from the same individual, using the two probes, implies that these are hybridizing with different subsets of minisatellite. We have no idea which minisatellite bands are associated with the many chromosomes of the house sparrow; however, it is remarkable that we detect 44 band fragments, representing 41 segregating loci, when the genome is believed (Bulatova *et al.* 1972) to consist of 38 pairs of chromosomes. This implies a wide dispersion of the minisatellite loci among the chromosomes, a conclusion supported by the observations of Jeffreys *et al.* (1987).

The data from the two autoradiographs can thus be merged for measurements of similarity. These combined data confirm the clustering of similarity coefficients determined for the data sets separately. The 14 full siblings resemble each other, with mean $F = 0.56$, so that they share approximately 50 per cent of their bands. This would be expected given the high degree of heterozygosity, since each progeny should acquire half of its bands from each parent, with a similar 50 per cent chance of the same band being transmitted to any two offspring. The resemblance between the male and these 14 progeny is 0.59. Since the bands are almost entirely unlinked, and only one is homozygous, we would expect 50 per cent of parental bands to be transmitted to each offspring. The observed 59 per cent inherited from the male is slightly higher than this, but he has rather more minisatellite bands than his mate (21 compared with 11), and so will pass more to the offspring. Conversely, the value of $F = 0.44$ for the comparison between the female and her 17 offspring is rather less than the expected 50 per cent because she has an apparent deficiency of the large scorable fragments.

The three 'mismatch' progeny compare closely with the female ($F = 0.523$) as their natural parent, but differ quite sharply from the male ($F = 0.214$). Their similarity to each other ($F = 0.300$) is close to that with the remainder of the broods ($F = 0.369$). Since they are known to be half-siblings of the other 14, it may be inferred that they are also half-siblings with each other, so that a different male fathered each one.

Using genetic fingerprinting the uniqueness of individuals can be quantified through the probability of band-sharing between tracks (Jeffreys *et al.* 1985*b*). We can compute this directly from the 14 siblings in Table 17.3. Considering the two autoradiographs separately, we can calculate the proportion of bands present in each individual that are also present in each of the others. The mean of these gives an estimate of the amount of band-sharing. While this is not the optimal method of analysis (Brookfield, in prep.) it is the best that can be attempted because we do not have detailed data relating to band frequencies in nature.

Since virtually all of the bands in Table 17.3 are unlinked, we can assume that they are independent of one another. Thus, the probability that two individuals will have the same genetic fingerprint is given as the probability that they share a band raised to the power of the average number of bands per individual (Jeffreys *et al.* 1985*b*). Thus, for 33.6, the probability of identical tracks is

$$0.614^{11.929} = 0.00297,$$

and for 33.15 it is

$$0.446^{3.929} = 0.0419.$$

The probability that two full siblings are identical with both probes is the product of these, 0.000125. In other words, the two adults would need to produce over 8000 progeny before we might expect two to have the same pattern with both probes.

There are four pairs of birds that are unrelated in this pedigree: the original adults form one pair and; the male is unrelated to the three mismatching progeny. However, since these three are half-siblings, they are not independent of one another. We have selected progeny 5 as representative of these, and so have two pairs that can be used to determine the amount of band-sharing between unrelated individuals. The data in autoradiograph 33.6 give a band-sharing probability of 0.090, with an average number of bands equal to 11.33. For probe 33.15 it is a little more difficult to estimate the amount of band-sharing, since the three birds in fact share none. We assume that each shares one-quarter of a band with each of the others, giving a total of one band shared, a mean band-sharing probability of 0.053, and an average of 4.67 bands per individual. The probability of identity of these unrelated birds is

$$0.090^{11.3} \times 0.053^{4.67} = 1.52 \times 10^{-18}.$$

Such an immense amount of variation implies an extremely high mutation rate, and Jeffreys *et al.* (1985*a, b*) estimated this as about 0.005 for humankind. In our present pedigree, we have two mismatching bands (indicated in bold type in Table 17.3). One of these occurs as band 10 in progeny 10 from the 33.6 autoradiograph; we have little idea with which parental band it is homologous, if indeed its progenitor was of a size to be detectable anyway. The second appears to be the same as a band inherited by progeny 7 from the extra-pair male; whether it is derived from the same locus, we do not know. However, it is very similar in size (within 1 mm in the autoradiograph) to band 9 in the male parent. If it is derived from this band, and most mutations tend to involve small changes in repeat number, it completes a pattern of segregation between bands 2 and 9 in the male parent. Among the 14 progeny, there are 221 bands that are faithfully transmitted from parent to offspring, and two that apparently are not. This suggests a mutation rate of approximately 2/220, which is reasonably close to those determined by Jeffreys *et al.* (1985*b*) and Burke and Bruford (1987).

The remarkable variation revealed by DNA fingerprinting—whether derived from the 'Jeffreys' probes, or using alternatives (Vassart *et al.* 1987; Fowler *et al.*, in press)—has great value to research that necessitates the recognition of individuality within a group of organisms. Apart from disentangling complex demographic problems (e.g. Wetton *et al.* 1987), it has obvious forensic applications (Jeffreys *et al.* 1985*c*; Gill *et al.* 1985). It has important implications for tissue recognition in biomedical science (Thein *et al.* 1987), and can be used to assist with chromosome mapping and the understanding of genetic disease (Jeffreys *et al.* 1986). Within the field of pest research, it has the potential to allow the recognition of different clones of parthenogenetically-reproducing aphids (Carter and Devonshire, unpublished), especially within species that are enzymically rather invariant. It seems likely that it will replace standard gel electrophoresis of proteins in situations where fine-scale recognition of individuality is required.

Acknowledgements

This research was supported by a Natural Environment Research Council postgraduate studentship to J.W., and a research grant to D.T.P. We are grateful to Mr Peter Benson and Brackenhurst College of Agriculture for permission to study the sparrows around their farm buildings, and to Dr T.A. Burke, who initiated the study of this population. We are especially grateful to Professor Alec Jeffreys and the Lister Institute for providing the DNA probes, and for help and advice at all stages of the study. Lastly, the manuscript was read and criticized by Roy Carter, Marian Hamshere, Jon Hesp, and Meng Anmeng. To all of them, we extend our thanks.

References

Bulatova, N.Sh., Radjabli, S.I., and Panov, E.N. (1972). Karyological description of three species of the genus *Passer*. *Experientia*, **28**, 1369–71.

Burke, T. and Bruford, M.W. (1987). DNA fingerprinting in birds. *Nature, Lond.*, **327**, 149–52.

Cavalli-Sforza, L.L. and Bodmer, W.F. (1971). *The genetics of human populations*. Freeman, San Francisco.

Chakraborty, R., Meager, T.R., and Smouse, P.E. (1988). Parentage analysis with genetic markers in natural populations. 1. The expected proportion of offspring with unambiguous paternity. *Genetics*, **118**, 527–36.

Cole, S.R. and Parkin, D.T. (1981). Enzyme polymorphism in the house sparrow, *Passer domesticus*. *Biol. J. Linn. Soc.*, **15**, 13–22.

Evans, P.G.H. (1987). Electrophoretic variability of gene products. In *Avian genetics* (ed. F.C. Cooke and P.A. Buckley). Academic Press, London.

Fowler, S.J., Werrett, D., and Higgs, D.R. (1988). Individual-specific DNA fingerprints from a hypervariable region probe: alpha globin 3′HVR. *Hum. Genet.* (in press).

Gill, P., Jeffreys, A.J., and Werrett, D.J. (1985). Forensic application of DNA fingerprints. *Nature, Lond.*, **318**, 577–9.

Jeffreys, A.J., Wilson, V., and Thein, S.L. (1985*a*). Hypervariable 'minisatellite' regions in human DNA. *Nature, Lond.*, **316**, 67–74.

Jeffreys, A.J., Wilson, V., and Thein, S.L. (1985*b*). Individual-specific 'fingerprints' of human DNA. *Nature, Lond.*, **316**, 76–79.

Jeffreys, A.J., Brookfield, J.F., and Semeonoff, R. (1985*c*). Positive identification of an immigration test-case using human DNA fingerprints. *Nature, Lond.*, **317**, 818–19.

Jeffreys, A.J., Wilson, V., Thein, S.L., Weatherall, D.J., and Ponder, B.A.J. (1986). DNA 'fingerprints' and linkage analysis in human pedigrees. *Am. J. hum. Genet.*, **39**, 11–24.

Jeffreys, A.J., Wilson, V., Kelly, R., Taylor, B.A., and Bulfield, G. (1987). Mouse DNA 'fingerprints': analysis of chromosome localization and germ-line stability of hypervariable loci in recombinant inbred strains. *Nucl. Acids Res.*, **15**, 2823–36.

Lansman, R.A., Shade, R.O., Shapira, J.F., and Avise, J.C. (1981). The use of restriction endonucleases to measure mitochondrial DNA sequence relatedness in natural populations. III. Techniques and potential. *J. molec. Evol.*, **17**, 214–26.

Nei, M. (1975). *Molecular population genetics and evolution*. North-Holland, Amsterdam.

Nevo, E. (1978). Genetic variation in natural populations: patterns and theory. *Theor. Pop. Biol.*, **13**, 121–77.

Parkin, D.T. and Cole, S.R. (1984). Genetic variation in the house sparrow (*Passer domesticus*) in the east midlands of England. *Biol. J. Linn. Soc.*, **23**, 287–301.

Sokal, R.R. and Rohlf, F.J. (1981). *Biometry* (2nd edn). Freeman, San Francisco.

Thein, S.L., Jeffreys, A.J., Govi, H.C., Flint, J., O'Connor, N.T.J., Weatherall, D.J., and Wainscoat, J.S. (1987). Detection of somatic changes in human cancer DNA by DNA fingerprint analysis. *Br. J. Cancer*, **55**, 353–6.

Vassart, G., Georges, M., Monsueir, R., Brocas, H., Lequarre, A.S., and Christophe, D. (1987). A sequence in M13 phage detects hypervariable minisatellite in human and animal DNA. *Science*, **235**, 683–4.

Wetton, J.H., Carter, R.E., Parkin, D.T., and Walters, D. (1987). Demographic study of a wild house sparrow population by DNA fingerprinting. *Nature, Lond.*, **227**, 147–9.

18 The population structure of *Deroceras reticulatum* (Mollusca: Pulmonata: Limacidae)

C.C. FLEMING

Department of Agricultural Zoology, The Queen's University of Belfast and Agricultural Zoology Research Division, Department of Agriculture for Northern Ireland, Newtownabbey, County Antrim, Northern Ireland, UK

Abstract

The use of biochemical techniques in the investigation of pest biology and systematics is well established. These techniques are particularly well suited to the study of molluscs, for which they can complement the more classical approaches to taxonomy and population biology.

Despite considerable efforts, much still has to be learned regarding the biology and ecology of slug pests. The present investigation attempts to demonstrate the efficacy of a biochemical approach to slug taxonomy and in assessing the population genetic structure of a major agricultural pest, *Deroceras reticulatum*. Allelic frequency data reveal great differences in the amounts of genetic variation in *D. reticulatum* populations and considerable genetic heterogeneity among populations. Much of the heterogeneity is the result of microgeographic variation and evidence of additional population substructuring is presented.

Introduction

Terrestrial gastropods occur in many regions of the world. However, it is usually only in cool, moist, temperate climates that they attain

Electrophoretic Studies on Agricultural Pests (ed. Hugh D. Loxdale and J. den Hollander), Systematics Association Special Volume No. 39, pp. 393–413. Clarendon Press, Oxford, 1989.

serious pest status. Such is the case in the British Isles, where slugs attack a wide range of crops, particularly potatoes and cereals (Port and Port, 1986).

Despite much effort, there are still taxonomic problems associated with many of the most common species. Much of the confusion in taxonomy results from the extreme plasticity and variation of the morphological characters used for identification (Evans 1985). This, combined with the presence of species complexes can, in many cases, make identification difficult and time-consuming. With interest growing in new approaches to slug control, for example using pheromones, pathogens, and plant resistance (Port and Port 1986), accurate species identification will become more important, as even closely related species sometimes differ greatly in ecology, behaviour, and susceptibility to different control measures.

Owing to the practical problems in working with slugs, and because of the localized and sporadic nature of serious pest outbreaks, research has in the past tended to be uncoordinated and much still has to be learnt regarding their biology and ecology. In particular, information on population size, dynamics, migration and homing, and the genetic structure of populations would be valuable to workers involved in control and in the construction of predictive models for damage and control. By its very nature and, perhaps more importantly, because of the subterranean habits of most slugs, such information is difficult to obtain using classical ecological methods. Biochemical techniques are now widely used in the study of agricultural pests and the approach is particularly amenable to the study of slugs, since many aspects of their taxonomy and population biology can be assessed using genetic data derived from the analysis of protein variation.

Although Gastropods have been the subject of numerous population genetic studies (Selander and Kaufman 1975; Johnson 1976; McCracken and Selander 1980; Nicklas and Hoffman 1981; Foltz *et al.* 1982), relatively little is known regarding the genetic population structures of the more important pest species in agricultural ecosystems.

In this chapter, the efficacy of electrophoresis, both as a tool in slug taxonomy and for obtaining information on the population and genetic structures of slug pests, is demonstrated with special reference to *Deroceras reticulatum* (Müller).

Materials and methods

1. *Species identification*

Samples of *Arion hortensis* (Ferrusac), *A. distinctus* (Mabelle) and A. *owenii* (Davies) (the *Arion hortensis* complex), were obtained from various locations in Northern Ireland and occasionally Britain. In addition, *A. circumscriptus* (Johnston), *A. intermedius* (Normand) and *D. reticulatum* were collected from Irish sites and used in the taxonomic analysis.

2. *Population structure analysis of D. reticulatum*

(a) Sampling strategy To assess the population structure of a species, it is important to devise an appropriate sampling strategy. Unfortunately, a sampling system that provides maximum information is only determined in the light of previous knowledge of the population structure of the target species — in other words, with the very information that is being sought. Consequently, in preliminary studies, it is usually necessary to use a less-efficient sampling programme. The data thus obtained can then be used to maximize the efficiency of subsequent studies.

Allendorf and Phelps (1981) described four measures of diversity that they considered important when designing a sampling programme. All can be estimated using allele frequency data:

(1) the amount of genetic variation in an average population;
(2) variability in the amount of genetic variation amongst populations;
(3) the amount of genetic differentiation between populations;
(4) the pattern of genetic similarity amongst populations.

Prior to obtaining such information, the practical aspects of sampling must also be considered. Initially, the total number of sample sites must be determined. Clearly, the greater the number the better; however, costs and time often restrict the number that can be studied. The distribution of sample sites is also important. Here the optimum distribution depends on the pattern of genetic differentiation within the species. When little is known about a particular species, a compromise must be made between sampling evenly across an area (providing the best description of broad geographic differences) and sampling clusters of local populations (assessing microgeographic variation). At this stage, the size of sample to survey at each site must also be decided upon. Marshall and Brown (1975) suggest that an investigation should aim at detecting at least 95 per cent of all alleles with a frequency of 0.05. Thus, for a di-allelic locus with one allele at a frequency of 0.05, at least 30 diploid individuals must be examined. Lastly, the actual number of loci screened electrophoretically must be determined. Ultimately, factors such as cost and the number of resolvable enzyme systems must limit the number of loci examined. However, in most cases, 20–30 loci are adequate, although species exhibiting low levels of polymorphism should be subjected to a more intensive analysis using as many loci as possible (Allendorf and Phelps 1981).

With *D. reticulatum*, it was decided that 30 should comprise the minimum sample size. At each sample site (discrete areas $\leqslant$ 10 m^2), individuals were either gathered by hand or trapped using tiles and bran bait. The geographic distribution of sample sites is shown in Fig. 18.1. Samples were taken from sites across Northern Ireland and, in addition, more intensive sampling was undertaken at two

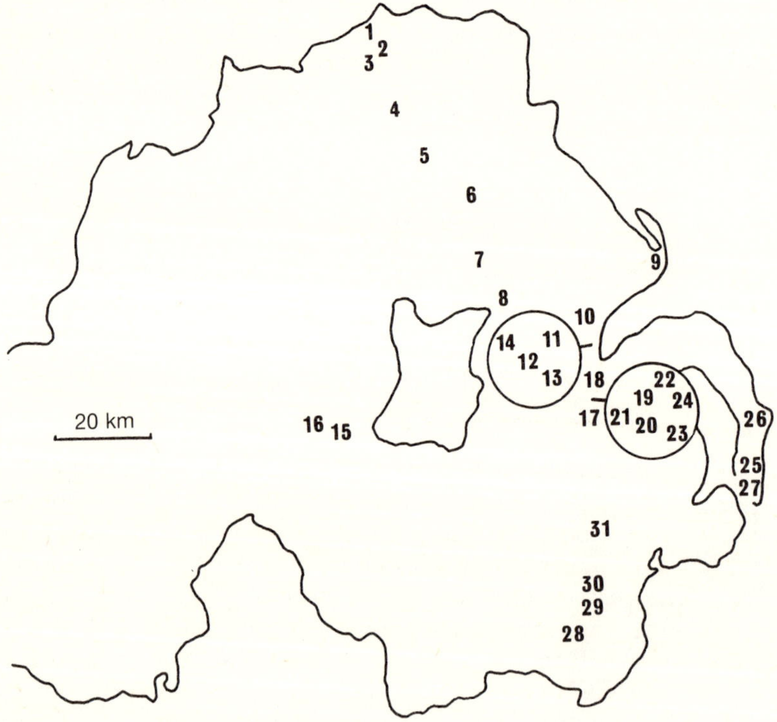

Fig. 18.1. Location of *D. reticulatum* sample sites in N. Ireland.

localities (Figs. 18.1 and 18.5) to assess the extent of the microgeographic variation. Sample sites were classified as four habitat types:

Disturbed (D)	recently excavated or infilled sites;
Cultivated (C)	agricultural land subject to regular cultivation, ploughing and/or molluscicide treatment;
Uncultivated (U)	agricultural land in permanent pasture;
Field margins (FM)	uncultivated field edges and hedgerows.

Details of samples are listed in Table 18.1.

Samples were also collected from an experimental site near the laboratory at Newtownabbey. This site comprised 17, each of 2 m^2, plots cultivated carefully for 10 years with no transfer of soil between plots. These samples were used to assess population structure and gene flow at the microgeographic level.

(b) Electrophoresis Slugs were prepared for starch gel electrophoresis as outlined by Dolan and Fleming (1988). Starch gels were prepared as described by Ferguson (1980), and electrophoresis of enzymes was performed as detailed by Dolan and Fleming (1988). Optimum resolution was obtained using three buffer systems:

Tris pH 8.6	Gel: 0.076 M tris, 0.005 M citric acid, 0.015 M boric acid, 0.005 M lithium hydroxide. Electrode: 0.3 M boric acid, 0.1 M lithium hydroxide.
TC pH 7.4	Gel: 0.009 M tris, 0.003 M citric acid. Electrode: 0.13 M tris, 0.043 M citric acid.
AM pH 6.1	Gel: 0.002 M citric acid adjusted to pH 6.1 with *N*-(3-aminopropyl)-morpholine. Electrode: 0.04 M citric acid adjusted to pH 6.1 with *N*-(3-aminopropyl)-morpholine.

Nine enzyme systems gave both good resolution and interpretable isoenzyme patterns: α-glycerophosphate dehydrogenase (α -GPD), esterase (EST), isocitrate dehydrogenase (IDH), leucine aminopeptidase (LAP), lactate dehydrogenase (LDH), peptidase (PEP), phosphoglucose isomerase (PGI), phosphoglucomutase (PGM), superoxide dismutase (SOD). Enzymes were stained using procedures described by Allendorf *et al.* (1977), with the exception of PEP which was stained as follows: 20 mg peptide, 5 mg amino acid oxidase, 5 mg peroxidase, 20 mg $MgCl_2$, 12 mg 3-amino-9-ethylcarbazole dissolved in 3 ml dimethyl sulphoxide. All ingredients were

Table 18.1 *D. reticulatum* samples from N. Ireland.

Region	Area	Sample no.	Sample site	N	Habitat[a]
North	N. Antrim	1	Bushmouth	36	U
		2	Bushleap	31	U
		3	Bushtown	30	U
		4	Dunloy	30	D
	Mid Antrim	5	Round	30	FM
		6	Roundun	30	FM
		7	Parkgate	32	C
		8	Kellswater	36	C
		9	Whitehead	30	D
	Belfast	10	Felden	53	C
		11	Castlepath	30	U
		12	Castlebrack	38	U
		13	Castlepark	49	U
		14	Castletop	36	U
	Tyrone	15	Parkanaur	30	U
		16	Dungannon	30	U
South	S. Antrim	17	Dunmurry	36	D
		18	Newforge	30	FM
		19	Scott 15	30	U
		20	Scott 16	30	U
		21	Scott 1	31	U
		22	Scott 11	33	U
		23	Scott 5	30	U
		24	Scott 8	27	U
	Ards	25	Portaferry	32	C
		26	Ballyhalbert	30	C
		27	Priestown	33	C
	S. Down	28	Hilltown	30	D
		29	Moneyroad	36	FM
		30	Moneyfield	31	C
		31	Ballynahinch	30	U

[a] U, uncultivated; D, disturbed; FM, field margins; C, cultivated

dissolved in 0.1 M Tris-HCl, pH ~ 7.4 and added to a 2 per cent agar overlay.

The nomenclature used for designating multiple loci and alleles is based on that of Allendorf and Utter (1979). These authors designate multiple loci by hyphenated numerals starting at the cathodal end of the gel and variant alleles (in parentheses) by the mobility of the homomeric band relative to a standard allele (usually the most common) designated 100. Statistical analysis of the electrophoretic data was performed using Biosys-1 (Swofford and Selander 1981).

Results

1. *Identification of slug species*

PEP and PGM patterns for the six slug species are shown in Fig. 18.2. The PEP patterns confirm the generic differences between *D. reticulatum* and the *Arion* species, the latter sharing a number of common alleles at both loci. While individual enzyme stains did not permit unequivocal identification of all *Arion* species, this was possible using a number of enzyme systems. The unreliability of

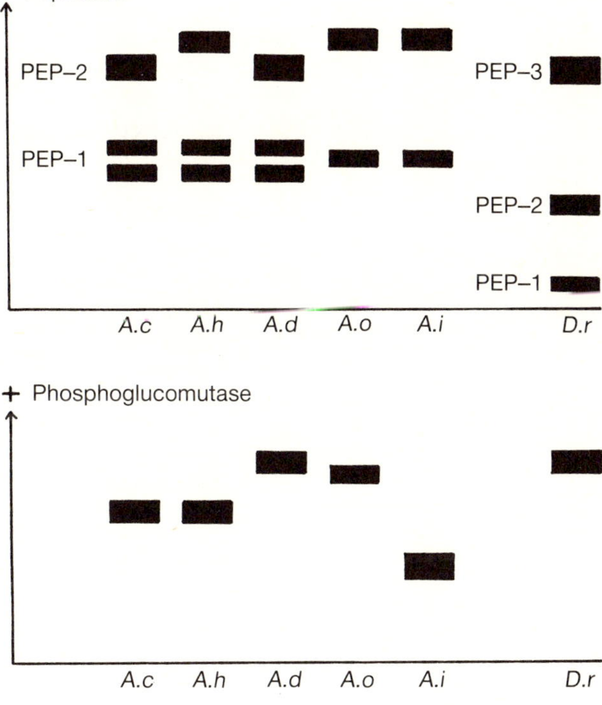

Fig. 18.2. Peptidase and phosphoglucomutase zymogram patterns from five *Arion* species and *D. reticulatum*. (Anodal end of the gel denoted by +). *A.c* = *Arion circumscriptus*; *A.h* = *A. hortensis*; *A.d* = *A. distinctus*; *A.o* = *A. owenii*; *A.i* = *A. intermedius*; *D.r* = *Deroceras reticulatum*.

Table 18.2 Matrix of genetic identity values (Nei 1978) for six slug species.

Species	*A.d*	*A.h*	*A.o*	*A.i*	*A.c*
A. distinctus					
A. hortensis	0.483				
A. owenii	0.438	0.500			
A. intermedius	0.185	0.182	0.400		
A. circumscriptus	0.365	0.500	0.286	0.546	
D. reticulatum	0.073	0.071	0.000	0.000	0.000

morphological characters was confirmed in the sense that a number of specimens identified (on the basis of external characters) as *A. hortensis* proved, following electrophoresis, to be *A. distinctus*. Quantitative assessment of genetic relationships amongst the six species was obtained using the data from all loci and computing genetic identity (*I*) values (Table 18.2). The relationships so obtained are presented in Fig. 18.3. The clustering of *A. hortensis*, *A. distinctus*, and *A. owenii* confirms their status as members of a species complex. It is also apparent that, despite their exhibiting very similar morphological characteristics, substantial genetic differences exist between the three species.

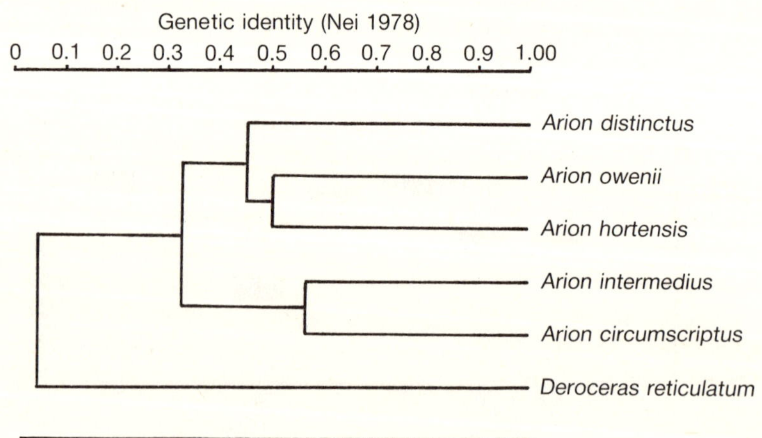

Fig. 18.3. Genetic relationships among *D. reticulatum* and five *Arion* species. Dendrogram (UPGMA; Ferguson 1980) based on mean genetic identity across 14 loci.

Table 18.3 Enzyme systems resolved in *D. reticulatum*.

Enzyme	No. of loci	No. of usable loci	Buffer
AGP	1	1	8.6
EST	5	2	6.1
IDH	2	1	6.1
LAP	1	1	7.4
LDH	2	2	8.6
PEP	3	2	7.4
PGI	1	1	6.1
PGM	2	2	7.4
SOD	2	1	8.6

2. *Polymorphic loci in* D. reticulatum

Table 18.3 shows the assumed number of loci coding for each enzyme system, the number of loci usable for population screening, and the buffer system providing best resolution. Seven of the usable loci proved polymorphic: Est-1, Est-5, Lap-1, Pep-2, Pep-3, Pgi-1 and Pgm-2.

3. *Distribution of genetic variation in* D. reticulatum

The amount of genetic variation in samples was assessed as the mean heterozygosity per locus (Ferguson 1980). Considerable differences in genetic variation were apparent among populations, values ranging from 0.04–0.22.

In Fig. 18.4 are shown the levels of genetic variation amongst populations separated according to habitat type. Populations from disturbed habitats exhibited particularly low levels of variation compared to other populations. Populations from cultivated sites also contained lower levels of genetic variation than those from field margins and uncultivated sites.

Hierarchical analysis (Wright 1978) was performed to assess the distribution of genetic variation throughout the species studied. Four hierarchical levels were considered in samples collected from Northern Ireland (see Table 18.1):

(1) N. Ireland as a whole;

(2) north and south regions;

(3) areas within regions;

(4) individual samples.

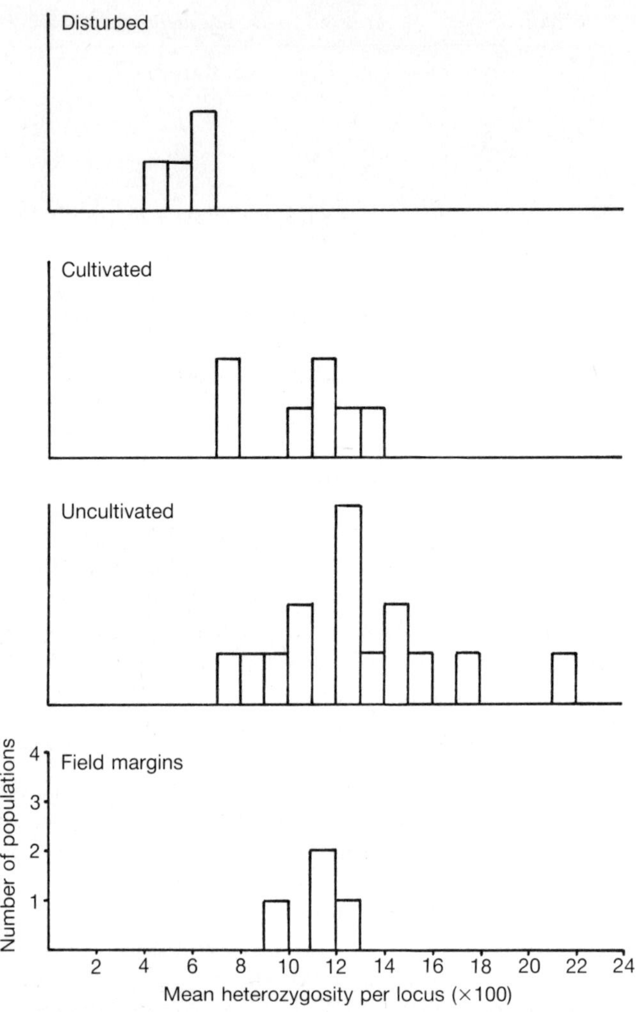

Fig. 18.4. Distribution of heterozygosity within 31 populations of *D. reticulatum* from N. Ireland.

Such hierarchic analysis allows determination of the partitioning of variation at the designated levels (Table 18.4).

Almost 65 per cent of the genetic variation resulted as variation within samples, and another 24 per cent by differences amongst samples. The remaining 11 per cent was due to differences amongst areas and between regions, indicating that macrogeographic variation was not marked within *D. reticulatum* from N. Ireland.

Table 18.4 Hierarchical analysis of genetic variation in *D. reticulatum*.

Level	Variance	Variance as percentage of total
Total	1.391	100.0
Within samples	0.900	64.7
Between areas within samples	0.332	23.9
Between areas within regions	0.111	8.0
Between regions within N. Ireland	0.048	3.4

4. Genetic variation in D. reticulatum

Table 18.5 shows the allelic frequencies at seven polymorphic loci in the 31 populations studied. As shown by variation in allele frequencies at the Lap-1 locus (Fig. 18.5), there was no clear evidence of macrogeographic differences in the allelic frequencies, thus confirming results of the hierarchical analysis. There was however, some evidence of local clusters of populations exhibiting similar allele frequencies.

Data from all loci were summarized by computing Nei's coefficient of genetic identity, I (Nei 1978), for all pairs of populations. The genetic relationships produced were then presented in the form of a dendrogram (Fig. 18.6). Despite the clustering of some local populations (e.g. Castlepath, Castletop, and Castlepark), there was generally little evidence of geographic clustering; local populations exhibited levels of genetic differentiation of the same magnitude as geographically more discrete populations. All of the observed genetic variation was above an I value of 0.95, a level typical of geographical populations (Ferguson 1980).

The significance of this interpopulation heterogeneity in allele frequencies was examined using a chi-square test (Workman and Niswander 1970) (Table 18.6). Significant heterogeneity was found at all polymorphic loci at the 0.001 level.

The data were also analysed using Wright's (1978) F statistics (Table 18.7). The mean F_{ST} statistic of 0.21 represents substantial genetic differentiation across samples. This heterogeneity was further highlighted by the large positive F_{IT} value, which indicated deviation from Hardy–Weinberg (H–W) proportions when all samples were pooled and is representative of substantial genetic

Table 18.5 Allele frequencies at seven loci in *D. reticulatum* from N. Ireland.

	EST-1		EST-5				LAP-1		PEP-2		PEP-3			PGI-1		PGM-2		
Sample	100	120	100	95	88	104	100	110	100	120	100	91	104	100	57	100	92	108
1	0.923	0.077	0.385	0.192	0.423	—	0.519	0.481	1.000	—	0.423	0.577	—	0.481	0.519	1.000	—	—
2	0.882	0.118	0.471	0.088	0.441	—	0.294	0.706	1.000	—	0.324	0.676	—	0.618	0.382	1.000	—	—
3	0.958	0.042	0.272	0.292	0.375	0.042	0.625	0.375	1.000	—	0.583	0.417	—	0.625	0.375	1.000	—	—
4	1.000	—	0.750	0.250	—	—	1.000	—	1.000	—	0.875	0.125	—	1.000	—	1.000	—	—
5	1.000	—	0.250	0.286	0.464	—	0.714	0.286	1.000	—	0.821	0.179	—	0.250	0.750	1.000	—	—
6	1.000	—	0.227	0.273	0.500	—	0.864	0.136	1.000	—	0.091	0.909	—	0.773	0.227	1.000	—	—
7	0.953	0.047	0.400	0.314	0.286	—	1.000	—	1.000	—	0.688	0.109	0.203	0.922	0.078	0.891	0.109	—
8	0.942	0.058	0.440	0.300	0.260	—	1.000	—	1.000	—	0.404	0.596	—	0.538	0.462	1.000	—	—
9	1.000	—	0.750	0.125	0.125	—	1.000	—	1.000	—	1.000	—	—	0.781	0.219	1.000	—	—
10	0.915	0.085	0.547	0.236	0.217	—	0.962	0.038	0.802	0.198	0.528	0.377	0.094	0.509	0.491	0.991	0.009	—
11	1.000	—	0.462	0.154	0.385	—	1.000	—	1.000	—	0.615	0.385	—	0.154	0.846	1.000	—	—
12	0.982	0.018	0.500	0.357	0.143	—	0.875	0.125	1.000	—	1.000	—	—	0.375	0.625	0.982	0.018	—
13	0.990	0.010	0.408	0.469	0.122	—	0.980	0.020	0.980	0.020	0.684	0.306	0.010	0.357	0.643	1.000	—	—
14	1.000	—	0.681	0.236	0.083	—	1.000	—	0.889	0.111	0.736	0.264	—	0.181	0.819	1.000	—	—
15	1.000	—	0.357	0.357	0.143	0.143	0.857	0.143	1.000	—	0.643	0.357	—	0.143	0.857	0.833	0.167	—
16	1.000	—	0.500	0.250	0.250	—	1.000	—	1.000	—	0.250	0.750	—	—	1.000	1.000	—	—
17	1.000	—	0.519	0.308	0.154	0.019	0.635	0.365	1.000	—	0.846	0.154	—	0.442	0.558	1.000	—	—
18	0.700	0.300	0.200	0.300	0.300	0.200	1.000	—	1.000	—	0.200	0.800	—	0.600	0.400	1.000	—	—
19	0.817	0.182	0.364	0.227	0.409	—	0.773	0.227	1.000	—	0.545	0.455	—	0.591	0.409	1.000	—	—
20	1.000	—	0.389	0.222	0.389	—	0.500	0.500	0.944	0.056	0.944	0.056	—	0.667	0.333	1.000	—	—
21	0.810	0.190	0.333	0.238	0.429	—	0.429	0.571	0.976	0.024	0.738	0.262	—	0.524	0.476	1.000	—	—
22	0.978	0.022	0.565	0.239	0.196	—	0.978	0.022	1.000	—	0.500	0.500	—	0.696	0.304	1.000	—	—
23	1.000	—	0.611	0.111	0.278	—	1.000	—	1.000	—	0.444	0.556	—	0.389	0.611	1.000	—	—
24	0.833	0.167	0.593	0.148	0.259	—	0.704	0.296	1.000	—	0.481	0.519	—	0.537	0.463	1.000	—	—
25	0.844	0.156	0.359	0.359	0.281	—	0.406	0.594	1.000	—	0.609	0.391	—	0.438	0.563	1.000	—	—
26	1.000	—	0.188	0.125	0.688	—	0.063	0.938	1.000	—	0.625	0.375	—	0.688	0.313	1.000	—	—
27	1.000	—	0.413	0.174	0.413	—	0.239	0.761	1.000	—	0.543	0.457	—	0.587	0.413	1.000	—	—
28	0.929	0.071	0.286	0.643	—	0.071	1.000	—	1.000	—	1.000	—	—	0.500	0.500	1.000	—	—
29	0.969	0.031	0.438	0.281	0.281	—	1.000	—	1.000	—	0.469	0.500	0.031	0.385	0.615	1.000	—	—
30	0.881	0.119	0.286	0.333	0.381	—	0.643	0.357	1.000	—	1.000	—	—	0.786	0.214	1.000	—	—
31	1.000	—	0.500	—	0.250	0.250	0.375	0.625	1.000	—	0.625	0.375	—	0.875	0.125	0.750	—	0.250

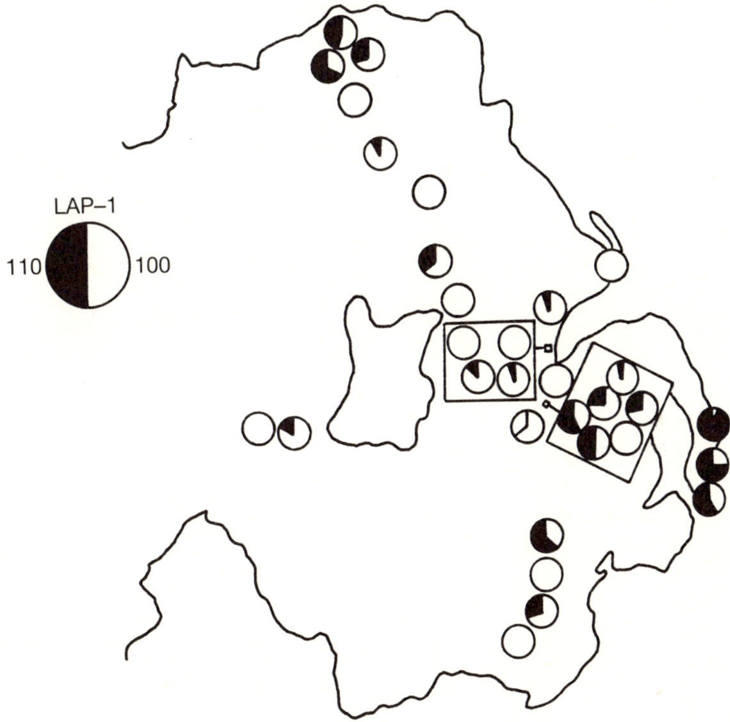

Fig. 18.5. Variation in allele frequencies at the *Lap-1* locus in *D. reticulatum* from N. Ireland.

Table 18.6 Heterogeneity chi-square test of allele frequencies in *D. reticulatum*.

Locus	No. of alleles	χ^2	df	*P*
Pgi-1	2	178.6	29	<0.001
Est-5	4	328.6	87	<0.001
Est-1	2	103.0	29	<0.001
Lap-1	2	449.1	29	<0.001
Pep-3	3	369.2	58	<0.001
Pep-2	2	154.9	29	<0.001
Pgm-2	3	400.8	58	<0.001
Total		1984.5	319	<0.001

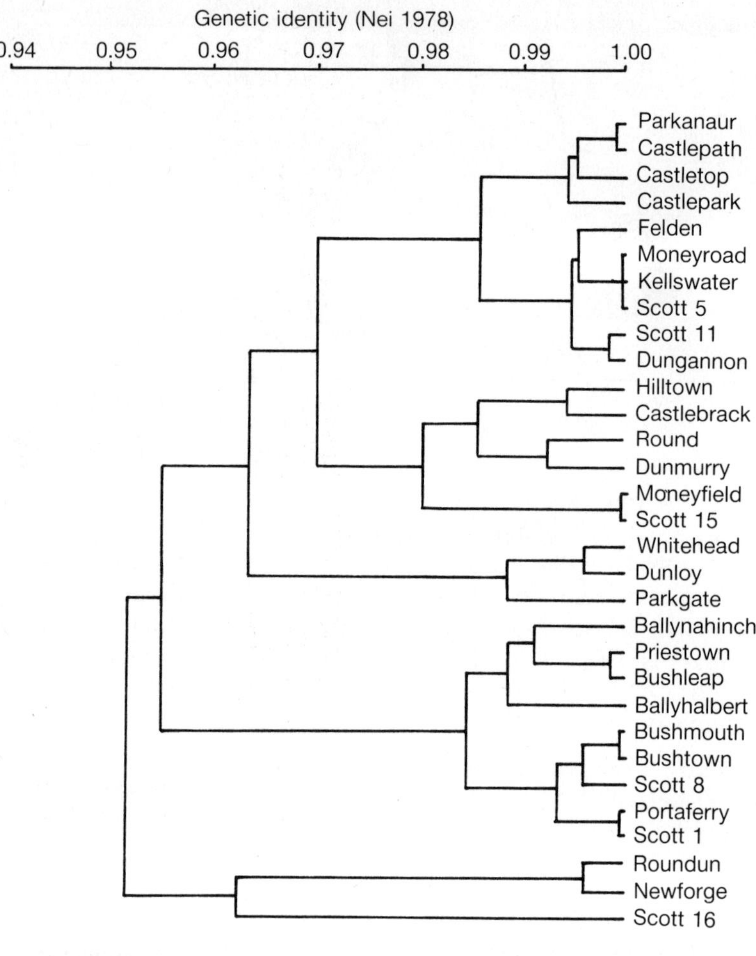

Fig. 18.6. Genetic relationships among 31 populations of *D. reticulatum* from N. Ireland. Dendrogram (UPGMA; Ferguson 1980) based on mean genetic identity across 13 loci.

Table 18.7 *F*-statistics for polymorphic loci in *D. reticulatum*.

Locus	F_{IS}	F_{IT}	F_{ST}
Pgi-1	0.076	0.247	0.185
Est-5	0.255	0.326	0.095
Est-1	0.219	0.431	0.271
Lap-1	0.053	0.435	0.403
Pep-3	0.252	0.427	0.233
Pep-2	0.213	0.311	0.124
Pgm-2	0.228	0.014	0.174
Mean	0.179	0.349	0.210

substructuring among samples. A mean F_{IS} value of ~ 0.18 indicated the possibility of inbreeding or further substructuring within the samples. This possibility was examined further by examining in detail the fixation indices at individual loci. Of 132 tests performed, 27 exhibited significant deviations from expected H–W proportions (Table 18.8). A positive fixation index, that is, a deficiency of heterozygotes, accounted for 26 of these.

Allelic frequency data from the experimental plot site provided additional evidence of genetic substructuring in *D. reticulatum*. At this site, four loci proved polymorphic and pooled genotype data revealed a significant deviation from H–W proportions ($P < 0.05$). Significant differences in allele frequencies across the 17 plots were detected at three of these loci (Table 18.9). Figure 18.7 shows the observed variation in allele frequencies at the Pep-3 locus. The distribution of allele frequencies indicates that the *D. reticulatum* population sample at this site did not represent a single panmictic unit. Single locus data from samples taken in individual plots revealed that of 14 cases exhibiting deviation from H–W expectations, 13 were caused by a deficiency of heterozygotes.

Discussion

Electrophoresis has proved a particularly useful tool in slug taxonomy, especially in cases where classical taxonomic techniques are time-consuming or inconclusive. Differentiating species of the *A. hortensis* complex can be a problem in the field, *A. hortensis* and *A. distinctus* often being indistinguishable without examination of their

Table 18.8 Fixation indices for polymorphic loci in *D. reticulatum* samples showing a deviation from Hardy–Weinberg proportions.

Sample no.	Locus	Fixation
1	*Est-1*	+0.458
1	*Pep-3*	+0.527
2	*Est-1*	+0.433
2	*Pep-3*	+0.597
3	*Pgi-1*	+0.822
5	*Pep-3*	+0.757
6	*Lap-1*	+0.614
8	*Pep-3*	+0.441
10	*Est-5*	+0.306
10	*Est-1*	+0.393
12	*Pgi-1*	+0.467
12	*Est-5*	+0.466
14	*Est-5*	+0.414
14	*Pep-2*	+0.438
19	*Pep-3*	+0.633
19	*Est-1*	−0.450
20	*Est-1*	+0.600
20	*Est-5*	+0.471
21	*Est-1*	+0.691
25	*Est-5*	+0.245
25	*Est-1*	+0.763
25	*Lap-1*	+0.611
26	*Est-5*	+0.475
28	*Est-5*	+0.429
29	*Pgi-1*	+0.350
30	*Est-5*	+0.568
30	*Lap-1*	+0.689

Table 18.9 Heterogeneity chi-square test of allele frequencies in *D. reticulatum* from 17 experimental plots.

Locus	χ^2	df	P
Pgi-2	19.9	16	NS
Pep-2	72.1	16	<0.001
Pep-3	48.9	32	<0.05
Est-5	67.4	32	<0.001

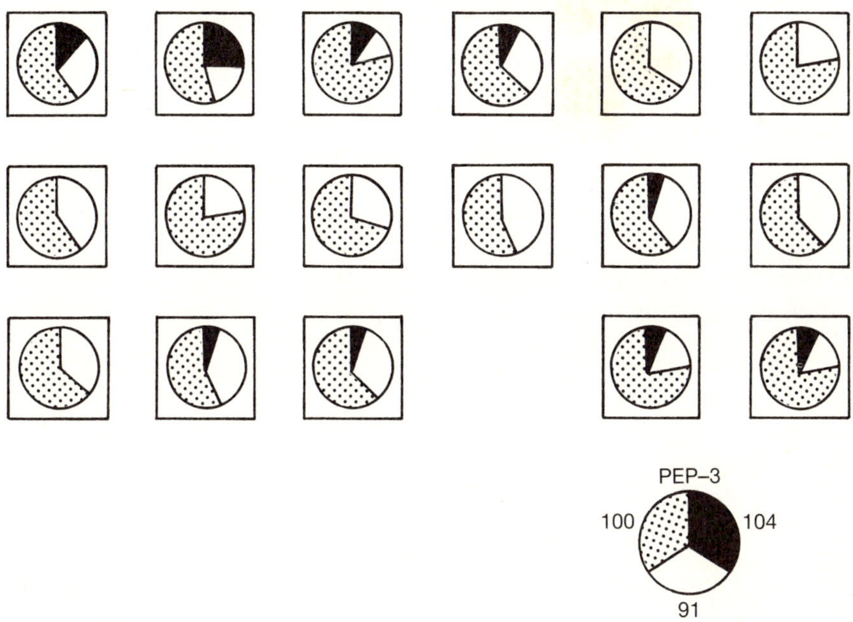

Fig. 18.7. Variation in allele frequencies at the *Pep-3* locus in *D. reticulatum* from 17 experimental plots.

internal anatomy (Davies 1979). Genetic identity values among the three species in the complex ranged from 0.43 to 0.50 and were of a magnitude expected for sibling species (Ferguson 1980). The degree of genetic differentiation within the complex provided numerous marker alleles useful for species identification, thus permitting large numbers of animals to be identified quickly and accurately. This was particularly valuable when juvenile slugs (often differing in coloration from adults) were examined. The application of electrophoresis to slug taxonomy should facilitate the resolution of other problem species complexes and help applied biologists to make routine and otherwise time-consuming identifications.

With the increased awareness by agriculturalists and horticulturalists of the pest status of slugs, integrated control measures will become more important as a means of minimizing their effects (Port and Port 1986). However, to maximize the efficacy of such measures, considerably more information regarding the slugs' ecology and population dynamics is required. The present study has demonstrated the use of the biochemical genetic approach in obtaining such information.

The population structure of a pest is an important consideration in any control strategy or population model: *D. reticulatum* exhibits a pattern of genetic substructuring that might be expected for a species with low dispersive powers. The wide range in heterozygosities among *D. reticulatum* populations probably reflects a number of factors that may be influenced by its population dynamics. The low heterozygosities in disturbed sites may result from large fluctuations in population size (creating genetic bottlenecks) or from the recent establishment of those populations via accidental introduction of a small number of individuals (founders) with correspondingly low levels of genetic variation. Bottlenecks may also be responsible for the apparent tendency towards lower heterozygosities in cultivated habitats, where, for example, seedbed production and molluscicide treatment can produce large decreases in population size (Port and Port 1986).

Whether reduced levels of genetic variation does affect the fitness or adaptability of slug populations is still a matter for speculation. However, some effects have been noted in other species, i.e. trout (Ryman and Ståhl 1980). Foltz *et al*. (1984) note that, in Britain, the major agricultural slug pests, including *D. reticulatum*, are highly heterozygous outcrossing species, whilst the less heterozygous species are uncommon in agricultural habitats. It is unclear whether this is due to the greater adaptability of the outcrossing species to novel habitat conditions.

D. reticulatum exhibits a high degree of genetic differentiation among populations — as evidenced by a high F_{ST} value (0.21) and heterogeneity. This was confirmed in the hierarchical analysis, which demonstrated that differentiation was due to microgeographic variation, local populations being as distinct from each other as geographically more separate ones. These findings suggest that *D. reticulatum* is a species with poor dispersive powers, spreading largely as a result of external factors such as human activity.

A high F_{IS} value provided evidence of additional sub-structuring or inbreeding within individual samples. Virtually all the observed deviations from H–W proportions resulted from a deficiency in the total number of observed compared to expected heterozygotes. This may be due to the presence in the samples of small groups of relatives (i.e. to inbreeding) or individuals from larger genetically distinct subpopulations (i.e. the 'Wahlund' effect). Either way, it may be inferred from the data that gene flow is restricted, even within quite small geographical areas.

The more intensive sampling regime within the plot site revealed a similar pattern of heterozygote deficiencies and substructuring, but on an even smaller scale. Study of the data revealed that, even at this level, the slug populations cannot be regarded as panmictic. Such evidence of restricted gene flow is at variance with the suggestions of Miles *et al.* (1931) and Carrick (1938), both of whom maintained that *D. reticulatum* exhibits considerable dispersal behaviour, migrations occurring between fields and from surrounding field margins. In contrast, the present data support the observations of South (1965), who described the spatial distribution of *D. reticulatum* in agricultural land and who noted that the species was aggregated in units of about 30 cm^2. There was also evidence that this aggregation was maintained by homing behaviour.

The apparent lack of extensive gene flow in *D. reticulatum* suggests that population build-up after, for example, a slug-control programme, may result largely from recruitment *in situ* rather than recolonization from field margins or adjacent fields. This finding thus supports the view of Barnes (1953), who showed that *D. reticulatum* was unable to recolonize a garden site until slugs were artificially introduced.

Conclusion

The present study has revealed some of the parameters possibly influencing the population dynamics of *D. reticulatum*. More importantly, it has demonstrated the way in which electrophoresis can complement a standard ecological approach to the study of slug population biology. As has been shown for many economically important animal groups, the efficient management of a species, either as a resource or as a pest, is only possible with an accurate knowledge of the population dynamics and genetic structure of the species in question (Ryman 1981; Menken and Ulenberg 1987). With slugs, the population data revealed by biochemical techniques can be used to maximize the effectiveness of control measures as well as to construct realistic population models for use as predictive tools.

Acknowledgements

I thank D. Craig and S. Dolan for their technical assistance.

References

Allendorf, F. W. and Phelps, S. R. (1981). Isozymes and the preservation of genetic variation in salmonid fishes. *Ecol. Bull.*, **34**, 37–52.

Allendorf, F.W. and Utter, F.M. (1979). Population genetics. In *Fish physiology*, Vol. 8 (ed. W.S. Hoar, D.J. Randall, and J.R. Brett), pp. 407–54. Academic Press, New York.

Allendorf, F.W., Mitchell, N., Ryman, N., and Ståhl, G. (1977). Isozyme loci in brown trout (*Salmo trutta* L.): detection and interpretation from population data. *Hereditas*, **86**, 179–90.

Barnes, H. F. (1953). The absence of slugs in a garden and an experiment in restocking. *Proc. Zool. Soc. Lond.*, **123**, 49–58.

Carrick, R. (1938). The life history and development of *Agriolimax agrestis* L., the grey field slug. *Trans. R. Soc. Edinb.*, **59**, 563–97.

Davies, S. M. (1979). Segregates of the *Arion hortensis* complex (Pulmonata: Arionidae), with the description of a new species, *Arion owenii*. *J. Conch., Lond.*, **30**, 123–7.

Dolan, S. and Fleming, C.C. (1988). Isoenzymes in the identification and systematics of terrestrial slugs of the *Arion hortensis* complex. *Biochem. Syst. Ecol.*, **16**, 195–8.

Evans, N.J. (1985). The use of electrophoresis in the separation of two closely related species of terrestrial slugs. *Biochem. Syst. Ecol.*, **13**, 325–8.

Ferguson, A. (1980). *Biochemical systematics and evolution*. Blackie, Glasgow.

Foltz, D.W., Ochman, H., Jones, J.S., Evangelisti, S.M., and Selander, R.K. (1982). Genetic population structure and breeding systems in arionid slugs (Mollusca: Pulmonata). *Biol. J. Linn. Soc., Lond.*, **17**, 225–41.

Foltz, D.W., Ochman, H., and Selander, R.K. (1984). Genetic diversity and breeding systems in terrestrial slugs of the families Limacidae and Arionidae. *Malacologia*, **25**, 593–605.

Johnson, M.S. (1976). Allozymes and area effects in *Cepaea nemoralis* on the Western Berkshire Downs. *Heredity*, **56**, 105–21.

Marshall, D. R. and Brown, A. H. D. (1975). Optimum sampling strategies in genetic conservation. In *Crop genetic resources for today and tomorrow* (ed. O.H. Frankel and J.G. Hawkes), pp. 53–80. Cambridge University Press, Cambridge.

McCracken, G.F. and Selander, R.K. (1980). Self-fertilisation and monogenic strains in natural populations of terrestrial slugs. *Proc. natn. Acad. Sci. USA*, **77**, 684–8.

Menken, S.B.J. and Ulenberg, S.A. (1987). Biochemical characters in agricultural entomology. *Agric. Zool. Rev.*, **2**, 305–53.

Miles, H.W., Wood, J., and Thomas, I. (1931). On the ecology and control of slugs. *Ann. appl. Biol.*, **18**, 370–400.

Nei, M. (1978). Estimation of average heterozygosity and genetic distance from a small number of individuals. *Genetics*, **89**, 583–90.

Nicklas, N. L. and Hoffman, R. J. (1981). Apomictic parthenogenesis in a hermaphroditic terrestrial slug, *Deroceras laeve* (Muller). *Biol. Bull.*, **160**, 123–35.

Port, C. M. and Port, G. R. (1986). The biology and behaviour of slugs in relation to crop damage and control. *Agric. Zool. Rev.*, **1**, 255–99.

Ryman, N. (1981). Conservation of genetic resources: Experiences from the brown trout (*Salmo trutta*). *Ecol. Bull.*, **34**, 61–74.

Ryman, N. and Ståhl, G. (1980). Genetic changes in hatchery stocks of brown trout (*Salmo trutta*). *Can. J. Fish. Aquat. Sci.*, **37**, 82–7.

Selander, R. K. and Kaufman, D. W. (1975). Genetic structure of populations of the brown snail (*Helix aspersa*). 1. Microgeographic variation. *Evolution*, **29**, 385–401.

South, A. (1965). Biology and ecology of *Agriolimax reticulatus* (Mull.) and other slugs: spatial distribution. *J. Anim. Ecol.*, **34**, 403–17.

Swofford, D. L. and Selander, R. B. (1981). BIOSYS-1: A FORTRAN program for the comprehensive analysis of electrophoretic data in population genetics and systematics. *J. Hered.*, **72**, 281–3.

Workman, P. L. and Niswander, J. D. (1970). Population studies on southwestern Indian tribes II. Local genetic differentiation in the Papago. *Am. J. Hum. Genet.*, **22**, 24–49.

Wright, S. (1978). *Evolution and the genetics of populations*, Vol. 4. The University of Chicago Press, London.

19 Characterization of populations of pathotypes of potato cyst nematodes

J. BAKKER and F.J. GOMMERS

Department of Nematology, Agricultural University, Wageningen, The Netherlands

Abstract

The sibling species *Globodera rostochiensis* and *G. pallida* were differentiated by 70 per cent of their polypeptides as revealed with two-dimensional gel electrophoresis (2-DGE) followed by a sensitive silver stain. These large differences indicate that these nematode species have accumulated protein differences during a period of millions of years without distinct changes in morphology.

The mosaic distribution of pathotypes of both *G. rostochiensis* and *G. pallida* in Europe impedes effective control by means of resistance. The five pathotypes of *G. rostochiensis* (Ro_1–Ro_5) and three pathotypes of *G. pallida* (Pa_1–Pa_3) were already present in Europe before resistant potato cultivars were grown. These spatial variations in virulence are mainly due to three processes: (1) the genetic structures of the initial populations introduced from South America, (2) random genetic drift, and (3) gene flow. The result of these colonizing processes was studied by 2-DGE by analysing intraspecific variations. Within *G. pallida*, 29 variant proteins, presumed to be encoded by 29 alleles at 11 loci, were recognized. Similarly, in *G. rostochiensis*, 20 variant proteins were found encoded by 9 presumptive loci. Similarity dendrograms constructed from distance matrices show that current pathotype classification is incapable of reflecting genetic relationships between populations of potato cyst nematodes. Interpopulation variations in both virulence and proteins are determined by the same three processes. Because these processes affect the whole gene pool of populations, similarities

Electrophoretic Studies on Agricultural Pests (ed. Hugh D. Loxdale and J. den Hollander), Systematics Association Special Volume No. 39, pp. 415–30. Clarendon Press, Oxford, 1989.

revealed by 2-DGE are also reflected at virulence loci, including those not yet resolved by the current pathotype scheme.

Introduction

Phytonematologists are confronted with a group of organisms whose nomenclature and taxonomic status continues to undergo change. The small size, uniform morphology, and relatively simple basic structure of nematodes create serious difficulties in establishing discriminating characters. The limited number of conspicuous morphological features has offered ample opportunity for personal judgement in delineating species and genera. Well-known examples are the cyst nematodes (Stone 1977). Initially, various populations were considered to be races of *Heterodera schachtii* (Filipjev and Schuurmans Stekhoven 1941), although eventually they were recognized as separate species, e.g. *H. schachtii, H. avenae, H. goettingiana*, and *H. rostochiensis*. The first sign that the potato cyst nematode, '*H. rostochiensis*', might not be a single species was the discovery of populations able to overcome certain host (potato plant) genes for resistance. In the early 1970s, it became apparent that not all pathotypes of *H. rostochiensis* interbreed freely and that two sibling species are present: *G. rostochiensis* and *G. pallida* (Jones *et al*. 1970). The late discovery of these species resulted from their morphological similarity. Most morphological characters are variable and overlap between the two species (Stone 1973, 1975).

Along with changes in the nomenclature of potato cyst nematodes, classification of their pathotypes has also been dominated by a state of uncertainty. As with many obligate plant parasites, the number of pathotypes increased with the discovery of more sources of resistance. Additional complications arose because of the fact that in several European countries, different systems were employed to distinguish field populations. In 1978, a number of European countries reached a consensus of opinion and proposed an international pathotype scheme (Kort *et al*. 1978). Currently, eight pathotypes are recognized in Europe, five within *G. rostochiensis* (Ro_1–Ro_5) and three within *G. pallida* (Pa_1–Pa_3). Resistance and virulence are thought to be mediated by genes operating on the basis of a gene-for-gene relationship (Jones *et al*. 1981; Turner *et al*. 1983). Unlike the case of resistance against various species of fungi, breeding programmes based on major genes can be profitable, since selection towards alleles for virulence is rather slow.

The international pathotype scheme (Kort *et al*. 1978) forms the basis on which decisions are presently made to control potato cyst nematode populations by means of resistance. However, the current pathotype classification has several drawbacks in characterizing the intraspecific diversity of potato cyst nematodes. First, measuring the virulence characteristics of potato cyst nematode populations is laborious: because no rapid assays are available, the multiplication factors of the population on different host clones ('differentials') have to be estimated in a traditional way by inoculating potato plants with cysts. Secondly, pathotypes are delineated in an arbitrary way. Populations are classified as virulent and avirulent for a certain differential if the multiplication factor is > 1 and $\leqslant 1$, respectively. Obviously, this way of classifying gives little information, because multiplication factors on the differentials may vary from 0 to 70. Thirdly, the number of differentials is too limited for a proper characterization of genetic diversity of potato cyst nematode populations in Europe. Supplementary differentials have already discriminated populations classified as identical according to the current international pathotype scheme. Fourthly, the expression of the nematode and host genotypes is rather variable, which often makes it difficult to decide whether a population should be classified as virulent or avirulent for a given differential.

Evidently, an accurate identification of potato cyst nematode populations is needed for a optimal control by means of host resistance. Cyst nematodes are major pests of potatoes in Europe and an improper identification or the inability to recognize distinct nematode populations can have drastic and costly consequences. Although traditional ways of nematode pathotyping and species identification will always retain their value, there is an increasing need for more efficient and accurate methods.

Compared to various other organisms, the molecular taxonomy of plant parasitic nematodes is poorly developed (Fox and Atkinson 1986; Platzer 1981). Obligate plant parasites are generally of microscopic size and difficult to rear in large quantities; consequently, it is difficult to use many of the available standard biochemical techniques. For example, because of the relatively large amounts of protein required, starch gel electrophoresis followed by specific enzyme stains, a routine technique for many organisms, has hardly been applied to plant parasitic nematodes (see Huettel *et al.* 1983). French (Berge *et al.* 1981; Dalmasso and Berge 1978) and German (Ohms and Heinicke 1983; Poehling and Wyss 1980) investigations tackled this problem by using micromethods for homogenizing and electrophoresing single specimens. Besides microelectrophoresis, nematologists have employed various other electrophoretic techniques to study nematode proteins. For example, the genetic variability of nematodes has been investigated using polyacrylamide gel electrophoresis at alkaline pH (Platzer 1981), isoelectric focusing (Fox and Atkinson 1984, 1985; Ohms and Heinicke 1983), immunoelectrophoresis (Fox and Atkinson 1985; Ohms and Heinicke 1983), sodium dodecyl sulphate electrophoresis (Pozdol and Noel 1984) and two-dimensional gel electrophoresis (2-DGE) (Bakker and Gommers 1982; Ferris *et al.* 1985) according to the method of O'Farrell (1975). 2-DGE combined with a sensitive silver stain (Merril *et al.* 1979; Oakley *et al.* 1980) seems — in theory — the most promising approach. A major advantage of 2-DGE over other techniques is that it is capable of resolving several hundred gene products on a single gel. In the first dimension, the proteins are separated according to their isoelectric points, and in the second dimension they are separated according to their molecular weights. This results in an almost unique position for each protein.

Genetic divergence of cyst nematode species

Using young female potato cyst nematodes, 2-DGE of 25 μg protein within the pH range 5–7 reveals an average of 245 polypeptides (Bakker and Bouwman-Smits 1988*b*). This figure can easily be increased by analysing a larger amount of protein and studying other pH regions as well. In this way, it is possible to study more than 700 gene products (Bakker and Bouwman-Smits 1988*a*). An accurate method of comparing the proteins of different species is to electrophorese a mixed sample. Analysing a protein pattern made from equal quantities of proteins from two species allows the detection of minute differences in isoelectric point and molecular weight. For example, the proteins A_1, B_1, and C_1 in *G. rostochiensis* differ only slightly in electrophoretic mobility from the corresponding proteins A_2, B_2, and C_2 in *G. pallida* (Figs. 19.1A, B, and C). The sibling species *Globodera rostochiensis* and *G. pallida* are differentiated at 70 per cent of their polypeptides. Similar results are obtained with *Heterodera glycines* and *H. schachtii*. These species are discriminated at 59 per cent of their polypeptides and yet are morphologically nearly indistinguishable.

A convenient measure for comparing the genetic differentiation of extant species using 2-DGE is given by Aquadro and Avise (1981),

$$F = \frac{2\,N_{xy}}{N_x + N_y}$$

in which N_x and N_y are the total number of proteins spots scored for population x and y, respectively, and N_{xy} is the number of spots shared by x and y. The genetic distance (D) is $1 - F$. The genetic distances between *G. rostochiensis* and *G. pallida* ($D = 0.70$), and between *H. glycines* and *H. schachtii* ($D = 0.59$) (Bakker and Bouwman-Smits 1988*b*) indicate that these species diverged a long time ago.

For comparison, 2-DGE studies on the sibling species *Drosophila melanogaster* and *D. simulans* revealed a genetic distance (D) of 0.19 (Ohnishi *et al.* 1983). Even two families of rodents have a smaller genetic distance ($D = 0.5$) (Aquadro and Avise, 1981) than these morphologically closely related cyst nematode species. Both 2-DGE distance data and calculations of the divergence time have been published for a number of species. 2-DGE studies on hominid primates indicate that a genetic distance (D) of approximately 0.2 corresponds to a divergence time of *ca.* 37 million years (Goldman *et al.* 1987). Although these values cannot be directly extrapolated to other organisms, it is evident that the closely related cyst nematode species diverged millions of years ago. Our data definitely exclude the possibility that *G. rostochiensis* and *G. pallida* speciated in South America as a result of independent potato cultivation by the Aymara and Quechua Indian tribes, as was suggested by Evans *et al.* (1975).

Contrasts between genetic and morphological similarities have been revealed in a wide variety of taxonomic groups ranging over bacteria, snails, fish, frogs, reptiles, and birds (Wilson 1976). These findings support the hypothesis that morphological evolution and structural gene evolution, measured by electrophoresis or other biochemical techniques, go on at virtually independent rates (King and Wilson 1975; Prager and Wilson 1975; Wilson *et al.* 1974). The evolution of proteins is thought to proceed at an approximately constant rate in all species (Kimura 1983), whereas the rate of morphological evolution is variable. Well-studied examples of rapid and slow morphological evolution include placental mammals and frogs, respectively (Wilson *et al.* 1974). Like frogs, cyst nematodes have accumulated protein differences over millions of years without any significant morphological changes. Another feature that cyst nematodes share indirectly with frogs, and in which both groups of organisms differ from organisms such as mammals, is the ability to produce viable interspecific hybrids in spite of large genetic distances. *H. schachtii* and *H. glycines* are able to produce fertile hybrids (Miller 1983), and matings between *G. rostochiensis* and *G. pallida* result in viable second-stage larvae (Mugniery 1979).

Slow morphological divergence is probably not rare throughout the phylum Nematoda. The large genetic distances revealed by

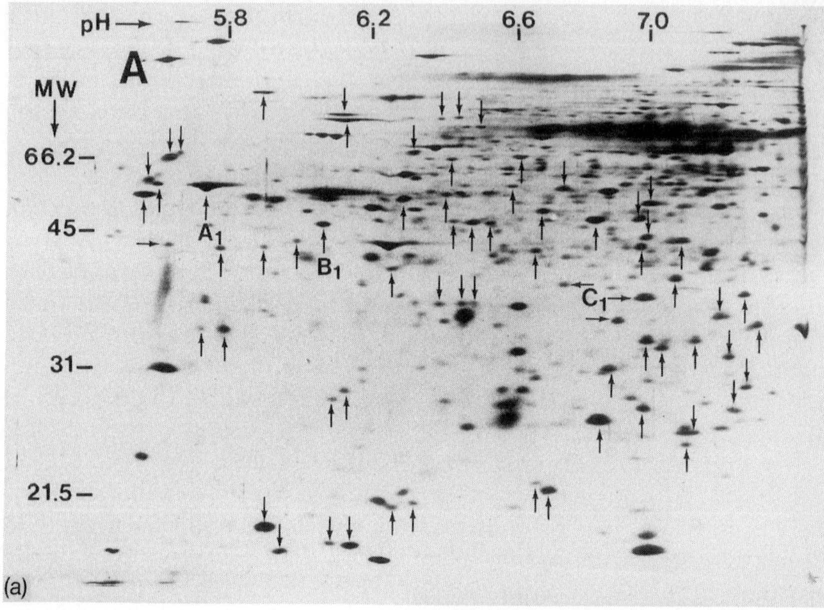

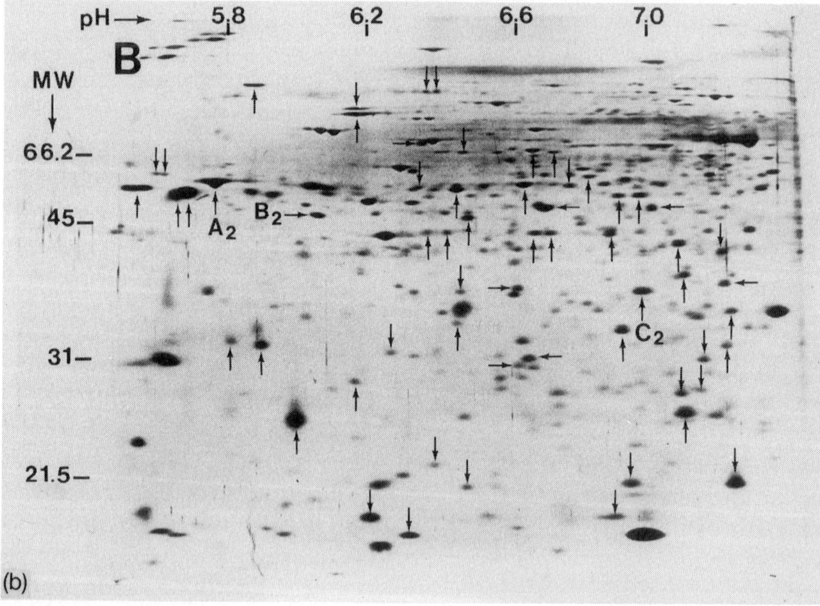

Fig. 19.1. 2-DGE protein patterns (25 μg) of females of *G. rostochiensis* population MIER (A) and *G. pallida* population HPL-1 (B) in which a number of major qualitative differences are marked with arrows. Common proteins are designated in a pattern (C) containing an equal quantities of protein of *G. rostochiensis* MIER (12.5 μg) and *G. pallida* HPL-1 (12.5 μg). Major qualitative

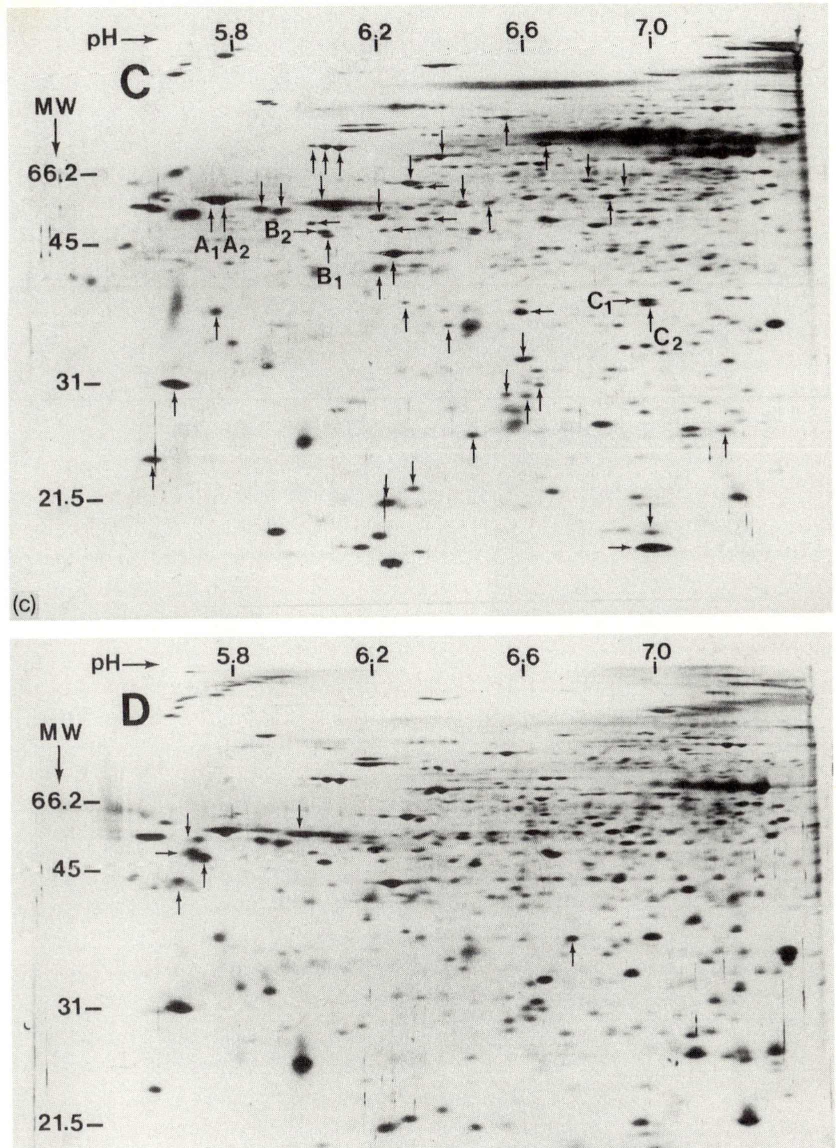

differences between *G. pallida* population HPL-1 (B) and *G. pallida* population 1337 (D) are indicated in (D). Proteins labelled with capitals and arabic numbers are referred to in the text. Molecular masses are given in kilodaltons. Protein patterns were made by electrophoresing a protein mixture from 100 adult females (Bakker and Bouwman-Smits, 1988*b*).

2-DGE are supported by the few studies in which starch gel electrophoresis has been applied to morphologically closely related nematode species (i.e. Butler *et al.* 1981; Huettel *et al.* 1983). Starch gel electrophoresis followed by specific enzyme stains is, unlike 2-DGE and the various other electrophoretic techniques used in plant nematology (Fox and Atkinson 1986), a suitable tool for estimating divergence times. For example, the morphologically nearly indistinguishable nematode species *Caenorhabditis elegans* and *C. briggsae* do not share any alleles at 22 of the 24 enzyme loci surveyed (Butler *et al.* 1981), from which it is inferred that these species diverged at least 10 million years ago.

Intraspecific variation in potato cyst nematodes

Genetic distances between conspecific populations of *G. rostochiensis* and *G. pallida* are usually less than 0.07 (Bakker 1987). Using 2-DGE, the large number of proteins surveyed and the number of variants detected is much larger than obtained with other electrophoretic techniques used to study intraspecific variation. For instance, isoelectric focusing of 23 enzymes of the *G. pallida* pathotypes Pa_1, Pa_2, and Pa_3 revealed only two allozymes displaying interpopulation variation (Fox and Atkinson 1984). Using 2-DGE, we detected 73 variant protein spots within *G. pallida* and 27 within *G. rostochiensis*.

The variant protein spots detected by comparison of conspecific populations can be divided into two classes: isoelectric-point variants (IP-variants) and non-isoelectric-point variants (NIP-variants) (Fig. 19.2) (Bakker and Bouwman-Smits 1988*a*). The IP-variants have the characteristics of variants that are the result of amino-acid substitutions that change the net charge of proteins. Corresponding IP-variants are assumed to be encoded by alleles at the same locus. The NIP-variants are the remaining variants for which no proper genetic interpretation has been proposed and which may include differences in regulatory and modifying genes. The IP-variants are the most reliable characters for estimating genetic relationships between conspecific populations, since homologous proteins are compared. Relationships based on NIP-variants may also be informative, but should be interpreted with care, as the number of genes involved is unknown and NIP-variants are more difficult to distinguish from artefacts owing to (for example) experimental variations and differences in physiological or developmental stages.

2-DGE is efficient when characterizing a large number of popula-

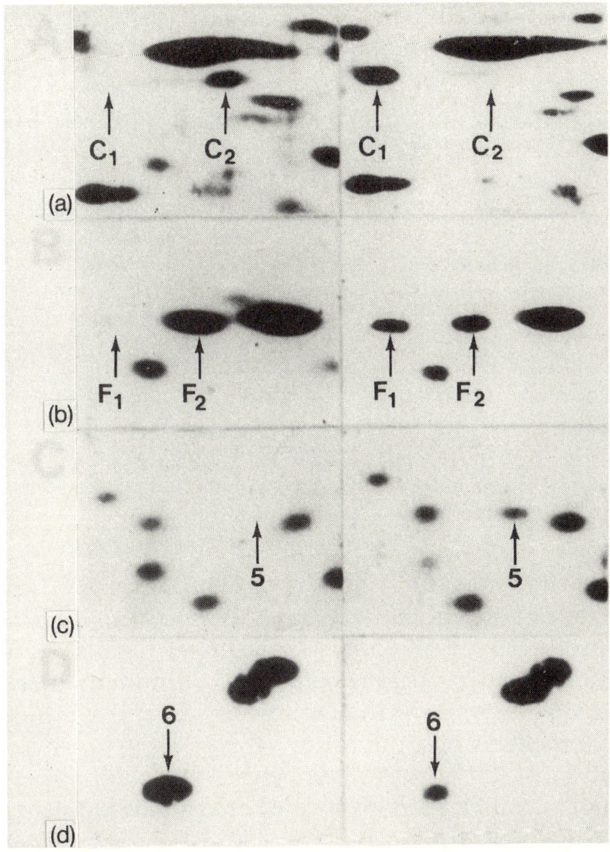

Fig. 19.2. Examples of the types of variant protein spots detected by the comparison of 2-DGE protein patterns of *G. ostochiensis* population Ro_1-M (left) and Ro_5-H (right). (A and B) Variations having the characteristics expected from amino-acid substitutions that alter the isoelectric point of a protein. The isoelectric point variants (IP-variants) are labelled with capital letters referring to the putative loci and numbers referring to the alleles. (C and D) Variants for which no putative corresponding allele product could be traced are labelled with arabic numbers. Differences expressed by presence/absence data (C) and differences in concentration (D) were combined within one class of variants, the non-isoelectric-point variants (NIP-variants). Protein patterns were made by electrophoresing a protein mixture from 100 adult females (Bakker and Bouwman-Smits 1988*a*).

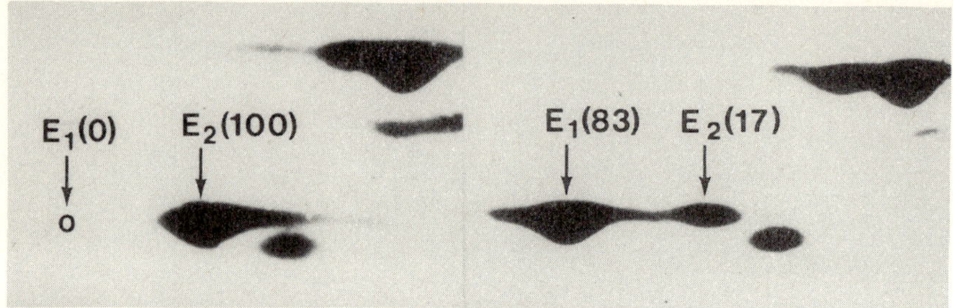

Fig. 19.3. Details of protein patterns of *G. pallida* population D-275 (left) and *G. pallida* population D.236 (right). Protein patterns were made by electrophoresing a homogenate of 100 adult females. The IP-variants are labelled with capitals and numbers referring to the putative loci and alleles, respectively. Estimates of the ratios between the protein quantities of the corresponding IP-variants, which were used as a measure of the allele frequencies (Bakker 1987) are given in parenthesis.

tions (Bakker 1987). 2-DGE patterns were made by electrophoresing a homogenate of 100 individuals and the ratio between the protein quantities produced by the alleles at a given locus was used as a measure for the allele frequencies (Fig. 19.3). Inherent to the staining techniques, such an approach is technically far more difficult in the case of enzyme electrophoresis.

Examination of 25 European *G. pallida* populations revealed 29 variant proteins, which seem to be the result of amino-acid substitutions changing net protein charge. These variant proteins were assumed to be encoded by 29 alleles at 11 loci. Comparison of 19 European *G. rostochiensis* populations revealed 20 IP-variant alleles at 9 loci (Bakker 1987). The allele frequencies at these loci were estimated as described above and used to construct similarity dendrograms. The dendrogram of the 25 *G. pallida* populations (Fig. 19.4) shows that the international pathotype scheme is incapable of reflecting the genetic diversity introduced into Europe. Genetic distances between populations classified as 'identical pathotypes' are often larger than those between populations classified as 'distinct pathotypes'. These data also show that identification of pathotypes using protein electrophoresis is far more difficult than has been suggested in various reports, either implicitly (Fox and Atkinson 1984; Wharton *et al.* 1983) or explicitly (Greet and Firth 1977; Ohms and Heinicke 1985).

In a previous report, we advanced a more realistic concept for exploiting electrophoretic data for pathotype management using

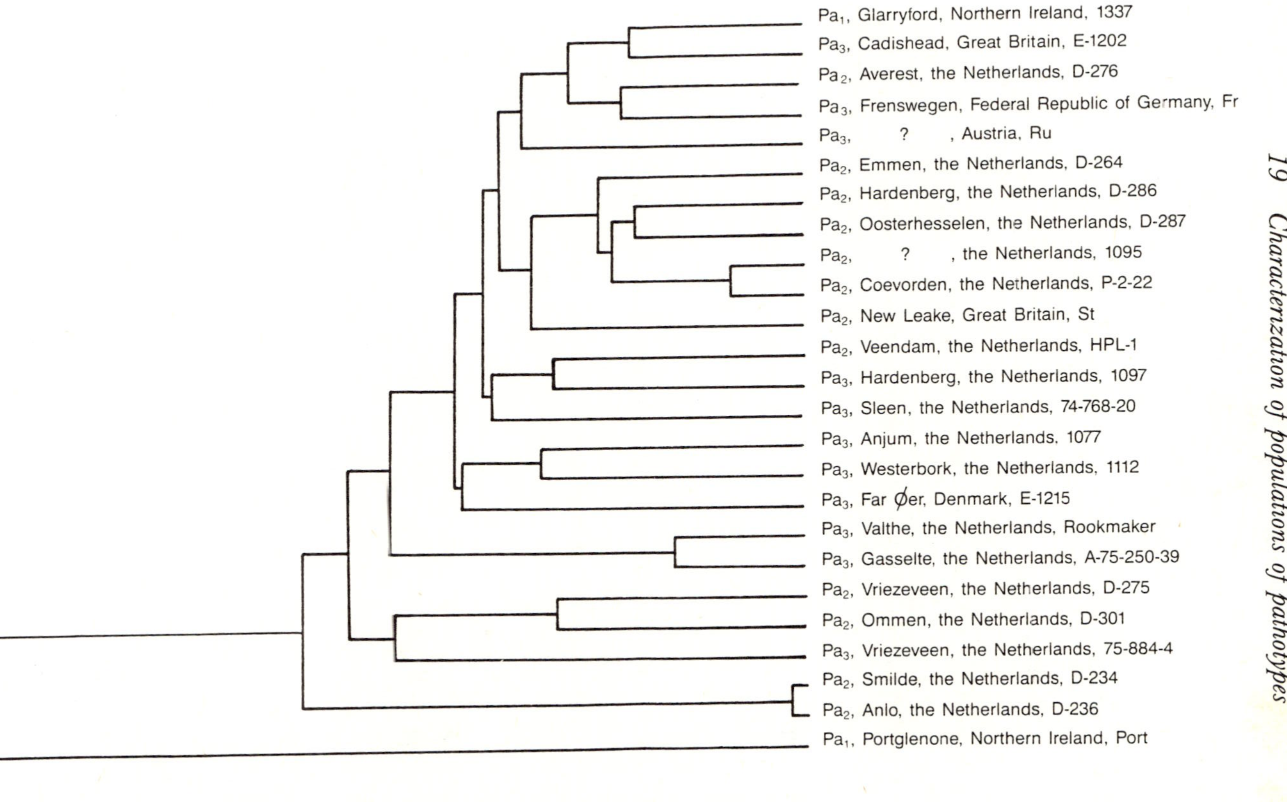

Fig. 19.4. Similarity dendrogram of 25 *G. pallida* populations constructed from the genetic distances at 11 IP-variant loci following the UPGMA method. Genetic distances were calculated according Rogers (1984). Invariant loci were not included (Bakker 1987).

differences in host resistance (Bakker 1987). It has been argued that, in the absence of the relevant host genes for resistance in Europe, both interpopulation variation in virulence and proteins of potato cyst nematodes are predominantly determined by the same three processes. First, spatial variations in virulence and proteins can in part be explained by the genetic structures of the primary founder nematodes introduced from South America, probably after 1850 (Jones and Jones 1974). Europe may have been infested from various native localities, resulting in the introduction of distinct primary founders in several areas. Random genetic drift is another important process and is inherent to the numerous secondary founding events in which Europe has been colonized by potato cyst nematodes. Fluctuations in allele frequencies between nematode generations may also occur when populations are nearly eradicated by control measures. The third process is gene flow. Inherent to the passive mode of spread, the extent of gene flow depends on factors such as soil transported by wind, farm machinery, and seed tubers. Selection is probably irrelevant in explaining the spatial variation found in proteins of potato cyst nematodes. At present, sufficient molecular data are available to support the hypothesis that selection is of minor importance in explaining intraspecific variation or that selection is too weak to be important in this context (Kimura 1983). Although no direct evidence is available, it seems reasonable to assume that selection of alleles for virulence and avirulence caused by adaptation to local environmental conditions is also of minor importance in explaining the mosaic distribution patterns of cyst-nematode pathotypes observed. An argument against strong selection of any type is that it would produce too large a genetic load on its spread and survival. Potato cyst nematodes produce only one generation in a growing season, and in a normal crop rotation the time between generations ranges from two to five years. The maximum multiplication factor is on average 25. Such considerations also argue that the contribution provided by gene mutation is irrelevant. Mutation, by itself, is a slow process and the time between the arrival of the first cyst nematode founders, i.e. after 1850, is too short for newly arisen mutants noticeably to have changed the allele frequencies observed.

The hypothesis that, in the absence of the relevant host genes for resistance in Europe, both interpopulation variation in proteins and virulence are predominantly caused by the genetic structures of the initial nematode populations and random genetic drift and gene flow affecting these, implies that 2-DGE data are valuable indicators for interpopulation variation in virulence. Because these three processes affect the whole gene pool of a nematode population, similarities and dissimilarities revealed by 2-DGE will also be reflected at virulence loci, including those loci not yet revealed by current pathotype schemes. Where nematode populations have large genetic distances at IP-variant loci, there is a fair chance that they behave differently towards various host resistance genes, e.g. because their native ancestors have been differentially exposed to genes for resistance. For example, *G. pallida* populations D-236 and D-264 (Fig. 19.4), both classified as Pa_2 in the current pathotype scheme, may be discriminated as distinct pathotypes by supplementary host genes for resistance. Populations that are identical or nearly so, e.g. *G. pallida* populations D-234 and D-236 (Fig. 19.4), will show resemblance at all virulence loci. Populations having relatively small genetic distances, e.g. populations 1095 and P-2-22 (Fig. 19.4), probably also have small genetic distances at their virulence loci.

2-DGE can be a valuable additional tool for plant breeders, because data on the genetic diversity of nematodes present in an area are indispensable for a successful control by means of host-plant resistance. Once a substantial number of populations in an area are investigated using 2-DGE, the affinities of populations can be used as a guide for a representative survey performed to estimate the proportion of virulent and avirulent nematode populations for any source of host resistance to be tested. Representatives of each group should be included. How to delineate groups in such a way that they represent useful entities, i.e. a manageable number of groups with an acceptable interpopulation variation in virulence, needs further investigation.

Lastly, although we proceed from stochastic processes in explaining the spatial variation in allele frequencies, we do not exclude other mechanisms. However, in the case or cases in which additional genetic mechanisms are operating, e.g. balanced polymorphisms or selection due to environmental conditions, it is very unlikely that such processes are strong enough and act on a sufficient number of loci to impair the value of 2-DGE data for planning and executing rational plant-breeding strategies.

References

Aquadro, C. F. and Avise, J. C. (1981). Genetic divergence between rodent species assessed by using two-dimensional electrophoresis. *Proc. natn. Acad. Sci. USA*, **78**, 3784–8.

Bakker, J. (1987). Protein variation in cyst nematodes. Ph.D. thesis. Agricultural University Wageningen, The Netherlands.

Bakker, J. and Bouwman-Smits, L. (1988*a*). Genetic variation in polypeptide maps of two *Globodera rostochiensis* pathotypes. *Phytopathology*, **78**, 894–900.

Bakker, J. and Bouwman-Smits, L. (1988*b*). Contrasting rates of protein and morphological evolution in cyst nematode species. *Phytopathology*, **78**, 900–4.

Bakker, J. and Gommers, F.J. (1982). Differentiation of the potato cyst nematodes *Globodera rostochiensis* and *G. pallida* and of two *G. rostochiensis* pathotypes by means of two-dimensional electrophoresis. *Proc. K. ned. Akad. Wet.*, **C85**, 309–14.

Butler, M.H., Wall, S.M., Luehrsen, K.R., Fox, G.E., and Hecht, R.M. (1981). Molecular relationships between closely related strains and species of nematodes. *J. Molec. Evol.*, **18**, 18–23.

Bergé, J.B., Dalmasso, A., Person, F., Rivoal, R., and Thomas, D. (1981). Isoesterases chez le nematode *Heterodera avenae*. 1. Polymorphisme chez differentes races francaises. *Revue Nematol.*, **4**, 99–105.

Dalmasso, A. and Bergé, J.B. (1978). Molecular polymorphism and phylogenetic relationship in some *Meloidogyne* spp.: Application to the taxonomy of *Meloidogyne*. *J. Nematol.*, **10**, 323–32.

Evans, K., Franco, J. and de Scurrah, M.M. (1975). Distribution of species of potato cyst-nematodes in South America. *Nematologica*, **21**, 365–9.

Ferris, V.R., Ferris, J.M. and Murdock, L.L. (1985). Two-dimensional protein patterns in *Heterodera glycines*. *J. Nematol.*, **17**, 422–7.

Filipjev, I.N. and Schuurmans Stekhoven, J.H. (1941). *A manual of agricultural helminthology*. E.J. Brill, Leiden.

Fox, P.C. and Atkinson, H.J. (1984). Isoelectric focusing of general protein and specific enzymes from pathotypes of *Globodera rostochiensis* and *G. pallida*. *Parasitology*, **88**, 131–9.

Fox, P.C. and Atkinson, H.J. (1985). Enzyme variation in pathotypes of the potato cyst nematodes *Globodera rostochiensis* and *G. pallida*. *Parasitology*, **91**, 499–506.

Fox, P.C. and Atkinson, H.J. (1986). Recent developments in the biochemical taxonomy of plant parasitic nematodes. *Agric. Zool. Rev.*, **1**, 301–31.

Goldman, D., Giri, P.R., and O'Brien, S.J. (1987). A molecular phylogeny of the homonoid primates as indicated by two-dimensional protein electrophoresis. *Proc. natn. Acad. Sci. USA*, **84**, 3307–11.

Greet, D.N. and Firth, J. (1977). Influence of host plant on electrophoretic protein patterns of some round-cyst nematode females and use of larvae to obtain less ambiguous results. *Nematologica*, **23**, 411–15.

Huettel, R.N., Dickson, D.W., and Kaplan, D.T. (1983). Biochemical identification of the two races of *Radopholus similis* by starch gel electrophoresis. *J. Nematol.*, **15**, 338–44.

Jones, F.G.W. and Jones, M.G. (1974). *Pests of field crops* (2nd edn). Edward Arnold, London.

Jones, F.G.W., Carpenter, J.M., Parrott, D.M., Stone, A.R., and Trudgill, D.L. (1970). Potato cyst nematode: one species or two? *Nature, Lond.*, **227**, 83–4.

Jones, F.G.W., Parrott, D.M., and Perry, J.N. (1981). The gene-for-gene relationship and its significance for potato cyst nematodes and their solanaceous hosts. In *Plant parasitic nematodes*, Vol. 3 (ed. B.M. Zuckerman and R.A. Rohde), pp. 23–36. Academic Press, New York.

Kimura, M. (1983). *The neutral theory of molecular evolution*. Cambridge University Press, Cambridge.

King, M.C. and Wilson, A.C. (1975). Evolution at two levels in humans and chimpanzees. *Science*, **188**, 107–16.

Kort J., Ross, H., Rumpenhorst, H.J., and Stone, A.R. (1978). An international scheme for identifying and classifying pathotypes of potato cyst-nematodes *Globodera rostochiensis* and *G. pallida*. *Nematologica*, **23**, 333–9.

Merril, C.R., Switzer, R.C., and Van Keuren, M.L. (1979). Trace polypeptides in cellular extracts and human body fluids detected by two-dimensional electrophoresis and a highly sensitive silver stain. *Proc. natn. Acad. Sci. USA*, **76**, 4335–9.

Miller, L.I. (1983). Diversity of selected taxa of *Globodera* and *Heterodera* and their interspecific and intergeneric hybrids. In *Concepts in nematode systematics* (ed. A.R. Stone, H.M. Platt, and L.K. Khalil), pp. 207–20. Academic Press, New York and London.

Mugniery, D. (1979). Hybridation entre *Globodera rostochiensis* (Wollenweber) et *G. pallida* (Stone). *Revue Nematol.*, **2**, 153–9.

Oakley, B.R., Kirsch, D.R., and Morris, N.R. (1980). A simplified ultrasensitive silver stain for detecting proteins in polyacrylamide gels. *Analyt. Biochem.*, **105**, 361–3.

O'Farrell, P.H. (1975). High resolution two-dimensional electrophoresis of proteins. *J. Biol. Chem.*, **250**, 4007–21.

Ohms, J.P. and Heinicke, D.H.K. (1983). Pathotypen des Kartoffelnematoden. 1. Schnellbestimmung der Artzugehörigkeit mittels Isoelektrofocussierung. *Z. PflKrankh. PflSchutz*, **90**, 258–64.

Ohms, J.P. and Heinicke, D.H.K. (1985). Pathotypen des Kartoffelnematoden. II. Bestimmung der Rassen von *Globodera rostochiensis* durch die Mikro-2D-Elektrophorese von Einzelzysten. *Z. PflKrankh. PflSchutz*, **92**, 225–32.

Ohnishi, S., Kawanishi, M., and Watanahe, T.K. (1983). Biochemical phylogenies of *Drosophila*: protein differences detected by two-dimensional electrophoresis. *Genetica*, **61**, 55–63.

Platzer, E.G. (1981). Potential use of protein patterns and DNA nucleotide sequences in nematode taxonomy. In *Plant parasitic nematodes* (ed. B.M. Zuckerman and R.A. Rohde), pp. 3–21, Academic Press, New York.

Poehling, H.M. and Wyss, U. (1980). Microanalysis of proteins in the aseptic host-parasite system: *Ficus carica — Xiphinema index*. *Nematologica*, **26**, 230–42.

Pozdol, R.F. and Noel, G.R. (1984). Comparative electrophoretic analyses of soluble proteins from *Heterodera glycines* races 1–4 and three other *Heterodera* species. *J. Nematol.*, **16**, 332–40.

Prager, E.M. and Wilson, A.C. (1975). Slow evolutionary loss of the potential for interspecific hybridization in birds: A manifestation of slow regulatory evolution. *Proc. natn. Acad. Sci. USA*, **72**, 200–4.

Rogers, J.S. (1984). Deriving phylogenetic trees from allele frequencies. *Syst. Zool.*, **33**, 52–63.

Stone, A.R. (1973). *Heterodera pallida* n. sp. (Nematoda: Heteroderidae), a second species of potato cyst nematode. *Nematologica*, **18**, 591–606.

Stone, A.R. (1975). Taxonomy of potato cyst-nematodes. *EPPO Bulletin*, **5**, 79–86.

Stone, A.R. (1977). Recent developments and some problems in the taxonomy of cyst nematodes, with a classification of the *Heteroderoidea. Nematologica*, **23**, 273–88.

Turner, S.J., Stone, A.R., and Perry, J.N. (1983). Selection of potato cyst-nematodes on resistant *Solanum vernei* hybrids. *Euphytica*, **32**, 911–17.

Wharton, R.J., Storey, R.M.J., and Fox, P.C. (1983). The potential of some immunochemical and biochemical approaches to the taxonomy of potato cyst nematodes. In *Concepts in nematode systematics* (ed. A.R. Stone, H.M. Platt, and L.K. Khalil), pp. 235–48. Academic Press, New York and London.

Wilson, A.C. (1976). Gene regulation in evolution. In *Molecular evolution* (ed. F.J. Ayala), pp. 225–34. Sinauer Associates, Inc. Sunderland, Mass.

Wilson, A.C., Maxson, L.R., and Sarich, V.M. (1974). Two types of molecular evolution. Evidence from studies of interspecific hybridization. *Proc. natn. Acad. Sci. USA*, **71**, 2843–7.

20 Isoelectric focusing techniques for the identification of plant-parasitic nematodes

M.P. ROBINSON

Department of Entomology and Nematology, Rothamsted Experimental Station, Harpenden, Hertfordshire, UK

Abstract

Effective control of plant-parasitic nematodes is to a large extent dependent on accurate identification of the species present in the field. The extensive morphological variation among and within species makes their identification difficult using morphometric methods. To overcome this, improvements and modifications to several applications of isoelectric focusing (IEF) have been made to aid the identification of nematode species.

Ultrathin polyacrylamide gels (0.15 mm thick) are being used routinely in conjunction with large voltage gradients and narrow pH ranges. Ultrathin gels have a number of advantages over conventional 2-mm thick gels: they are less expensive, easier to cast, require shorter processing times, and show better resolution with smaller sample loads. Mini-IEF apparatus has proved as effective as the more conventional systems, but is considerably less expensive. Silver staining has been used to enhance the sensitivity of detection. The variability of conventional ampholyte pH gradients can be reduced by replacing the ampholytes with a mixture of amphoteric buffers. These eliminate cathodal drift and do not complex with the proteins being separated. Emphasis is placed on developing rapid, simple, and inexpensive methods of accurately identifying economically important species of plant-parasitic nematodes.

Electrophoretic Studies on Agricultural Pests (ed. Hugh D. Loxdale and J. den Hollander), Systematics Association Special Volume No. 39, pp. 431–42. Clarendon Press, Oxford, 1989.

Introduction

The correct identification of plant-parasitic nematodes to species and sometimes subspecies level is necessary for their effective control. This is because control strategies, in particular non-chemical methods, are usually directed at specific nematode species. In most cases, these species have been described by morphometric and morphological data, with additional evidence from geographical distribution (Sturhan 1983). However, many populations of plant-parasitic nematodes show a range of variation in morphology that makes their identification less reliable. The difficulty in distinguishing intraspecific from interspecific differences in animals of such small size has always been a major problem for taxonomists. Nematologists have therefore attempted to support conclusions from morphological data by using a variety of biochemical techniques to help accurately identify plant-parasitic nematode species (Hussey 1979).

The majority of biochemical systematic studies on plant-parasitic nematodes have relied on the technique of electrophoresis to detect species differences (Fox and Atkinson 1986). Recently, isoelectric focusing (IEF) of soluble nematode proteins has been shown to be a practical and effective technique for the identification of closely related species and members of species complexes. Most of the work has been done on cyst-nematodes (Fleming and Marks 1982, 1983; Fox and Atkinson 1984*a*,*b*; Ohms and Heinicke 1983; Podzol and Noel 1984), but the IEF technique has also been effective in separating populations of root-knot nematodes, *Meloidogyne* spp. (Lawson *et al.* 1984).

However, there are several problems associated with conventional IEF procedures used in these studies. Among the more serious are lack of sensitivity, poor resolution, and band distortion. This has precluded the use of the technique for separation of nematode species with only limited protein heterogeneity. The need for specialized equipment and the high cost of conventional IEF techniques may also limit their use as routine tools for nematode identification. This chapter describes modifications to the conventional IEF procedure that overcome many of these problems and demonstrates the effectiveness of new IEF methods in separating species of the genus *Meloidogyne*.

Materials and methods

Nematode populations

Populations of *M. incognita* Race 1 (North Carolina State University, USA, Stock #78), *M. javanica* (NCSU #72), *M. arenaria* Race 1 (NCSU #351), *M. hapla* (NCSU #482), *M. oryzae* (ex Surinam, 1980) and *M. graminicola* (ex Bangladesh, 1980) were raised on tomato plants *cv*. Moneymaker or rice plants *cv*. IR36 (*M. graminicola* and *M. oryzae*). After 50–55 days, nematode eggs and juveniles were extracted by standard methods (McClure *et al.* 1973).

Sample preparation

Isoelectric focusing is intolerant of high concentrations of salt in protein preparations (Allen *et al.* 1984, p. 77). Therefore, water-soluble proteins from purified and washed second-stage juveniles and eggs were extracted in either deionized water, 2 per cent ampholyte or 1 per cent glycine containing 50 μg ml^{-1} L-1-tosylamide-2-phenylethyl-chloromethyl ketone (TPCK), 50 μg ml^{-1} *N*-α-tosyl-L-lysine-chloromethyl ketone (TLCK) and 10 mM phenylmethylsulphonyl fluoride (PMSF, Sigma Chemical Co.) in isopropanol as inhibitors of proteolysis. Nematode suspensions were homogenized at 4°C in 0.1 ml glass homogenizers (Jencons Ltd, UK), centrifuged at 10 000g for 10 min at 4°C and the soluble fraction has retained. Care was taken to ensure that the protein samples were free of precipitates and lipid. The protein concentration of each sample was measured spectrophotometrically (Bradford 1976), adjusted to 1 mg ml^{-1} and stored at − 45°C until required.

Ultrathin gel casting

Polyacrylamide gels of 1–2 mm thickness are usually employed in analytical electrofocusing using commercial systems. In the present study, 150-μm thick gels (5 per cent T, 3 per cent C_{BIS}) containing 4 per cent w/v Ampholine carrier ampholytes (LKB-Pharmacia, Sweden) were prepared on Gelbond polyester films (LKB-Pharmacia) by the flap technique (Radola 1980). This technique involves binding the Gelbond support film with the hydrophilic side upwards on to a glass plate using a few drops of water. Self-adhesive tape (150 μm thick) is used to cover the edges of a second glass plate that has been silanized (Repel-Silane, LKB-Pharmacia) to prevent binding of acrylamide to this surface. The polymerization mixture is poured along the centre of the silanized plate and spread evenly over the surface by lowering the support plate on to it. After polymerization (1 hour), the glass plates are separated, leaving the ultrathin polyacrylamide gel bound to the polyester support film.

Mini-IEF gels

Polyacrylamide gels (5 per cent T, 3 per cent C_{BIS}) containing 5 per cent glycerol and 2 per cent Resolyte carrier ampholytes (BDH Ltd, UK) were prepared on to Gelbond support films using a capillary casting tray (BIORAD Laboratories, Richmond, Calif., USA). Final gel dimensions were 125 × 65 × 0.4 mm.

Buffer electrofocusing

As an alternative to using pH gradients established with synthetic carrier ampholytes (e.g. Ampholine, Resolyte, Pharmalyte, Biolyte), a linear gradient was produced over the pH range 3–10 using a mixture of 47 amphoteric and non-amphoteric buffers (Cuono and Chapo 1982; Poly/Sep 47, Polysciences Inc, Warrington, USA). The buffer concentrate was mixed with a 40 per cent acrylamide stock solution and catalysts in a 6 :1 :1 ratio to provide a gel monomer solution of 5 per cent T, 3 per cent C_{BIS}. Gels were cast using the flap technique with 0.8-mm spacers.

Procedures for electrofocusing

Ultrathin gels, covalently bound to Gelbond support films, were placed directly onto the cooling plate of standard electrofocusing apparatus (Multiphor II, LKB-Pharmacia). A thin film of paraffin between the gel and the cooling plate was used to improve thermal conductivity. Electrode strips (LKB-Pharmacia) were soaked in anolyte (1 per cent acetic acid) and catholyte (1 per cent ethanolamine) and placed along the longer dimensions of the gel. Two μl of each sample were applied in a 5-mm wide band directly onto the gel surface 1 cm from the cathodic wick. Platinum electrodes (7 cm apart) were placed on the wicks and separations were carried out under constant power with stepped increases when voltages reached 300, 500, 1000, 1500, and 2000 V (Table 20.1). This protocol is derived from the cold focus conditions used by Allen *et al.* (1984, p. 122). Final field strength and total volt-hour products were approximately 400 V cm^{-1} and 800 V h.

Table 20.1 Electrofocusing conditions for high voltage gradient IEF using 150-μm thick gels. Starting voltage levels at each step are set under constant power. Temperature is that of the cooling plate setting.

Time (min)	Voltage (V)	Temperature (°C)
10	100	6
Load samples		
3	300	6
5	500	6
10	1000	4
10	1500	2
5	2000	2
3	2500	0

Samples for mini-IEF were applied onto the gel surface as above. The gels were then inverted and placed directly onto two, 0.95-cm diameter graphite electrodes with an inter-electrode distance of 5 cm (BIORAD). This system eliminates the need for electrode buffers and wicks and does not require associated cooling apparatus. Focusing was carried out under constant voltage conditions at 100 V for 15 min, 200 V for 15 min, and 450 V for 60 min.

Buffer electrofocusing was carried out at 5°C using flat-bed apparatus as for ultrathin gels. Electrode strips were saturated with catholyte (1 M KOH) or anolyte (1 M oxalic acid) and gels were prefocused at constant power of 10 W for 40 min. Sample solutions (20 μl) were then absorbed onto 5 × 10-mm strips of Whatman GF/B glass fibre and placed onto the gel surface 3 mm from the leading edge of the catholyte strip. Electrofocusing was conducted at constant power of 20 W until the potential reached 1000 V, and continued at constant field strength of 1000 V for a total of 2 hours (Cuono and Chapo 1982).

After electrofocusing, all gels were placed in fixative (20 per cent trichloroacetic acid in double-distilled water) and the protein bands were vizualized with non-diamine oxidative silver staining (Merril *et al.* 1981).

Results

The techniques here described were developed and tested to overcome many of the problems encountered with conventional isoelectric focusing. Common artefacts caused by unstable pH

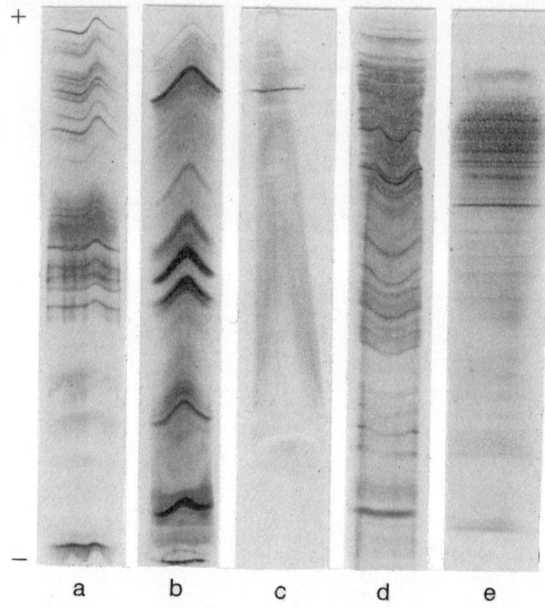

Fig. 20.1. Common problems resulting in artifacts in isoelectric focusing. (a) and (b) Uneven contact between gel and electrode wicks caused by air bubbles or partial drying. (c) Lipid in sample. (d) Sample applied to the anodic end of the gel, where most proteins focus. (e) Ideal result with sharp, straight band patterns.

gradients and impurities in the sample are shown in Fig. 20.1. For comparison, a track from mini-IEF (e) is included, which demonstrates the sharp, straight banding patterns required for detailed comparisons of nematode species.

The effect of different ampholyte pH ranges on the separation of water-soluble nematode proteins is shown in Fig. 20.2. Most nematode proteins have an acidic isoelectric point and therefore focus in a tightly-packed region at the acidic end of the pH 3.5–9.5 gel (Fig. 20.2(c)). Greater separation is achieved by using narrower pH ranges that bracket this region (Fig. 20.2(a) and (b)).

The results of high voltage gradient, ultra-thin IEF (HVGIEF) are shown in Fig. 20.3. The degree of resolution is demonstrated with a densitometric trace (Hoeffer GS300) of proteins from *M. graminicola* (Fig. 20.4). Over 40 easily distinguishable bands are resolved, a number of these being species-specific. *M. graminicola* and *M. oryzae* can occur on the same host and, therefore, this technique could be

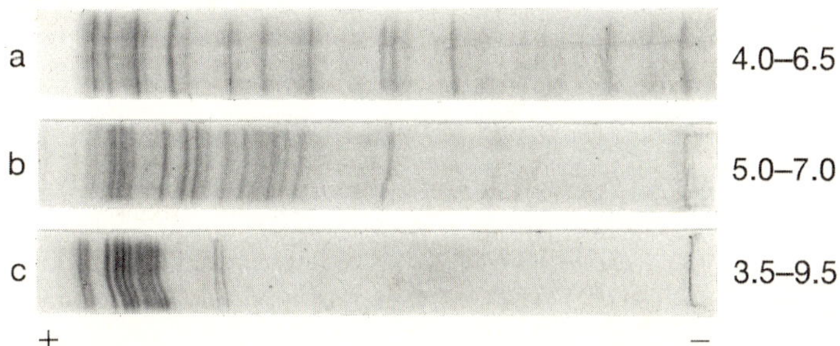

Fig. 20.2. Selected ampholyte pH ranges available for isoelectric focusing. (a) pH 4.0–6.5 (LKB-Pharmacia); (b) pH 5.0–7.0 (BDH); (c) pH 3.5–9.5 (LKB-Pharmacia).

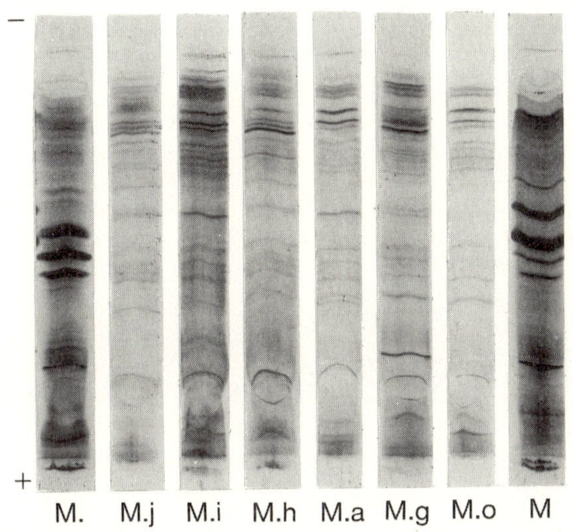

Fig. 20.3. Ultrathin gel, high voltage gradient isoelectric focusing of *Meloidogyne* species over the pH range 4.0–6.5. *Abbreviations*: M, isoelectric point markers; M.j, *M. javanica*; M.i, *M. incognita* Race 1; M.h, *M. hapla*; M.a, *M. arenaria*; M.g, *M. graminicola*; M.o, *M. oryzae*.

used to confirm morphological identification of the nematode species.

Much more detailed examination of HVGIEF tracks is required to separate the remaining four *Meloidogyne* spp. shown in Fig. 20.3. A number of bands are consistently useful in identifying each species. However, the most reliable method of identification was to run

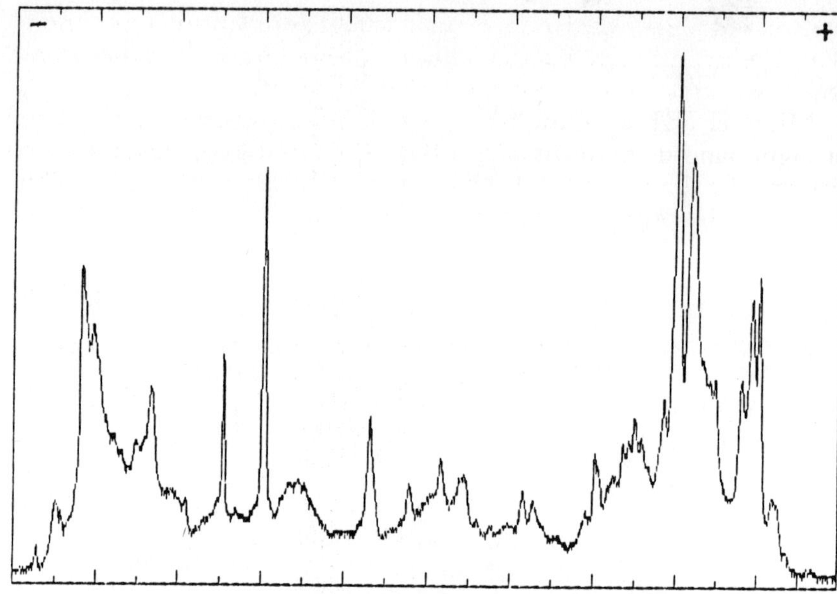

Fig. 20.4. Densitometric trace of *Meloidogyne graminicola* in Fig. 20.3.

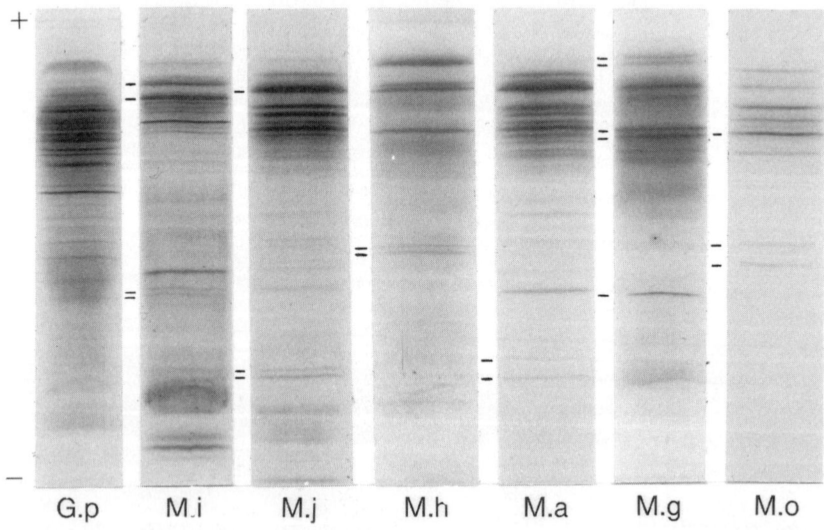

Fig. 20.5. Mini-gel isoelectric focusing of *Meloidogyne* spp. and *Globodera pallida* (G.p) proteins over the pH range 4.0–8.0. (Abbreviations as in Fig. 20.3). Reproducible species-specific bands are marked to the left of each track.

control samples of known species adjacent to unidentified specimens. The profiles, especially in the acidic region of the gel, were then found to be reproducible for a number of populations.

Mini-gel IEF over the pH range 4.0–8.0 consistently produced straight banding patterns (Fig. 20.5). The number of bands resolved was smaller than with HVGIEF but this was compensated for by the clarity of the profiles obtained. A number of reproducible, species-specific bands are shown. Species separation was not judged by the intensity of the bands obtained but by the presence or absence of a particular band. For assessment of intergeneric differences, a preparation from *Globodera pallida*, a potato cyst nematode, was run on the same gel as *Meloidogyne* spp. It is evident that very few proteins are conserved between these two nematode genera (Fig. 20.5).

Buffer electrofocusing, used as an alternative to synthetic carrier ampholytes, resulted in good separation of *M. graminicola* from *M. oryzae* (Fig. 20.6). The conductivity of the buffers was one-tenth that

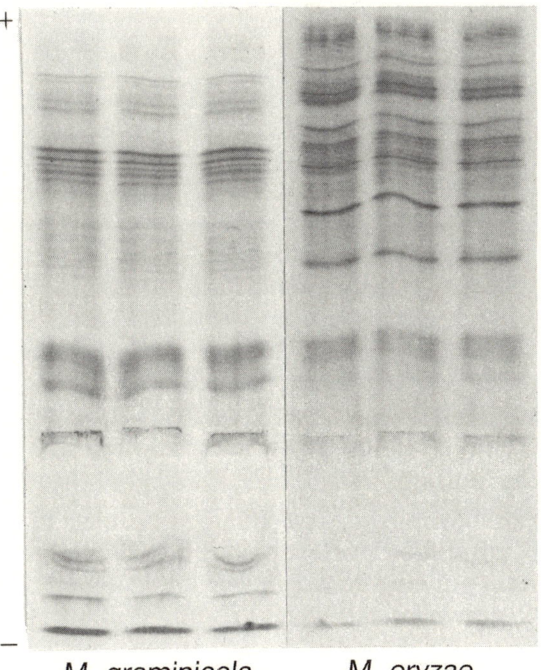

Fig. 20.6. Buffer electrofocusing over the pH range 3.0–10.0 of proteins from *Meloidogyne graminicola* and *M. oryzae*.

of ampholytes, resulting in a high degree of resolution. The gradient was stable for extended periods and there was no evidence of cathodal pH gradient drift — a common problem with conventional IEF (Righetti and Drysdale 1971).

Discussion

Ultrathin HVGIEF, mini-gel IEF, and buffer electrofocusing have not been used before for plant-parasitic nematode identification. This work has shown that these approaches have a number of advantages over conventional IEF. With HVGIEF, a higher field strength with increased resolution potential may be employed, which results in focusing times of only *ca.* 30 min. This effectively eliminates the possibility of 'cathodal drift', a phenomenon caused by extended focusing times. HVGIEF is only possible with the inherent improved cooling characteristics of ultrathin gels. With these gels it is also possible to double the concentration of ampholytes, which improves gradient production but would normally result in overheating with thicker gels. Ultrathin gels require smaller total amounts of reagents and thus the costs of IEF analyses are one-third those of conventional gels. Processing times of the gels after focusing have been reduced, enabling silver staining to be used routinely for nematode identification. This leads to improved sensitivity and increases the number of bands detected. Currently, the equivalent of 400 eggs and second-stage juveniles or one adult female (2 μg of protein) is required for each track.

Mini gel IEF, with the system used here, also has a number of advantages. As the gel is run upside-down, directly in contact with the electrodes, the need for electrode buffers and wicks is eliminated. This overcomes the common problem of uneven contact between gel and electrode wick caused by air bubbles and partial drying. The result is reproducible, straight band profiles that are ideal for routine identification (Fig. 20.5). The system is also considerably cheaper than HVGIEF and conventional IEF, since neither specialized cooling apparatus nor a power pack are required.

With buffer electrofocusing, the advantages are a high degree of resolution and the lack of batch variation encountered with synthetic carried ampholytes. An additional advantage is the fact that these low-molecular-weight amphoteric and non-amphoteric substances do not complex with proteins (Cuono and Chapo 1982), which can lead to inconsistent results.

The techniques presently described have been found to be

applicable to a number of plant parasitic nematode genera, including *Aphelenchoides*, *Globodera*, *Heterodera*, *Pratylenchus*, *Hexatylus*, *Anguina*, and *Bursaphelenchus* (Robinson, unpublished results). The work reported is part of a continuing programme of research into the biochemical taxonomy of *Meloidogyne* species.

Acknowledgements

This work was performed in collaboration with Dr S.B. Jepson, Southampton University, during the tenure of grant R4101 from the Overseas Development Administration.

References

Allen, R.C., Saravis, C.A., and Maurer, H.R. (1984). *Gel electrophoresis and isoelectric focusing of proteins*. Walter de Gruyter, Berlin.

Bradford, M. (1976). A rapid and sensitive method for the quantitation of microgram quantities of proteins utilizing the principle of protein-dye binding. *Analyt. Biochem.*, **72**, 248.

Cuono, C.B. and Chapo, G.A. (1982). Gel electrofocusing in a natural pH gradient of pH 3–10 generated by a 47-component buffer mixture. *Electrophoresis*, **3**, 65–75.

Fleming, C.C. and Marks, R.J. (1982). A method for the quantitative estimation of *Globodera rostochiensis* and *G. pallida* in mixed species samples. *Rec. Agric. Res.*, **30**, 67–70.

Fleming, C.C. and Marks, R.J. (1983). The identification of the potato cyst nematodes *Globodera rostochiensis* and *G. pallida* by isoelectric focusing of proteins on polyacrylamide gels. *Ann. appl. Biol.*, **103**, 277–87.

Fox, P.C. and Atkinson, H.J. (1984*a*). Glucose phosphate isomerase polymorphism in field populations of the potato cyst nematodes, *Globodera rostochiensis* and *G. pallida*. *Ann. appl. Biol.*, **104**, 503–6.

Fox, P.C. and Atkinson, H.J. (1984*b*). Isoelectric focusing of general proteins and specific enzymes from pathotypes of *Globodera rostochiensis* and *G. pallida*. *Parasitology*, **88**, 131–9.

Fox, P.C. and Atkinson, H.J. (1986). Recent developments in the biochemical taxonomy of plant-parasitic nematodes. *Agric. Zool. Rev.*, **1**, 301–31.

Hussey, R.S. (1979). Biochemical systematics of nematodes — a review. *Helminthol. Abstr.*, **48**(B), 141–8.

Lawson, E.C., Carter, G.E., and Lewis, S.A. (1984). Application of isoelectric focusing to the taxonomic identification of *Meloidogyne* spp. *J. Nematol.*, **16**, 91–6.

Merril, C.R., Goldman, D., Sedman, S.A., and Ebert, M.H. (1981). Ultrasensitive stain for proteins in polyacrylamide gels shows regional variation in cerebrospinal fluid proteins. *Science*, **211**, 1437.

McClure, M.A., Kruk, T.H., and Misaghi, I. (1973). A method of obtaining quantities of clean *Meloidogyne* eggs. *J. Nematol.*, **5**, 230.

Ohms, J.P. and Heinicke, D.H.K. (1983). Pathotypen des kartoffelnematoden. 1. Schnellbestimung der Artzugehörigkeit mittels Isoelektrofocussierung. *Z. PflKrankh. PflSchutz.*, **90**, 258–64.

Podzol, R.F. and Noel, G.R. (1984). Comparative electrophoretic analyses of soluble proteins from *Heterodera glycines* Races 1–4 and three other *Heterodera* species. *J. Nematol.*, **16**, 332–9.

Radola, B.J. (1980). Ultrathin-layer isoelectric focusing in 50–100 μm polyacrylamide gels on silanized glass plates or polyester films. *Electrophoresis*, **1**, 43–56.

Righetti, P.G. and Drysdale, J.W. (1971). Isoelectric focusing in polyacrylamide gels. *Biochim. biophys. Acta*, **236**, 17.

Sturhan, D. (1983). The use of the subspecies and the superspecies categories in nematode taxonomy. In *Concepts in nematode systematics* (ed. A.R. Stone, H.M. Platt, and L.F. Khalil), pp. 41–53. Academic Press, London and New York.

21 The use of electrophoresis in the study of hymenopteran parasitoids of agricultural pests

W. POWELL

Entomology and Nematology Department, Rothamsted Experimental Station, Harpenden, Hertfordshire, UK

and M.P. WALTON

School of Pure and Applied Biology, University of Wales College of Cardiff, Cardiff, Wales, UK

Abstract

Hymenopteran parasitoids can cause considerable mortality in populations of many agricultural pests. Consequently they have been used extensively in biological control and integrated pest-management programmes throughout the world. However, there are a number of problems involved with their use that often hinder the success of such control programmes. These include taxonomic difficulties such as the recognition of sibling species and distinct biotypes that may have varying biological characteristics; monitoring problems such as the accurate and at the same time rapid assessment of levels of parasitism in pest populations; and difficulties with the mass rearing of parasitoids that may involve fast rates of genetic drift leading to reduced variability in cultured populations.

The potential of electrophoresis for helping to solve or alleviate these problems is discussed.

Introduction

Hymenopteran parasitoids form one of the most important groups of natural enemies attacking agricultural pests. They have conse-

Electrophoretic Studies on Agricultural Pests (ed. Hugh D. Loxdale and J. den Hollander), Systematics Association Special Volume No. 39, pp. 443–65. Clarendon Press, Oxford, 1989.

quently been widely used in biological and integrated control programmes throughout the world (Greathead 1986; Powell 1986). However, the taxonomy of many parasitoid groups is difficult, often requiring specialist expertise. This can seriously hinder their successful exploitation in pest-control programmes; misidentifications may lead to the selection of an inappropriate parasitoid species and result in wastage of time, effort, and money. The use of electrophoresis as an aid in systematics can alleviate some of these problems. In addition, electrophoresis shows considerable potential as a sampling technique that can improve the efficiency of parasitoid monitoring — both in basic ecological research and in the assessment of their performance as biological control agents. In electrophoretic studies of the hosts of parasitoids, the recognition of endoparasitoid banding patterns is essential to avoid erroneous interpretations of gels. For example, parasitized individuals may be mistaken for new genetic variants, perhaps associated with important biological traits such as resistance to pesticides.

In this review we outline the ways in which electrophoresis can aid the study of hymenopteran parasitoids, placing emphasis on their role as natural enemies of agricultural pests. Consequently, we have not attempted to cover work in which electrophoresis is being used to elucidate the effects of endoparasitism on the physiology and biochemistry of host insects, but information on this aspect may be found in the following papers: Smilowitz (1973), Smilowitz and Smith (1977), Dahlman and Greene (1981), Ferkovich *et al.* (1983), Ferkovich and Dillard (1986), and Beckage *et al.* (1987). Some examples of studies involving electrophoretic work on hymenopteran parasitoids are listed in Table 21.1.

Monitoring parasitism in host populations

1. Assessment of parasitization levels

Evaluation of entomophagous parasitoids for use in integrated pest-management or classicial biological-control programmes normally involves some assessment of their impact on the population dynamics of the host species in field situations. This is most often expressed as percentage parasitism, defined in terms of the parasitoid and host developmental stages sampled. Data on parasitism levels are also useful in the development and implementation of pest-forecasting procedures.

The main sampling techniques currently employed to assess

Table 21.1 Examples of studies involving electrophoresis of hymenopteran parasitoids

Parasitoid	Host	Purpose	Reference
Aphidiidae			
Aphidius ervi	*Acrythosiphon pisum*	To evaluate the importance of genetic drift in laboratory populations	Unruh *et al*. (1983)
A. ervi	*A. pisum, Sitobion avenae*	To supplement behavioural data on host biotypes	Cameron *et al*. (1984).
A. ervi	*A. pisum* *Microlophium carnosum*	To investigate genetic polymorphism in field populations and host biotypes	Němec and Starý (1983*a*, *b*, 1984*a*, 1985*b*) Starý *et al*. (1985)
A. ervi *Aphidius matricariae* *Aphidius picipes* *Aphidius rhopalosiphi*	Various aphid hosts	To clarify taxonomic status and aid species identification	Pungerl (1986)
A. ervi *A. picipes* *Aphidius pisivorus* *Aphidius eadyi* *Aphidius smithi* *Aphidius* sp. nr *smithi*	*A. pisum* *A. kondoi*	To clarify species and intraspecific relationships	Unruh *et al*. (1986)
A. picipes *Aphidius uzbekistanicus* *Ephedrus plagiator* *Praon volucre* *Toxares deltiger*	*S. avenae*	To detect parasitoid larvae within hosts and differentiate between species involved	Castañera *et al*. (1983)

Table 21.1 *Continued*

Parasitoid	Host	Purpose	Reference
A. ervi *A. rhopalosiphi* *A. picipes* *E. plagiator* *P. volucre*	*S. avenae*	To assess levels of parasitism in field populations of the host and to differentiate between species involved	Walton (1986)
Aphidius rosae *Ephedrus* sp.	*Macrosiphum rosae*	To assess levels of parasitism in host populations	Tomiuk and Wöhrmann (1980)
A. matricariae	*Myzus persicae*	To detect parasitism in the host	Wool *et al*. (1978)
Diaeretiella rapae	*Brevicoryne brassicae* *M. persicae* *Hayhurstia atriplicis*	To investigate genetic diversity of host biotypes	Němec and Starý (1984*b*)
Lysiphlebus fabarum *L. cardui* *L. confusus*	*Aphis fabae* *A. farinosa*	To investigate genetic diversity of field populations	Němec and Starý (1985*c*)
Praon abjectum *Trioxys angelicae*	*A. sambuci*	To investigate genetic diversity of field populations	Starý and Němec (1986)
Aphelinidae			
Encarsia lutea' *Eretmocerus mundus*	*Bemisia tabaci*	To detect parasitism in the host	Wool *et al*. (1984)
Aphelinus abdominalis	*Sitobion avenae*	To assess levels of parasitism in field populations of the host	Walton (1986)
Ichneumonidae			
Glypta fumiferanae	*Choristoneura occidentalis*	To detect parasitism in field	Castrovillo and Stock

Diadegma armillata *Itoplectis maculator* *Mesochorus vittator* *Pimpla turionellae* *Triclistus yponomeutae* *Trieces tricarinatus*	Various *Yponomeuta* spp.	To differentiate between host and parasitoid zymograms	Menken (1982)
Encyrtidae			
Ageniaspis fuscicollis	Various *Yponomeuta* spp.	To differentiate between host and parasitoid zymograms	Menken (1982)
Eulophidae			
Tetrastichus evonymellae	Various *Yponomeuta* spp.	To differentiate between host and parasitoid zymograms	Menken (1982)
Pteromalidae			
Spalangia endius *S. cameroni* *S. nigroaenea* *Muscidifurax raptor* *M. zarapter*	*Musca domestica* *Stomoxys calcitrans*	To clarify taxonomic status and aid species indentification	Propp (1986)
Trichogrammatidae			
Trichogramma evanescens *T. maidis*	*Ephestia kuehniella*	To identify a new species and to distinguish between sibling species	Pintureau and Voegelé (1980)
T. evanescens *T. maidis*	*E. kuehniella* *Pieris brassicae* *Ostrinia nubilalis*	To characterize laboratory strains	Pintureau and Babault (1981)
T. evanescens group (10 spp.)	Various Lepidoptera	To clarify taxonomic relationships	Pintureau and Babault (1980)
Trichogramma (14 spp.)	*Heliothis virescens*	To investigate biosystematics and genetics of related species	Hung and Huo (1985)

percentage parasitism are host dissection, host rearing until parasitoid emergence, and the counting of parasitized individuals if these are readily identifiable (e.g. mummified aphids). However, all these methods have their drawbacks (Wool *et al.* 1978; Stock 1981; Castrovillo and Stock 1981; Walton 1986). First, dissection of the host is time-consuming and often inaccurate, and the efficiency of different individuals doing the dissections can vary greatly. Parasitoid eggs in particular can be difficult to detect and, if more than one species of parasitoid is involved, larval identification may be a problem. Rearing of the host allows identification of emerging adult parasitoids but necessitates a delay between field sampling and obtaining the data. Rearing is also expensive in terms of labour and space and may lead to biased results, since some mortality factors, such as climate and predation, are removed, whilst others, resulting from handling and rearing conditions, are introduced (Wool *et al.* 1978; Walton 1986). In the case of aphid parasitoids, 'mummy' counts are the most commonly used method of assessing parasitism in field populations. Parasitoid pupation occurs within or beneath the dead aphid host, which thereby acquires a distinctive appearance and is known as a parasitoid mummy. Assessing percentage parasitism from the proportion of mummified aphids in a field population is quick and relatively easy but can provide very misleading results for several reasons. First, aphids containing parasitoid eggs or larvae are ignored. In addition, mummies may be easily dislodged from plants (e.g. by strong winds) and some parasitized aphids leave their host plant before mummification (Stary 1970; Powell 1980). Furthermore, at certain times of the year, many parasitoids enter diapause in the mummy stage, leading to their accumulation in the crop, thereby biasing subsequent samples.

Electrophoresis offers an alternative to these sampling techniques that has the potential to overcome many of their drawbacks. This was first appreciated by Wool *et al.* (1978), who used polyacrylamide slabgel electrophoresis to detect *Aphidius matricariae* Haliday (Aphidiidae) in a glasshouse population of the aphid *Myzus persicae* (Sulz.). They looked at several enzyme systems and obtained the most promising results with esterase (EST) (EC 3.1.1.1), malate dehydrogenase (MDH) (EC 1.1.1.37), and malic enzyme (ME) (EC 1.1.1.40). These all gave distinct banding patterns for aphids containing parasitoid larvae, allowing easy separation of parasitized and healthy individuals. MDH was also used on starch gels to assess levels of parasitism in field populations of the rose aphid *Macrosiphum rosae* (L). (Tomiuk and Wohrmann, 1980). Castañera *et al.* (1983) investigated detection of the parasitoid *Aphidius uzbekistanicus* (Luzh.)

(= *A. rhopalosiphi* (De Stef.)) in laboratory cultures of the grain aphid *Sitobion avenae* (F.) using 14 enzyme systems. They agreed with Wool *et al.* (1978) that the most useful systems were EST, MDH, and ME, but they also obtained reasonable results with glutamate-oxaloacetate transaminase (GOT) (EC 2.6.1.1.). Two of the systems tested, phosphatase (PHOS) (EC 3.1.3.) and peptidase (PEP) (EC 3.4.11.), failed to detect the parasitoids. This work was followed up by Walton (1986) who used EST to monitor percentage parasitism in glasshouse and field populations of *S. avenae*.

Electrophoresis has also been used to detect parasitoids in insects other than aphids. Four out of the 15 enzyme systems tested proved useful for detecting the endoparasitoid *Glypta fumiferanae* (Vier.) (Ichneumonidae) in larvae of the western spruce budworm *Choristoneura occidentalis* Freeman (Castrovillo and Stock 1981). These were EST, aconitase (ACON) (EC 4.2.1.3.), leucine aminopeptidase (LAP) (EC 3.4.13.), and 6-phosphoglucose dehydrogenase (6PGDH) (EC 1.1.1.44.). ACON and 6PGDH produced monomorphic host bands, whereas EST and LAP were polymorphic and it was suggested that for the routine screening of host populations in order to assess percentage parasitism, enzyme systems giving single host bands would be quicker and easier to interpret (Castrovillo and Stock 1981).

Wool *et al.* (1984) looked at six enzyme systems for detecting the tiny endoparasitoids *Encarsia lutea* (Masi) and *Eretmocerus mundus* (Mercet.) (Aphelinidae) in the whitefly *Bemisia tabaci* (Genn.). Because of the difficulty of homogenizing such tiny insects, the whitefly were individually squashed with a plastic rod in the middle of a narrow strip of filter paper, which was then placed in a pocket of the acrylamide gel. EST produced the best results, although weakly staining bands were also obtained using MDH and aldehyde oxidase (AO) (EC 1.2.3.1). Enzyme activity was not detected for either host or parasitoids with alkaline phosphatase (AP) (EC 3.1.3.1), acid phosphatase (ACP) (EC 3.1.3.2), or α-glycerophosphate dehydrogenase (α-GPD) (EC 1.1.1.8).

To assess the value of electrophoresis for estimating percentage parasitism in cereal aphid field populations, Walton (1986) compared results using electrophoresis with those from mummy counting and host rearing. Electrophoretic estimates were very similar to those from host rearing. This was not unexpected, since both methods sample the same parasitoid developmental stages and both involve the removal of live aphids from the field and hence from the risk of further parasitization.

In order to interpret electrophoretic estimates of percentage parasitism, it is necessary to know which developmental stages of the parasitoid can be reliably detected by the technique. When establishing the best enzyme systems for detecting parasitism of *S. avenae*, Castañera *et al*. (1983) deliberately used aphids containing final-instar larvae, but they stressed the importance of determining the earliest detectable stage for the purpose of field monitoring. In an attempt to do this, Walton (1986) set up a series of laboratory trials in which he exposed pots of third-instar *S. avenae* nymphs to female parasitoids for 24 hours. These were sampled every day thereafter, half the aphids being dissected to establish the progress of parasitoid development and half being run on acrylamide gels and stained for EST activity. Three parasitoid species, *A. rhopalosiphi*, *Praon volucre* (Hal.), and *Ephedrus plagiator* (Nees), were investigated, and in all species EST staining failed to detect the egg stage. This may have been due to the absence of esterase in the eggs, or more probably to their small size. Other enzyme systems were not tested. The failure to detect eggs with EST staining may explain why host rearing sometimes gave slightly higher estimates of percentage parasitism than did electrophoresis in Walton's field surveys.

2. *Identification of parasitoid species present*

Many pests are attacked by more than one species of parasitoid and this introduces a further complication when monitoring parasitism levels in field populations. Separation of the different parasitoids attacking a particular host species may be straightforward when they utilize different developmental stages of the host or when they are widely separated taxonomically. However, groups of closely-related species are sometimes associated with the same host species. For example, in a single cereal field at Rothamsted Experimental Station in England, six aphidiid species, including three species in the single genus *Aphidius*, were recorded attacking *S. avenae* (Powell, unpublished data). A primary parasitoid from another family (Aphelinidae) plus five hyperparasitoid species were also present, giving a total of 12 hymenopteran parasitoids associated with the same host in the same place at the same time.

Taxonomic determination of parasitic Hymenoptera using conventional morphological characters is often difficult for non-specialists, and if a time delay is necessary in order to rear out adult parasitoids for identification, then the value of the data for monitoring and forecasting purposes will be considerably reduced. Electrophoresis offers a solution to these difficulties, provided that characteristic allozyme banding patterns can be obtained for the parasitoids involved. Distinctive banding patterns were obtained, using malate dehydrogenase, for two parasitoids — *Aphidius rosae* Hal. and *Ephedrus* sp. — attacking the rose aphid *Macrosiphum rosae* (Tomiuk and Wohrmann 1980). Castañera *et al.* (1983) examined adults of five aphidiid species — *A. rhopalosiphi, Aphidius picipes* (Nees), *Ephedrus plagiator, Praon volucre,* and *Toxares deltiger* (Hal.) — attacking *S. avenae* and established that they were readily distinguished from each other on the basis of their EST banding patterns. Patterns for three of these species are illustrated in Fig. 21.1.

The potential of polyacrylamide gel electrophoresis for distinguishing larval stages of the primary parasitoids of *S. avenae* was investigated by Walton (1986). He examined fumarase (FUM) (EC 4.2.1.2.), PEP, ME, MDH, and GOT in addition to EST and confirmed that EST was the most useful enzyme system for species separation but identified two problems. First, the two closely related species, *A. rhopalosiphi* and *Aphidius ervi* Hal., could not readily be distinguished from one another because of the very high degree of polymorphism shown by both at the most variable EST locus. Secondly, there was a possibility that some individuals of *E. plagiator* and of *Aphelinus abdominalis* Dalm. would be overlooked during the screening of aphid samples because the single EST bands produced by their larvae were occasionally obscured by host bands. When

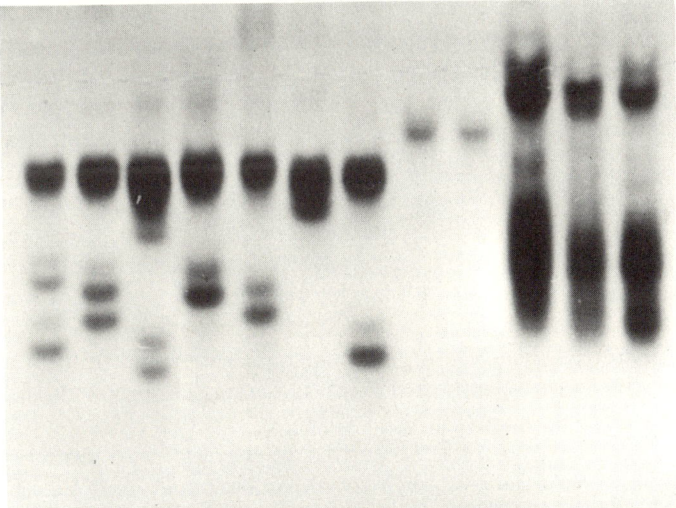

Fig. 21.1. Esterase isoenzyme patterns of three parasitoid species attacking the grain aphid, *Sitobion avenae*. Left to right: 1–7, *Aphidius picipes*; 8–9, *Ephedrus plagiator*; 10–12, *Praon volucre*.

using polyacrylamide gel electrophoresis to detect endoparasitoids of the whitefly *Bemisia tabaci*, Wool *et al*. (1984) could not discern any differences in the EST banding patterns for the two parasitoids *Encarsia lutea* and *Eretmocerus mundus*. The authors point out that the identical, single EST bands shown by the two parasitoids could either be produced by the parasitoids themselves or by the host in response to the presence of either species.

Obviously, the ability to recognize endoparasitoid allozyme banding patterns can be of great value in biological control and parasitoid population research. However, it can also be important in studies concerned primarily with the host species. For example, during electrophoretic work on the systematics of small ermine moths (Yponomeutidae), the banding patterns produced by one dipteran and eight hymenopteran endoparasitoids were determined in order to avoid erroneous interpretations of these as allelic variations of the host (Menken 1982).

3. *Advantages and disadvantages of electrophoresis for sampling parasitoids*

As a sampling tool for the assessment of parasitism in pest populations, electrophoresis has a number of obvious advantages over the more conventional methods of host rearing and dissection. Parasitoids can be detected at an early stage in their development and the species involved can usually be identified without the need to rear through to the adult. This allows objective estimates of parasitization levels to be obtained very rapidly, which greatly enhances their value in forecasting and pest control decision making. Large numbers of host individuals can be processed within a short time and this is likely to increase as progress is made in the design of commercial equipment. Walton (1986), whilst screening cereal aphid populations, was able to test around 400 aphids per day when staining for esterase activity only; in their evaluation of electrophoresis for the detection of endoparasitoids in western spruce budworms, Castrovillo and Stock (1981) established that a single technician could process 150 budworm larvae per day.

The principal disadvantage of electrophoresis is the requirement for specialized equipment and the associated technical expertise. The establishment of appropriate facilities, at a series of laboratories, for the routine electrophoretic screening of specific crop pests is seldom likely to be cost-effective. However, Castrovillo and Stock (1981) suggest that it may become possible to subcontract such routine screening work as the number of laboratories carrying out electrophoretic research increases. If results are not required quickly, samples can be stored deep-frozen for several months without any apparent detrimental effects, allowing optimal use of personnel, equipment, and laboratory space (Wool *et al.* 1978; Castañera *et al.* 1983).

Parasitoid population studies

1. *Field populations*

The efficient exploitation of both indigenous and introduced parasitoid species in pest control requires a sound knowledge of parasitoid population dynamics. The development of electrophoresis as a technique for sampling insect endoparasitoids has provided new opportunities for the study of their populations in the field. Although parasitoids commonly feature as mortality factors in field population studies of pest insects, they are seldom the subject of population studies themselves. This is regrettable, since there is an urgent need for a better understanding of parasitoid ecology within agricultural and forest ecosystems, encompassing both crop and non-crop habitats, in order to facilitate efforts to manipulate them and so increase their impact on crop pests. For example, it is important to determine how and when parasitoids are utilizing available resources (e.g. hosts, adult food sources, shelter) in different habitats within the agroecosystem. How far do they move in response to resource availability? How much intraspecific genetic heterogeneity is there between different populations separated both in space and, in the case of oligophagous species, on different hosts? Electrophoresis could help in the study of such problems.

Němec and Starý (1983*a*) have emphasized the need to study aphid parasitoid populations in entire agroecosystems. They examined esterase banding patterns on polyacrylamide disc-gels as a means of measuring population diversity in several aphid parasitoid species in Czechoslovakia (Němec and Starý 1983*a*, 1984*b*, 1985*a,b*). Populations of oligophagous parasitoids attacking different host species differed in the relative frequencies of allozymes at polymorphic loci, and in some cases certain allozymes were absent in populations on particular hosts. These observations led to the concept of the 'population diversity centre', a term applied to the host on which a parasitoid species shows the greatest degree of genetic diversity and which is therefore regarded as its main host (Němec and Starý 1984*b*, 1986). On this basis, *Brevicoryne brassicae* (L.) and *Acyrthosiphon pisum* (Harr.) have been designated as the population diversity centres of *Diaeretiella* rapae (McInt.) and *A. ervi*, respectively. It is assumed that oligophagous and evolutionarily young parasitoid species will show the greatest genetic variability, whilst monophagous and evolutionarily ancient species will show the least (Němec and Starý 1983*b*).

The concept of population diversity centres is important with respect to parasitoid collection for classical biological control programmes, as it is often desirable to tap the most genetically diverse source populations. Also, knowledge of host preferences is pertinent to the use of alternative hosts to build up field populations in and around crops susceptible to pest aphids. However, although esterases are considered to be the most highly polymorphic enzymes (Wagner and Selander 1974), they are only one indicator of genetic variability and the population diversity centre theory should be tested further by examining a range of enzyme systems showing polymorphic loci to see whether the greatest allozyme diversity always occurs on the same host species.

Studies of allozyme frequencies in naturally occurring parasitoid populations should prove invaluable, not only for identifying genetic variability of potential benefit in the selection and screening of biological control agents, but also for the detection of genetic markers indicative of specific biological traits or geographical origins (Mackauer 1972; Unruh *et al.* 1983). Populations of a parasitoid from distinct geographical regions often differ in biological traits that may prove highly pertinent to their success in classical biological control programmes. Propp (1986) analysed isoenzymes in field and laboratory populations of three species of *Spalangia* and two species of *Muscidifurax* (Hymenoptera: Pteromalidae), all parasitizing house flies (*Musca domestica* L.) and stable flies (*Stomoxys calcitrans* L.). Although variability was generally low amongst conspecific populations, some unique alleles were detected, and the author suggested that these could be used as electrophoretic markers to identify introduced exotic strains and to monitor the incorporation of new genes into the indigenous gene pool.

Electrophoresis should also be a useful tool for following the fate of parasitoids after inoculative or inundative releases (Propp 1986). Rates of spread from release areas and interactions with indigenous populations, including hybridization, could be monitored (Gonzalez *et al.* 1979). Changes in allozyme frequencies in populations following release, indicating genotypes most likely to persist, will provide useful information pertinent to the selection of strains or biotypes for future releases.

2. *Laboratory populations*

Laboratory populations of insect parasitoids are maintained for two main purposes; to provide material for research into their basic biology and behaviour and to mass-rear insects for release in biological control programmes. In either case, it is important to be aware that laboratory cultures are usually much less variable genetically than wild populations. Unruh *et al.* (1983) recorded considerably more allozyme variability in field populations of the aphid parasitoid *A. ervi* than was evident in laboratory populations, as did Walton (1986) for both *A. ervi* and *A. rhopalosiphi*. The low variability often noted in laboratory cultures is sometimes the result of small founder numbers or of the population passing through a 'bottleneck' owing to culturing difficulties. However, genetic drift also appears to be important. Unruh *et al.* (1983) monitored the decline in allozyme variability in a series of *A. ervi* populations of different sizes, maintained at three different temperature regimes. They estimated that, at the observed rate of decline, heterozygosity would be halved in 18 generations under the normal rearing conditions in their laboratory. This was equivalent to less than nine months in culture.

In view of this, it is obviously unwise to assess the biological and behavioural traits of a parasitoid using only individuals from small or long-established laboratory populations. For example, distinct host preferences were displayed by *A. rhopalosiphi* when in culture, and these differed between different laboratory populations. Furthermore, these host preferences were not apparent in field populations (Powell and Wright 1988). Thus, genetic drift can result in the inadvertent selection of laboratory strains that display certain biological traits and this can lead to biased or conflicting conceptions of the characteristics of the species as a whole. This has important implications when biological control decisions need to be made. When examined electrophoretically, some field populations of *Spalangia* spp. and *Muscidifurax* spp. had heterozygosities as low as those in laboratory colonies, but this does not rule out the possibility that adaptive traits had been lost from the latter (Propp 1986). As Unruh *et al.* (1983) point out, 'Even though much of the genetic variation detected by electrophoresis may be effectively neutral, it does provide a reliable measure of genetic drift which may act to reduce genetic variation in both adaptive and neutral traits.'

Greater use of electrophoresis needs to be made in order to monitor the quality of laboratory cultures and to detect the development of major deviations in genetic composition from that existing in field populations. Electrophoretic data could thereby help to establish the optimum size and life-span for a laboratory population and appropriate times to replace cultures or to introduce fresh material from the field. Quality control is also important during the importation and mass-rearing of parasitoids for release against agricultural pests.

Unruh *et al.* (1983) discuss the necessity of considering 'effective population size' during culturing. This is normally smaller than the actual population size, since not all individuals contribute to reproduction to the same degree. A further possible benefit of electrophoretic monitoring is the detection of cross-contamination between cultures of sibling species (Menken and Ulenberg 1987).

Parasitoid systematics

1. Species separation

The value of electrophoresis as a tool in systematics is being increasingly recognized by taxonomists (Avise 1975; Menken and Ulenberg 1987). However, there have been few published studies concerning insect parasitoids. This is surprising, considering the serious problems that can arise in biological control attempts as a result of misidentifications and unrecognized sibling-species complexes. Electrophoresis can be particularly useful for the recognition of sibling species, which often differ significantly in their protein composition (Menken and Ulenberg 1987) or in allozyme frequencies (Berlocher 1984), especially in the case of enzymes such as esterases that show considerable intraspecific variability.

Propp (1986) used starch gel electrophoresis to examine the inter- and intraspecific isoenzyme variation in three species of *Spalangia* (*S. cameroni* Perk., *S. endius* Wlk., *S. nigroaenea* Curt.) and two *Muscidifurax* species (*M. raptor* Gir., *M. zaraptor* Kog.), pteromalid parasitoids attacking the pupae of muscoid flies. The three *Spalangia* species were readily distinguishable at 8 of the 15 loci examined and the two *Muscidifurax* species at 3 of 10 loci. Calculation of Rogers' Similarity Index (Rogers 1972) from these data gave values below 0.3 for the *Spalangia* species, but a value of 0.65 for the *Muscidifurax* species, indicative of a genetic similarity associated with sibling species (Propp 1986). The electrophoretic results were in agreement with existing morphological and biological characters used to identify these particular species.

The most extensive enzymatic study of parasitic Hymenoptera has been done on *Trichogramma*, a genus in which species identification using morphological characteristics is frequently difficult. Examination of enzyme banding patterns, especially of esterases, from numerous cultures has allowed clarification of inter- and intraspecific relationships. This includes recognition of new species and the provision of evidence of species hybridization (Jardak *et al.* 1979; Pintureau and Babault 1980, 1981, 1982; Pintureau and Voegele 1980; Voegele and Berge 1976; Hung 1982; Hung and Huo 1985). Voegele and Berge (1976) first demonstrated the value of gel electrophoresis for separating closely related *Trichogramma* species, based on esterase banding patterns. Referring to differences at nine esterase loci, Jardak *et al.* (1979) separated *Trichogramma oleae* sp.n. from *T. evanescens* Westw., whilst Pintureau and Babault (1981) provided enzymatic data to support the separation of *Trichogramma maidis* sp.n. from *T. evanescens*. Esterases, with supplementary data from malate dehydrogenase, readily permitted recognition of other related species of *Trichogramma*, but tetrazolium oxidase distinguished only distant species (Pintureau and Babault 1982).

Much of the work on *Trichogramma* has utilized laboratory cultures and strains. This introduces the risk that interpretations of isozyme data could be biased as a result of inbreeding and inadvertent selection (Berlocher 1984). Because of this risk, Hung and Huo (1985) recommended that isoenzyme data in systematic studies of *Trichogramma* should be obtained from at least six laboratory cultures, which should be established from females collected in the same locality and from the first five generations only. It would be wise to use similar precautions in other electrophoretic studies of parasitic Hymenoptera and to make use of naturally occurring populations as far as it is practicable.

Sound taxonomy is fundamental to the successful use of insect parasitoids in biological and integrated pest control. However, the parasitic Hymenoptera in general present many species-identification problems. In order to alleviate these problems, greater use should be made of electrophoresis as a systematics aid, but at present the potential of this technique is largely unrealized.

2. *Biotypes*

The recognition of intraspecific variation can be as crucial for the success of biological control programmes as is sound species determination. Unfortunately, there exists considerable confusion regarding infraspecific terminology, with terms such as subspecies, variety, race, strain, and biotype all being widely used. The problems this poses in biological control work have been discussed by Gonzalez *et al.* (1979), who proposed that the term 'biotype' alone should be used to designate biological characteristics of parasitoid populations. Populations of the same species that are separated spatially are termed geographical biotypes, and the term host biotype is applied to populations of the same parasitoid existing on separate host species. Different geographical and host biotypes often can be distinguished electrophoretically (Pintureau *et al.* 1980; Němec and Starý 1985*a*; Unruh *et al.* 1986; Walton 1986; Smith and Hubbes 1986). Gonzalez *et al.* (1979) advocated the use of electrophoresis in conjunction with hybridization and behaviour studies in order to clarify the status of parasitoid populations being used in research.

One strategy for the exploitation of oligophagous parasitoids in biological control involves the use of alternative hosts, to improve synchrony between parasitoid and target host, to improve parasitoid distribution, and to reduce intraspecific competition in the parasitoid population (van den Bosch and Telford 1964). The success of this strategy depends upon the ability and willingness of parasitoids to switch between host species. This in turn depends upon the extent to which the relevant host biotypes are distinct.

Laboratory attempts to transfer the aphid parasitoid *A. ervi* from the pea aphid, *Acyrthosiphon pisum*, to one of its alternative hosts, the nettle aphid *Microlophium carnosum* (Buckt.), met with little success (Powell and Wright 1988). However, the parasitoid readily switched hosts in the opposite direction, from nettle aphid to pea aphid. Examination of the esterase banding patterns for the two laboratory cultures used in these trials revealed greater variability in the nettle aphid biotype than in the pea aphid biotype (Powell and Loxdale,

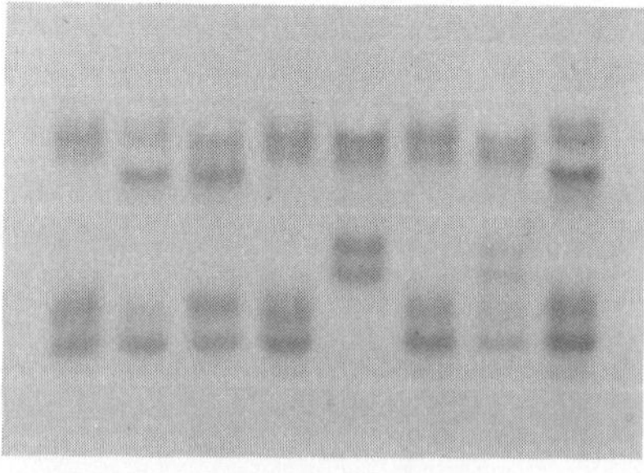

Fig. 21.2. Esterase isoenzyme patterns of *Aphidius ervi* females, showing the two main alleles at EST-3. Left to right: 1–4, 6, 8, homozygous for the fast allele; 5, homozygous for the slow allele; 7, heterozygous.

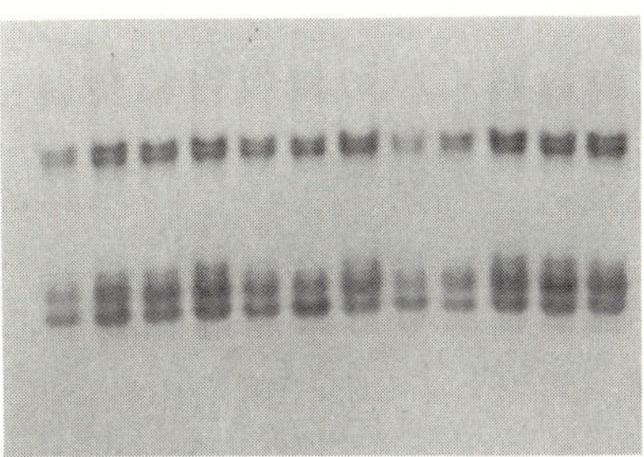

Fig. 21.3. Esterase isoenzyme patterns of *Aphidius ervi*, laboratory culture of a pea aphid (*Acrythosiphon pisum*) biotype. All homozygous for the fast allele at EST-3.

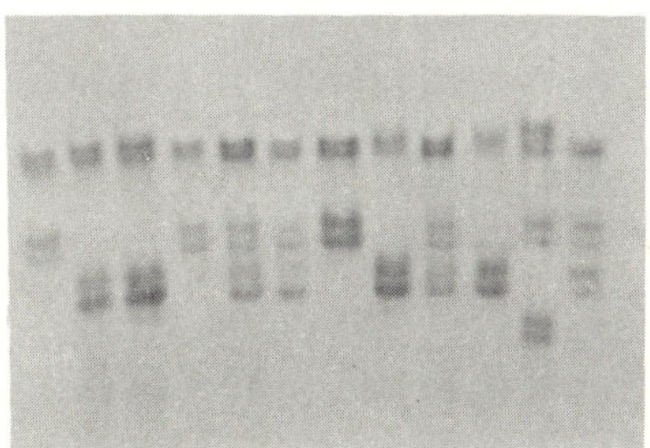

Fig. 21.4. Esterase isoenzyme patterns of *Aphidius ervi*, laboratory culture of a nettle aphid (*Microlophium carnosum*) biotype. Left to right: 1, male, hemizygous for the slow allele at EST-3; 2, 3, 8, 10, females, homozygous for the fast allele; 4, 7, females, homozygous for the slow allele; 5, 6, 9, 12, females, heterozygous for the two main alleles; 11, female, heterozygous for the slow allele and a third, very fast allele.

unpublished data). The most mobile of the esterase isoenzymes on *A. ervi* gels (designated as EST-3 by Cameron *et al.* (1984) and Walton (1986) and as EST-1 by Unruh *et al.* (1983)), appears to be a monomer coded by a locus with two main alleles, a slow and a fast allozyme, each represented by a double band (doublet) (Fig. 21.2). Because *A. ervi* males are haploid, they show either the slow or the fast doublet, whereas females are either homozygous for the slow or fast or are heterozygous. Individuals from the pea aphid biotype culture showed only the fast doublet at EST-3 (Fig. 21.3), whereas heterozygous females, and hence both fast and slow males, were present in the nettle aphid biotype culture (Fig. 21.4). Crossing experiments between the two cultures revealed that when a male nettle aphid biotype carrying the slow allele was mated with a female

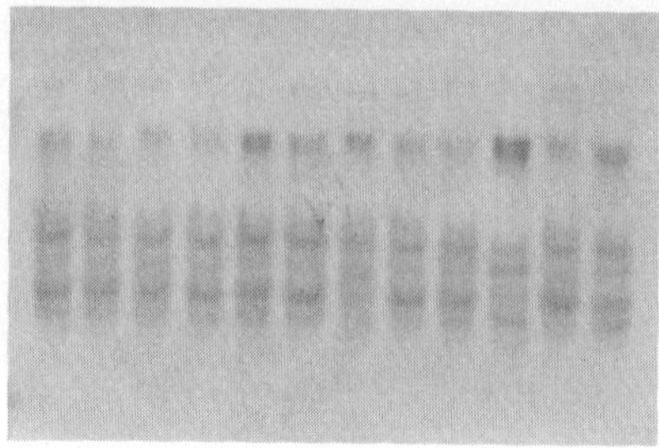

Fig. 21.5. Esterase isoenzyme patterns of female *Aphidius ervi* produced by crossing a female pea aphid biotype (homozygous for the fast allele at EST-3) with a male nettle aphid biotype carrying the slow allele. All female offspring heterozygous.

pea aphid biotype (homozygous for the fast allele), their female offspring (all heterozygous at EST-3) (Fig. 21.5) readily switched from the pea aphid to the nettle aphid (Powell and Wright 1988). In an earlier series of experiments, using different laboratory cultures, a pea aphid biotype was transferred to *S. avenae* (Cameron *et al.* 1984). Although, in this case, both the fast and slow alleles were present in the culture on pea aphid, the fast allele always seemed to be lost when the parasitoid was switched to *S. avenae*, suggesting selection of a laboratory host biotype on *S. avenae* from a more diverse parental genotype on the original host. A similar phenomenon was noted by Němec and Starý (1983*b*): transfer of *A. ervi* from an oligomorphic (as regards EST banding patterns) nettle aphid biotype to pea aphid gave rise to a monomorphic population on the new host.

The esterase enzyme system is probably not related to behavioural or biological traits governing host preference and/or acceptability. Nevertheless, the observations on host switching by *A. ervi* do emphasize the potential role of electrophoretic markers for identifying distinct biotypes and for labelling biological traits in parasitoid populations. This could be extremely important in connection with the choice of material for collection, mass-rearing, and release of parasitoids in pest-control programmes. Electrophoretic labels for biotypes may also aid the evaluation of parasitoid performance following release against specific pests. Pintureau *et al.* (1980) used esterase isoenzyme differences to monitor the success of six geographical biotypes of *Trichogramma embryophagum* Hart. following their release for control of the codling moth, *Cydia pomonella* (L.), in France.

Concluding remarks

It is clear from the work reviewed above that electrophoresis can be used in a variety of ways to advance our understanding of parasitoid biology and ecology, and thereby improve our ability to exploit them for the control of agricultural pests. Related techniques such as isoelectric focusing also offer exciting opportunities for the future that have yet to be fully realized.

Acknowledgements

We would like to thank Dr Hugh Loxdale for helpful discussion during the preparation of this paper and for permission to reproduce Fig. 21.1, and Cliff Brookes and Ann Wright for technical assistance with the parasitoid work at Rothamsted.

References

Avise, J. C. (1975). Systematic value of electrophoretic data. *Syst. Zool.*, **23**, 465–81.

Beckage, N. E., Templeton, T. J., Nielsen, B. D., Cook, D. I., and Stoltz, D. B. (1987). Parasitism-induced hemolymph polypeptides in *Manduca sexta* (L.) larvae parasitized by the braconid wasp *Cotesia congregata* (Say). *Insect Biochem.*, **17**, 439–55.

Berlocher, S. H. (1984). Insect molecular systematics. *A. Rev. Ent.*, **29**, 403–33.

Cameron, P. J., Powell, W., and Loxdale, H. D. (1984). Reservoirs for *Aphidius ervi* Haliday (Hymenoptera: Aphidiidae), a polyhagous parasitoid of cereal aphids (Hemiptera: Aphididae). *Bull. ent. Res.*, **74**, 647–56.

Castañera, P., Loxdale, H. D., and Nowak, K. (1983). Electrophoretic study of enzymes from cereal aphid populations. II. Use of electrophoresis for identifying aphidiid parasitoids (Hymenoptera) of *Sitobion avenae* (F.) (Hemiptera: Aphididae). *Bull. ent. Res.*, **73**, 659–65.

Castrovillo, P. J. and Stock, M. W. (1981). Electrophoretic techniques for detection of *Glypta fumiferanae*, an endoparasitoid of western spruce budworm. *Entomologia exp. appl.*, **30**, 176–80.

Dahlman, D. L. and Greene, J. R. (1981). Larval hemolymph protein patterns in tobacco hornworms parasitized by *Apanteles congregatus*. *Ann. ent. Soc. Am.*, **74**, 130–3.

Ferkovich, S. M. and Dillard, C. R. (1986). A study of uptake of radiolabelled host proteins and protein synthesis during development of eggs of the endoparasitoid, *Microplitis croceipes* (Cresson) (Braconidae). *Insect Biochem.*, **16**, 337–45.

Ferkovich, S. M. Greany, P. D., and Dillard, C. (1983). Changes in hemolymph proteins of the fall armyworm *Spodoptera frugiperda* (J. E. Smith), associated with parasitism by the braconid parasitoid *Cotesia marginiventris* (Cresson). *J. Insect Physiol.*, **29**, 933–42.

Gonzalez, D., Gordh, G., Thompson, S. N., and Adler, J. (1979). Biotype discrimination and its importance to biological control. In *Genetics in relation to insect management* (ed. M. A. Hoy and J. J. McKelvey), pp. 129–36. The Rockefeller Foundation, New York.

Greathead, D. J. (1986). Parasitoids in classical biological control. In *Insect parasitoids*, Symposium of the Royal Entomological Society of London, No. 13 (ed. J. Waage and D. J. Greathead), pp. 289–318. Academic Press, London.

Hung, A. C. F. (1982). Chromosome and isozyme studies in *Trichogramma*. *Proc. Ent. Soc. Wash.*, **84**, 791–6.

Hung, A. C. F. and Huo, S. (1985). Malic enzyme, phosphoglucomutase and phosphoglucose isomerase isozymes in *Trichogramma* (Hym.: Trichogrammatidae). *Entomophaga*, **30**, 143–49.

Jardak, T., Pintureau, B., and Voegelé, J. (1979). Mise en évidence d'une nouvelle espèce de *Trichogramma* (Hym.: Trichogrammatidae).—Phénomène d'intersexualité; étude enzymatique. *Annls Soc. ent. Fr. (N. S.)*, **15**, 635–42.

Mackauer, M. (1972). Genetic aspects of insect production. *Entomophaga*, **17**, 27–48.

Menken, S. B. J. (1982). Enzymatic characterization of nine endoparasite species of small ermine moths (Yponomeutidae). *Experimentia*, **38**, 1461–2.

Menken, S. B. J. and Ulenberg, S. A. (1987). Biochemical characters in agricultural entomology. *Argric. Zool. Rev.*, **2**, 305–353.

Němec, V. and Starý, P. (1983*a*). Genetic polymorphism in *Aphidius ervi* Hal. (Hym.; Aphidiidae), an aphid parasitoid on *Microlophium carnosum* (Bckt.). *Z. angew. Ent.*, **95**, 345–50.

Němec, V. and Starý, P. (1983*b*). Elpho-morph differentiation in *Aphidius ervi* Hal. biotype on *Microlophium carnosum* (Bckt.) related to parasitization on *Acyrthosiphon pisum* (Harr.) (Hym., Aphidiidae). *Z. angew. Ent.*, **95**, 524–30.

Němec, V. and Starý, P. (1984*a*). Utilization of isozyme analysis in the research on population diversity of aphid parasitoids (Hym., Aphidiidae). *Z. angew. Ent.*, **98**, 150–9.

Němec, V. and Starý, P. (1984*b*). Population diversity of *Diaeretiella rapae* (McInt.) (Hym., Aphidiidae), an aphid parasitoid in agroecosystems. *Z. angew. Ent.*, **97**, 223–33.

Němec, V. and Starý, P. (1985*a*). Genetic diversity and host alternation in aphid parasitoids (Hymenoptera: Aphidiidae). *Ent. Gener.*, **10**, 253–8.

Němec, V. and Starý, P. (1985*b*). Genetic diversity of the parasitoid *Aphidius ervi* on the pea aphid, *Acyrthosiphon pisum* in Czechoslovakia (Hymenoptera, Aphidiidae; Homoptera, Aphididae). *Acta ent. bohemoslov.*, **82**, 88–94.

Němec, V. and Starý, P. (1985*c*). Population diversity in dueterotokous *Lysiphlebus* species, parasitoids of aphids (Hymenoptera, Aphidiidae). *Acta ent. bohemoslov.*, **82**, 170–4.

Němec, V. and Starý, P. (1986). Population diversity centers of aphid parasitoids (Hym.: Aphidiidae): a new strategy in integrated pest management. In *Ecology of aphidophaga.* (ed. I. Hodek), pp. 485–8. Academia, Prague.

Pinureau, B. and Babault, M. (1980). Comparison des estérases chez 19 souches de *Trichogramma* (Hym. Trichogrammatidae) appartenant au groupe d'espèces evanescens. *Arch. Zool. exp. gén.*, **121**, 249–60.

Pintureau, B. and Babault, M. (1981). Characterzation enzymatique de *Trichogramma evanescens* et de *T. maidis* (Hym.: Trichogrammatidae): Etude des hybrides. *Entomophaga*, **26**, 11–22.

Pintureau, B. and Babault, M. (1982). Comparaison des enzymes chez 10 souches de *Trichogramma* (Hym.: Trichogrammatidae). In *Les Trichogrammes*, Colloquium INRA (France), No. 9, pp. 31–44.

Pintureau, B. and Voegelé, J. (1980). Une nouvelle espèce proche de *Trichogramma evanescens*: *T. maidis* (Hym.: Trichogrammatidae). *Entomophaga*, **25**, 431–40.

Pintureau, B, Goujet, R., Martouret, D., and Voegelé, J. (1980). Étude des esterases chez *Trichogramma embryophagum* Hartig (Hym.: Trichogrammatidae) choix d'un souche pour lutter contre *Laspeyresia pomonella* L. (Lep.: Tortricidae) dans la region parisienne. *Bull. Soc. ent. Mulhouse* **1980**, 17–24.

Powell, W. (1980). *Toxares deltiger* (Haliday) (Hymenoptera: Aphidiidae) parasitizing the cereal aphid, *Metopolophium dirhodum* (Walker) (Hemiptera: Aphididae), in southern England: a new host-parasitoid record. *Bull. ent. Res.*, **70**, 407–9.

Powell, W. (1986). Enhancing parasitoid activity in crops. In *Insect parasitoids*, Symposium of the Royal Entomological Society of London, No. 13 (ed. J. Waage and D.J. Greathead), pp. 319–40. Academic Press, London.

Powell, W. and Wright, A.F. (1988). The abilities of the aphid parasitoids *Aphidius ervi* Haliday and *A. rhopalosiphi* De Stefani Perez to transfer between different known host species and implications for the use of alternative hosts in pest control strategies. *Bull. ent. Res.*, **79**, 683–93.

Propp, G.D. (1986). Characterization of intra- and interspecific isoenzyme variation in *Spalangia* spp. and *Muscidifurax* spp. (Chalcidoidea: Pteromalidae), parasitoids of synanthropic Diptera. In *Biological control of muscoid flies*, Miscellaneous Publications of the Entomological Society of America, No. 61 (ed. R.S. Patterson and D.A. Rutz), pp. 164–74.

Pungerl, N.B. (1986). Morphometric and electrophoretic study of *Aphidius* species (Hymenoptera: Aphidiidae) reared from a variety of aphid hosts. *Syst. Ent.*, **11**, 327–54.

Rogers, J. S. (1972). Measures of genetic similarity and genetic distance. In *Studies in genetics VII*, University of Texas Publications, No. 7213, pp. 145–53.

Smilowitz, Z. (1973). Electrophoretic patterns in hemolymph protein of cabbage looper during development of the parasitoid *Hyposoter exiguae*. *Ann. ent. Soc. Am.*, **66**, 93–9.

Smilowitz, Z. and Smith, C. L. (1977). Hemolymph proteins of developing *Pieris rapae* larvae parasitized by *Apanteles glomeratus*. *Ann. ent. Soc. Am.*, **70**, 447–54.

Smith, S.M. and Hubbes, M. (1986). Isoenzyme patterns and biology of *Trichogramma minutum* as influenced by rearing temperature and host. *Ent. exp. Appl.*, **42**, 249–58.

Starý, P. (1970). *Biology of aphid parasites with respect to integrated control*. Dr. W. Junk, The Hague.

Starý, P. and Němec, V. (1986). Common elder, *Sambucus nigra*, as a reservoir of aphids and parasitoids (Hymenoptera, Aphidiidae). *Acta ent. bohemoslov.*, **83**, 271–8.

Starý, P., Pospíšil, J., and Němec, V. (1985). Integration of olfactometry and electrophoresis in the analysis of aphid parasitoid biotypes (Hymn. Aphidiidae). *Z. angew. Ent.*, **99**, 476–82.

Stock, M.W. (1981). Isozyme studies of forest insect populations. In *USDA Forest Service General Technical Report*, No. PSW-48, pp. 23–6.

Tomiuk, J. and Wöhrmann, K. (1980). Population growth and population structure of natural populations of *Macrosiphum rosae* (L.) (Hemiptera, Aphididae). *Z. angew. Ent.*, **90**, 464–73.

Unruh, T.R., White, W., and Gonzalez, D., Gordh, G., and Luck, R.F. (1983). Heterozygosity and effective size in laboratory populations of *Aphidius ervi* (Hymn.: Aphidiidae). *Entomophaga*, **28**, 245–58.

Unruh, T.R., White, W., Gonzalez, D., and Luck, R.F. (1986). Electrophoretic studies of parasitic hymenoptera and implications for biological control. In *Biological control of muscoid flies*, Miscellaneous Publications of the Entomological Society of America, No. 61 (ed. R.S. Patterson and D.A. Rutz), pp. 150–63.

van den Bosch, R. and Telford, A.D. (1964). Environmental modification and biological control. In *Biological control of insect pests and weeds* (ed. P. DeBach), pp. 459–88. Chapman and Hall, London.

Voegelé, J. and Berge, J.B. (1976). Les Tichogrammes (Insectes, Hymenop., Chalcidens, Trichogrammatidae), caracteristiques isoesterasiques de deux especies: *Trichogramma evanescens* Westw. et *T. achaeae* Nagaraja et Nagarkatti. *C.R. Acad. Sci., Paris*, **283**, 1501–3.

Wagner, R.P. and Selander, K. (1974). Isozymes in insects and their significance. *A. Rev. Ent.*, **19**, 117–38.

Walton, M.P. (1986). The application of gel-electrophoresis to the study of cereal aphid parasitoids. Ph.D thesis. The Hatfield Polytechnic, UK.

Wool, D., van Emden, H. F., and Bunting, S. W. (1978). Electrophoretic detection of the internal parasite, *Aphidius matricariae* in *Myzus persicae*. *Ann. appl. Biol.*, **90**, 21–6.

Wool, D., Gerling, D., and Cohen, I. (1984). Electrophoretic detection of two endoparasite species, *Encarsia lutea* and *Eretmocerus mundus* in the whitefly, *Bemisia tabaci* (Genn.) (Hom., Aleurodidae). *Z. angew. Ent.*, **98**, 276–9.

22 The use of electrophoresis for determining patterns of predation in arthropods

R.A. MURRAY, M.G. SOLOMON, and J.D. FITZGERALD

AFRC Institute of Horticultural Research, East Malling, Maidstone, Kent, UK

Abstract

Ecological studies of invertebrates often require determination of their predator–prey interrelationships. Since the quantity of prey material ingested by the predator(s) is usually very small, immunological methods have been used for detecting prey protein within the gut of the predator. These methods, however, require preparation of antisera from quantities of prey material that often present major difficulties in collection, particularly for small prey organisms such as mites. This problem prompted the development of an alternative method of analysing predator diets.

The method detects prey enzymes within the gut of a predator by polyacrylamide gradient gel electrophoresis and subsequent staining for enzyme activity. It is relatively rapid and sensitive, and does not require collection of large numbers of prey. The enzymes usually selected are esterases, since these are readily detected by a simple, sensitive staining procedure.

The method requires the prey species to have at least one distinctive esterase, which can be isolated as a band (or bands) at a characteristic position(s) on the gel, enabling it to be distinguished from esterases arising from the predator or from other possible prey species. If this requirement is not fulfilled, alternative enzyme markers may be selected.

Electrophoretic Studies on Agricultural Pests (ed. Hugh D. Loxdale and J. den Hollander), Systematics Association Special Volume No. 39, pp. 467–83. Clarendon Press, Oxford, 1989.

In almost all relationships examined, the method using esterases has provided a sensitive method for detecting predation, and has been used in a variety of laboratory and field studies. Thus, in addition to studies on predator–prey feeding relationships in orchard environments, it has also been employed to study Antarctic mites, notonectids, and carabids.

Introduction

Ecological studies of invertebrates often require determination of their predator–prey interrelationships. Trophic relationships may be quite limited as in, for instance, an Antarctic soil-living arthropod community. Alternatively, they may be quite complex and extensive, as found in apple orchards, where many of the animals present may be potential predators of several species and in turn potential prey for several other species. Knowledge of the relative importance of the components of such a feeding web is required to identify the most effective predators and their food sources, and hence to assist in the rational design of a programme of integrated pest management.

Such information may be acquired by direct observation in the field or in the laboratory, but this approach is limited by its labour-intensive nature, the artificiality of the laboratory environment, and its non-applicability to, for example, soil arthropods (Lister *et al.* 1987). In particular instances, recognizable prey remains may be identified in the gut of predators that chew their food (Sunderland 1975; Phillips 1981), or a semiquantitative assessment of predation can be made by observing coloured prey remnants in the gut of a predator with transparent cuticle (Rabbinge 1976). Until recently, however, the most widely applicable and most widely used method was based on immunological detection using the precipitin test (reviewed in Boreham and Ohiagu 1978), either to investigate predation by a range of predators on a particular prey species or, using the Ouchterlony plate technique, to detect a range of different prey proteins within the gut of a single predator species (Pickavance 1970). More recently, an enzyme-linked immunosorbent assay (ELISA) has been used (Sunderland *et al.* 1987). Such methods can be very sensitive, but they require the collection of sufficient prey material (sometimes as much as 1 g) to enable preparation of an antiserum with adequate titre. This can be a major difficulty with small prey organisms such as mites, which may weigh less than 1 μg, particularly for a complex feeding web when several such antisera might be required.

In mosquitoes, electrophoretic methods have been described either for the identification of the species of vertebrate host (Marinkelle 1969) or for the detection of multiple human blood meals (Boreham and Lenahan 1976). However, the particular methods used were most suited to arthropods that extract large blood meals from their prey, mainly owing to the relatively high limit of detectability of the host blood haptens. The problems associated with all these methods prompted the development of an electrophoretic technique for analysing diets of small invertebrate predators.

Any method developed had to be capable of detecting the presence of a marker characteristic of the prey species within the gut of the predator. Since the total quantity of prey material ingested by a predatory mite is very small (several micrograms for a phytoseiid mite weighing typically 7 μg), the method developed had to be very sensitive yet relatively rapid in operation so as to allow large numbers of samples to be analysed, and had not to require collection of large numbers of prey individuals. Such requirements were met by polyacrylamide gradient slab gel electrophoresis and subsequent

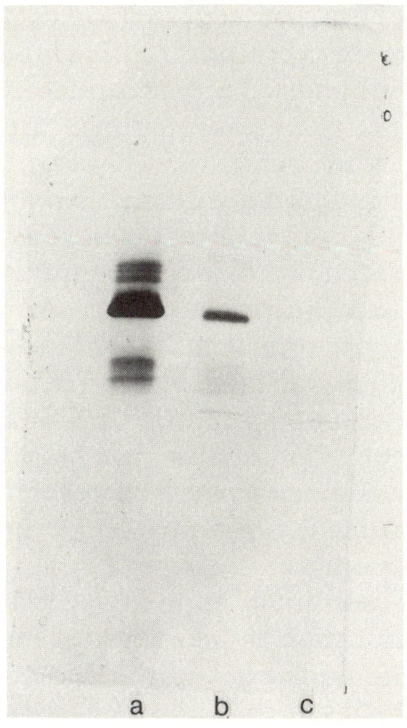

Fig. 22.1. Electrophoresis gel, stained for esterase activity, demonstrating predation on phytophagous mites. Samples, from individual mites, are (a) *Panonychus ulmi* (prey); (b) *Typhlodromus pyri* (predator) fed on *P. ulmi*; (c) *T. pyri* starved for 48 hours.

detection, by histochemical techniques, of enzymes from homogenates of predator and prey individuals (Murray and Solomon 1978).

Recognition of at least one enzyme (subsequently called the marker enzyme) that has the same electrophoretic mobility in the prey and fed predator, but which is not present in the starved predator, is an indication of feeding (Fig. 22.1). To allow full interpretation of the electrophoretic patterns, homogenates of individual predators that have been starved and fed in the laboratory, and homogenates of all potential prey species, should be analysed to demonstrate the applicability of the method for each particular predator/prey pair. Such a study would be a prerequisite for the electrophoresis of homogenates of field-collected predators and the subsequent search for the marker enzyme in each potential prey species.

Electrophoretic techniques

Polyacrylamide gradient gel electrophoresis was selected during method development because of its ability to focus the migrating proteins into narrow bands. This maximizes the resolution of the separation and the sensitivity of detection, since the limited quantity of marker enzyme is concentrated. It also offers the facility for modification of the slope of the gradient so that the absolute separation of particular bands in the region of the gel where the prey marker is found may be increased. In this way, identification of the prey band and its subsequent quantification are easier. An additional advantage of gradient gel electrophoresis is that if the proteins are run to 'pore limit', their migration distance is essentially independent of minor variations in running conditions. Consequently, very good reproducibility of separation is obtained between gels, particularly for those gels within a given batch. If batches of gels are prepared under carefully controlled, reproducible conditions then interbatch variability of protein migration is also small. This minimizes the requirement for standards to be applied to gels and permits easier gel scanning and quantification by automatic data processing systems. The ability to prepare batches of gels which may then be stored for several weeks before use is also a great convenience when large numbers of samples (50–100) are being analysed daily.

The use of concentration-gradient gels is not a prerequisite for successful application of the principles of the method; predation by notonectid bugs has been studied by electrophoresis on a disc electrophoresis system with homogeneous polyacrylamide rod gels (Giller 1982, 1984). In Giller's study, however, the problems that can arise with gel variability are apparent—in particular, the differences in absolute band migration distance on different gels. This may be of little consequence when the marker enzymes for two or more potential prey species are widely separated from each other, as was the case in his study, but when they have similar migration distances (as, for example, has been observed for the apple aphids *Dysaphis plantaginea* (Pass.) and *Rhopalosiphum insertum* (Walk.)) (Murray, pers. obs.) reproducibility of migration distance is essential. Even then, the addition of marker proteins on each gel, either run as separate samples or added to each of the samples under investigation as internal calibrations, may be advisable (Lister 1984*a*).

An alternative electrophoresis system which would seem to retain the advantage of sharp bands is isoelectric focusing. In this technique, in which proteins are separated according to their isoelectric point, very fine resolution is attainable. Additionally, as with gradient gels, the final separation is theoretically dependent only on the nature of the gel and is relatively independent of running conditions. However, the isoelectric focusing buffers are quite expensive and the thin-layer gels that would minimize the cost per gel are not readily made in batches for storage and convenient immediate availability when large numbers of samples are to be analysed.

To overcome the problem of low enzyme activity in some species, such as the mite *Typhlodromus pyri* (Scheuten), and the consequent difficulty in detection of individual mites, thin-layer cellulose acetate has been used as the support medium (van der Geest and Overmeer 1985). The use of these thin layers allowed very small aliquot portions of sample homogenate to be analysed, implying a marked increase in sensitivity of enzyme detection.

The limit of detectability of a band will be a function not only of its maximum absorbance (itself a function of band sharpness for a given quantity of enzyme), but also of the background non-specific staining of the gel. A criterion commonly adopted for detectability is that maximum peak absorbance should be twice that due to the background. It would seem that, for a given quantity of enzyme present in a band on an electrophoresis gel, the maximum sensitivity of detection will be attained when the stain resulting from that enzyme is concentrated beneath the minimum possible surface area of gel. This surface area will be minimized not only when the band is as narrow as possible but also when it has the shortest possible length, i.e. when the width of the sample applicator is as small as possible. The maximum sensitivity should thus be obtained by using very narrow sample applicators. On 3-mm slab gels, useful gains in sensitivity are obtained in this way, but the full potential is not realized because the initially narrow bands widen through lateral diffusion as they migrate down the gel (Lister 1984*a*, p. 888).

This problem is eliminated by the use of microcapillary gradient gels (Neuhoff 1973). In such gels, prepared in 0.45-mm diameter microcapillaries, the band focusing inherent on gradient slab gels is retained, while lateral diffusion is eliminated. Consequently, sharp, visible bands can be obtained from very small samples. Such gels, however, are not easy to prepare reproducibly, and are probably not suited to routine investigation of large numbers of samples. They may, nonetheless, often be of value in establishing relationships between predators and prey that are not amenable to study by the other techniques described. Such limited studies are facilitated by the fact that small numbers of gels can be prepared quickly and easily without recourse to special equipment. Gels can then be visually graded for consistency; matched gels can be loaded with the samples and then run in easily fabricated apparatus of the type described by Solomon and Murray (1986).

Sample preparation

To realize the maximum sensitivity of detection offered by the electrophoretic techniques described above, especially in microcapillary or thin-layer systems, it is advantageous to homogenize the sample in a minimum volume of buffer. The maximum sample proportion can then be applied to the electrophoretic support medium in a realistically achievable volume. In the case of small insects and mites, the volume of homogenization buffer can be as little as 2 μl for microcapillary electrophoresis. This presents problems in high recovery of the homogenate from the homogenization vessel. It is also important that the sample is reproducibly homogenized, particularly for quantitative assessments. Additionally, when large numbers of such samples are to be analysed, it is convenient to prepare each one in a separate disposable vessel, which, in addition, minimizes risk of sample cross-contamination.

These requirements are met by a microhomogenizer described by Murray and Solomon (1985) which can be used to prepare homogenates of individual mites weighing as little as 5 μg in 2–10 μl of buffer. The equipment is a microscale version of a conventional Potter–Elvehjem tissue grinder in which the disposable homogenization vessels are fabricated from sections of standard melting-point capillaries (ca. 1.25 mm in diameter) by sealing one end in a Bunsen-burner flame. Matching, close fitting, pestles are fabricated from stainless-steel wire.

The method is equally applicable to larger samples such as individual bugs, which would be homogenized in a larger volume of buffer (up to 20 μl). Subsamples could then be analysed, permitting multiple enzyme staining if, for example, evidence was being sought for predation on a range of potential prey that could not be detected by a single enzyme stain.

Alternative methods of homogenization are discussed by Murray and Solomon (1985). An additional method described by Brookes and Loxdale (1985) is particularly suited to the processing of large numbers of insect samples.

Sample detection: staining methods

Successful detection of the very small quantities of proteins ingested by predators of mites requires highly sensitive histochemical staining. Additionally, such staining should ideally reveal only a limited number of bands on an electrophoretic gel so as to minimize coincidence of bands derived from a predator and its potential prey. These requirements militate against the use of non-specific protein stains such as Coomassie blue R-250, and favour the use of a stain that detects enzyme activity.

Enzyme detection is inherently more sensitive than simple protein staining because of the different nature of the chromogenic reactions involved. In simple protein staining with, for example, Coomassie blue R-250, each protein molecule binds to a defined maximum number of dye molecules that depends on the number of active sites on the protein available for binding. This stoichiometric relationship defines the maximum stain intensity achievable for a given quantity of protein. Similar considerations limited the sensitivity of immunological detection in previous work. In contrast, enzyme detection relies on a non-stoichiometric enzymic reaction with, at least in theory, no upper limit to the stain intensity. If small quantities of enzyme from a prey species are to be detected, it is only necessary to extend the period of staining. In practice, the maximum staining period will be limited by non-specific background staining. An investigation of candidate enzyme-detection systems, taking into consideration diversity of the enzyme, ease and speed of staining, sensitivity of detection, and cost of reagents, led Murray and Solomon (1978) to select esterases as the marker enzyme of choice.

Consideration of similar criteria has led others to the same conclusion (Giller 1984; Lister 1984*b*; van der Geest and Overmeer 1985).

Esterase staining is capable of higher sensitivity than many other enzymes because the chromophore generated by esterases from the substrate and stain most commonly used for that class, namely 1-naphthyl acetate and Fast blue RR, is an azo dye that has a high extinction coefficient and is readily detected at low concentration. Using this staining technique permits detection of predatory mite meals of as little as 1.5 μg of *Panonychus ulmi* (Koch) (Fitzgerald, unpubl. obs.) or 2 μg of the springtail, *Cryptopygus antarcticus* (Willem) (Lister *et al.* 1987). The chromophore is also generated directly in a single staining step, which is particularly convenient because it allows the staining time to be optimized by direct visual observation.

Occasionally, esterases are unsuitable as marker enzymes, either because of coincidence of electrophoresis bands in predator and prey species, as found by Lister *et al.* (1987) for the mites *Gamasellus racovitzai* (Trouessart) and *Friesia* spp., or because of low esterase

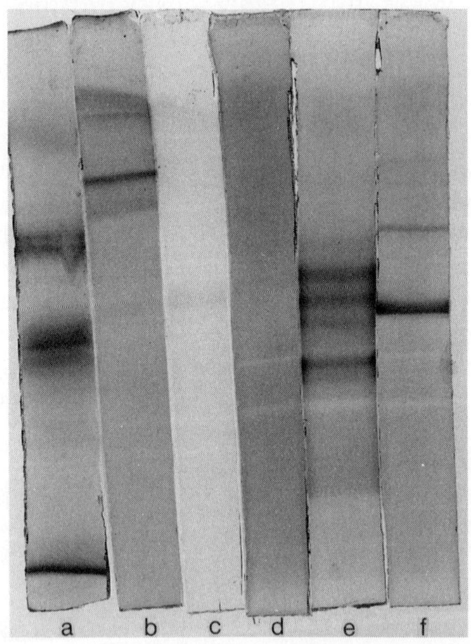

Fig. 22.2. Aliquot samples of *Phorodon humuli* showing the different banding patterns produced when separated by electrophoresis and stained for activity of (a) esterase; (b) alkaline phosphatase; (c) acid phosphatase; (d) α-glycerophosphate dehydrogenase; (e) malate dehydrogenase; (f) malic enzyme.

activity of the prey species, as observed for *T. pyri* (Murray and Solomon 1978; van der Geest and Overmeer 1985). In such cases, other enzymes may be considered as markers. Although at least 17 alternatives have been investigated, only phosphoglucoisomerase (van der Geest and Overmeer 1985; Heitmans *et al.* 1986), malate dehydrogenase (Lister 1984*b*), or alkaline phosphatase, malic enzyme, and malate dehydrogenase (Murray, unpubl. obs.) have generally shown sufficient activity to be potentially usable (Fig. 22.2). When sodium 1-naphthyl phosphate is used as the substrate for alkaline phosphatase, the advantages of esterase staining are retained.

Interpretation of band patterns

Valid interpretation of electrophoretic band patterns of predator and prey enzymes requires confidence that the band selected as the prey marker should have a reproducible migration distance on the gel. The significance of the selection of the electrophoresis system in this context has already been discussed. Other factors may need to be considered, notably the uniformity (quantitative and qualitative) of the marker enzyme between various life stages of the prey and between various individuals of a single stage. Different esterase patterns may be obtained if a phytophagous species is raised on different host plants, as has been observed for the citrus red mite *Panonychus citri* (McGregor) on citrus and pear (Osakabe 1984), although no such differences are observed between *P. ulmi* individuals reared on apple, myrobolan, or pear (Fitzgerald, unpubl. obs.). Variability may also arise if samples are collected from different geographical locations (Osakabe 1984) or from a population within which there is genetic variation. Such observations do not necessarily preclude the successful application of the technique, as Lister *et al.* (1987) have shown. In their investigation of Antarctic mites, although some prey species showed apparent pairs of closely-spaced electromorphs of certain bands, these electromorphs were sufficiently characteristic that presence of either form in the predator demonstrated predation. Such variation has not been observed between adult individuals in the mite species *P. ulmi* or *Tetranychus urticae* (Koch) (Murray and Solomon 1985), nor has any qualitative or quantitative difference been observed between various larval stages of *P. ulmi* (Fitzgerald, unpubl. obs.). None the less, it would be prudent to establish at an early stage of an investigation whether any variation in electrophoretic band patterns occurs.

Another factor that might cause difficulty in interpretation is modification of the prey marker protein in the predator gut so that its electrophoretic mobility is altered. This is unlikely if enzyme activity is being detected, since it would imply partial digestion with retention of enzyme activity, although this might occur if a polymeric isoenzyme were broken down into its subunits. A modified enzyme whose position did not correspond to its precursor in the original prey might still be acceptable as a marker if its modified position were consistent.

Confusion in interpretation may also arise from the predator and prey having a common food source (usually the host plant), giving rise to a common electrophoretic band not associated with predation. Related to this is the possibility of undigested food in the prey gut being detected in the predator following its feeding. Such undigested food should not inadvertently be selected as a marker band. Parasitism of the prey species should also be recognized as a potential cause of misinterpretation for a similar reason (cf. Powell and Walton, this volume).

Selective feeding of predatory bugs (Hemiptera) can lead to absence of some prey bands from a fed predator sample. The larva of the alder sawfly *Fenusa dohrnii* (Tisch.), and the alder aphid *Pterocallis alni* (de Geer) each show a number of sharp, well-resolved bands (Murray and Solomon 1978). However, analysis of the gut contents of the Hemipteran predator *Anthocoris nemoralis* (F.) that has fed on either of the two prey species reveals only a single band from each. This is attributed to the selective feeding of predatory bugs by extraction from their prey of haemolymph that contains only the esterase ultimately detected in the predator gut. A predator that consumed tissue other than, or in addition to, haemolymph would be expected to abstract a different combination of enzymes from its prey.

All such possibilities should be detected and considered in the preliminary feeding experiments designed to identify suitable marker enzymes.

Period of detectability of predation

In any detailed study of a feeding web, a prerequisite is assessment of the period of detectability of predation after the feeding event. In its simplest form, this can be estimated by starving individual predators for a period predetermined to eliminate residual prey protein from the gut. The starved predators are then allowed to feed on a single prey organism until satiated, then subsequently starved for a range of times before homogenization and analysis. Quantification by densitometry of the residual prey enzyme in each of the predators then demonstrates the period of detectability following a single meal. In this way, Murray and Solomon (1978) demonstrated that predation by *T. pyri* on *P. ulmi* could be detected after at least 31 hours (at 20°C). Similarly, the mite *Aculus schlechtendali* (Nalepa) can be detected in the gut of *T. pyri* for at least 40 hours at 25°C (Dicke and de Jong 1986) and *C. antarcticus* in the predatory mite *G. racovitzai* for 19 days at 5°C (Lister 1984*a*).

An investigation of temperature dependence of the period of detectability of the corixid waterboatman *Sigara falleni* (Fieb) and of the cockroach *Periplaneta americana* (L.) by the predatory waterbug (Backswimmer) *Notonecta glauca* (L.) (Giller 1984) confirmed that, as one might expect, the rate of digestion decreases as temperature falls. Thus, at 12°C, the period of detectability was 48 hours, but increased to 96 hours at 4°C. Such temperature dependence may also account in part for the extended period of detectability in the Antarctic mites at 5°C.

The study of predation by notonectid waterboatman also demonstrated that, for this group of predators at least, a large meal (represented by two corixids) is more persistent than a small meal (one corixid). This increased persistence is attributed to some food from the large meal remaining longer in the foregut than does the small meal, and not being subjected to the action of digestive enzymes until it is moved down to the midgut.

Quantitative assessment of predation

Simple qualitative analysis of predation can be valuable in the identification of the main predator and prey interactions in a feeding web. More detailed studies designed to assess, for example, the relative effectiveness of various predators in maintaining pest species at acceptably low levels, and the importance of non-pest species as alternative food sources, require more detailed interpretation of the electrophoresis gels. This can be limited to the determination of the proportion of each predatory species feeding on each prey species (e.g. Dicke and de Jong 1986).

A more detailed analysis of gels involves quantitative measurement of the intensity of staining of the prey marker band. Fitzgerald *et al.* (1986) investigated various factors influencing the accuracy of quantitative assessment of staining intensity and subsequent estimation of the quantity of prey remaining in the predator gut. They emphasized the need to determine the relationship between stain intensity (estimated from peak areas of gel scanning densitometer traces) and sample size using a series of standard prey homogenates. This is important because the response curve for some esterases is not linear at high sample concentrations or long staining time. This is attributed in part to masking of esterase activity in the centre of the gel following substantial depletion, at high esterase concentration, of the inwardly-diffusing substrate. This results in non-uniform staining through the gel thickness. If such esterases are selected as marker enzymes, then sample dilution may be required to bring the concentration of prey in the gut sample within the linear, steeper portion of the curve to allow accurate quantification. Calibration curves of stain intensity corresponding to known weights of prey taken by a predator can be prepared by weighing the predator and/or the prey before and after feeding. If the digestion rate of the prey within the predator gut is also determined from laboratory feeding experiments, these results can be incorporated into predation models and used to estimate predation rates (Lister *et al.* 1987; Solomon and Fitzgerald, unpubl. obs.).

Range of animals studied

The techniques using enzyme markers for electrophoretic analysis of predation were developed to overcome the difficulties of antiserum preparation with small prey species. They have subsequently been applied in several small arthropod studies, but have also have useful with larger species.

Predatory mites

Much of the earlier work on the development and assessment of the method was based on predation on the spider mites *P. ulmi* and *T. urticae*, major pests in apple orchards and glasshouse crops, respectively (Murray and Solomon 1978). Predation on these phytophagous species by the mites *T. pyri* and *Amblyseius fallacis* (Garman) was detected, demonstrating the sensitivity of the method, with which feeding by a single predatory mite on a single phytophagous mite could be detected.

Subsequent studies on other predatory phytoseiids in orchard habitats have been reported, namely, on *Amblyseius potentillae* (Garman) (Solomon *et al.* 1985) and *Amblyseius finlandicus* (Oudemans) (van der Geest and Overmeer 1985; Dicke and de Jong 1986). The range of prey species studied has been extended to include *Eotetranychus pruni* (Oudemans), *Tetranychus viennensis* (Zacher) (van der Geest and Overmeer 1985) and the rust mites *A. schlechtendali* and *Aculus fockeui* (Nalepa et Trouessart) (Dicke and de Jong 1986, 1988). Because the rust mites are so small (a single *A. schlechtendali* weighs *ca.* 0.15 μg), detection was only possible when the sample contained 20–30 individuals and, consequently, only when a predator had consumed at least this number. This relative insensitivity of detection in terms of numbers of *A. schlechtendali* is not a serious problem in the context of analysing feeding relationships. Rust mite population densities are typically far higher than those of spider mites, so that for a predator to have a significant impact on the population dynamics of a rust mite species it would need to consume large numbers. Dicke and de Jong (1986, 1988) detected *A. schlechtendali* and *P. ulmi* proteins in 2 per cent and 65 per cent, respectively, of *T. pyri* collected from apple leaves on which the absolute numbers of the two phytophagous mite species varied widely, but where the relative numbers were typically about 10 to 1 . This result demonstrated that over the (relatively high) range of spider mite densities studied, *T. pyri* consumed few *A. schlechtendali*. This contrasted with the findings for the predatory mite *A. finlandicus* on cherry, where, over a range of densities of the rust mite *A. fockeui* and the spider mite *P. ulmi*, the two phytophagous species were detected in 29 per cent and 26 per cent, respectively, of the predators examined.

In an electrophoretic study of an Antarctic microarthropod community (Lister 1984*a, b*), the only potential prey of a single species of predatory gamasid mite, *G. racovitzai*, were two species of collembola and six species of mite. Sampling over a 12-month period detected food traces in 82 per cent of the predators collected during the summer, but virtually none in winter. The three largest and most abundant potential prey species — two springtails, *C. antarcticus* and *Parisitoma octooculata* (Willem), and one mite *Alaskozetes antarcticus* (Michael) — were identified by esterase detection within the predators, offering no clear evidence of selective feeding; prey species were evidently taken in proportion to their relative availabilities (a reflection of abundance and susceptibility to capture and consumption). However, it is conceded that the underrepresentation of *A. antarcticus* in the detected diet of *G. racovitzai* may not reflect a low feeding frequency in this species, but rather a short period of detectability for the prey proteins in the gut of the predator. This study underlines the need to determine the persistence of the prey enzymes in the predator gut.

Predatory insects

In the study of predation by insects, it may be possible, because of their greater size, to prepare a sample from the excised gut rather than the entire insect. In this way, most of the proteins from other tissues of the predator are discarded, and the chances of predator proteins masking prey marker enzymes are minimized. This practice also reduces the possible inclusion of endoparasites in the sample.

Using samples prepared from the abdomen or excised gut of such predators, Murray and Solomon (1978) demonstrated that feeding by anthocorid and mirid bugs (Hemiptera) on a range of mites and insects could be detected. A study of the feeding of *Anthocoris nemoralis* collected from alder trees revealed that the leaf-mining larvae of the alder sawfly *Fenusa dohrnii* did not form a significant proportion of the diet of *A. nemoralis*, and that the predator fed mainly on the aphid *Pterocallis alni*.

Skinner (1983) used the electrophoretic technique in an investigation of the feeding of anthocorids on aphids and psyllids in apple orchards. The study showed that while the aphid *Rhopalosiphum insertum* was consumed by both *Anthocoris nemorum* (L.) and *A. nemoralis*, it appeared to be the preferred prey of the former. *A. nemoralis*, on the other hand, was shown to be feeding preferentially on *Psylla mali* (Schmid.) Neither species of anthocorid appeared to feed on the woolly aphid *Eriosoma lanigerum* (Haus.) despite its abundance in the experimental orchard.

In a feasibility study on predatory carabid beetles (*Pterostichus* spp., *Trechus* spp., and *Nebria brevicollis* (Fab.)), the same electrophoretic technique has been shown to offer potential for detecting feeding in a range of prey (D. Sutton, pers. comm.).

In an investigation of the feeding behaviour of the predatory anthocorid *Orius vicinus* Ribaut in apple trees in The Netherlands, Heitmans *et al.* (1986) used cellulose acetate plate electrophoresis and stained for phosphoglucoisomerase. They showed that this predator was feeding almost exclusively on the apple rust mite *Aculus schlechtendali*, which was very abundant on most of the trees examined. Only when rust mite numbers were below 50 per leaf was *O. vicinus* found to be consuming other prey. In this circumstance, bands were seen that were common to the predatory mite *T. pyri*, tydeid mites, and a leafhopper (*Typhlocyba* sp). Laboratory feeding tests indicated that tydeids and *Typhlocyba* sp. did not form part of the diet of *O. vicinus*, so it was inferred that the bands in question signified feeding on *T. pyri*.

A feasibility study by Giller (1982) using a disc electrophoresis system investigated the potential for analysis of predation by five species of notonectid water bugs on a range of prey species. In all examples of laboratory-fed predators studied, it was possible to detect the presence of a known prey species in the predator's gut utilizing one or more marker esterases. This initial study was extended to investigate the relationship between predator gut state, assessed by direct observation of the excised gut, and marker esterase detectability (Giller 1984). It demonstrated that prey esterases can almost always be detected while some food remains in the foregut. Detection may still be possible while the midgut is very full, but is not possible for any subsequent gut states.

In a further study of the diet of two species of notonectid in the laboratory and field, Giller (1986) identified characteristic bands from 33 potential aquatic prey types. Of these, 18 types were detected in the gut of field-collected notonectids. The results showed that the smaller instars specialized on small crustacean prey, whereas fourth instars fed on a mixture of crustacea, mayflies, and dipteran larvae, and fifth instar and adults fed mainly on mayflies.

Conclusion

The application of electrophoretic techniques coupled with enzyme staining is of particular value in ecological studies of predation on small prey for which antiserum preparation is difficult. It is ideally suited to the investigation of communities in which there is a large number of different potential prey species. It has also been found useful for those situations where, even though neither of these restrictions necessarily applies, it is inconvenient to maintain small mammals for the preparation of antisera.

The method has been successfully applied to a wide range of taxa for direct diet analysis of predators ranging from small mites weighing 5 μg to large insects. The ability to quantify predation offers the potential not only to determine the existence of a feeding relationship but also to assess the impact of particular predators on a prey population. It is a valuable tool that has a place in the investigation of many trophic relationships.

References

Boreham, P.F.L. and Lenahan, J.K. (1976). Methods for detecting multiple blood-meals in mosquitoes (Diptera, Culicidae). *Bull. ent. Res.*, **66**, 671–9.

Boreham, P.F.L. and Ohiagu, C.E. (1978). The use of serology in evaluating invertebrate prey-predator relationships: a review. *Bull. ent. Res.*, **68**, 171–94.

Brookes, C.P. and Loxdale, H.D. (1985). A device for simultaneously homogenizing numbers of individual small insects for electrophoresis. *Bull. ent. Res.*, **75**, 377–8.

Dicke, M. and de Jong, M. (1986). Prey preference of predatory mites: electrophoretic analysis of the diet of *Typhlodromus pyri* Scheuten and *Amblyseius finlandicus* (Oudemans), collected in Dutch orchards. In *IOBC/WPRS Bulletin*, Vol. 9, part 4, pp. 62–7.

Dicke, M. and de Jong, M. (1988). Prey preference of the phytoseiid mite *Typhlodromus pyri*. 2. Electrophoretic diet analysis. *Exp. appl. Acarol.*, **4**, 15–25.

Fitzgerald, J.D., Solomon, M.G. and Murray, R.A. (1986). The quantitative assessment of arthropod predation rates by electrophoresis. *Ann. appl. Biol.*, **109**, 491–8.

Geest, L.P.S. van der and Overmeer, W.P.J. (1985). Experiences with polyacrylamide gradient gel electrophoresis for the detection of gut contents of phytoseiid mites. *Mededelingen van Faculteit Landbouwwetenschappen Rijksuniversiteit Gent*, **50**, 469–71.

Giller, P.S. (1982). The natural diets of waterbugs (Hemiptera–Heteroptera): electrophoresis as a potential method of analysis. *Ecol. Ent.*, **7**, 233–7.

Giller, P.S. (1984). Predator gut state and prey detectability using electrophoretic analysis of gut contents. *Ecol. Ent.*, **9**, 157–62.

Giller, P.S. (1986). The natural diet of the Notonectidae: field trials using electrophoresis. *Ecol. Ent.*, **11**, 163–72.

Heitmans, W.R.B., Overmeer, W.P.J., and van der Geest, L.P.S. (1986). The role of *Orius vicinus* Ribaut (Heteroptera; Anthocoridae) as a predator of phytophagous and predacious mites in a Dutch orchard. *J. appl. Ent.*, **102**, 391–402.

Lister, A. (1984*a*). Predation in an Antarctic micro-arthropod community. In *Acarology VI*, Vol. 2 (ed. D.A. Griffiths and C.E. Bowman), pp. 886–92. Ellis Horwood, Chichester.

Lister, A. (1984*b*). Studies on the Antarctic predatory mite *Gamasellus racovitzai*. D. Phil. thesis. University of York, UK.

Lister, A., Usher, M.B., and Block, W. (1987). Description and quantification of field attack rates by predatory mites: an example using an electrophoresis method with a species of Antarctic mite. *Oecologia* (*Berlin*), **72**, 185–91.

Marinkelle, C.J. (1969). Host identification of arthropod blood meals by agar electrophoresis. *Mitteilungen der Instituto Columbo-Aleman de Investigaciones Cientificas*, **3**, 29–48.

Murray, R.A. and Solomon, M.G. (1978). A rapid technique for analysing diets of invertebrate predators by electrophoresis. *Ann. appl. Biol.*, **90**, 7–10.

Murray, R.A. and Solomon, M.G. (1985). A microtechnique for preparing homogenates of biological samples. *Analyt. Biochem.*, **151**, 400–2.

Neuhoff, V. (1973). Micro-electrophoresis on polyacrylamide gels. In *Micromethods in molecular biology* (ed. V. Neuhoff), pp. 1–83, Springer Verlag, Berlin.

Osakabe, M. (1984). Esterase zymogram of the citrus red mite, *Panonychus citri* (McGregor), on citrus and pear. *Jap. J. appl. Ent. Zool.*, **28**, 1–4.

Phillips, M.L. (1981). The ecology of the common earwig *Forficula auricularia* in apple orchards. Ph.D. thesis. University of Bristol, UK.

Pickavance, J.R. (1970). A new approach to the immunological analysis of invertebrate diets. *J. Anim. Ecol.*, **39**, 715–24.

Powell, W. and Walton, M.P. (1989). The use of electrophoresis in the study of hymenopteran parasitoids of agricultural pests. This volume, pp. 443–65.

Rabbinge, R. (1976). *Biological control of fruit-tree red spider mite*. Pudoc, Wageningen.

Skinner, R.N. (1983). The biology of apple aphids and their predators. Ph.D. thesis. University of London, UK.

Solomon, M.G. and Murray, R.A. (1986). Apparatus for electrophoresis in microcapillary polyacrylamide gels. *J. Chromat.*, **355**, 281–4.

Solomon, M.G., Murray, R.A., and van der Geest, L.P.S. (1985). Analysis of prey by means of electrophoresis. In *Spider mites. Their biology, natural enemies and control; World crop pests*, Vol. 1B (ed. W. Helle and M.W. Sabelis), pp. 171–3. Elsevier, Amsterdam.

Sunderland, K.D. (1975). The diet of some predatory arthropods in cereal crops. *J. appl. Ecol.*, **12**, 507–15.

Sunderland, K.D., Crook, N.E., Stacey, D.L., and Fuller, B.J. (1987). A study of feeding by polyphagous predators on cereal aphids using ELISA and gut dissection. *J. appl. Ecol.*, **24**, 907–33.

Index

Systematics Association Publications

1. Bibliography of key works for the identification of the British fauna and flora, *3rd edition* (1967)
   Edited by G. J. Kerrich, R. D. Meikle and N. Tebble
2. Function and taxonomic importance (1959)
   Edited by A. J. Cain
3. The species concept in palaeontology (1956)
   Edited by P. C. Sylvester-Bradley
4. Taxonomy and geography (1962)
   Edited by D. Nichols
5. Speciation in the sea (1963)
   Edited by J. P. Harding and N. Tebble
6. Phenetic and phylogenetic classification (1964)
   Edited by V. H. Heywood and J. McNeill
7. Aspects of Tethyan biogeography (1967)
   Edited by C. G. Adams and D. V. Ager
8. The soil ecosystem (1969)
   Edited by H. Sheals
9. Organisms and continents through time (1973)†
   Edited by N. F. Hughes

Published by the Association (out of print)

Systematics Association Special Volumes

1. The new systematics (1940)
   Edited by J. S. Huxley (Reprinted 1971)
2. Chematoxonomy and serotaxonomy (1968)*
   Edited by J. G. Hawkes
3. Data processing in biology and geology (1971)*
   Edited by J. L. Cutbill
4. Scanning electron microscopy (1971)*
   Edited by V. H. Heywood
   Out of print
5. Taxonomy and ecology (1973)*
   Edited by V. H. Heywood

6. The changing flora and fauna of Britain (1974)*
   Edited by D. L. Hawksworth
   Out of print
7. Biological identification with computers (1975)*
   Edited by R. J. Pankhurst
8. Lichenology: progress and problems (1976)*
   Edited by D. H. Brown, D. L. Hawksworth and R. H. Bailey
9. Key works to the fauna and flora of the British Isles and north-western Europe, *4th edition* (1978)*
   Edited by G. J. Kerrich, D. L. Hawksworth and R. W. Sims
10. Modern approaches to the taxonomy of red and brown algae (1978)*
    Edited by D. E. G. Irvine and J. H. Price
11. Biology and systematics of colonial organisms (1979)*
    Edited by G. Larwood and B. R. Rosen
12. The origin of major invertebrate groups (1979)*
    Edited by M. R. House
13. Advances in bryozoology (1979)*
    Edited by G. P. Larwood and M. B. Abbot
14. Bryophyte systematics (1979)*
    Edited by G. C. S. Clarke and J. G. Duckett
15. The terrestrial environmental and the origin of land vertebrates (1980)*
    Edited by A. L. Panchen
16. Chemosystematics: principles and practice (1980)*
    Edited by F. A. Bisby, J. G. Vaughan and C. A. Wright
17. The shore environment: methods and ecosystems (2 Volumes) (1980)*
    Edited by J. H. Price, D. E. G. Irvine and W. F. Farnham
18. The Ammonoidea (1981)*
    Edited by M. R. House and J. R. Senior
19. Biosystematics of social insects (1981)*
    Edited by P. E. Howse and J. -L. Clément
20. Genome evolution (1982)*
    Edited by G. A. Dover and R. B. Flavell
21. Problems of phylogenetic reconstruction (1982)*
    Edited by K. A. Joysey and A. E. Friday
22. Concepts in nematode systematics (1983)*
    Edited by A. R. Stone, H. M. Platt and L. F. Khalil
23. Evolution, time and space: the emergence of the biosphere (1983)*
    Edited by R. W. Sims, J. H. Price and P. E. S. Whalley

24. Protein polymorphism: adaptive and taxonomic significance (1983)*
    Edited by G. S. Oxford and D. Rollinson
25. Current concepts in plant taxonomy (1983)*
    Edited by V. H. Heywood and D. M. Moore
26. Databases in systematics (1984)*
    Edited by R. Allkin and F. A. Bisby
27. Systematics of the green algae (1984)*
    Edited by D. E. G. Irvine and D. M. John
28. The origins and relationships of lower invertebrates (1985)‡
    Edited by S. Conway Morris, J. D. George, R. Gibson and H. M. Platt
29. Infraspecific classification of wild and cultivated plants (1986)‡
    Edited by B. T. Styles
30. Biomineralization in lower plants and animals (1986)‡
    Edited by B. S. C. Leadbeater and R. Riding
31. Systematic and taxonomic approaches in palaeobotany (1986)‡
    Edited by R. A. Spicer and B. A. Thomas
32. Coevolution and systematics (1986)‡
    Edited by A. R. Stone and D. L. Hawksworth
33. Key works to the fauna and flora of the British Isles and northwestern Europe, *5th edition* (1988)‡
    Edited by R. W. Sims, P. Freeman and D. L. Hawksworth
34. Extinction and survival in the fossil record (1988)‡
    Edited by G. P. Larwood
35. The phylogeny and classification of the tetrapods (2 Volumes) (1988)‡
    Edited by M. J. Benton
36. Prospects in systematics (1988)‡
    Edited by D. L. Hawksworth
37. Biosystematics of haematophagous insects (1988)‡
    Edited by M. W. Service
38. The chromophyte algae: problems and perspectives (1989)‡
    Edited by J. C. Green, B. S. C. Leadbeater, and W. Direr
39. Electrophoretic studies on agricultural pests (1989)‡
    Edited by Hugh D. Loxdale and J. den Hollander

* *Published by Academic Press for the Systematics Association*
† *Published by the Palaeontological Association in conjunction with the Systematics Association*
‡ *Published by the Oxford University Press for the Systematics Association*